Geophysical Logging
for
Mineral and Engineering Applications

Geophysical Logging
for
Mineral and Engineering Applications

JAMES K. HALLENBURG

PennWell Books
PennWell Publishing Company
Tulsa, Oklahoma

Library of Congress cataloging in publication data

Hallenburg, James K.
Geophysical logging for mineral and engineering applications.

Includes bibliographical references and index.
1. Geophysical well logging. I. Title.
TN269.H33 1983 622.1 83-4136
ISBN 0-87814-235-5

Printed in the United States of America

1 2 3 4 5 • 87 86 85 84 83

Contents

PREFACE vii

ACKNOWLEDGMENTS viii

INTRODUCTION ix

CHAPTER 1 **History 1**

References 3

CHAPTER 2 **Formation Types 4**

References 10

CHAPTER 3 **Borehole Effects and Mud 11**

Cuttings 13

References 17

CHAPTER 4 **Electric Logging Systems 18**

Resistance, resistivity, and conductivity 18

Spontaneous potentials 50

Dipmeters 76

Induced polarization 79

References 88

CHAPTER 5 **Radioactivity Methods 90**

Radioactivity 90

Gamma-ray logging 112

Scattered gamma-ray systems 130

Neutron logging 146
References 162

CHAPTER 6 **Magnetic Methods 165**
Magnetic susceptibility 165
Magnetic field strength 166
Surface and airborne magnetic methods 167
References 168

CHAPTER 7 **Mechanical Methods and Sampling Methods 170**
Acoustic methods 170
Calipers 176
Borehole deviation methods 182
Temperature measurements 186
Sampling 194
Gas measurements 196
Electric cone penetrometer 200
References 200

CHAPTER 8 **Analog and Digital Systems 202**

CHAPTER 9 **Interpretation Methods 206**
Initial data treatment 207
Single curves 208
Multiple curves and log suites 222
General considerations 237
References 244

APPENDIX 246
GLOSSARY 249
INDEX 251

Preface

This text is a logging manual that is a combination of a description of borehole geophysical and petrophysical logging devices and methods of using them. A large portion of the useful logging and interpretation techniques has been included. It is an attempt to put together the information available for the use of geophysical borehole logs for nonpetroleum use. The purpose is to make the information available to the mineral industry, engineering firms, and persons engaged in the use of geophysical logs for scientific purposes.

This manual is oriented toward mineral, engineering, scientific, and other nonpetroleum uses because of a previous lack of such a text. There is literature published for petroleum logging, but little exists that is oriented to nonpetroleum use.

Inevitably, such a manual as this involves petroleum phases of logging. Much of the equipment and many of the techniques in the nonpetroleum logging industries have been borrowed from petroleum logging. Recently, however, developments aimed specifically at mineral and engineering logging have appeared. Geothermal and other nonpetroleum, nonmineral uses have been developing rapidly. This situation will probably continue to grow as the nonpetroleum logging industries become more and more self-supporting.

One should remember that petroleum and nonpetroleum methods are not likely to be the same. A person looks for different materials in different situations and with different characteristics. Petroleum logging people should be especially careful of the distinction whenever they apply their valuable experience and expertise to nonpetroleum projects.

The logging industry in the past has been small enough that one had personal contact with most of the active persons in this field. Therefore, if one had a question about the use of a method or about the way to proceed, he asked his peers. Now, however, the logging industry and logging methods are complex and diverse enough that a basic description of them is valuable.

At the rate the industry is evolving, it is impossible to cover all methods and systems thoroughly in an up-to-date manner. The most common and some of the more exotic methods have been included. There are also some methods that are not being used but that should be or that may become valuable methods in the near future. These too have been included.

It is surprising that there is a paucity of information concerning the history of logging. This is an area that someone could cover in detail. The few bits such as Allaud and Martin's book about Schlumberger are fascinating. One comes to the conclusion that, in contrast to many disciplines, geophysical methods have evolved with little serendipity. Virtually all of the methods, systems, and tools were developed deliberately to perform the function for which they were ultimately used.

When examining the history of logging, one is struck by the typical S-shaped growth curve. It appears that nonpetroleum logging is in the steep part of the curve. If this is true, one wonders what interesting developments will appear in the near future.

It is hoped that this volume will be useful. It is also hoped that the various logging fields and other geological and geophysical disciplines will work more closely together and that this volume will help promote that end.

Acknowledgments

I wish to acknowledge the courtesy and kind permission of several persons and organizations who have allowed me to use their illustrations in this book:

L. Allaud and M. Martin, from their book, *Schlumberger, the History of a Technique* (New York: John Wiley and Sons, 1977).

American Petroleum Institute, figures from API RP33, "Recommended Practice for Standard Calibration and Format for Nuclear Logs," (April 1974).

Birdwell Division of the Seismograph Service Corporation (E.J. Schonblom).

Century Geophysical Corporation.

William H. Curry III, from the *Earth Science Bulletin,* volume 12, No. 2 (June 1979).

Dresser Atlas Division of Dresser Industries, from "Relating Diplogs to Practical Geology."

Gearhart Industries, from the "Formation Data Handbook."

Geometrics, from "Applications Manual for Portable Magnetometers."

Gulf Oil Company, from "A Review of Little Knife Field Development," *Earth Science Bulletin,* volume 11, No. 4 (1978).

The IRT Corporation.

George V. Keller, from the *Colorado School of Mines Quarterly,* volume 57, No. 2 (April 1962).

G.V. Keller and F. Frischknecht, from *Electrical Methods in Geophysical Prospecting* (Pergamon Press).

The Log Analyst, from "A Practical Method for Determining Cementation Exponents," by Sr. Gomez-Rivero; "A Technique for Interpreting Depositional Environments of Sandstones from the SP Log Using a Computer." by W.A. Ruoff, vol. XVII, No. 4; and "Formation Density Logging/Gamma Gamma Principles and Practice," J. Tittman and J. Wahl, SPWLA Reprint Volume *Gamma Ray, Neutron, and Density Logging* (March 1978).

Owl Technical Associates Inc.

Sylvain Pirson, from *The Handbook of Well Log Interpretation* (Englewood Cliffs: Prentice-Hall, 1963).

Princeton Gamma Tech.

Schlumberger Well Services.

Society of Petroleum Engineers of the AIME, from "Introduction to Induction Logging" and "Applications to Logging Wells Drilled with Oil Base Mud" by H.G. Doll, tp 2641 (June 1949); "The SP Log in Shaly Sands" by H.G. Doll, tp 2912 (1949); "Coal Analysis from Well Logs" by J.J. Kowalski and M.E. Holter, SPE5503 (1975); "Temperature Surveys in Gas Producing Wells," *JPT,* volume 205 (1955); and *JPT* (1978).

Welex, A Halliburton Company, paper no. 906-15-c.

U.S. Department of Energy, "Uranium Industry Symposium" (1970).

Bruce Rubin, from "Uranium Roll Front Zonation in the Southern Powder River Basin, Wyoming," *WGA Earth Science Bulletin,* volume 3, No. 4 (December 1970).

Ertec Western Inc.

Also, I would like to thank Eunice Mohan and Jacquelyn Hallenburg for their many hours of help.

Introduction

Logging is a term meaning to make a record. Geophysical and petrophysical logging are usually used in the context of making a measurement and a record in a borehole in the earth. The record may be of any parameter, but geophysical logging is done for the purpose of defining the formations and their structural characteristics. Petrophysical logging records the characteristics of the rocks themselves.

We use geophysical and petrophysical logging for many purposes, some of which are described in this manual. Each of the uses to which logging is put is a little different from the others, even though the techniques of one have been used and adapted from another.

Nonpetroleum logging techniques and instruments have been borrowed from the petroleum logging industry. However, petroleum logging and the various types of nonpetroleum logging are not the same thing, even though they resemble each other. Tools and techniques designed to locate and assess a petroleum reservoir are used with only moderate to poor success for uranium development, unless one understands the differences. Thus, while petroleum and nonpetroleum methods are similar, one should remember that they are separate and distinct.

Nonpetroleum logging is literally any logging not used for petroleum exploration, development, or production. It primarily involves mineral logging. Besides uranium and coal, nonpetroleum logging covers other mineral fields that are becoming important. These are oil shale, potash, and trona exploration and development. Hard-rock mineral explorations are using logging, and water is being considered a mineral resource to a greater extent daily. Logging for engineering purposes, before building a dam or an office building or before mining, can be tremendously valuable in terms of cost, time, and reliability. Plate tectonic theories have been substantiated largely through logging. It is probable that this will soon be as large of a field as petroleum logging.

Few logs measure the parameter desired directly. This is an advantage in a way, as it forces an examination and understanding of the method. A suite of logs designed for one purpose may not be suited for other purposes or another locality. A uranium suite of logs picked for development logging in New Mexico, Colorado, and Wyoming sedimentary environments that may contain uranium compounds is not appropriate for coal or oil-shale logging. It is not even appropriate for other types of uranium-bearing environments, such as hydrothermal deposits. Even though a contract logging company may offer a uranium suite or a coal suite of logs, one should choose the proper combination of logs for the purpose. To a large extent a person should design a suite of logs for the purpose intended. Tables A–1 to A–3 in the Appendix list some hypothetical environments and the associated minerals with the combination of suites that may be best suited to them.

It is necessary to cover some of the instrumentation problems of the various logging methods in this text. This is because instrument design and response enter into interpretation to a large degree. This is true not only of the sensors themselves but also of the circuits. For example, the time-constant circuit of an analog gamma-ray system plays a great part in determining the logging speed, distortion, appearance, and accuracy of the measurement. On the other hand, the time-constant concept is not used in digital systems for interpretation. Other features are similarly involved in other circuits.

In a like manner, designing an instrument to operate in a hostile environment, such as a borehole, has a profound effect on the response of that tool. The hostile environment of one type of logging may be the normal environment for another type.

Mathematics is an integral part of geophysical measurements. It is a powerful tool in understanding the results of a log, but few people are mathematicians. Therefore, the mathematical handling in this text is held to a minimum. Where it does occur, it is explained or referenced.

Finally, the pattern of this text is to explain the theory and response of a system and then explain how that system can be used. It should be kept in mind, however, that an error-free, effective analysis can seldom be made from one curve. For this reason, a separate handling of multiple-curve usage is an important part of this text. Also, for this reason information about cores, cuttings, and surface methods is included along with the downhole explanations.

Combinations of logs and methods, such as downhole and surface or physical and chemical, may become more important in the years to come. The ultimate effectiveness can be gained from combinations of methods.

CHAPTER 1 History

Geophysical borehole logging apparently began in 1869, when Lord Kelvin (of temperature studies fame) logged a water well with a temperature tool using a platinum-resistance thermometer circuit. Borehole geophysics remained an academic study until the Schlumberger brothers, Marcel and Conrad, started in the business of open-hole geophysical logging for oil exploration in 1927.

In 1921 Marcel Schlumberger logged a borehole to determine the in situ resistivity of the formation in the Beseges coal basin near Molieressur-Ceze. This hole was 2,500 ft deep and was cased to 1,500 ft. Between March 20 and 22, several resistivity measurements over a few feet at the bottom of the hole did reflect the variations in the nature of the formations.

On April 28, 1927, Conrad Schlumberger outlined the principle of Electrical Coring in a note entitled "Reserches Electriques dans les Sondages" (Electrical Research in Boreholes). Henri Doll was given the responsibility for equipment design and testing. One of the opinions held at this time was that below a certain depth all geological formations became compacted and infinitely resistive.

The first operation by Doll was on September 5, 1927, at Diefenbach in well 2905, rig 7. A reproduction of that log is shown in Fig. 1–1. A photograph of the logging equipment used is shown in Fig. 1–2. From this beginning, the logging industry has grown to a needed and reliable service.

In 1931 the spontaneous potential measurement was added. Gamma-ray logging was started in 1938. The induction log was introduced in 1949. The induction log was so successful in solving the problem of measuring resistivity in the undisturbed formation that it replaced the electric log within a year.

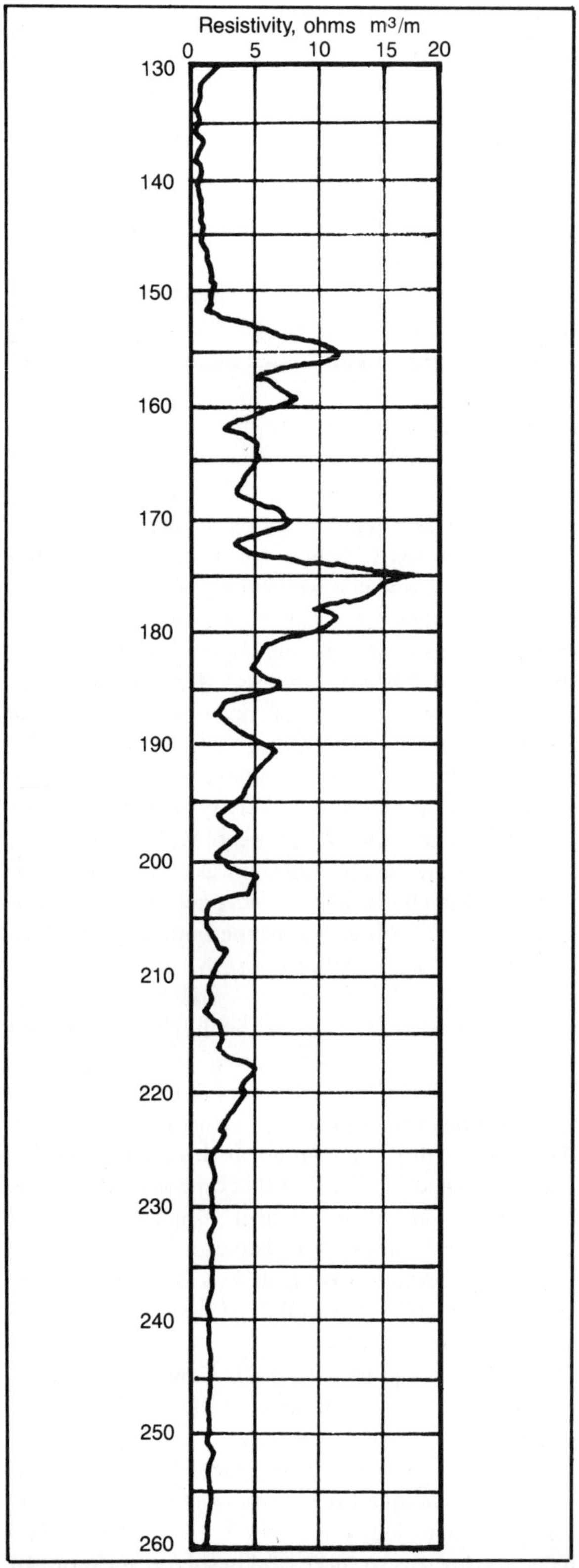

FIG. 1–1

A copy of the first Schlumberger commercial log

(after Allaud and Martin, *Schlumberger, The History of a Technique*, courtesy John Wiley and Sons, 1977)

New methods came rapidly at that time. The MicroLog was introduced in 1949. Neutron porosity logging (c. 1950) added an independent means of measuring formation porosity. The acoustic logs (c. 1956) became the first good porosity measurement. The density tool (c. 1962) was able to make very accurate measurements. It also gave a third means, other than resistivity, to determine porosity. Focused resistivity measurement methods were revived (from early work by Doll) for salty mud environments (c. 1960).

Following this prolific period of introduction of new methods (1931–1962), much work was done to refine and improve these methods. The compensated neutron system reduced adverse borehole effects. Investigation of neutron generators began in 1962.

Concurrently with the refinement of methods there was intense examination of existing methods and systems. The spontaneous potential (SP) was examined. The selective/SP and static SP systems were

FIG. 1–2

The winch, cable, and pickup truck used in the beginnings of electrical coring at Pechelbronn

(after Allaud and Martin, *Schlumberger, The History of a Technique*, courtesy John Wiley and Sons, 1977)

tried in 1949. Focused resistivity measurements were adapted to sidewall or micro systems and the Microlaterolog and Microguard log methods were successful (c. 1962). Cable characteristics were studied and depth measurement accuracy was improved.

From 1960–1970 oil wells became deeper and hotter. This prompted the development of high-temperature explosives for perforating. The small diameter and the high temperature gamma-ray-neutron systems were developed about 1963–1965.

Probably the most significant event that affected logging was the introduction in 1960 of the silicon solid-state devices. Early silicon transistors cost $85 each, and 15–29 were used in a gamma-ray-neutron tool. Their low energy requirements and reliability resulted in a rapid adoption. These devices not only improved reliability, but made combination tools practical. The various combinations, such as the dual induction-Laterolog 8, the Microlaterolog-dual Laterolog, and many others have been used since.

Computer methods appeared about 1970. Coincident with this were investigations to facilitate the transmission of data. At first, logged data were transmitted to the headquarters computer facility and then returned in finished form. This method was superceded with the introduction of the minicomputer. Logging trucks capable of processing data in the field became common. One mineral logging contractor had a complete fleet of 150 trucks in the field by 1979. These were completely digital in nature.

The next step was to put some of the digital control into the downhole tool. This is a continuing process. Also, the logging truck is being used for other types of data processing than merely the processing of logs and logged data. They are becoming mobile geophysical laboratories, capable of gathering and processing data from any source.

In 1978, the petroleum business logged approximately 200 MM ft of borehole, while 67 MM ft additional were logged in the mineral industry. In addition to measuring resistivities in sediments, borehole geophysical logging measures almost every conceivable physical parameter of the formations for exploration, development, and production purposes in petroleum, uranium, coal, copper, heavy metals, evaporates, engineering, and scientific fields.

REFERENCES

Allard, Louis, and Martin, Maurice. *Schlumberger, The History of a Technique*. New York: John Wiley and Sons, 1977.

Houston Chapter of the Society of Professional Well Log Analysts. *The Art of Ancient Log Analysis*. 1979.

CHAPTER 2 Formation Types

There are several major classifications of environment types as far as logs are concerned. We can classify the environments as sedimentary, hard rock, and fluid, for example. No attempt is made to describe these various formations geologically. These descriptions are primarily geophysical.

If we use the three classifications listed in the first paragraph, the sediments can be subdivided into sands, carbonates, and shales. The hard-rock environments can be subdivided into metamorphics and igneous rocks. Perhaps both the metamorphic rocks and igneous environments can be grouped as granites, basalts, quartz, and feldspars.

Fluids, of course, divide into liquids, such as oil, water, and mud, and gases, such as hydrocarbon and others. In between, one must fit foams and emulsions. Sediments usually contain large amounts of water, both as free water and bound water. Hard-rock environments usually contain little water. Fluids move through both, but the format is different for each.

Petroleum logging is almost entirely concerned with porous, sedimentary formation types. Petroleum is found in the formation interstices. Therefore, virtually all interest in petroleum logging is centered around porous sandstones, porous limestones and dolomites, shales, and, occasionally, evaporites. When loggers talk about sandstone, we often mean clastic rocks. When we talk about shales, we mean shales or clays. Hard-rock or tight formations have low porosity.

True sandstones are characterized as granular in nature. The grain sizes of sands vary over a large range: from about 0.0625–2 mm diameter. Mixed with these sizes of sand can be material up to boulder size and down through clay particle sizes. The predominant sizes fall in the sand range, however.

Sandstones may be cemented by various substances, such as calcareous or siliceous compounds. This usually results in a reduction of porosity and perhaps a modification of the porosity type from granular. They typically have some degree of silt and clay in them, which also reduces the porosity but does not substantially alter the porosity type. The shale (silt and clay) may coat the sand grains or may occur interstitially. Sandstones are usually consolidated to some degree. However, they may occasionally be quite unconsolidated in the shallow, nonpetroleum-type boreholes.

Virtually all of the electrical conductivity in a sandstone occurs in the interstitial spaces. Therefore, the sandstone electrical resistivities are determined to a major extent by the fluid, clay, shale, and other material filling the interstices and the amount and type of interconnected pore space—thus, the cementation. Clean, shale-free sandstone resistivities may run from as low as 0.08–0.09 ohmm (35% porosity filled with saturated salt water) to several thousand ohmm in a highly cemented, freshwater-filled sand. If shale is present, either of these extremes may be changed (Table 2–1). Except for clay, resistivity is virtually independent of rock type.

Typical porosities in the shallow mineral and engineering formations vary from about 22–40%. These sands also have substantial amounts of silt, clay, and shale in them. In deeper sands, typical of oil exploration, porosities range from about 7–25 vol%. Cementation is less common in the shallow sands than in the deep ones.

Radiation from sandstones is about 25–40 API units of gamma radiation originating almost entirely from the uranium and thorium daughter products with some emission (about 20%) from potassium 40 (Fig. 2–1). This occurs primarily in the feldspathic components. Densities, acoustic velocities, and neutron scattering depend on the sand material and the amount and type of pore material. Densities range from 2.0–2.65 g/cc. Acoustic travel times range from 57 micro- to 100 μsec/ft. Neutron scattering indicates up to about 35% porosity (Table 2–2).

The quality of sandstone varies continuously from clean sand to shaly sand, sandy shale, and shale. The point of differentiation between shaly sand and sandy shale corresponds approximately with the appearance or disappearance of permeability.

The presence of shale, clay, or silt lowers the real porosity of a sand since it fills the interstitial spaces,

Table 2–1

Formation and Minerals Resistivity Catalog

Formation/Mineral	Resistivity-ohms, m^2/m (at stp unless otherwise stated)*
Sulfur	10^9–10^{16} (1.9×10^{15} at 20°C, 9.5×10^9 at 115°C)
Petroleum	10^9–10^{16}
Biotite	10^{14}–10^{15}
Sylvite	10^{14}–10^{15}
Mica	$10^{14}<$
Halite (common salt)	10^4–$10^{14}<$
Quartz	10^{12}–10^{14}
Muscovite	10^{11}–10^{12}
Calcite	10^7–10^{12}
Cinnabar	10^6–10^{10}
Sheelite	10^6–10^{10}
Anhydrite	10^4–10^{10}
Limonite	10^6–10^8
Sphalerite	10^5–10^7
Anthracite coal	10^{-3}–5 (most below 1 ohm-m^2/m)
Bauxite	10^2–10^6
Hematite	2.1×10^{-3}–1×10^6
Bituminous coal	10–10^6
Basalt	8×10^2–10^5
Gabbro	8×10^2–10^5
Diabase	8×10^2–10^4
Gneiss	10^2–10^4
Subbituminous coal	10^2–10^4
Dolomite	1–7×10^3
Limestone (dense)	80–6×10^3
Lignite	4×10^3
Conglomerates	up to 3,000
Siderite	10–1,000
Sand	up to 1,000
Slate	up to 1,000
Marl	up to 600
Siltstone	up to 300
Peat	10–300
Argillite	1–300
Braunite	0.01–100
Ilmenite	0.01–100
Marcasite	0.01–100
Shale	up to 15
Pyrolusite	1–10
Chalcopyrite	0.001–0.1
Sulfides	0.001–0.1
Pyrite	0.0001–0.1
Magnetite	0.0001–0.01
Bornite	0.000001–0.01
Galena	0.00001–0.001
Pyrrhotite	0.00001–0.0001
Graphite	0.000001–0.0001
Native metals	less than 0.000001

*The resistivity of most bedded formations varies with the direction of measurement.

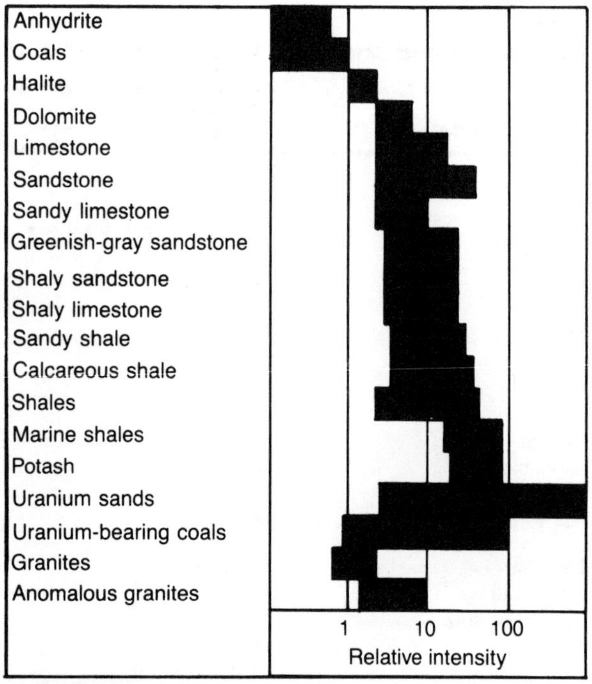

FIG. 2–1

Relative natural gamma-ray intensities

(after Russell, "Well Logging by Radioactivity," *AAPG Bulletin*, volume 5, No. 9, 1941)

displacing the pore fluid. Because the electrical, radiological, and mechanical properties of clay are so different from those of the quartz grain materials of sand, the presence of shale profoundly alters the logged characteristics of a sand. The electrical characteristics of fine-grained quartz also depart from those of coarser-grained material because of grain surface and edge effects.

The resistivities of shales and clays are lower than would be expected from the water that can be extracted from them because of the cation exchange capacity (CEC) of the contained clays and silts. Clays have the capacity to absorb calcium, magnesium, hydrogen, potassium, sodium, and ammonia ions and retain them in an exchangable state. The exchange capacities of some common clays are as follows:

	milliequivalents/100 g of clay
Kaolinite	3–15
Halloysite $2H_2O$	5–10
Halloysite $4H_2O$	40–50
Montmorillonite	80–150
Illite	10–40
Vermiculite	100–150
Chlorite	10–40
Attapulgite	20–30

Fine-grained minerals, such as fine-grained quartz, silts, and zeolites, exhibit cation exchange capacities because of unsatisfied crystal bonds along the large number of edges of the material grains. Therefore, resistivities of silt, clay, and shale-bearing sands are lower than expected.

The potassium, uranium, and thorium contents of clays contribute an appreciable amount of gamma radiation to clays and shales. A typical midcontinent shale has a gamma radiation of about 100 API gamma units. The potassium content of many common materials is shown in Table 2–3. Marine shales may have radiations up to 150 API units.

The acoustic velocities in shales and clays are low. The velocities of the compressional wave in various geologic materials are shown in Table 7–1. Gamma-ray readings, spontaneous potential readings, acoustic velocity, and resistivity values furnish clues as to the shale, clay, and silt contents of shaly sands. They are particularly effective when used in crossplots.

Limestones and dolomites generally have lower porosities than sandstones. Usually, the porosities in this type of formation are less than 25 vol%. The porosity, too, takes a somewhat different form, which may affect some of the logged results. When carbonates are oolitic, the porosity affects logged results in much the same way as that of sandstones or clastics. Vugular porosity may, however, be composed of vugs large enough and poorly distributed enough that a nonrepresentative measurement is obtained by the logging system.

Obviously the shape of the pore spaces affects physical measurements. This is reflected in the variation of the value of m with cementation in the formation resistivity factor expression:

$$F = a\phi^{-m} \tag{2.1}$$

Also, very small pore spaces affect both electrical and mechanical measurements (Fig. 2–2).

Carbonates are characterized in logging by high resistivities, high acoustic velocities, moderately high densities, and low radioactivities. These are, respectively, 100 to >10,000 ohmm, about 20,000 ft/sec, 2.4–2.9 g/cc, and about 10–20 API units.

To a great extent, the rock or formation types discussed so far are common to petroleum, uranium, and coal logging. In hard-rock mineral logging, however, we are beginning to log in harder, denser, lower-porosity rocks. Granite, quartz, basalt, and other massive rock-type formations are commonly encountered in hard-rock uranium exploration. Porosity is found mainly in fracture systems. Conductivities

Table 2–2

Formation and Mineral Densities, Z/A Ratios, and Thermal Neutron Capture Cross Sections*

Material	Z/A Ratio	Matrix Density	Apparent Density**	Σ_{ma}, 10^{21} barns/cc
Lead Pb	0.3953	11.34	8.97	5.61
Uraninite UO_2	0.4000	8.25 (6.5–10.8)	6.60	49.69
Cinnabar H_gS	0.4143	8.1 (8.0–8.2)	6.71	7,981.16
Iron Fe	0.4687	7.87	7.38	214.90
Galena PbS	0.4093	7.5 (7.4–7.6)	6.14	12.47
Wulfenite $PbMoO_4$	0.4187	6.9 (6.7–7.0)	5.78	32.50
Arsenopyrite FeAsS	0.4605	6.1 (5.9–6.2)	5.62	165.22
Cobaltite CoAsS	0.4517	6.1 (6.0–6.3)	5.51	936.73
Chalcocite Cu_2S	0.4610	5.65 (5.5–5.8)	5.21	173.56
Hematite Fe_2O_3	0.4787	5.26 (4.9–5.3)	5.04	100.47
Magnetite Fe_3O_4	0.4774	5.18 (4.97–5.18)	4.95	112.10
Bornite Cu_5FeS_4	0.4643	5.15 (4.8–5.4)	4.78	145.63
Pyrite FeS_2	0.4850	5.06 (4.95–5.17)	4.91	89.06
Illmanite $FeTiO_3$	0.4757	4.75 (4.5–5.0)	4.52	158.23
Zircon $ZrSiO_4$	0.4691	4.69 (4.2–4.86)	4.40	5.42
Stibnite Sb_2S_3	0.4436	4.57 (4.52–4.62)	4.05	17.82
Pyrrhotite Fe_5S_6	0.4812	4.55 (4.58–4.64)	4.40	90.52
Barite $BaSO_4$	0.4454	4.45 (4.30–4.60)	3.96	19.40
Chromite $FeCr_2O_4$	0.4753	4.45 (4.30–4.60)	4.23	102.20
Rutile TiO_2	0.4756	4.20 (4.15–4.25)	3.80	202.75
Chalcopyrite $CuFeS_2$	0.4751	4.2 (4.1–4.3)	3.99	80.95
Corundum Al_2O_3	0.4904	4.02 (3.95–4.10)	3.94	11.04
Carnotite $K_2O \cdot 2UO_3 \cdot V_2O_5 \cdot 2H_2O$	0.4350	4.0_+	3.48_+	56.21
Rhodocrosite $MnCO_3$	0.4793	4.0 (3.5–4.0)	3.84	278.93
Sphalerite ZnS	0.4720	4.0 (3.9–4.1)	3.78	38.33
Siderite Fe_2CO_3	0.4797	3.88 (3.0–3.88)	3.72	68.81
Limonite $2Fe_2O_3 \cdot 3H_2O$	0.4897	3.8 (3.51–4.0)	3.72	74.10
Dunite (4 samples)	0.4978	3.3 (3.24–3.74)	3.29	17.03
Olivine $(Mg, Fe)_2SiO_4$	0.4892	3.3 (3.27–3.37)	3.23	31.74
Magnesite $MgCO_3$	0.4992	3.1 (3.0–3.2)	3.1	1.48
Norite (11 samples)	0.4970	2.984 (2.720–3.020)	2.97	12.88
Diabase (6 samples)	0.4954	2.98 (2.96–3.05)	2.95	17.12
Gabbro (27 samples)	0.4938	2.976 (2.850–3.120)	2.94	21.47
Anhydrite $CaSO_4$	0.4995	2.95 (2.89–3.05)	2.95	12.30
Aragonite $CaCO_3$	0.4995	2.94 (2.85–2.94)	2.94	8.12
Muscovite $KAl_2(AlSi_3)O_{10}(OH)_2$	0.4966	2.93 (2.76–3.1)	2.91	17.30
Biotite $H_2K(Mg, Fe)_3Al(SiO_4)_3$	0.4900	2.90 (2.65–3.1)	2.84	25.20
Dolomite $CaMg(CO_3)_2$	0.4994	2.85 (2.80–2.99)	2.85	4.78
Illite $KAl_5Si_7O_{20}(OH)_4$	0.4954	2.84 (2.60–3.0)	2.81	39.90
Diorite (13 samples)	0.4964	2.839 (2.721–2.960)	2.82	14.33
Langbeinite $K_2Mg_2(SO_4)_3$	0.4961	2.83	2.61	78.87
Polyhalite $2CaSO_4 \cdot MgSO_4 \cdot K_2SO_4 \cdot 2H_2O$	0.5013	2.78	2.79	21.00
Synite (24 samples)	0.4971	2.757 (2.630–2.899)	2.74	16.43
Granodiorite (11 samples)	0.4963	2.716 (2.668–2.785)	2.696	11.33
Chlorite $(Mg, Al, Fe)_{12}(Si, Al)_8O_{20}(OH)_{16}$	0.5056	2.71 (2.60–3.22)	2.74	17.56
Calcite $CaCO_3$	0.4996	2.71 (2.71–2.72)	2.71	7.48
Aluminum Al	0.4818	2.70	2.60	13.99
Plagioclase feldspar $xNaAlSi_2O_8$, $yCaAl_2Si_2O_8$	0.4925	2.69 (2.62–2.76)	2.65	6.99
Limestone (average of 345 samples)	0.5000	2.69 (2.66–2.74)	2.69	8.72
Granite (155 samples)	0.4969	2.667 (2.516–2.809)	2.65	11.62
Quartz SiO_2	0.4993	2.65 (2.65–2.66)	2.65	4.36

(Courtesy of Gearhart Industries Inc.)

*Density at room temperature and 1 atm pressure unless otherwise stated

**Based on tool calibration using a Z/A ratio of 0.5.

Table 2–2 *continued*

Material	Z/A Ratio	Matrix Density	Apparent Density**	Σ_{ma}, 10^{21} barns/cc
Sandstone (average of 12 samples)	0.4990	2.655 (2.59–2.84)	2.655	8.66
Kaolinite $(OH)_8Al_4Si_4O_{10}$	0.5103	2.63 (2.40–2.68)	2.68	13.06
Albite $NaAlSi_3O_8$	0.4885	2.62 (2.61–2.65)	2.56	6.77
Orthoclase feldspar $KAlSi_3O_8$	0.4958	2.57 (2.55–2.63)	2.55	16.00
Kieserite $MgSO_4 \cdot H_2O$	0.4724	2.57	2.43	12.77
Concrete		2.35 (1.98–2.35)		
Montmorillonite $(OH)_4Si_8Al_4O_{20} \cdot nH_2O$ (n = 1)	0.5009	2.35 (2.00–3.00)	2.35	8.10
Gypsum $CaSO_4 \cdot 2H_2O$	0.5111	2.32 (2.30–2.35)	2.37	19.40
Glauconite KMg (FeAl) $(SiO_3)_6 \cdot 3H_2O$	0.4998	2.30 (2.20–2.80)	2.30	16.80
Graphite C	0.4995	2.22 (2.09–2.23)	2.22	0.38
Serpentine $Mg_3Si_2O_5(OH)_4$	0.5062	2.20	2.23	8.80
Halite NaCl	0.4799	2.16 (2.135–2.165)	2.07	752.36
Nahcolite $NaHCO_3$	0.4905	2.20	2.16	
Kainite $MgSO_4 \cdot KCl \cdot 3H_2O$	0.5140	2.13 (2.1–2.13)	2.19	196.13
Trona $Na_2CO_3HNaCO_3 \cdot 2H_2O$	0.5043	2.125 (2.11–2.15)	2.14	16.21
Sulfur, orthorhombic (below 95.4°C)	0.4990	2.07 (2.05–2.09)	2.07	19.06
Potash $K_2CO_32H_2O$	0.5049	2.04	2.06	39.70
Sylvite KCl	0.4829	1.99 (1.97–1.99)	1.92	570.68
Cement (32 samples)		1.99		about 13
Sulfur, monoclinic (above 95.5°C) at 1 atm; above 150°C at 1.304 atm) S	0.4990	1.96	1.96	18.05
Kernite $Na_2B_4O_7 \cdot 4H_2O$	0.5026	1.91	1.92	12,793.69
Carnallite $KMgCl_3 \cdot 6H_2O$	0.5095	1.61 (1.60–1.61)	1.64	370.92
Anthracite coal 0.9350(C) 0.0281(H) 0.0097(N) 0.0272(O)	0.5134	1.60 (1.32–1.80)	1.64	1.08
Bituminous coal 0.8424(C) 0.0555(H) 0.0152(N) 0.0869(O)	0.5201	1.35 (1.15–1.7)	1.40	1.54
Lignite		1.10 (0.5–1.5)	1.16	
Water (300,000 ppm NaCl)	0.5325	1.219	1.298	146.22
(250,000 ppm NaCl)	0.5363	1.1825	1.268	122.55
(200,000 ppm NaCl)	0.5401	1.146	1.238	100.08
(150,000 ppm NaCl)	0.5438	1.109	1.206	78.75
(100,000 ppm NaCl)	0.5476	1.073	1.175	58.69
(50,000 ppm NaCl)	0.5513	1.0365	1.143	39.02
(30,000 ppm NaCl)	0.5528	1.022	1.130	32.56
(pure water)	0.5551	1.00	1.11	22.08
NaCl solution density at STP $\cong$ 1 + (.00000073 $\times$ ppm)				
oil $n(CH_2)$, 10°API, stp	0.5703	1.00	1.14	28.02
30°API, stp		0.88	1.00	25.89
40°API, stp		0.85	0.97	24.22
50°API, stp		0.78	0.85	22.23
(C_8H_{18}), 70°API, stp	0.5778	0.70	0.81	22.12
n-Pentane C_5H_{12}, stp	0.5823	0.626	0.733	20.80
200°F, 7,000 psi		0.603	0.702	20.02
n-Hexane C_6H_{14}, stp	0.5803	0.659	0.765	21.38
200°F, 7,000 psi		0.628	0.739	20.37
n-Heptane C_7H_{16}, stp	0.5778	0.684	0.790	21.80
200°F, 7,000 psi		0.657	0.759	20.84
n-Octane C_8H_{18}, stp	0.5778	0.703	0.812	22.12
200°F, 7,000 psi		0.673	0.778	21.12
n-Nonane C_9H_{20}, stp	0.5768	0.718	0.828	22.37
200°F, 7,000 psi		0.686	0.791	21.37
n-Decane $C_{10}H_{22}$, stp	0.5763	0.730	0.841	22.55
200°F, 7,000 psi		0.701	0.808	21.65
n-Undecane $C_{11}H_{24}$, stp	0.5759	0.740	0.852	22.71
200°F, 7,000 psi		0.713	0.821	21.87
Methane CH_4, stp	0.5703	0.000677	0.00076	0.028
200°F, 7,000 psi		0.2189	0.2497	10.88

Table 2–2 *continued*

Material	Z/A Ratio	Matrix Density	Apparent Density**	Σ_{ma}, 10^{21} barns/cc
Ethane C_2H_6, stp	0.5986	0.001269	0.00015	0.051
200°F, 7,000 psi		0.4104	0.4913	16.34
Propane C_3H_8, stp	0.5896	0.00186	0.0022	0.067
n-Butane C_4H_{10}, stp	0.5850	0.00246	0.0029	0.085
Helium He, stp	0.4997	0.00017	0.00017	0.0000
Carbon dioxide CO_2, stp	0.4999	0.001858	0.001857	0.0001
Nitrogen N_2, stp	0.4998	0.001182	0.001185	0.004
Oxygen O_2, stp	0.5000	0.001350	0.001350	0.00001
Hydrogen sulfide H_2S, stp	0.5281	0.001438	0.001519	0.029
Air (dry) (N, 78%; O, 21%; A, 1%)	0.4997	0.001224	0.001223	
Argon A, stp	0.4859	0.001688	0.00144	0.017
Average natural gas, stp	0.5735	0.0007726	0.000886	
200°F, 7,000 psi		0.252	0.289	
Hydrogen H	0.9921	0.00009		
Oxygen O	0.5000	0.001429		
Nitrogen N	0.4998	0.00125		
Carbon C	0.4995	3.52		
Calcium Ca	0.4990	1.5		
Sulfur S	0.4990	2.07		
Silicon Si	0.4985	2.4		
Magnesium Mg	0.4975	1.74	1.73	

**Based on tool calibration using a Z/A ratio of 0.5.

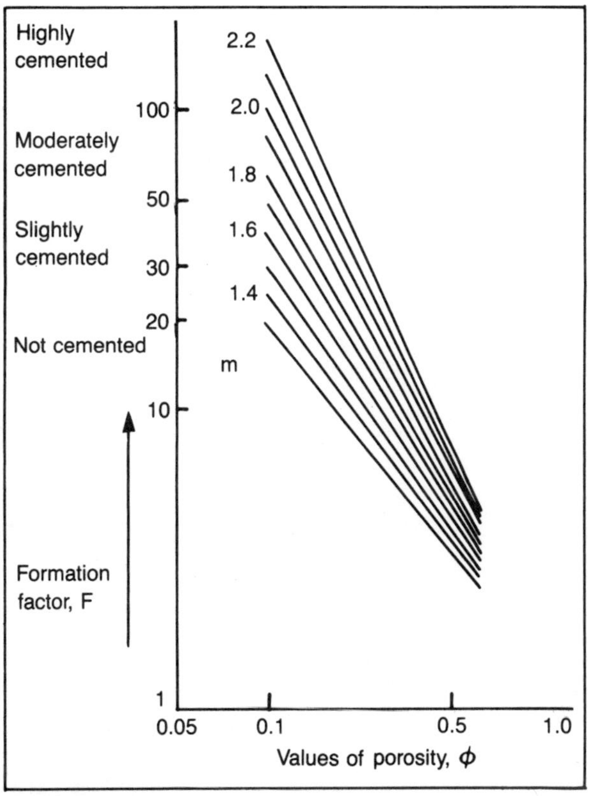

FIG. 2–2

Formation factor vs. formation porosity

(per Archie and Humble formulas, after Pirson, *Oil Reservoir Engineering,* courtesy McGraw-Hill, 1958)

are low (less than 1 mmho-m) and acoustic velocities are high (more than 20,000 ft/sec).

Monzonites and diorites are common in heavy metal prospects, and basalts may be found in many locations. These types are characterized by high acoustic velocities, extremely low and poorly distributed and often fractured porosity, low to high radioactivities, extremely high resistivity, and structural competence. Resistivity and potential logging systems are almost useless in this type formation. Guard log systems are occasionally useful. On the other hand, density, gamma-ray, neutron, and acoustic velocity systems are definitive.

Conductive solids such as pyrites and other sulfides frequently occur with hard-rock materials. These generally have low resistivities, high densities, and a uniform structure. Evaporites are usually classed with the sediments. They have characteristics that cover a broad range. Usually, they are highly resistive. Typically, they cause sudden, marked drops in the resistivity of the drilling liquid. Their radioactive characteristics range from low gamma emission, because of their crystalline structure, to fairly high gamma output, because of their potassium content. In density they range from very light to as heavy as over 3 g/cc.

Hard-rock environments require a careful setting of scales because parameters such as density vary frequently, over a limited range. The characteristic

Table 2–3

Potassium Content of Various Formation Materials*

Material	Potassium Content, wt % Average	Range
Sylvite	54	
Potash	44.9	
Langbeinite	20	
Microcline	16	
Kainite	15.1	
Carnallite	14.1	
Orthoclase	14	
Polyhalite	12.9	
Muscovite	9.8	
Biotite	8.7	
Illite	5.2	3.51–8.31
Arkose (sandstone)	4.6	4.4–5.1
Synite	4.53	
Glauconite	4.5	3.2–5.8
Granite	4.0	2.0–6.0
Norite	3.3	
Granodiorite	2.90	
Shale	2.7	1.6–9.0
Igneous rock	2.6	
Graywacke (sandstone)	1.8	1.2–2.1
Diorite	1.66	
Basalt	1.3	
Sandstone	1.1	0–5.1
Gabbro	0.87	
Diabase	0.75	
Kaolinite	0.63	0–1.49
Limestone	0.27	0–0.71
Montmorillonite	0.22	0–0.60
Orthoquartzite (sandstone)	0.08	0–0.12
Dolomite	0.07	0.03–0.1
Dunite	0.04	
Sea water	0.035	

*This table shows the effects of formation potassium content on gamma-ray response. About 0.012% of all natural potassium is radioactive (i.e., K-40) giving off a 1.5 MeV gamma ray on disintegration. Approximately 20% of the gamma-ray emissions from shale are caused by the isotope K-40. The remaining radiation from shales is generally caused by uranium and thorium series elements. The average sandstone contains about 12% feldspar. Even orthoquartzite sands may contain 10% feldspar. "Although any kind of feldspar may be present the acid feldspars, particularly potash-bearing (orthoclase) varieties are most common" (Pettijohn). A number of sandstones from California have been reported to contain an average of 50% feldspar.

values can be discerned. However, calibrations must be done with precision, and scales must be chosen with some care.

Many special formation materials and minerals exist. These cases include coal, oil shale, tar sands, water, and organic sediments. Before logging for a certain type of material, we should define its physical parameters and log responses. Table 2–2 shows typical response ranges of some of the common materials and logging systems.

REFERENCES

Gearhart-Owen Industries Inc. *Log Interpretation Reference Data Handbook*. Fort Worth: Gearhart-Owen Industries Inc., 1972.

Keller, G.V., and Frischknecht, F.C. ''*Electrical Methods in Geophysical Prospecting*. Elmsford: Pergamon Press.

LeRoy, L.W., and LeRoy, D.O. *Subsurface Geology*. Golden: Colorado School of Mines.

Stokes, W.L., and Varnes, D.J. *Glossary of Selected Geologic Terms*. Golden: Colorado School of Mines.

CHAPTER

3 Borehole Effects

Borehole geophysical logging requires the presence of a borehole, by which the probe and instruments gain access to the formation for detailed measurements. The fluids in the borehole provide the means for the measuring device to contact the formation. These fluids also distort the measurements. Therefore, the presence of the borehole, its size, its fluid, its shape, and its direction are all vitally important to the borehole measurement.

The borehole is also a profoundly disturbing element to the formation environment. The portions of the formations immediately around the borehole are not the same as those portions farther away from the borehole. The borehole has introduced foreign fluids and stresses into the formation and represents a missing portion of them. Fig. 3–1 illustrates a typical mineral logging setup.

During the drilling of a borehole, the drill bit may be a rotary type or a percussive type. It may be a blind or plug type, a coring type, or a reaming type. It may be a standard or an upreaming system. In all of these systems, portions of the formation are cut away by mechanical force and removed from the borehole. Diagrams of some bit types are shown in Fig. 3–2.

When the formation is cut away by the drill bit by rotary scraping (drag or coring bit) or by hammering (percussion and cone bits), the formation is subject to severe mechanical stress. This stress is not always confined to the immediate vicinity of the cutting zone. The stress is distributed through some volume, the size of which depends on the character of the formation. A slightly consolidated sand formation may confine the resulting strain more closely than a hard formation such as a dense limestone or a basalt. The

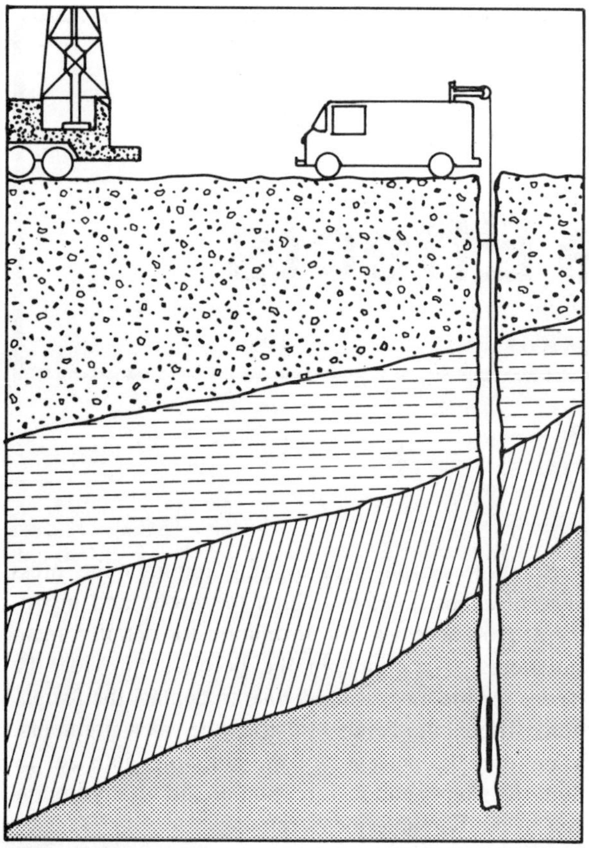

FIG. 3–1
Mineral logging setup

strain may show up as distortion or compaction in a loosely consolidated formation or as micro or macrofractures in a hard formation. The strain also may be dissipated with little or no damage to the formation if the elastic limits of the formation are not exceeded.

The work done in cutting away parts of the formation shows up as heat at the cutting interface. The amount of heat depends on the rate of cutting and on the degree of consolidation of the formation. This heat usually is not a serious problem since one of the purposes of the drilling fluid is to remove it. The drilling fluid may be air, water, a mud mixture, oil, an oil-water emulsion, or a foam. Regardless of the type, the drilling fluid has a great effect on the formation and on the logged measurement.

Drilling mud of any type, especially water-based mud (the most common type), performs several important functions:

- Cools and lubricates the bit
- Cleans the bottom of the hole
- Carries cuttings to the surface
- Permits cuttings removal from the mud at the surface
- Deposits an impermeable mud cake
- Overcomes formation fluid pressures
- Prevents caving of borehole walls
- Avoids damage to or protects productive zones
- Allows good electric logs (and other types) to be obtained
- Protects drillpipe against corrosion

One of the functions of the drilling mud is to prevent caving of the borehole walls. This is an especially severe problem in mineral boreholes because of the nature of the holes.

Except for solution mining projects, mineral boreholes are only drilled to obtain information about subsurface conditions and composition. This is in contrast to the petroleum industry, where the hole

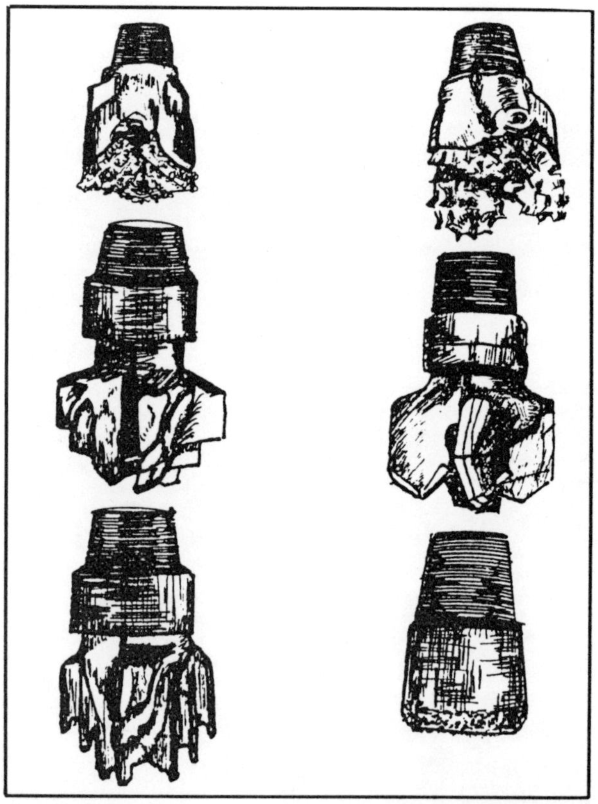

FIG. 3–2
Bit types
(courtesy Barry Hallenburg)

is drilled to produce the petroleum. The cost of the hole must be kept as low as possible, consistent with obtaining good information. This last must be given attention. It is of little use to drill holes on a mineral exploration location or an engineering project if the information cannot be obtained because of poor borehole conditions.

Evidence of borehole problems shows up before logging starts. Large quantities of shale in the returns, sticking or torquing up of the drillstring, bridges, and filling of the hole all indicate hole problems or incipient hole problems. To these can be added many other problems, such as water entry, lost circulation, crooked holes, poor circulation, and changing mud characteristics. All of these indicate that the hole may be unstable. In most cases the fault may be attributed to the drilling mud.

If the formation zone is porous or clay/shale, it contains fluids. The fluid may be liquid or gas and is usually at a normal hydrostatic pressure. Most often the fluid is water with some amount of dissolved salts. If the hydrostatic head of the mud is lower than that of the formation fluids, the pressure of the formation fluids forces fluids and fragments of formation into the borehole. The hydrostatic head of the mud must be kept higher than that of the formation components. Water-loss and weighting materials must be used if indicated.

However, a higher mud hydrostatic head than the formation pressure can result in excessive filtration into the formations and, perhaps, lost circulation. These must be controlled by water-loss additives, lost-circulation materials, and mud weight control. Excessive invasion affects all of the logs, but it affects the electric logs the most.

Borehole instability is more likely to occur if the fluid in the hole is moving violently at the wall. This can be caused by too high drilling rates (rotational speed and mud velocity), the rheology of the mud, and an irregular geometry of the hole (caving). The rheology can be controlled by adjusting the plastic viscosity, the yield point, or the pump rate of the mud. The problem of water loss and filtration affects hole stability severely and also affects the log quality in ways that are difficult to evaluate. The filtrate quantity and quality both have a bearing on the hole stability and log quality. The quantity of filtrate obviously affects electric logs. Some filtration is necessary with a water-based mud to build up a protective mud cake. This quantity can be very small, with a resulting improvement in the resistivity and spontaneous potential measurements. Low filtration quantity (low water loss) can reduce the shale caving problem also. This results in a marked improvement of the quality of all logs, as well as reducing the incidence of stuck tools.

Filtrate quality is determined in water-based muds by the amount and type of dissolved solids. Many organic materials, both natural and added, tend to disperse the clays and thin the muds. This enhances the disintegration of the shales. Organics often slow or prevent the setting of cement. Inorganic salts, such as sodium chloride, retard the disintegration process or even prevent it. The rate of retardation depends on the salt type and concentration. Note that potassium salts in the mud may affect the radioactive logs.

Crooked holes and poor circulation are usually the result of hurrying. They invariably result in logging problems, stuck drillpipe, and loss of the hole. Changing mud characteristics are the result of insufficient circulation or fluid entry into the hole. A borehole should always be well circulated before logging. If required, weighting materials or water-loss materials must be added during circulation. Usually, a good bentonite is enough. Changing or spotty mud characteristics affect most of the logs in ways that are difficult or impossible to evaluate.

CUTTINGS

Drill cuttings are a valuable source of information in mineral, petroleum, and other types of boreholes. Samples must be collected at frequent, regular intervals. They must be laid out in a logical manner. Examination is the duty of the project geologist. Sloppy collection is inexcusable on the part of the driller and is considered shoddy practice. Collection and presentation of the cuttings are not the places for practical jokes.

Cuttings are laid out on the ground as a function of drill depth. The particles, however, take a finite amount of time or lag to reach the surface from the drill bit. The amount of lag depends on the velocity of the mud in the annulus and its viscosity, as well as on the cross-sectional areas and densities of the particles.

After cuttings depths have been corrected for lag, they can indicate many things: the lithology, the effect of the mud on the shale, mineral occurrences, depositional sequences, depositional sources, direction of exploration, and the redox state. These cuttings should all be logged carefully and compared to the indications on the core and wire-line logs. In some cases it is even possible to assay the cuttings, although this is usually a last resort.

Water-based drilling mud is composed of water that contains dissolved salts and other compounds and solids, which are clays, cuttings, and silts. It is circulated down through the central drillpipe (usually), out at the drill bit, around the cutting surfaces, and up the annulus, carrying away cuttings and formation water. At the surface, it flows into a pit where the cuttings drop out. Additives are put in at this point, if required. The mud pump picks up the mud and pumps it down the drillpipe.

Salts and other dissolved materials are in the original water from which the mud is made. Additives may have been used to give the mud particular properties, and dissolved materials are picked up from the formations traversed by the borehole. The solid portion of a water-based mud is comprised of clays, cuttings, sand, and silt picked up by the passage of the drill bit through the formation. Clay and other solids may be added to maintain viscosity and other properties. Sand and heavy solids are dropped out into the mud pit.

When the drilling mud is forced into the formation by the hydrostatic pressure of the drilling mud circulating system, the restricted permeability of the hole wall separates the mud into its two components. The solid portion is deposited on the wall of the hole at a permeable zone as a mud cake. The liquid portion invades the formation as a mud filtrate (Fig. 3–3).

The thickness of the mud cake depends primarily on the water-loss of the mud. To a lesser extent it depends on the permeability of the formation. Low water-loss muds generally have thin, tough mud cakes and are considered to be a better quality mud. Mud cakes may be a small fraction of an inch to several inches thick. It is probable that many of the hole blockages encountered are really mud cake and not caves or ledges.

The depth of invasion of the formation by the mud filtrate depends on the permeability and porosity of the formation. A low porosity or a high permeability formation together with a high water-loss mud result in a deep invasion by the filtrate. A high porosity, a low permeability, or a low water-loss mud restrict the depth of invasion by the mud filtrate. Invasion depth may be a few inches with a low water-loss mud to several feet with a high water-loss mud.

As the mud filtrate invades the formation, it displaces the formation (connate) water. If there is liquid hydrocarbon in the formation, the invading fluid does not displace all of the hydrocarbon because

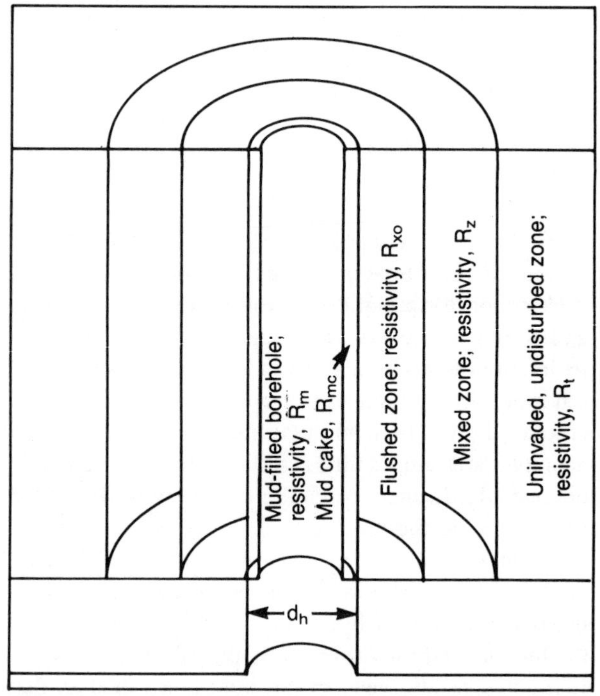

FIG. 3–3

Diagram of an invaded formation

the hydrocarbon preferentially wets the sand grains. If there is interstitial gas, the gas may or may not be displaced by the invading fluid. If the invading fluid is less saline than the formation water and there are clays in the formation, the clay tends to adsorb and absorb the filtrate. This tendency may also be accompanied by swelling. This formation is the so-called *heaving shale*. It swells and sluffs off, which results in permeability changes. If the invading fluid is more saline, the clay loses water and shrinks in volume. If it is less saline, the clay may swell. Thus, there are permeability changes.

Since the invading mud filtrate differs electrically, physically, and chemically from the interstitial fluid, it changes the character of the formation and of any cores taken. The mud filtrate has a resistivity, R_{mf}, that may be different from the resistivity of the formation water, which results in an electrochemical potential at the interface. This is a portion of the spontaneous potential of the formation. The movement itself generates a diffusion potential or electrofiltration potential.

The electrical resistivity of the mud filtrate, R_{mf}, is probably different from the formation water resistivity, R_w. Therefore, the invaded zone resistivities,

R_{xo} and R_z, are different from the undisturbed formation resistivity, R_t (Fig. 3–4).

It is sometimes helpful to use fluid other than water in the borehole. If ground water is not plentiful or if the permeability is low, the hole may be drilled with air. In this case a large compressor furnishes air to the drillpipe to blow the cuttings out of the hole and cool the drill bit. Using a water truck and driver or an expensive mud system can be saved. Often, reverse circulation—up the drillpipe instead of the annulus—is used. This allows a higher velocity for chip removal.

If possible, an air-drilled hole should be logged with radioactive tools and an induction log. Acoustic velocity tools are useless, and electric logs are unsatisfactory. Electric logs are often attempted with scratcher or contact electrodes, but these are invariably noisy.

When a small amount of ground water is present and the permeability is low, foam is often used. This is a water-based mud mixed with a detergent or other foaming agent and foamed with air. High air velocities are not needed to get cuttings out of the hole with foam. The characteristics of the foams vary greatly, so it is difficult to predict how an electric log will act. Spontaneous potential curves are often unreliable. Resistivity curves may be surprisingly good. Radioactive logs operate almost as if the hole is air filled. Foam resistivities should be measured as if the foam were a liquid mud. Do not wait until the foam collapses. In general, resistivities of foam are low, and invasion is negligible.

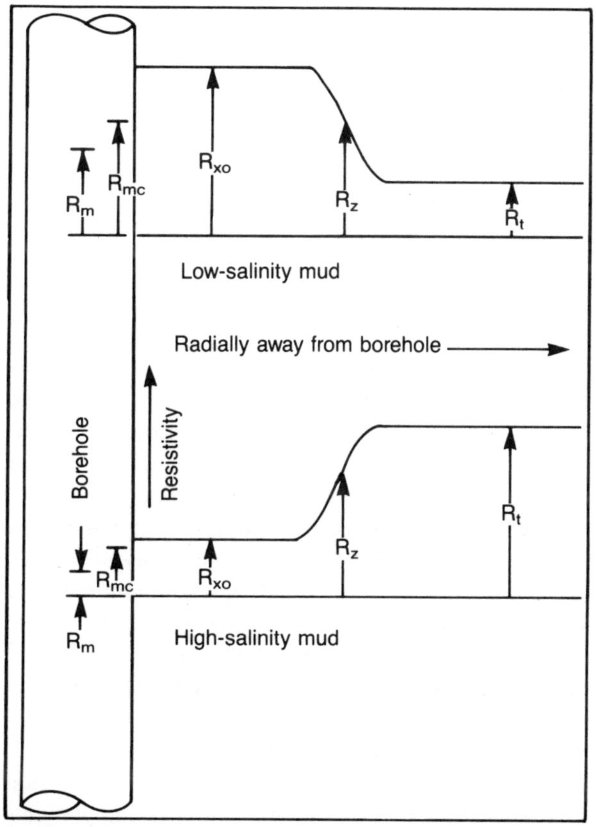

FIG. 3–4

Resistivity of invaded zone

Problems with swelling clay and damaged permeability sometimes are combated with an oil-based emulsion or a brine mud. Oil-based muds and oil-phase emulsions have electrical characteristics similar to an air-filled hole, except for radioactivity logs. Radiation is attenuated as with a water-based mud. Electric logs are less reliable than in air holes. The induction log is excellent, and acoustic velocity systems are usable.

Water-phase emulsions are as unpredictable as foams. In general, electric logs are unsatisfactory in oils and emulsions. Radioactive logs should be handled as in water-based muds. Induction logs and acoustic velocity logs are usable.

Brines usually adversely affect the recording of electric logs, induction logs, and neutron logs. Allowances must be made for the salinity, but often, the systems are quite usable. Spontaneous potential curves are reversed, and resistivity readings are low. Neutron counting rates are low and hole conditions are usually good.

In general, when a drilling program—especially the mud program—is set up, the logging is not considered. In mineral and engineering logging projects there usually is no mud program. This is a serious mistake. The main purpose for drilling an exploratory hole in mineral, engineering, and scientific projects is to get geophysical log data. Even in a petroleum hole, data gathering is of major importance. A mud program must be designed with logging in mind. Muds must be used to extend the life of the borehole and to increase the ease and reliability of interpretation. Money and time are wasted otherwise. Expensive equipment may be damaged, and faulty information may be gathered. It is false economy to save money on drilling by cutting down the usefulness of the logs and risking lost probes.

The borehole itself affects measurements of most logging systems because it is part of the measured volume. This is true of the omnidirectional tools. The directional and sidewall tools have usually been designed to minimize or eliminate borehole effects.

An omnidirectional probe, unless it is deliberately centered, is assumed to be against the wall of the hole. This is probably a valid assumption, except when the borehole is nearly vertical (< 2° slant) and the logging speed is high (> about 20–30 ft/min, 1,200–1,800 ft/hr) in water or mud. Then, the Bernoulli effect tends to float the probe away from the hole wall. This problem should be cured by using eccentering devices on the sonde. In this case the borehole is an unsymmetrical part of the measured volume. However, usually there is an advantage to keeping the probe eccentric.

If an omnidirectional probe is centered in the borehole, the borehole is a symmetrical part of the measured volume of the tool. In this case the effect of the borehole is usually easy to calculate. This is the analysis we usually see in a mathematical study.

The borehole forms a zero portion of the measured volume when it is the same diameter as the sonde. When the diameter of the borehole is increased, the borehole becomes a larger and larger part of the measured volume, until the borehole finally fills the entire volume of a centered tool and the detector is measuring nothing but borehole. If the probe is eccentered, the borehole has no effect if it is the same diameter as the sonde. This case is identical to the centered case. As the borehole diameter is increased, however, with the probe remaining in contact with the wall of the hole, there is always some formation within the measurement volume. The end case, as the diameter of the borehole approaches infinity, is that the borehole wall becomes a flat surface, and the borehole fills only one-half of the measured volume (approximately) with formation filling the other half. Therefore, the eccentered situation is usually preferred. These cases are illustrated in Fig. 3–5.

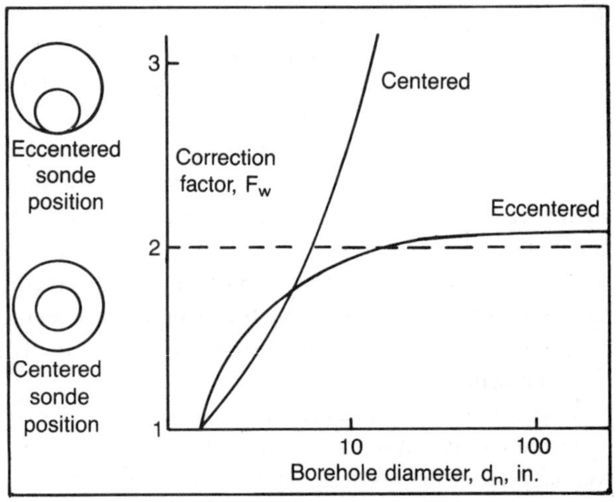

FIG. 3–5

Water-filled hole correction factor

As each system is covered, note the effect of the borehole. It can be negligible or it may be severe. The borehole, hole fluid, and drilling method can have a great effect on the measurement from the borehole. These effects usually must be corrected. Therefore, it is vital that borehole, mud, and drilling information be noted on the log heading. These important data are often left off. Items such as mud resistivity, bit sizes, rig type (rotary, core, cable tool, etc.), mud additives, casing, date, and temperature must be noted on the heading.

Often the borehole is cased to prevent its collapse. If it is to be logged over a period of time and then abandoned, the casing has a minimum of effort and money expended upon it. It is often plastic and is neither centered nor cemented. If the hole is to be used for production of oil, gas, water, uranium, sulfur, or evaporites, the casing is usually steel and is frequently centered and cemented in place. These two situations can produce different effects on any logging done in and through the casing (Fig. 3–6).

Individual peculiarities of each system that is used to log through casing are treated in detail with each type. There are some general things to keep in mind, however. Casing of any type, if it is not centered in the hole, can be a severe problem. The casing often supports its own weight. This causes the casing to have undulations and spirals. A probe in such a casing may be at the wall of the borehole or it may be at the center; there is no way of knowing which. The effect upon the gross-count gamma-ray tool is one of the least severe. Even in this case, however, if we assume the tool is eccentered (the usual case) in a 6-in. diameter hole but the casing position actually has it centered, the signal is 14% in error (Fig. 3–7).

Casing may be made of PVC (polyvinylchloride), ABS (acrylic resin), polyethylene, polypropylene, fiberglass-polyester, fiberglass-epoxy, aluminum, or steel. The chlorine in PVC and the boron in the borosilicate glass of fiberglass have large cross sections to thermal neutrons and reduce the counting rates of the neutron porosity tools and the delayed fission neutron (DFN). Steel and aluminum have a severe attenuating effect on gamma rays. Correction charts and factors are available. Steel casing prevents the use of a magnetic deviation tool. An inertial device must be used in logging through steel casing.

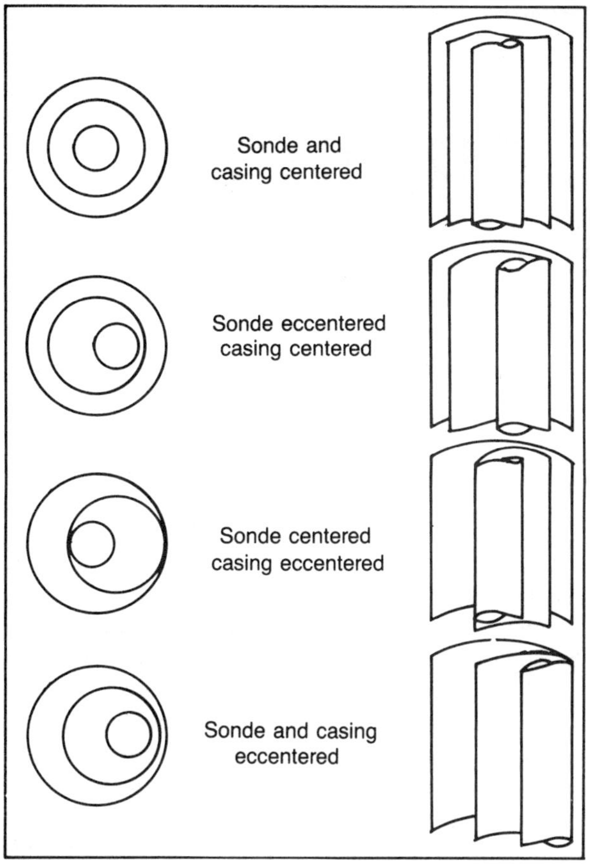

FIG. 3–6

Tool position within casing

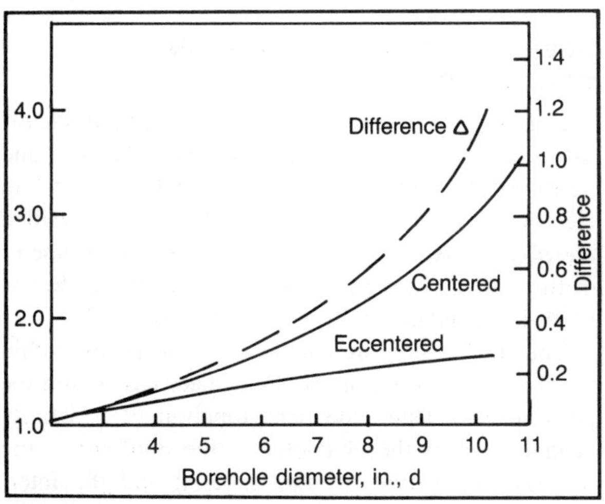

FIG. 3–7

Gamma-ray correction factors in a water-filled hole (after Czubeck, *Quantitative Interpretation of Gamma Ray Logs,* courtesy Institute of Nuclear Research, Krakow, Poland, 1962)

Any casing prevents recording an electric log, but only steel and aluminum casing prevent the use of an induction log.

All casing and drillpipe have collars at their joints that affect radioactivity logs. These are periodic attenuations that appear every 20–30 ft. Drill collars have a severe effect on the radioactivity logs because of their thick steel walls. Often, the driller does not know the wall thickness; he chooses the collar by weight. If he does know the wall thickness, it is probably the new thickness. If the collar is worn, the wall thickness is smaller. The thickness should be measured by the logger. In other respects the drill collars should be corrected for in the same way as the casing.

It is worthwhile to determine whether drill collars have been used during drilling or if a pull-down has been used. Most of the small, mineral-type, portable, rotary drilling rigs use pull-downs in place of drill collars. With this device, part or sometimes all of the weight of the drilling rig is put on the top of the drillstring by a mechanical or a hydraulic system. However, drillpipe is flexible, and this system can give extremely crooked holes, especially if the driller is in a hurry. These holes can give severe problems.

Drill collars are usually used in holes deeper than 200–300 m. These holes are straight and nearly vertical, especially if stabilizers were used. Straight, vertical holes can result in unusual logging problems because the logging tool may not ride on the wall of the hole. These holes are generally better and safer to log than many of the shallower ones.

REFERENCES

Czubek, Jan A. *Quantitative Interpretation of Gamma Ray Logs*. Krakow, Poland: Institute of Nuclear Research, 1962.

Dakhnov, V.N. "Geophysical Well Logging." *Quarterly of the Colorado School of Mines,* volume 57, No. 2, Golden: Colorado School of Mines, April 1962.

Pirson, Sylvain J. *Handbook of Well Log Analysis*. Englewood Cliffs: Prentice-Hall, 1963.

Vennard, John K. *Elementary Fluid Mechanics*. New York: John Wiley & Sons, 1961.

CHAPTER

4 Electric Logging Systems

Electric logging systems are those that measure natural and induced electrical phenomena from the borehole. The name *electric logging* should correctly be applied only to those systems that measure electrical phenomena and is so applied in this text.

The factors that affect the electrical properties of rocks, in decreasing order of their importance, are: porosity and water content, water chemistry, degree of alteration and mineralization, amount of evaporites, amount of humic acids, temperature, and others. Electric logging includes resistivity, conductivity, spontaneous potential measurements, and induced polarization measurements. Dipmeters have been included because they are usually resistivity measurements at the present time.

RESISTANCE, RESISTIVITY, AND CONDUCTIVITY

If a material that contains free charged particles is subjected to a voltage gradient, the particles flow and transport their charges. The material conducts an electrical current. The impedance to this flow is called *electrical resistance* and is a function of the geometry of the flow and the resistivity of the material. Resistivity is a fundamental property of a material.

The current flow in a sediment is primarily within the fluids and some of the solid materials within the pore spaces. Thus, the determination of resistivity is indicative of the character of the pore space and material. Resistivity of the material and the interstitial fluid can be related to the pore volume. Resistivity curves are useful for correlation, determination of stratigraphy, and depositional history. Various tools have been designed for specific types of measurements: shallow, deep, symmetrical, and surface. Some are focused and others are not. Each system has advantages and disadvantages.

Resistivity and conductivity measurements are made by causing a current to flow in the earth. The resulting voltage drop is measured, and the resistivity, resistance, or conductivity is calculated. These measurements are used to identify and determine lithology, stratigraphy, salinity, porosity, and permeability. Since the conductivity through the rock structure of a sedimentary formation is usually lower than that of the fluids filling the pore space, virtually all of the measurement is made in the pore material.

Resistivity measurements reflect the following:

- The bulk resistivity of the formation material
- The degree of water saturation or the relative amount of water within the interconnected pore space of the material
- The resistivity of the waters filling the interconnected pore space of the formation material
- The amount of clay or other conductive mineral within the formation volume
- The resistivity and thickness of the wall mud cake
- The resistivity and diameter of the borehole
- The geometry of the conductive fluids and solids filling the formation volume

Resistivity measurements are affected by and must be corrected for:

- Electrode spacing
- System geometry
- Borehole resistivity
- Borehole size
- Bed thickness

Errors in measurement can be caused by:

- Thin beds
- Exposed metal on the sonde
- Casing
- Fish in the borehole
- Insufficient distance to the return electrodes

Physical Description

If a material contains free, charged particles, such as ions or electrons, and is subjected to a voltage gradient or a voltage field, the charged particles move under the influence of the field and transport their charges through the field. The amount of electrons or equivalent particles flowing is called the *current*. The impedance to their flow is the resistance.

Such a situation is encountered if one connects the two terminals of a battery (a current source) with a wire (a material containing free electrons) (Fig. 4–1). There is a voltage drop or voltage gradient, E, along the wire. We can measure E with a voltmeter, V. If a current meter is put into the circuit, the flow of electrons, I (current), can be measured. If the voltage drop is in volts and the current is in amperes, the relationship tells the resistance, r, in ohms:

$$r = E/I \tag{4.1}$$

This relationship is Ohm's Law. The resistance is a measure of the impedance to the flow of current by the wire and is in ohms. If another piece of material, such as carbon, rock, or metal, is put into the circuit, as in Fig. 4–2, the resistance of that piece of material may be determined.

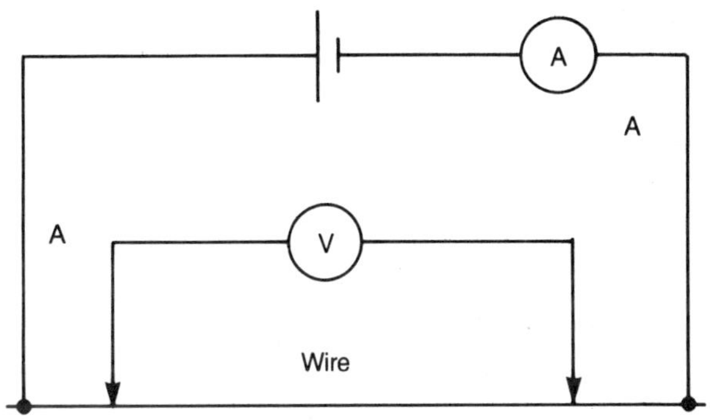

FIG. 4–1
Voltage drop in the wire

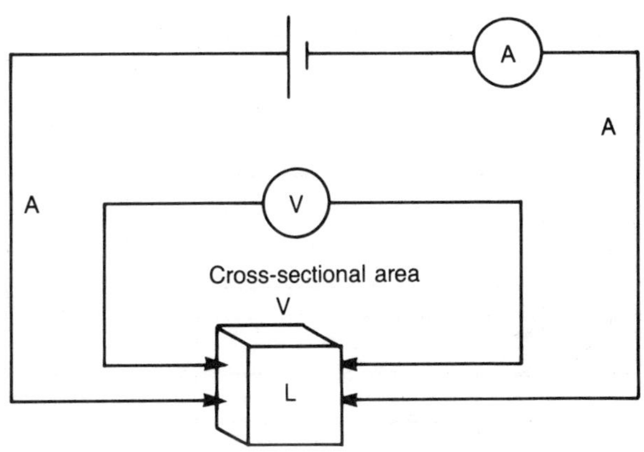

FIG. 4–2
Voltage drop in the block

If two identical blocks of material as in Fig. 4–3 are placed in series, the resistance is twice that of one block. If three are placed in series, the resistance is three times that of one block. Thus, the resistance is some function of the length, L, of the blocks:

$$r = f_1(L) \qquad (4.2)$$

Similarly, if we put the blocks in parallel as in Fig. 4–4, we find that two blocks in parallel have half the resistance of one block and three have one-third the resistance of one block. (The conductance is the reciprocal of the resistance.) Thus, we say that the resistance, r, is also a function of the reciprocal of the cross section of the blocks, (1/A):

$$r = f_2(1/A) \qquad (4.3)$$

Therefore, the resistance r is a function of the geometry of the material.

Eqs. 4.2 and 4.3 can be combined into a single equation with a conversion factor, R:

$$r = R(L/A) \qquad (4.4)$$

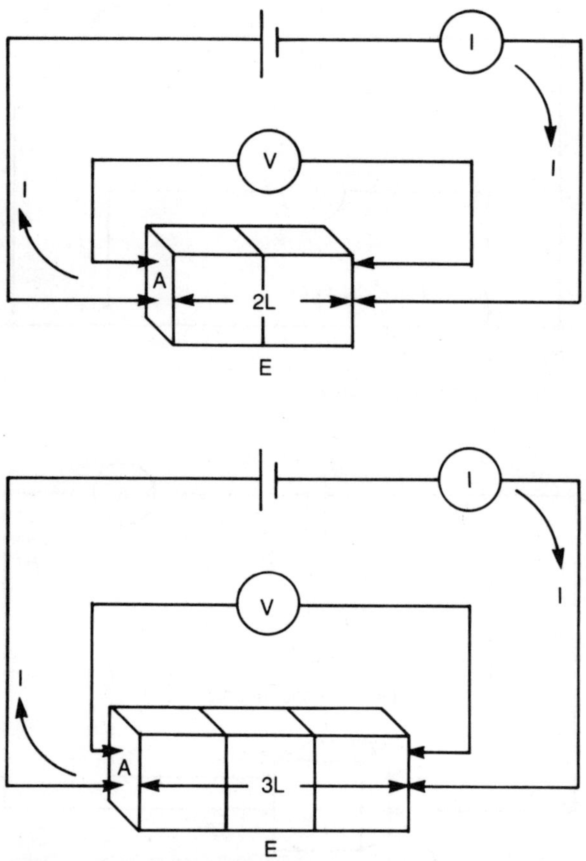

FIG. 4–3

Resistance of blocks in series

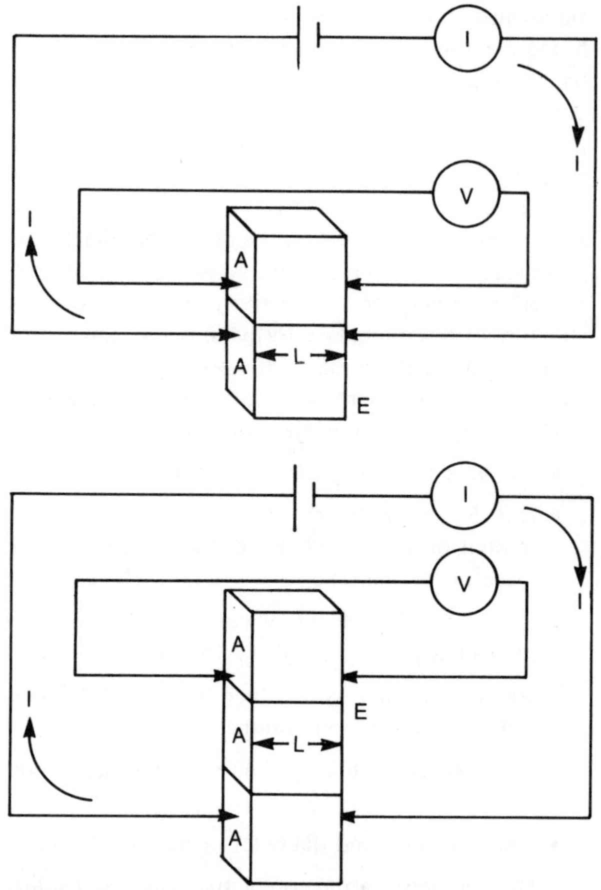

FIG. 4–4

Resistance of blocks in parallel

The resistivity, R, is a fundamental property of the material and is in ohmmeters. Thus, if we can determine the resistivity, R, we have a means of identifying a property of the material:

$$R = r(A/L)$$

This property is independent of the geometry of the material. It is of the same nature as the color, smell, taste, and density. The reciprocal of R is conductivity and is in mhos per meter. Now, substitute a general geometric factor, K, for the particular one (A/L):

$$R = Kr = K(E/I) \qquad (4.5)$$

Eq. 4.5 is the fundamental relationship used for resistivity measurements. The problem for the designer is to determine the form, value, and variations of the constant, K.

Within the formation, the flow of current is primarily within the fluid filling the pore space of a sedimentary formation. Table 4–1 lists the resis-

Table 4–1
Resistivities of Geologic Materials

Material	Resistivity Range, ohmm
Sulfur	10^9–10^{16}
Petroleum	
Biotite	
Sylvite	
Mica	
Quartz	
Muscovite	
Halite	10^4–10^{14}
Calcite	10^7–10^{12}
Cinnabar	10^6–10^{10}
Sheelite	
Limonite	
Anhydrite	10^4–10^{10}
Sphalerite	10^5–10^7
Bauxite	10^2–10^6
Bituminous coal	10–10^6
Basalt	10^3–10^5
Gabbro	
Diabase	10^3–10^4
Gneiss	10^2–10^4
Subbituminous coal	
Hematite	10^{-3}–10^6
Dolomite	1–10^4
Limestone	80–6 × 10^3
Lignite	5–10^3
Conglomerates	up to 3 × 10^3
Siderite	10–10^3
Sand	up to 10^3
Slate	
Marl	up to 600
Siltstone	up to 300
Peat	
Argillite	
Braunite	0.01–100
Ilmenite	
Marcasite	
Shale	up to 15
Pyrolusite	1–10
Chalcopyrite	10^{-3}–10^{-1}
Sulfides	
Pyrite	10^{-4}–10^{-1}
Magnetite	10^{-4}–10^{-2}
Bornite	10^{-6}–10^{-2}
Galena	10^{-5}–10^{-2}
Pyrrhotite	10^{-5}–10^{-4}
Graphite	10^{-6}–10^{-4}
Native metals	10^{-6}

tivities of common geological materials. Notice that most of the common rock materials, such as quartz, calcite, and other sedimentary materials, have extremely high resistivities. Their resistivities range from more than 10^3 ohmm to over 10^{12} ohmm. The major exceptions are the shales and the clay minerals.

The fluids filling the pore spaces are usually solutions of salts and other compounds (mostly ionizable) in water. These compounds are often dissociated and, therefore, electrically conductive. A solution of sodium chloride—for example NaCl, the most common solute—may have a resistivity as low as a fraction of an ohmmeter.

When the pore water and the quartz sand are pictured as resistors in parallel, virtually all of the current of a resistivity measuring system flows in the medium with the lower resistivity, e.g., the pore water solution, rather than in the sand grains, (Fig. 4–5). If the water, for example, has a resistivity of 1 ohmm and the sand grains 1 of 10^{14} ohmm, then 10^{14} times as much current will flow in the fluid as in the sand itself. Thus, the resistivity measurement in a clean sand, or any other shale-free sedimentary formation, is primarily a function of the electrical effects of the pore space.

Shale in the formation is a solid material filling part or all of the pore space in a sand. But, it is very conductive compared to the sand material. In fact, it has a conductivity or resistivity comparable to that of the pore fluid. Therefore, when a formation contains shale, the shale content must be taken into account in any analysis. A clean or shale-free sand is considered first. Then a sand contaminated with shale is considered. Once the resistivity of a sedimentary material has been determined, it may be used to examine the character of that material. The ratio of the resistivity of the formation material, R_o, to

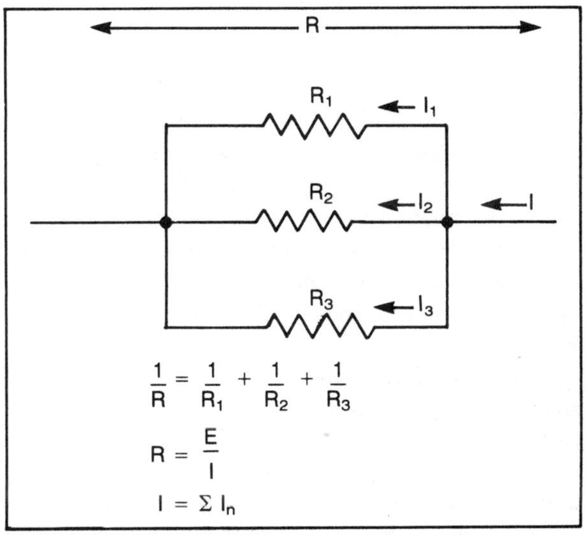

FIG. 4–5
Current flow in parallel resistances

that of the fluid filling the material pore space, R_w, is called the formation resistivity factor, F, or simply, the formation factor:

$$F = R_o/R_w \tag{4.6}$$

If the formation is shale free, has a granular porosity, and is completely water saturated, then the formation factor, F, is:

$$F = a\phi^{-m} \tag{4.7}$$

Where:

a and m = Constants whose values depend on the degree of cementation of the formation

If the formation is not completely water saturated, the amount of saturation, S_w, is

$$S_w = (R_o/R_t)^{1/n} \tag{4.8}$$

Where:

R_t = True resistivity of the formation as measured
R_o = Resistivity of the 100% water saturated formation
n = Saturation exponent

The value of n may vary from 1.5–2.5 and is usually about 2.0; n is similar to m.

The apparent value of the formation factor of the deep, undisturbed formation can be determined using a deep investigation device, such as a focused resistivity or conductivity system. The spontaneous potential curve can be used to calculate a value for R_w. Thus:

$$F = R_o/R_w$$
$$F\ (\text{apparent}) = R_t/R_w \tag{4.9}$$

The relationship in Eq. 4.7 is Archie's Relation:

$$F = a/\phi^m \tag{4.7}$$

Eq. 4.7 was proposed a number of years ago by G.E. Archie of Shell Oil. The exponent, m, of Archie's Relation is called the cementation exponent or constant. The relationship between any value of m, the formation factor, F (ratio of measured resistivity to the water resistivity), and the fractional porosity, ϕ, of the material under consideration is shown in Fig. 2–2.

The value of the cementation constant reflects the shape of the pore volume, or tortuosity. It may vary from a low value (1.3 in an unconsolidated sand) to more than 2.2 in a tight, low porosity limestone or dolomite. The more tortuous the electrical path becomes, the higher the value of m will be.

The constant, *a*, has no real physical meaning. It has been called the compaction constant, but that

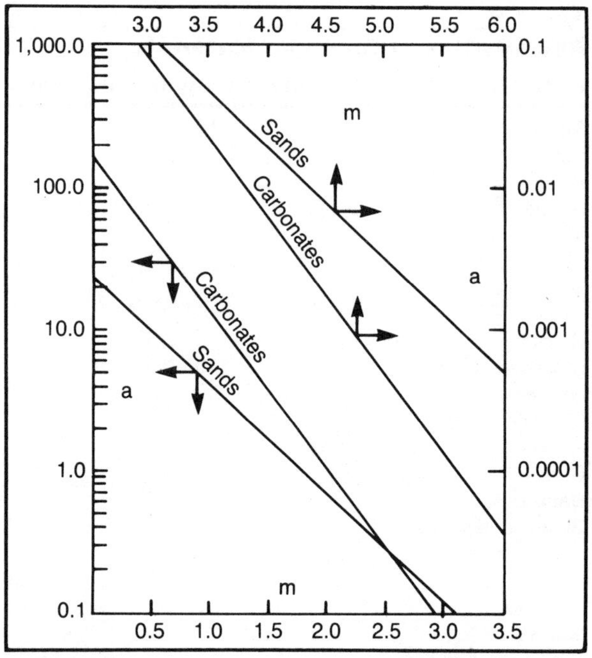

FIG. 4–6

Chart for determining *a* or m

(after Gomez-Rivero, "A Practical Method for Determining Cementation Exponents," SPWLA, *The Log Analyst*, volume XVII, No. 5, 1976)

name is without validity. It is dependent on the value of m, as shown in Fig. 4–6.

Obviously, there is an uncertainty possible in the determination of the values of m and, thus, *a*. There are several possible ways to determine m:

1. If the value of the porosity is determined by analysis of a sufficiently representative sample from cores, log values of the measured resistivities, $R_{measured}$, and the fluid resistivities, R_{fluid}, may be applied in Archie's Relation and the chart of Fig. 4–6:

$$m = \frac{\log(a\ R_{fluid}/R_{measured})}{\log \phi} \tag{4.7a}$$

Where:

log = Logarithm to the base 10

2. The resistivity and porosity values of the core samples (instead of the log values) may be used, as in method 1. This method is probably a little more reliable than the first.
3. Neutron, density, or acoustic devices may be used to determine porosity. Then, method 1 or 2 may be applied.
4. Several resistivity devices may be used to solve for the three unknowns in Archie's Relation.

5. The values of m and a may be determined empirically in a given zone in one field. Simply use the values that experience shows work best.
6. Finally, any combination of these methods obviously may be used.

Since a is a dependent variable, there have been suggestions to eliminate it by setting it equal to one and modifying the value of m:

$$a/\phi^{m} = a\phi^{-m} = \phi^{-m_b} \quad (4.7b)$$

This m_b becomes a new cementation constant whose value is porosity dependent:

$$m_b = m - \frac{\log a}{\log \phi} \quad (4.7c)$$

Raiga-Clemenceau advances a good argument coupling the value of m to the permeability, k, of a sand:

$$m_k = 1.28 + \frac{2}{\log k + 2} \quad (4.7d)$$

Where:
k = md

Thus:

$$F = \phi^{-m_k} \quad (4.7e)$$

This relation appears to fit the presented data quite well. It has the tremendous advantage of suggesting a method for determining permeability or explaining some variations of m:

$$a\phi^{-m} = \phi^{-m_b}$$

$$\log a - m \log \phi = -m_b \log \phi$$

$$\frac{\log a - m \log \phi}{\log \phi} = m_b$$

$$\frac{m \log \phi}{\log \phi} - \frac{\log a}{\log \phi} = m_b$$

$$m - \frac{\log a}{\log \phi} = m_b$$

Water Resistivity

The resistivity values and other effects measured in electric logging depend primarily on the solution resistivities involved. These generally are the mud resistivity, R_m, occasionally the mud-cake resistivity, R_{mc}, and the formation-water resistivity, R_w. These are all basically water-solution resistivities, since most of the other components (the matrix, gas, hydrocarbons, etc.) have much greater resistivities. The major exception to the latter is the shale or clay resistivity, which will be discussed separately.

The conductivity (1/R) of any material is a measure of its ability to carry or conduct an electrical current (Eq. 4.5). When electrical conductors are in parallel, the fraction of the total current carried by each is in direct proportion to its conductivity. A sand and water mixture is such a parallel system (Fig. 4–7). Water solution conductivities may run from about 5 mmho/m to many thousands. Most rock materials, on the other hand, have conductivities much lower than 1 mmho/m. Quartz, for example, has a conductivity range from 1×10^{-9} to 1×10^{-11} mmho/m. Thus, in any mixture of quartz, sand, and water, most of the electrical current (and thus, most of the measurement) flows through the water by a factor of 10^9 or more. With a 50% mixture of quartz and water, which has a conductivity of 1,000 mmho/m (1 ohmm resistivity), 0.5×10^{-11} part of the current would flow through the quartz and 0.99999999995 part of the current would flow through the water solution. Thus, with the normal resolution of our logging instruments (approximately ± 1%), essen-

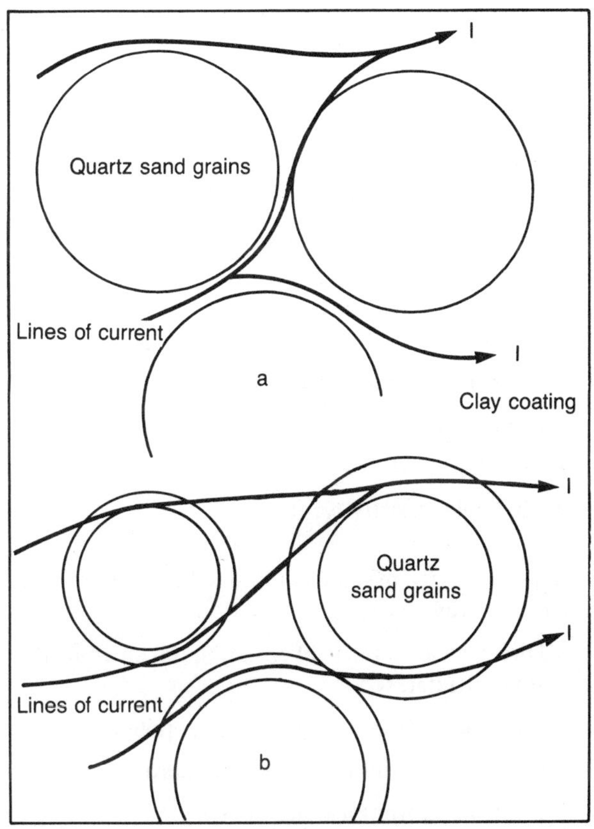

FIG. 4–7
Sand and water parallel system

tially none of the electrical current used in measurement of resistivity in sediments flows through the rock matrix. Therefore, with the exception of shales, clays, and conductive solids (pyrites, etc.), the resistivity and spontaneous potential measurements are independent of the chemical composition of the rock type.

The resistivity of a water solution depends on the types of ions involved, the concentrations of the ion types, and the temperature of the solution. This resistivity is R_f:

$$R_f = \frac{1}{\Sigma(C_a L_a + C_c L_c)} \qquad (4.8a)$$

Where:

C = Ion concentration
L = Equivalent conductance
Subscripts *a* and c = Anion and cation types, respectively

If the resistivity, R_{f1}, is known at any temperature, T_1, the resistivity, R_{f2}, at any other temperature, T_2, may be calculated:

$$\frac{R_1}{A_1} = \frac{R_2}{A_2} \qquad (4.8b)$$

Where:

$$A = \frac{1}{1 + 0.025\,(T_c - 18°C)} = \frac{1}{1 + 0.014\,(T_F - 64.4°F)} \qquad (4.8c)$$

The water resistivity may be encountered in several forms:

- Formation water resistivity, R_w
- Mud resistivity, R_m
- Mud-filtrate resistivity, R_{mf}
- Mud-cake resistivity, R_{mc}

The resistivity may be determined in one or more of several ways:

- Direct measurement of a sample taken from the zone of interest
- Calculation from the spontaneous potential
- Calculation from resistivity and porosity data
- Conversion from a water analysis
- Information from catalogues

Direct Measurement from a Sample

A sample of mud may be measured directly for a value of R_m. It may be separated with a mud press to obtain a sample of filtrate (the liquid portion of the mud) for a value of R_{mf} and the mud cake (the solid portion of the mud) for a value of R_{mc}. Many times, the driller can obtain a sample of water from a permeable, water-saturated zone if the circumstances are good. This is usually true in a water well but can be done often in other boreholes.

The sample is measured with a mud tester, a resistivity-measuring circuit with a set of electrodes designed for the purpose (a mud cup). A standard resistivity-measuring circuit in a logging unit can be easily adapted for this purpose.

Calculation from the Spontaneous Potential

The procedure for calculating the spontaneous potential is covered later. It suffers from the problem that total dissolved solids (TDS) in equivalent sodium chloride units (ppm eNaCl) is determined. If the relative ionic concentrations of the dissolved solid content is known or can be estimated, a good correction from the resistivity of the equivalent sodium chloride solution to the actual resistivity can be made:

$$\log(\text{ppm eNaCl}) = \Sigma C_a D_a + \Sigma C_c D_c \qquad (4.9)$$

Where:
C = Ionic concentration in ppm
D = Dunlap multiplier for that ion type

Fig. 4–8 is a chart showing the Dunlap multipliers for the several common ion types. The resistivity at any temperature can then be found from Fig. 4–52. The equivalent sodium chloride concentration in ppm is:

$$\log(\text{ppm}) \simeq \log\frac{4 \times 10^5}{°F} - 1.058 \log R_f^* \qquad (4.9a)$$

In Eq. 4.9a the term log indicates logarithm to the base 10. In the centigrade scale:

$$\log(\text{ppm}) = \log\frac{2.22 \times 10^5}{C° + 17.8} - 1.058 \log R_f \qquad (4.9b)$$

Calculation from Resistivity and Porosity Data

The preceding expressions (Eqs. 4.6 through 4.9) can be used to determine the fluid resistivity, R_w or R_{mf}, if the porosity is known from an independent source, such as a core analysis, a density log, a neutron log, or an acoustic log.

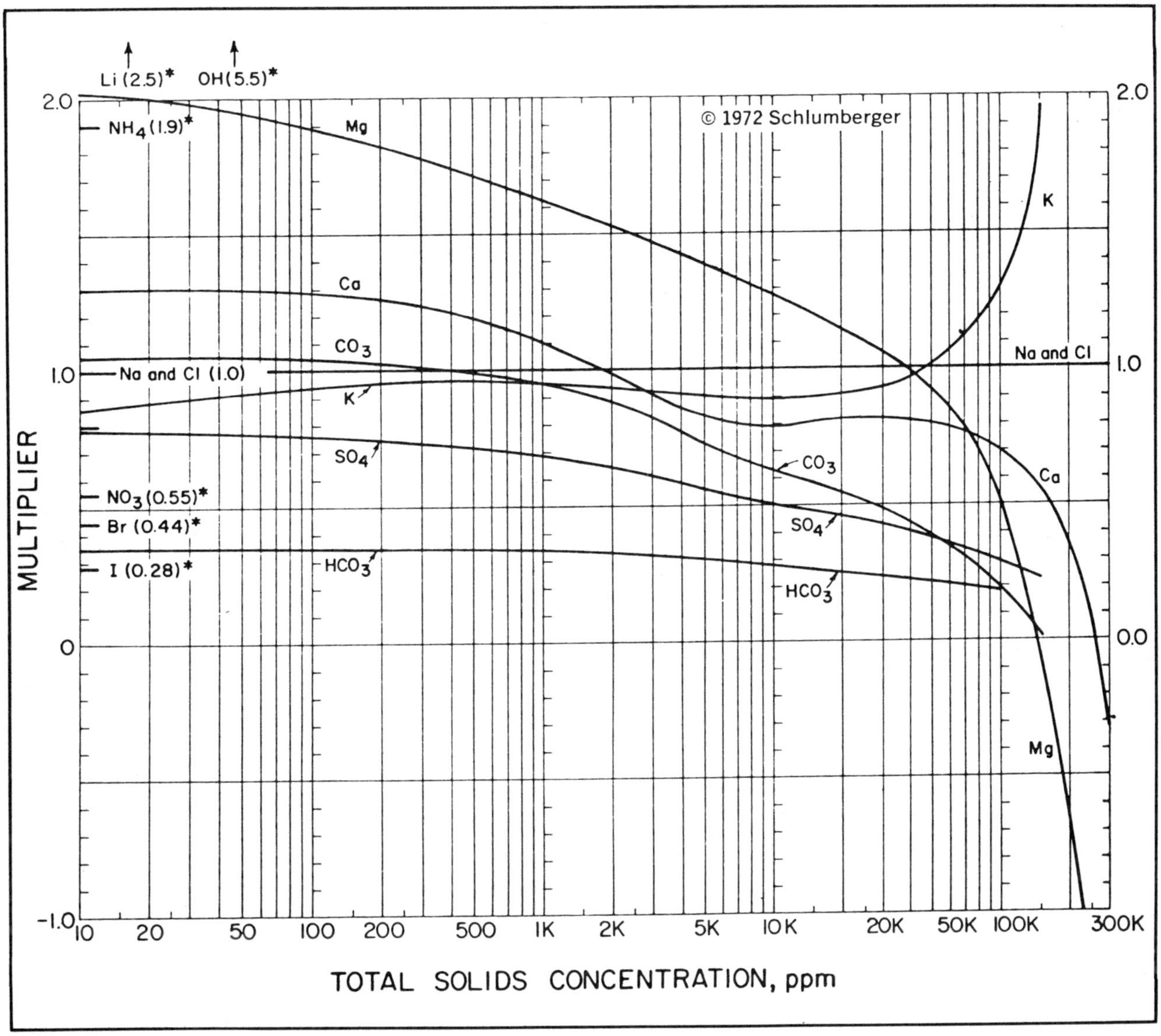

FIG. 4–8

Resistivities of solutions (courtesy Schlumberger Well Services Inc., 1972)

Conversion from Water Analysis

Conversion from water analysis is essentially the reverse of the spontaneous potential determination method. For example, a water analysis gives the following TDS content:

Ion Type	*Concentration*	*Multiplier*	*ppm eNaCl*
Na	2,000	1.0	2,000
Cl^-	1,500	1.0	1,500
$SO_4^{=}$	200	0.6	120
HCO_3^-	100	0.31	31
	3,800 ppm		3,651

From Eq. 4.9a:

$$\log 3{,}651 = \log \frac{4 \times 10^5}{57^\circ} - 1.058 \log R$$

$$R = 1.43 \text{ ohmm at } 75^\circ F$$

Information from Catalogs

Most state geologists and state engineers as well as numerous other organizations catalog water resistivities for well-known water-bearing formation members. Unfortunately, these do not help determine mud-component resistivities, nor are they valid at shallow depths near outcrops. If the values are used, they must be converted to the resistivity at formation temperature.

Actual resistivity measurements are always to be preferred, but the resistivity of a water sample can be estimated from its chemical analysis. An equivalent sodium chloride concentration can be determined by the use of Fig. 4–9.

For total solids concentrations above 1,000 ppm, the chart is constructed from data given by Desai and Moore in their paper, "Equivalent NaCl Concentrations from Ionic Concentrations," *The Log Analyst,* May–June 1969. For lower concentrations, data from the Variable Dunlap Method *(1968 Schlumberger Log Interpretation Chart Book)* are included. Multipliers that do not vary appreciably for low concentrations (less than 10,000 ppm) are shown at the left margin of the chart.

The chart is entered in abscissa with the total solids concentration of the sample in ppm to find the weighting multipliers for the various ions present. The concentration of each ion is multiplied by its weighting multiplier, and the products are summed to get equivalent NaCl concentration.

Similarly, if a shallow device such as a short, normal, single point or a microdevice is used, the resistivity of the flushed zone, R_{xo}, may be measured. Since the flushed zone is filled with the fluid portion of the drilling mud, the mud filtrate resistivity value, R_{mf}, of this fluid may be measured on the surface, eliminating one unknown. In this case:

$$F_{xo} = R_{xo}/R_{mf} \tag{4.10}$$

The value of F_{xo} should equal the value of F if the values are correct.

Oil, gas, or air have resistivities that are on the order of those of the grain material. If the deep, undisturbed zone contains one or more of these, the value of R_t is larger than the resistivity from a zone not containing oil or gas but that is otherwise similar (the value R_o). A comparison allows a calculation of the amount of oil or gas in the formation pore space. This is usually given in terms of the water saturation, S_w, which is one minus the oil saturation. The values of R_o and R_t can be used in expression 4.8. If the value of n is not known, the approximate value of 2 can be used:

$$S_w{}^n = F/F_a \tag{4.11}$$

$$R_o = R_w F$$

$$S_w = (R_w F/R_t)^{1/n} \tag{4.12}$$

In a like manner if the shallow and deep investigation resistivity devices are both available, a comparison of the apparent value of F of the deep zone with that of the shallow zone gives the value or relative amount of movable oil, since some of the oil is left in the shallow zone when it is flushed. The value of F in the shallow zone may be determined with any of the shallow porosity devices, such as the neutron porosity, density, or acoustic velocity systems. Merely solve Eq. 4.7 for F using the porosity determined by one of the other porosity devices.

The value of S_w is actually the ratio of the value of the apparent porosity, ϕ_a, as determined by a deep resistivity or conductivity device to the value of the real porosity, ϕ:

$$S_w = \frac{\phi_a}{\phi} \tag{4.12a}$$

Shales and Fine-Grained Rocks

The presence of a shale and a fine-grained material in a sand or a carbonate usually lowers the resistivity of the horizon and results in apparent porosities that are too large. The presence of this material must be accounted for.

The decrease in resistivity in fine-grained materials is a function of the ion-exchange properties of the included clays and by surface conduction through the water of the double layer along dielectric grain surfaces. The shale correction may take several forms. One of the common methods is proposed by Alger and is now in general use to determine water-saturation values in shaly formations. For dispersed shales, q is the fraction of solid material or matrix occupied by shale:

$$\rho_b = (1 - \phi_z)\rho_m + \phi_z[q\rho_{sh} + (1 - q)\rho_f]$$

Where:

ρ = Density
b = Bulk response
ϕ_z = 1 − matrix volume, V_m
m = Matrix
sh = Shale
f = Free fluid in the pore space
q = Fraction of ϕ_z occupied by shale

Also:

$$q = \frac{\phi_s - \phi_\rho}{\phi_s} = \frac{aV_{sh}}{\phi_s}$$

Where:

ϕ = Calculated or apparent porosity
S = Sonic measurement
V_{sh} = Fractional volume of shale
$a = x_2 - x_1$

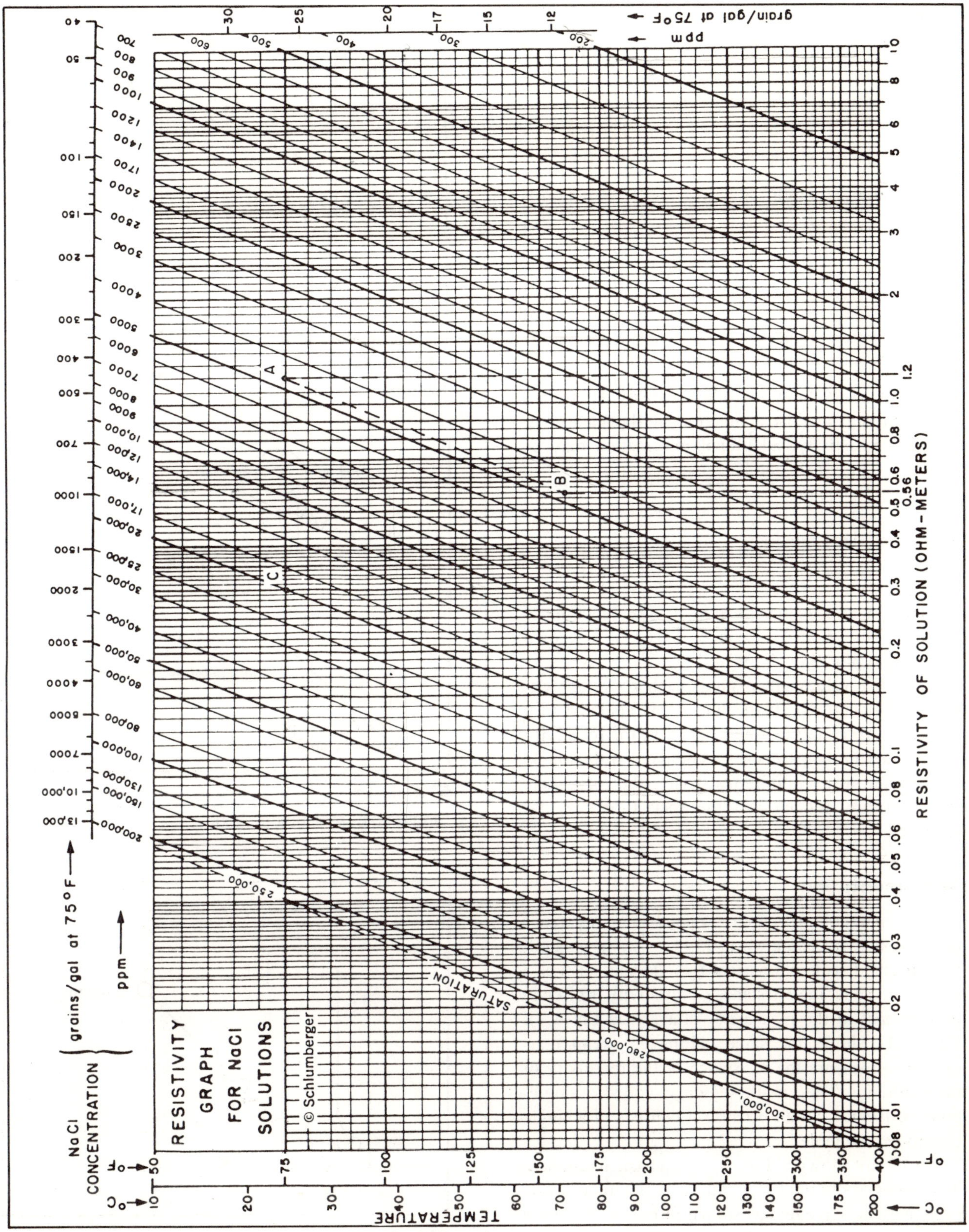

FIG. 4–9 Schlumberger NaCl solution resistivity, salinity chart (courtesy Schlumberger Well Services Inc., 1979)

$$x_1 = \frac{\rho_m - \rho_{sh}}{\rho_m - \rho_f}$$

$$x_2 = \frac{\Delta t_m - \Delta t_{sh}}{\Delta t_m - \Delta t_f}$$

Where:
Δt = Sonic travel time

The water saturation, S_w, in a shaly formation is then:

$$S_w = \frac{[(F_{\Delta t} R_w/R_t) + q^2/4]^{0.5} - q/2}{1 - q}$$

Where:
$F_{\Delta t}$ = Apparent formation resistivity factor determined from the sonic log apparent porosity

If the porosity from any other porosity determination is plotted against a function of the measured, corrected resistivity, additional information may be obtained. The y axis can be scaled as:

$$y = [(aR_w)/R_t]^{1/m}$$

As a generalized relation:

$$y = [(aR_{fluid})/R_{measured}]$$

Then the x axis can be scaled in any function of porosity (i.e., g/cc Δ_t, or ϕ_n). This plot is known as a Hingle plot (Fig. 9–38).

When log values are plotted, each rock type exhibits a linear trend. If the line is extrapolated to a value corresponding to 100% porosity, the resistivity of the water in the pore space is defined. If the line is extrapolated to an infinite resistivity, the rock characteristics are defined and the rock type may be identified.

Instrumentation

The general ohmmeter circuit to measure resistance or the resistivity is shown in Fig. 4–10. In general, the current, I, is kept constant by a device, such as a regulator. Thus, the resistance or resistivity is directly proportional to the measured voltage (see Eq. 4.1). Fig. 4–11 shows the same ohmmeter circuit rearranged to measure the resistance or the resistivity of the earth, r_g. The circuit shown is called a single-point resistance circuit because it consists of a single electrode downhole.

The circuit shown in Fig. 4–11 is a basic resistance-measuring or ohmmeter circuit. The electrode, however, may be separated as shown in Fig. 4–12 into a current electrode, e_a, and a measure electrode, e_m. Similarly, the return electrode may be separated.

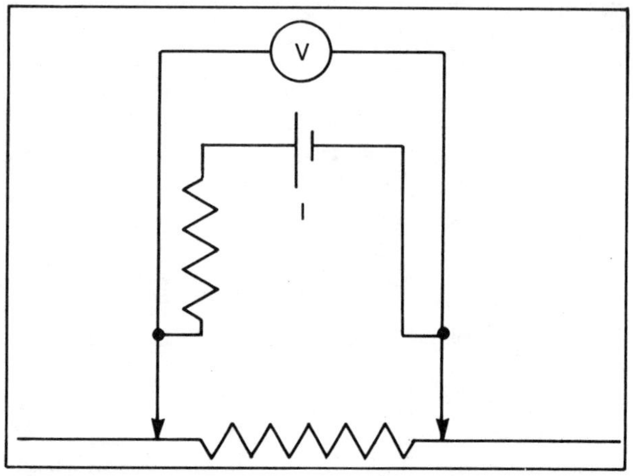

FIG. 4–10
Ohmmeter circuit

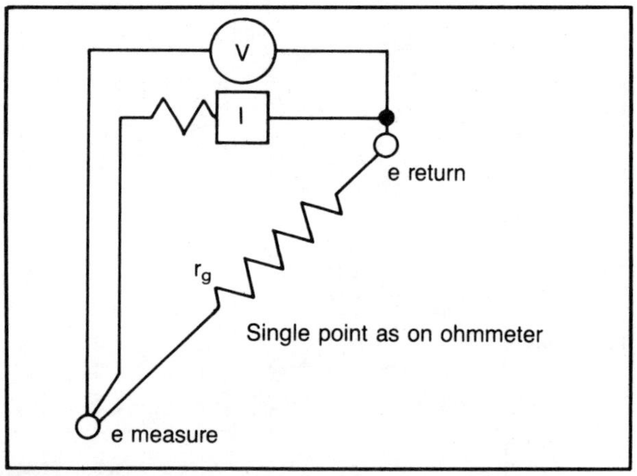

FIG. 4–11
Single-point resistance circuit

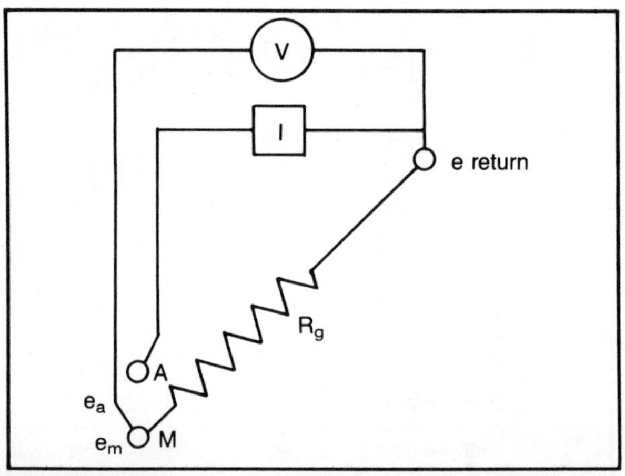

FIG. 4–12
Normal resistivity circuit

These are the variations that constitute normal and lateral arrays.

If the measure circuit uses a separate electrode, the system is known as a normal array or a normal resistivity circuit. It may have two, three, or four electrodes downhole (Fig. 4–13). A lateral or inverse circuit always has three or four electrodes downhole (Fig. 4–14).

The current source is usually an alternating type. However, occasionally (in surface work particularly), a direct current source may be used. The calculations remain the same.

An alternating current source is used because the electrical current flow causes chemical reactions to take place at the electrode surface. These reactions interfere with the flow of current but are short term and reversible. Therefore, the periodic reversal of the flow of an alternating current tends to nullify the effects of the chemical reactions.

With these exceptions, all of the resistance and resistivity circuits are basically as shown in Figs. 4–12 to 4–14. There are some minor exceptions, but the principle remains the same. An electrical current is caused to flow through the subject material, the voltage drop along the current path is measured,

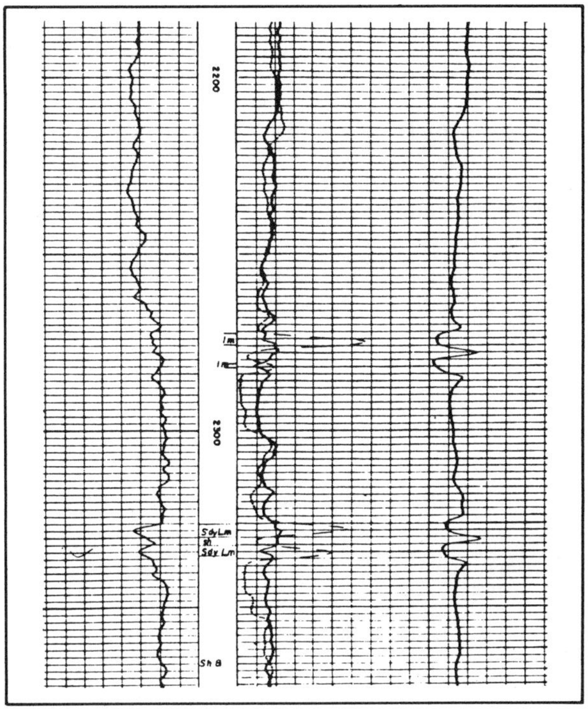

FIG. 4–13

Electric log—SP, 16 in. N, 18 ft Lateral, 64 in. N
(after Guyod, *Resistivity Determination from Electric Logs,* 1958)

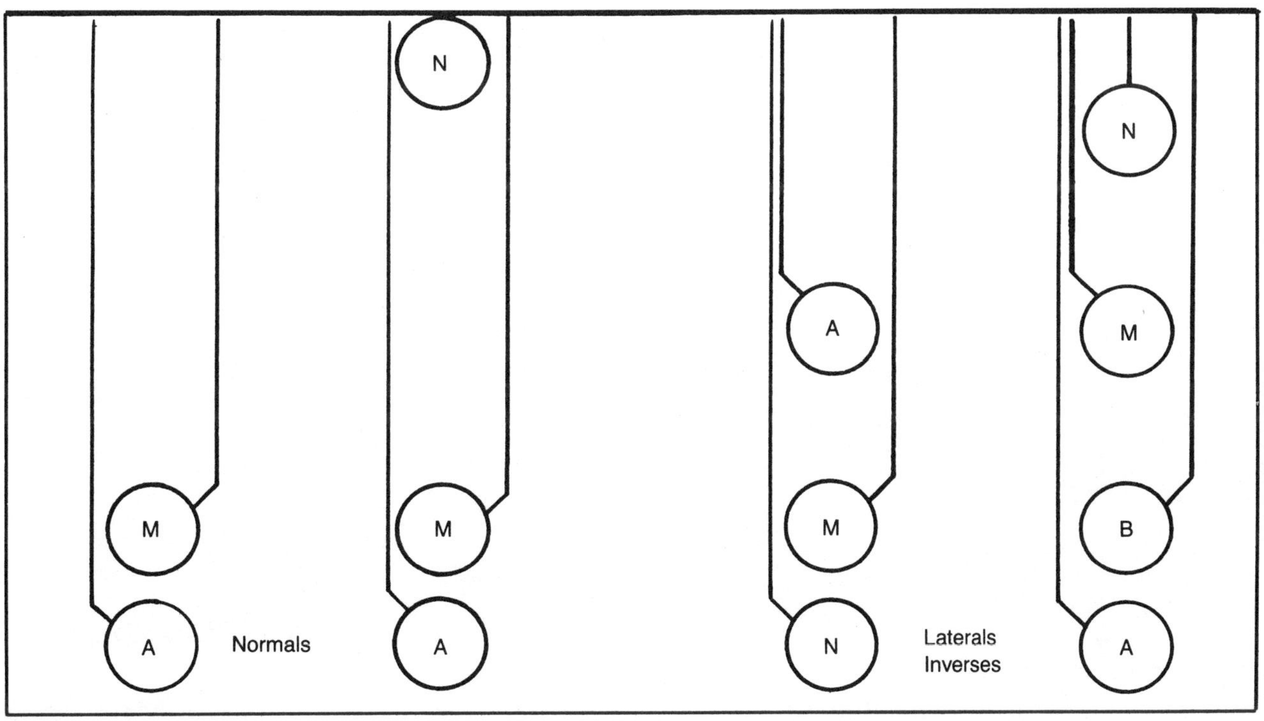

FIG. 4–14

Electrode arrays

and the resistance or resistivity is calculated. The electrode configuration determines the measurement geometry.

Single-Point Resistance and Resistivity

If a spherical electrode is placed in a homogeneous medium, as shown in Fig. 4–15, the resistance of the medium is called the *grounding resistance*, r_g, of the electrode. The electrode potential E_e is given with respect to an infinitely distant return by the expression:

$$E_{e\infty} = \frac{RI}{4\pi(d_e/2)} \qquad (4.13)$$

Or:

$$R = r_g K$$

Where:
$K = 4\pi(d_e/2)$
d_e = Electrode diameter

The grounding resistance of the electrode is:

$$r_g = \frac{E_{e\infty}}{I} = \frac{R}{4\pi(d_e/2)} \qquad (4.14)$$

Where:
R = Resistivity of the medium
I = Current
d_e = Electrode diameter

Thus, the grounding resistance is directly proportional to the resistivity, R, of the medium. This is similar to the concept of a pole in surface geophysics.

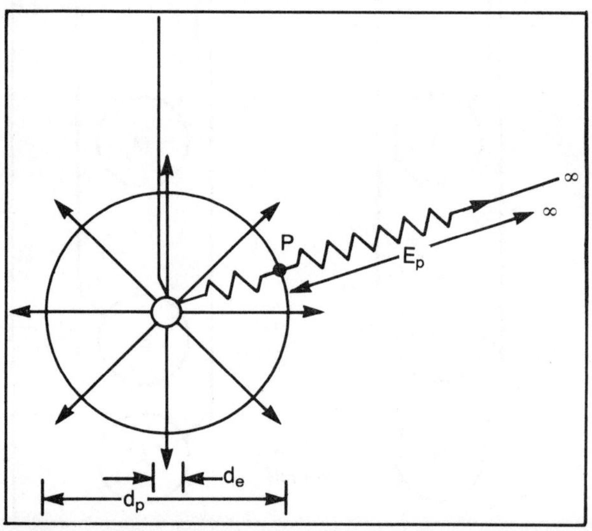

FIG. 4–15
Spherical electrode in homogeneous medium

The potential at any point, p, in the medium also is similar (Fig. 4–16):

$$E_{p\infty} = \frac{RI}{4\pi(d_p/2)} \qquad (4.15)$$

Where:
d_p = Diameter of the equipotential sphere that has point p on its surface

Therefore, the resistance outside a sphere in the medium surrounding the electrode is:

$$r_p = \frac{E_{p\infty}}{I} = \frac{R}{4\pi(d_p/2)} \qquad (4.16)$$

Subtracting Eq. 4.16 from Eq. 4.14 determines the size of the sphere of investigation of the single-point system for any fraction of its response.

If the sphere, d_p, has a radius that contains a volume contributing 90% of the observed signal, then:

$$E_{p\infty} = 0.1\, E_{e\infty}$$

$$\frac{RI}{4\pi(d_p/2)} = 0.1 \frac{RI}{4\pi(d_e/2)}$$

$$1/d_p = 0.1\,(1/d_e)$$

$$d_p = 10\, d_e$$

Thus, a 1-in. diameter electrode receives 90% of its signal from a sphere 10 in. in diameter in a homogeneous medium. Anything outside of about five times this distance is essentially an infinite distance away. This is a shallow investigation device, and the

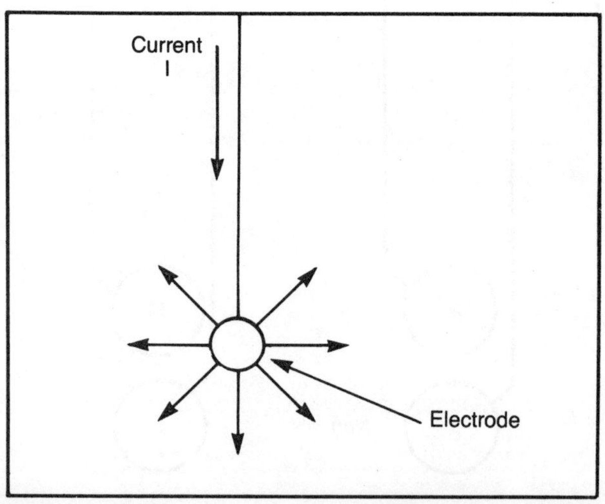

FIG. 4–16
Single point in a homogeneous medium

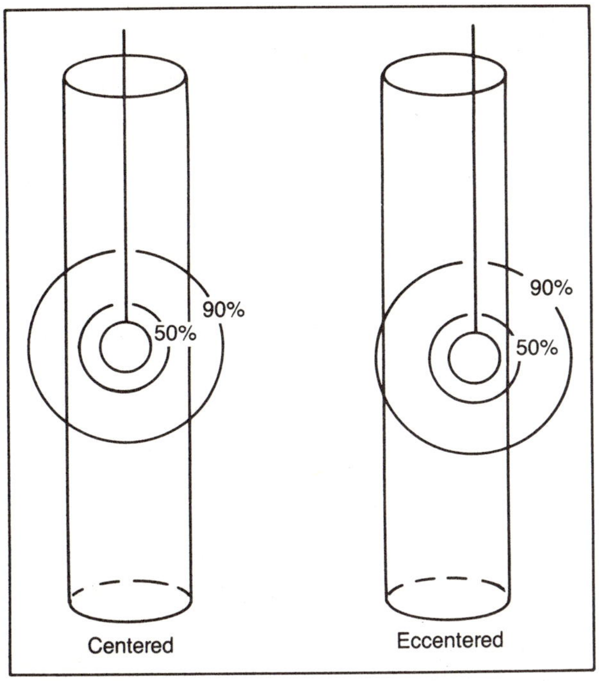

FIG. 4–17
Single-point investigation

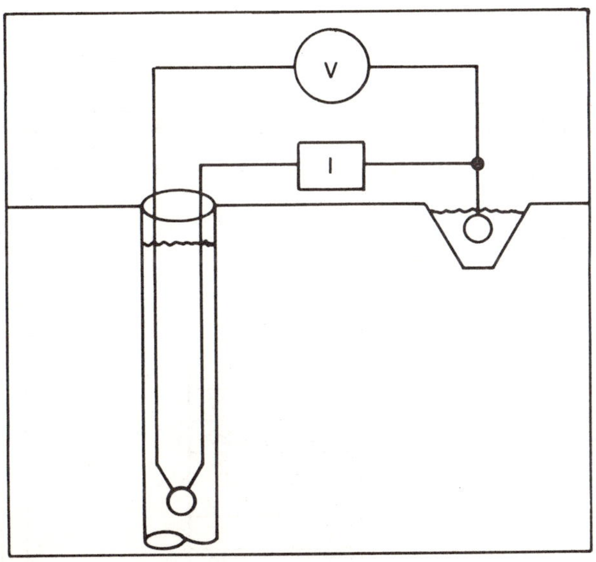

FIG. 4–18
Single-point resistivity

contribution of the borehole is a major factor. This is illustrated in Fig. 4–17.

A similar procedure can be followed to determine the influence of the return electrode (Fig. 4–18). If we assume that an accuracy of measurement of ±1% is needed, the influence of the position of the return electrode must be less than 1%. Solving for d_p in this case:

$$d_p = 100\ d_e \tag{4.17}$$

Therefore, with a 1-in. diameter electrode, the return electrode position has less than 1% influence on the measurement if it is more than 50 in. away from the measure electrode. The total circuit, including the high impedance measuring circuit, is shown in Fig. 4–19.

A spherical electrode is difficult to incorporate into a downhole tool mechanically. Therefore, the electrode is usually made cylindrical. However, except for the expression for the grounding resistance, the effects are the same as for spherical electrodes.

The expression for the resistivity measurement of a cylindrical electrode is:

$$R = (4\pi\ Le/\ell n(2L_e/d_e))r \tag{4.18}$$

Where:
L_e = Length of the electrode
ℓn = Logarithm to the base e (the natural logarithm)

Notice in Eq. 4.18 that both the length and diameter of the electrode must be known in order to solve for the resistivity, R. Since the logarithm of d_e is involved, its value is not as critical as L_e. However, in order to use the single-point curve quantitatively, the electrode dimensions must be known and noted on the log. What is probably better, although seldom done by contractors, is to scale the log in resistivity

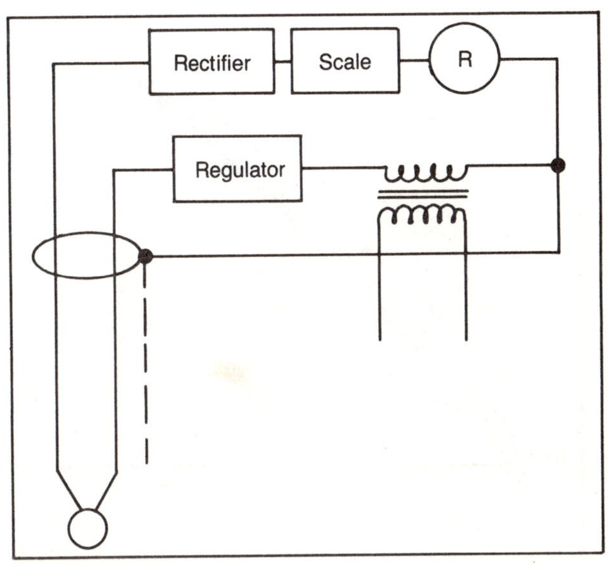

FIG. 4–19
Single-point resistivity circuit

and adjust the circuit to correct to resistivity. This is done routinely for all other arrays.

The combination of an extremely shallow depth of investigation, a linear dependence on resistivity, and a response symmetry make the single-point resistance curve quite useful for many purposes:

$$R = rK_s \quad (4.19)$$

$$K_s = 4\pi \frac{Le}{\ell n \frac{2Le}{d}}$$

There are no unsymmetrical relationships in these factors. The constant, K_s, describes an ellipsoid of revolution. The major axis of the ellipsoid is the electrode length.

Several things are indicated by Eq. 4.18. These are:

- The response of the single-point system is symmetrical and, therefore, excellent for correlation. Single-point signatures are quite representative of formation character. Bed boundaries are at curve inflection points.
- The resistance deflection, the resolution, and the curve smoothness all depend on the electrode length. Increasing the electrode length results in a decreased deflection, an increased curve smoothness and depth of investigation, and a decreased resolution.
- The shallow depth of investigation might be used to examine the flushed zone around the borehole and to determine R_{xo}.

The determination of R_{xo} depends on being able to separate the effects of the borehole from the effects of the formation. This relationship has the appearance of a pair of parallel resistors:

$$\frac{1}{R} = \frac{b}{R_m} + \frac{(1-b)}{R_{xo}} \quad (4.20)$$

Fig. 4–20 shows the probable relationship of the value of b as a function of hole size and the ratio of formation resistivity to mud resistivity (departure curves). Notice the great variation of reading with hole size. Also, notice the greater linearity at low resistivities.

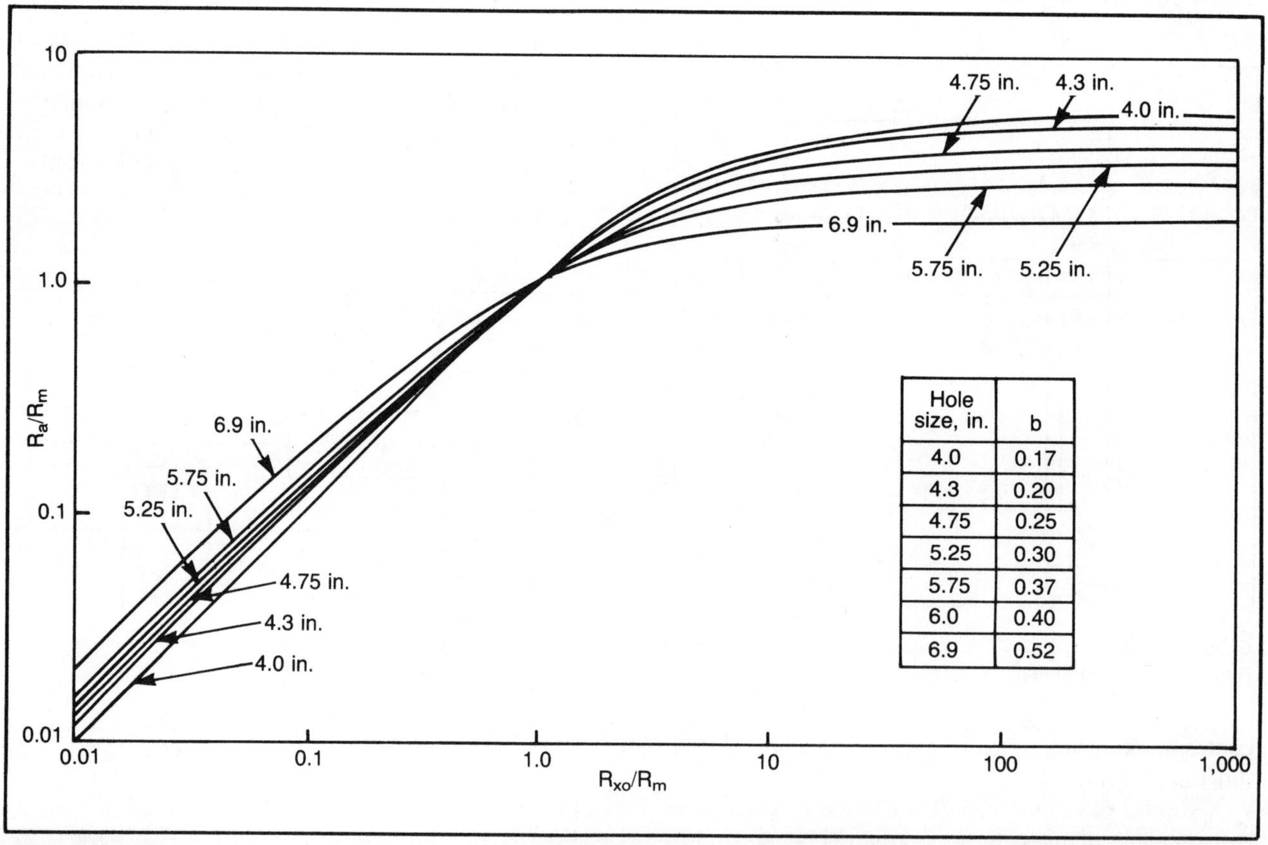

Hole size, in.	b
4.0	0.17
4.3	0.20
4.75	0.25
5.25	0.30
5.75	0.37
6.0	0.40
6.9	0.52

FIG. 4–20

Single-point resistivity departure and hole size correction curves

As far as is known, charts of this type have not been verified by any of the logging contractors using the single-point resistivity system.

Fig. 4–21 shows the current distribution around the singe-point electrode under various conditions of resistivity. Notice that the current distribution can depart radically from a radial one. Consequently, the equipotential surfaces depart severely from the spherical one described by the conversion constant, K_s. This problem is also illustrated by the departure curves (Fig. 4–20) in the flattening of the curves at the resistivities greater than R_m.

This last problem, that of a changing current distribution around the current electrode (the geometry), is the major impediment to using the single point quantitatively. While the relationship shown in Fig. 4–20 and in Eq. 4.20 is valid, there is an extreme loss of resolution when the formation resistivity is more than twice the mud resistivity. Therefore, the major usage of the single-point curve has been qualitative. It can be used, if care is taken. It is a poor curve for coals and limestones, however.

To this point the influence of the return electrode has been ignored. The assumption is that it is so far away that it has no influence on the measurement. It is also assumed that it has a negligible grounding resistance. While both of these factors usually are correct, one or both of them may not be negligible.

The measurement of the resistance includes the grounding resistance of the measure electrode and that of the return electrode. They are in series; therefore, their grounding resistances must be added. As

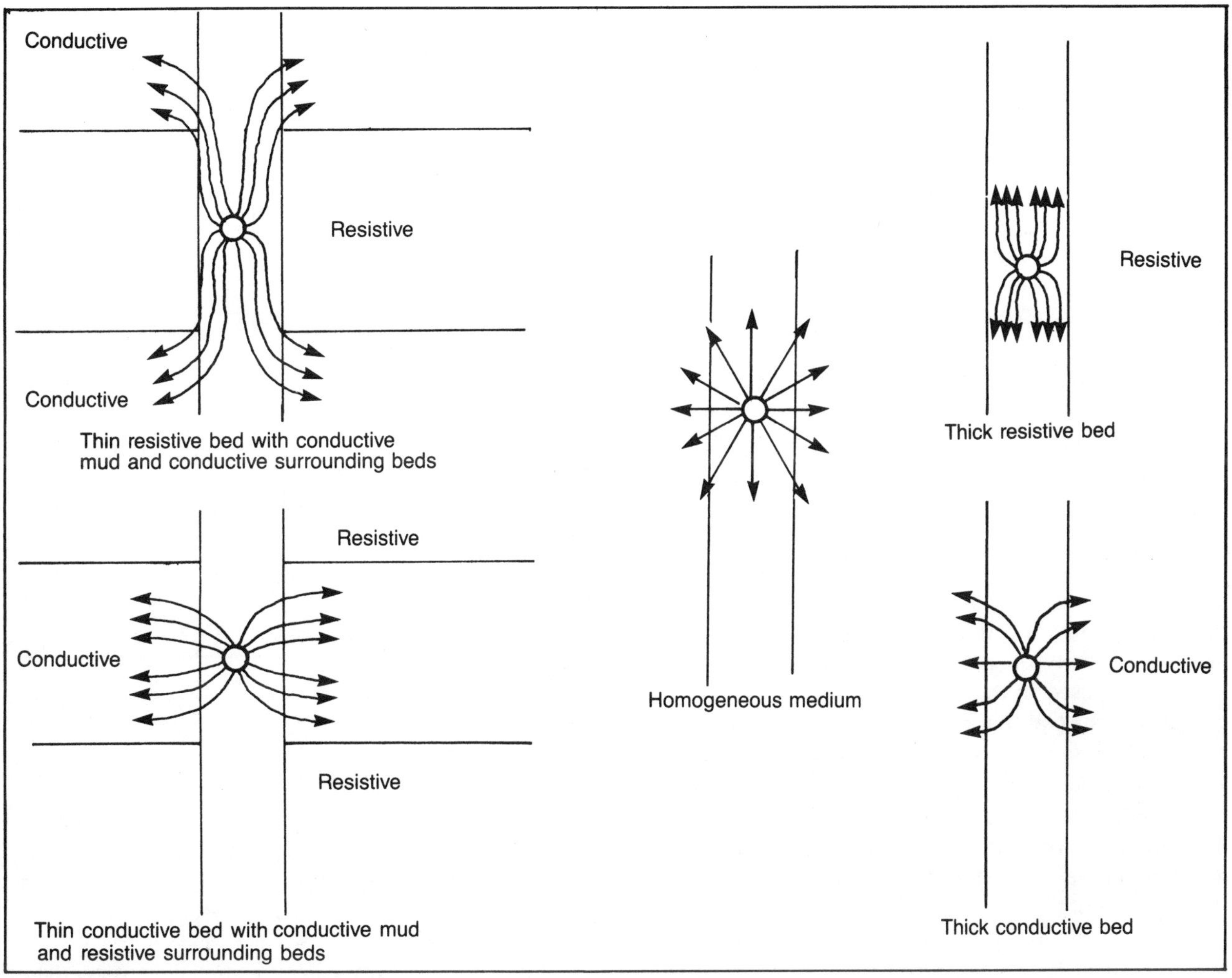

FIG. 4–21

Electrical current flow from an electrode in a borehole

long as the curve is used qualitatively and the return electrode resistance remains constant, there is little problem from this. However, if the curve is to be used quantitatively, these factors must be considered.

There are two major options for handling the return electrode problem. First, we may make the electrode a length that results in a significant grounding resistance but that holds its dimensions constant in a constant environment. Then this value need only be subtracted from the total reading on the log. Therefore, it must be kept in a constant environment, such as the mud pit. Sometimes the mud pit is such an environment, but often it is not. Freezing, drying, or nonexistent mud pits are a problem in the arid climates of many mineral projects because of the high resistance of the surface layers.

It is common practice to return the resistance or resistivity current circuit to the cable armor. This puts the return electrode into the borehole fluid. From Fig. 4–22, it can be seen that this will result in a return electrode that is normally extremely long. Thus, the grounding resistance of such an electrode is very low—so low as to be negligible. For example, 100 m of 3/16-in. cable has a grounding resistance of about 10% of the mud resistivity in ohms.

Using the cable as a return electrode is an excellent idea, except when the sonde approaches the surface of the earth, the surface of the mud in the hole, or a thick, high-resistivity zone, such as a thick, massive limestone or coal. In this case as the sonde approaches the surface, the return electrode becomes shorter until its grounding resistance is no longer negligible or constant. This is called the *Delaware effect* and is present in other arrays as well as the single point.

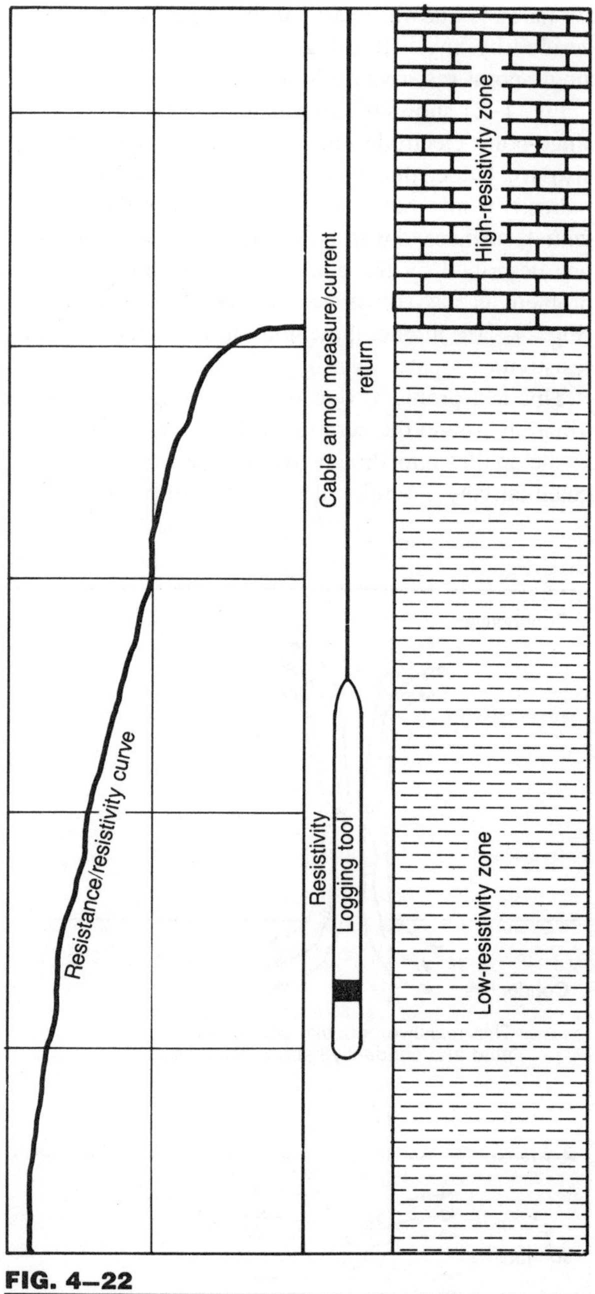

FIG. 4–22

Delaware effect

■ Normal and Lateral Resistivity

The problems inherent in the single-point system, the shallow investigation, sensitivity to hole size, and the difficulty of quantitization, have resulted in it not being used in the petroleum logging industry, except for special purposes. In its place, normal resistivity systems of several different spacings or arrays have been employed. Although many of the problems of the single point are present in the normals, the normals have been extensively investigated and departure curves are available from any of the major logging contractors for many different arrays.

A normal resistivity device is one where the current electrode, A, is much closer to the measure electrode, M, than either is to the return electrodes, N and B. This is illustrated in Fig. 4–23. The measure-return electrode, N, and the current-return electrode, B, are at near infinite distances from the measure and current electrodes, M and A, respectively. This is similar to a pole-pole array in surface work.

In equation 4.15, it can be seen that the potential of the point, p, is the same anywhere on that equipotential surface. Therefore, an electrode, M, can be used to measure the potential, E, at a constant distance, $d_p/2$, from the current electrode on that equipotential surface. The potential, E, is caused by the flow of current from the current electrode, A, and the resistivity of the medium surrounding the array. This is a normal device of spacing $\overline{AM}$. The resistivity measured by this system in a homogeneous medium is:

$$R = rK_N = \frac{E_M}{I_A} 4\pi \, \overline{AM} \tag{4.21}$$

Where:
E_M = potential of the measure electrode, M
I_A = Current through the current electrode, A.

If the normal device is in an infinite, homogeneous medium, the current distribution around the current electrode is radial, which is described by the factor 4π in Eq. 4.21. The resulting equipotential surfaces, which are always at right angles to the current, are spheres. These shapes are subject to the same variations and distortions as with the single-point system. The departure curves for the normal devices have been published and verified. Also, the depth of investigation is great enough that the nonlinearities must be large to be significant.

Since the equipotential surface through the measure electrode is a sphere of radius $\overline{AM}$, applying Eq. 4.17 shows that approximately 50% of the signal measured by M comes from a volume whose radius is $2\overline{AM}$. Thus, the depth of investigation of the normal device (50%) is $2\overline{AM}$. The volume of investigation (50%) is a sphere whose radius is $2\overline{AM}$.

Because of the substantially larger volume of investigation of the usual normal device, compared to the single point, the resolution of the normal is proportionally less. However, for the same reason the effects of the borehole are proportionally less with the usual normal device.

The A electrode and the M electrode are physically separated on the sonde body thus, within the measurement volume. Because of this, a bed boundary or any other change has little effect on the signal value while the boundary is between the two electrodes.

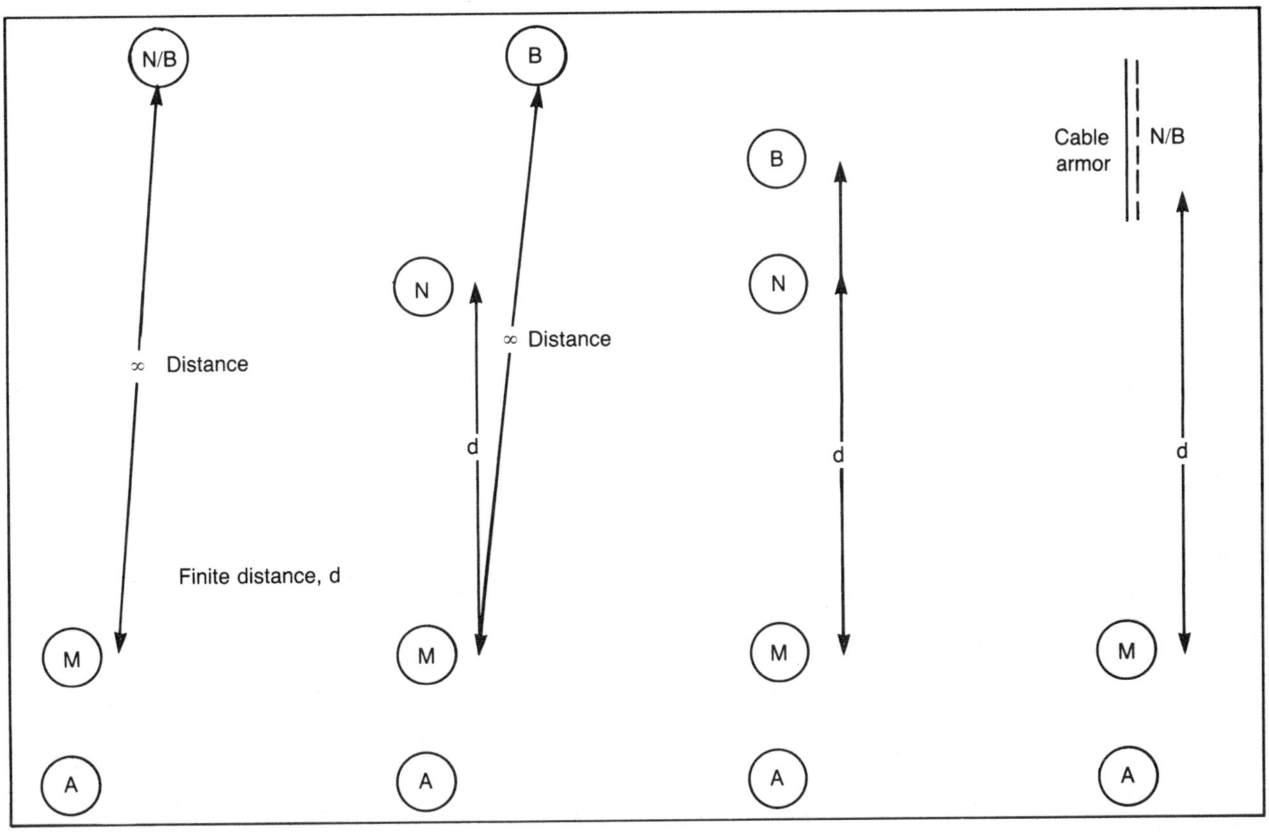

FIG. 4–23
Normal resistivity configurations

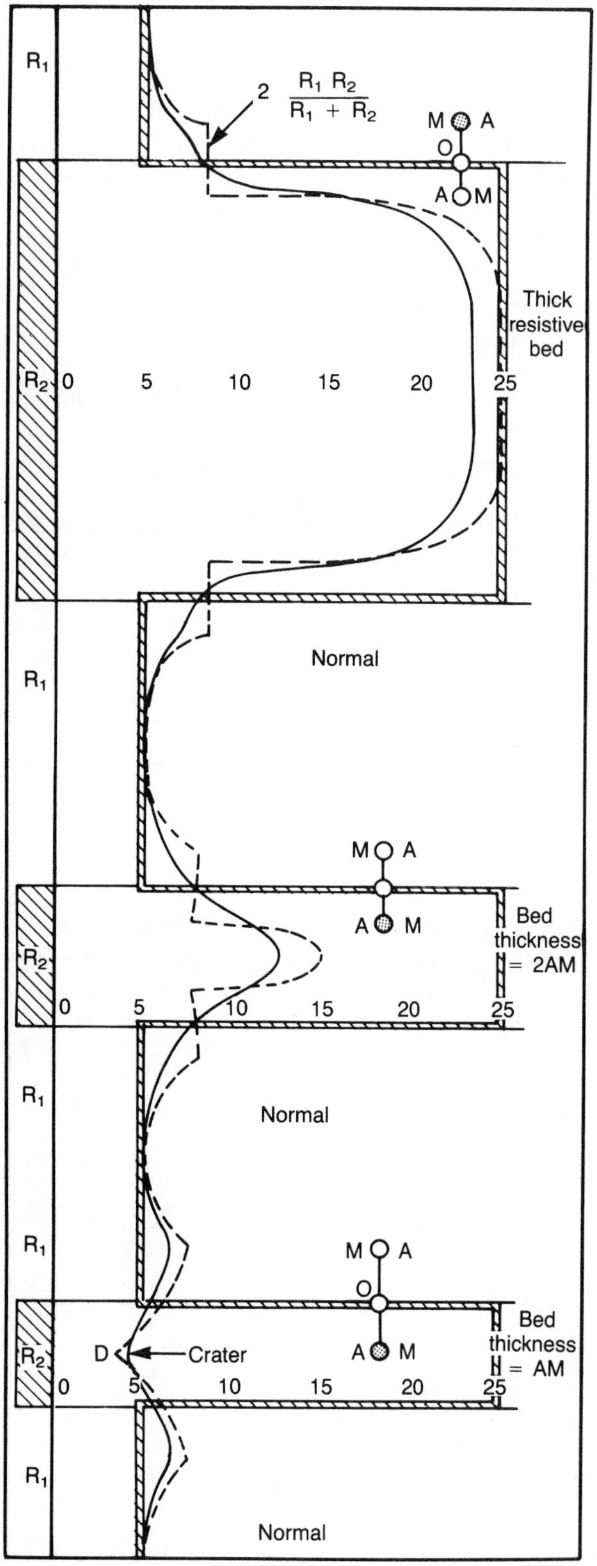

FIG. 4–24a

Normal curve shapes

(after Chombart, *Factors Involved in Practical Electric Log Analysis*, courtesy Schlumberger Well Services Inc., 1949)

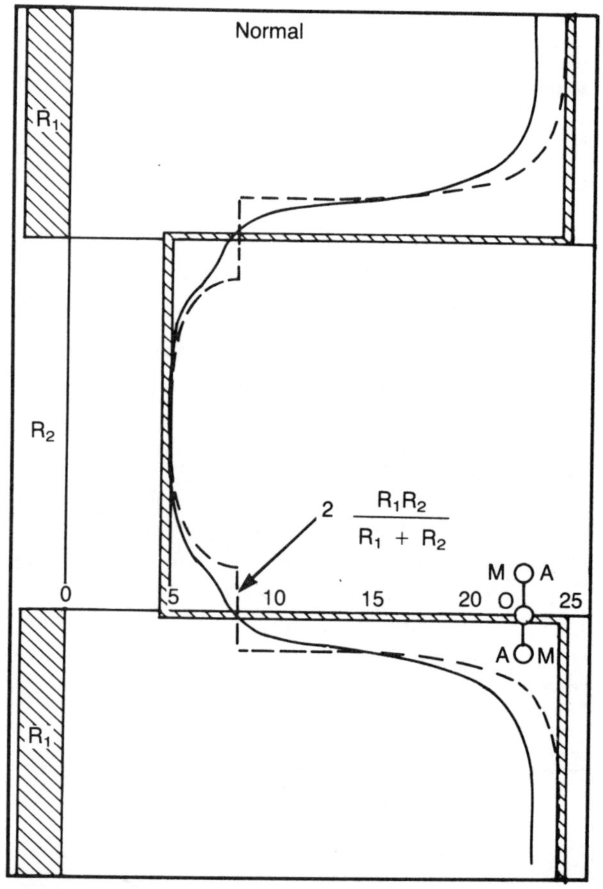

FIG. 4–24b

Normal curve shape in a thick conductive bed

(after Chombart, *Factors Involved in Practical Electric Log Analysis*, courtesy Schlumberger Well Services Inc., 1949)

The resulting curve shapes are shown in Figs. 4–24a, 4–24b, and 4–24c. Notice that there are distortions at the bed boundary that make it extremely difficult to determine the location of the bed boundary accurately. In nonpetroleum logging, more than in petroleum logging, this is a detriment to the use of the normal systems because of the greater interest in thin beds. When a normal device is run, a single point probably should accompany it.

Since the depth of investigation can be controlled and determined easily, a common practice in petroleum logging is to run a combination of two or more normal spacings. This allows one to assess the effects of the invasion of the borehole fluids into the permeable formation. This is illustrated in Fig. 4–25.

Besides the bed boundary effects of the normal device, the response in thin beds (thinner than $\overline{AM}$) makes its use difficult in nonpetroleum logging. Fig.

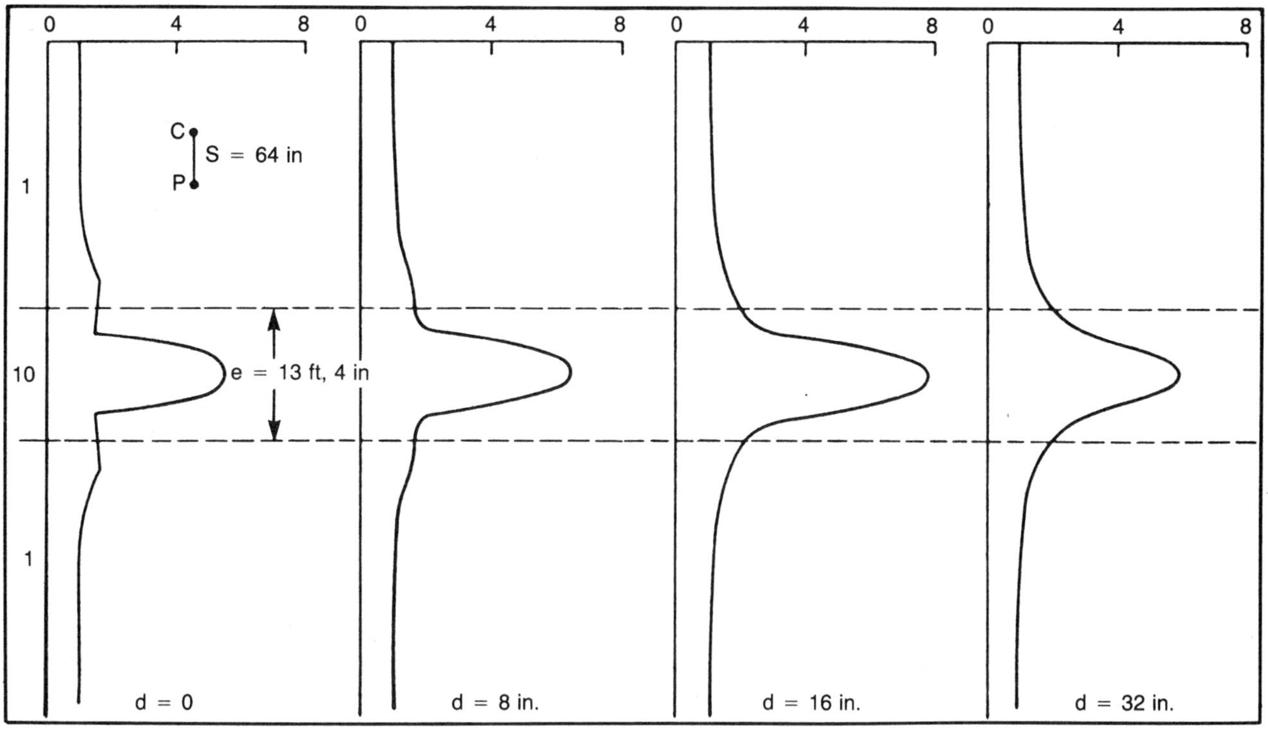

FIG. 4–24c

Comparison of normal curves for four different hole diameters (after Guyod, *Resistivity Determination from Electric Logs,* 1958)

4–24a illustrates this problem. One can see that it is possible to mistake one thin bed for two beds. If the bed thickness is exactly equal to $\overline{AM}$, there is a very ambiguous deflection.

Departure curves for the 16-in. and 64-in. normal are shown in Figs. 4–26 and 4–27. Fig. 4–28 shows a comparison of the sample sizes of the single point and the two normals.

On lateral devices the A electrode is closer to the B electrode than it is to the M electrode. This is shown in Fig. 4–29. This arrangement results in a deep penetration device. On the other hand, the response curves shown in Fig. 4–30 illustrate the extremely unsymmetrical nature of the curve. For this reason, the lateral or inverse systems are seldom used. One occasionally runs across them in log libraries. This is a pole-dipole in surface work.

The response of the lateral or inverse device is:

$$R = K_L r$$

$$K = 4\pi \frac{(\overline{MN})(\overline{AB})(\overline{AB} - \overline{MN})}{(\overline{MN})(\overline{AB}) + (\overline{AB} - \overline{MN})} \qquad (4.22)$$

Eq. 4.22 is the general form that may be used for any four-electrode device.

Because of the deep depth of investigation of the lateral or the inverse or double dipole (which are the same thing), this device was used with the early electric logs for determining R_t or R_o. Fig. 4–30 illustrates how to pick the probable best value of the formation resistivity from this curve. Because of the difficulty of using the lateral curve, it is probably best to use it only as a last resort.

Focused Resistivity Systems

The correct readings of the nonfocused devices depend on correcting for the departure of the measurement from linearity, or on making the resolution so gross that interfering effects are negligible, in the case of the lateral. It is also possible to stabilize the measurement electronically and retain resolution. This class of focused systems is gaining importance in the mineral business. These are also known as guard logs and Laterologs (Schlumberger trademark).

An equipotential surface around a single point electrode is an ellipsoid of revolution. The length of the major axis of the ellipsoid depends on the length of the electrode. If the length of the electrode is

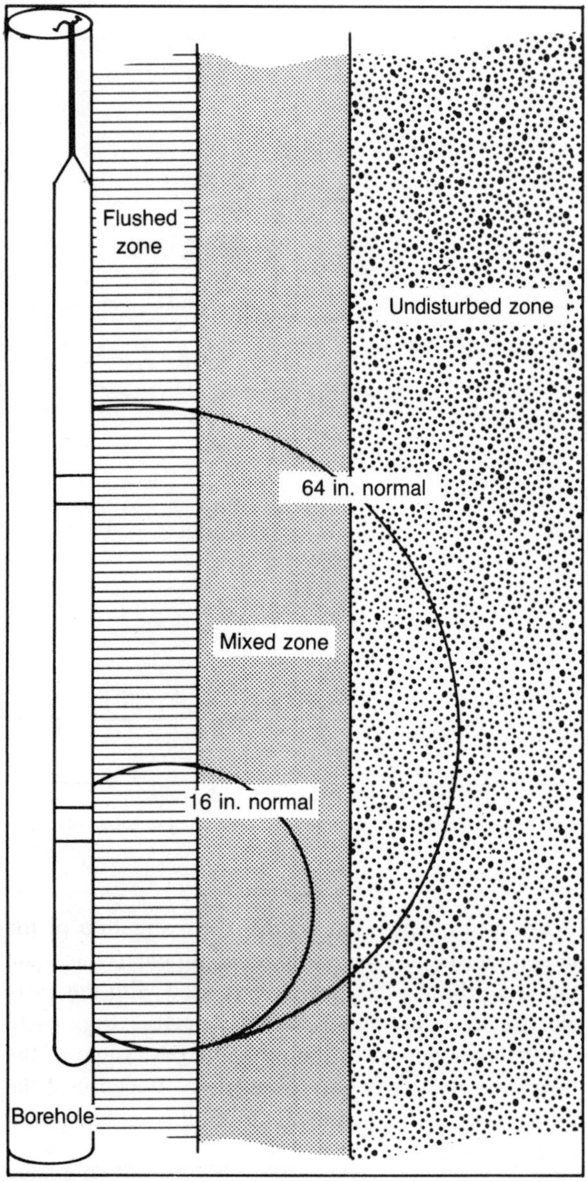

FIG. 4–25

Depth of investigation determined by spacing of electrodes

extended (Fig. 4–31), the flow of current near the center of the electrode is nearly parallel and that portion of the equipotential surface becomes very nearly a cylinder.

The electrode can be separated into three pieces and a measurement made from the center piece. In a homogeneous medium, the current flows parallel into the formation. Most of the measurement is from a cylinder whose height is the length of the measure electrode and whose diameter depends on the length of the two end pieces or guard electrodes. Furthermore, if the guard electrodes are electronically kept at the same potential as the measure electrode, the size and shape of the measure cylinder remain constant, even in conditions of nonhomogeneity. In a typical system an electronic circuit compares the potential of the measure electrode at constant current with the guard electrodes potential and forces the guard electrodes to the measure potential by adjusting the guard current. As a result, the geometry of such a system is stable and constant. Such a system is shown in Fig. 4–32. The response of signal to resistivity is linear over a wide range. The depth of investigation can be deep. The resolution is symmetrical and can be designed to any value from fine to coarse.

Several variations of this device are in use. The guard electrodes may be separated far from the measure electrode, and monitor electrodes may be used to monitor the potential differences. This is a seven-electrode system (Laterolog 7) that is popular in petroleum logging. It is used because of its deep penetration and because it can allow the recording of a spontaneous potential curve. This device is shown in Fig. 4–33.

The eight-electrode system and the dual-guard log systems have current returns that are not at an infinite distance as they are in the seven-electrode system. This results in a shallower investigation (Fig. 4–34).

The beam or current sheet of most of the guard log or focused systems is thin enough and deeply penetrating enough that borehole effects are minimal or negligible. This is even true with respect to the deeper penetration devices in the invaded zone of high-porosity formations. Therefore, these devices often measure accurately the value of R_t. They need a minimum amount of correction.

The use of these devices in petroleum work is to obtain a deeply penetrating, symmetrical response when the value of R_m is much lower than R_t. Neither the nonfocused devices nor the induction log are at their best in this situation. In mineral logging the focused devices are used to obtain a high, thin bed resolution while retaining a symmetrical response and minimal borehole effects. Because they can be designed for a wide range of linear response, they make an excellent quantitative resistivity tool in coals, limestone, and hard-rock environments. A typical response to a thin bed is shown in Fig. 4–35.

A typical mineral-type focused resistivity device is a three-electrode array, 5 cm in diameter and

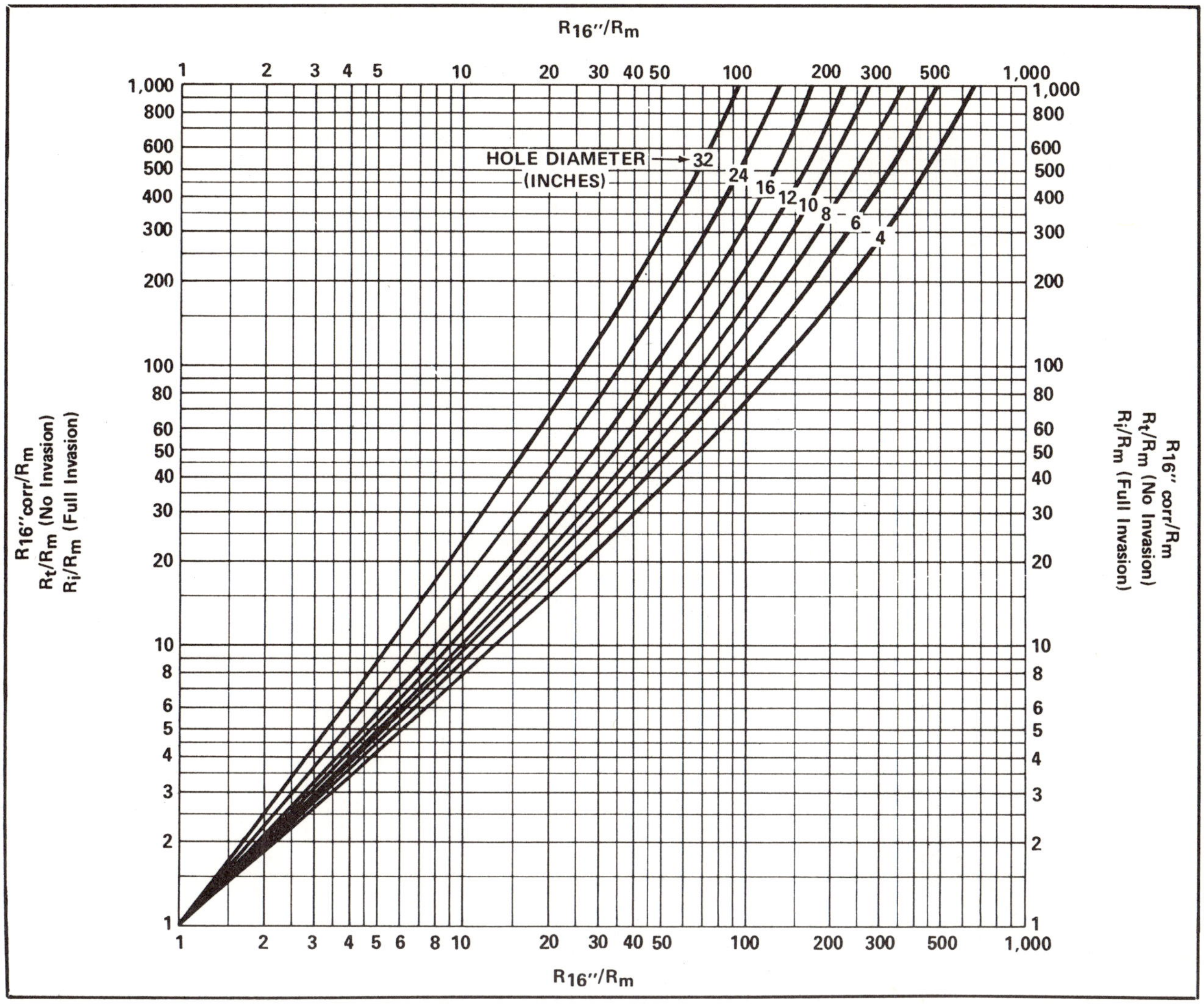

FIG. 4–26

Borehole correction for 16-in. normal readings, thick beds, no invasion or full invasion (courtesy Gearhart Industries Inc.)

122 cm long with a 10–13 cm long measure electrode. Petroleum tools may be three, seven, or more electrode systems, often recording more than one focused curve simultaneously. They are usually 8.9–10.0 cm in diameter and may have arrays from 50–300 cm long.

Fig. 4–36 shows a comparison of a focused log with a standard, nonfocused set. Fig. 4–37 shows the adjacent bed effect of one type of focused tool. When the resistivity contrast between a thin bed and the adjacent beds is great, there is danger of exceeding the design capability of the guard electrode control circuits. Then a distortion is present, which must be taken into account.

Induction Logs

The induction logging systems were developed for petroleum logging in oil-based muds. They need no electrical contact with the formation and can be used effectively in oil, water, or gas-filled boreholes. They depend on using magnetic fields to induce electrical currents in the surrounding formation proportional to the formation conductivity.

If a magnetic field moves across a conductive medium, it induces an electrical current in the medium. Therefore, if a coil of wire with the solenoid axis parallel to the borehole has an alternating current flowing in it, an alternating magnetic field

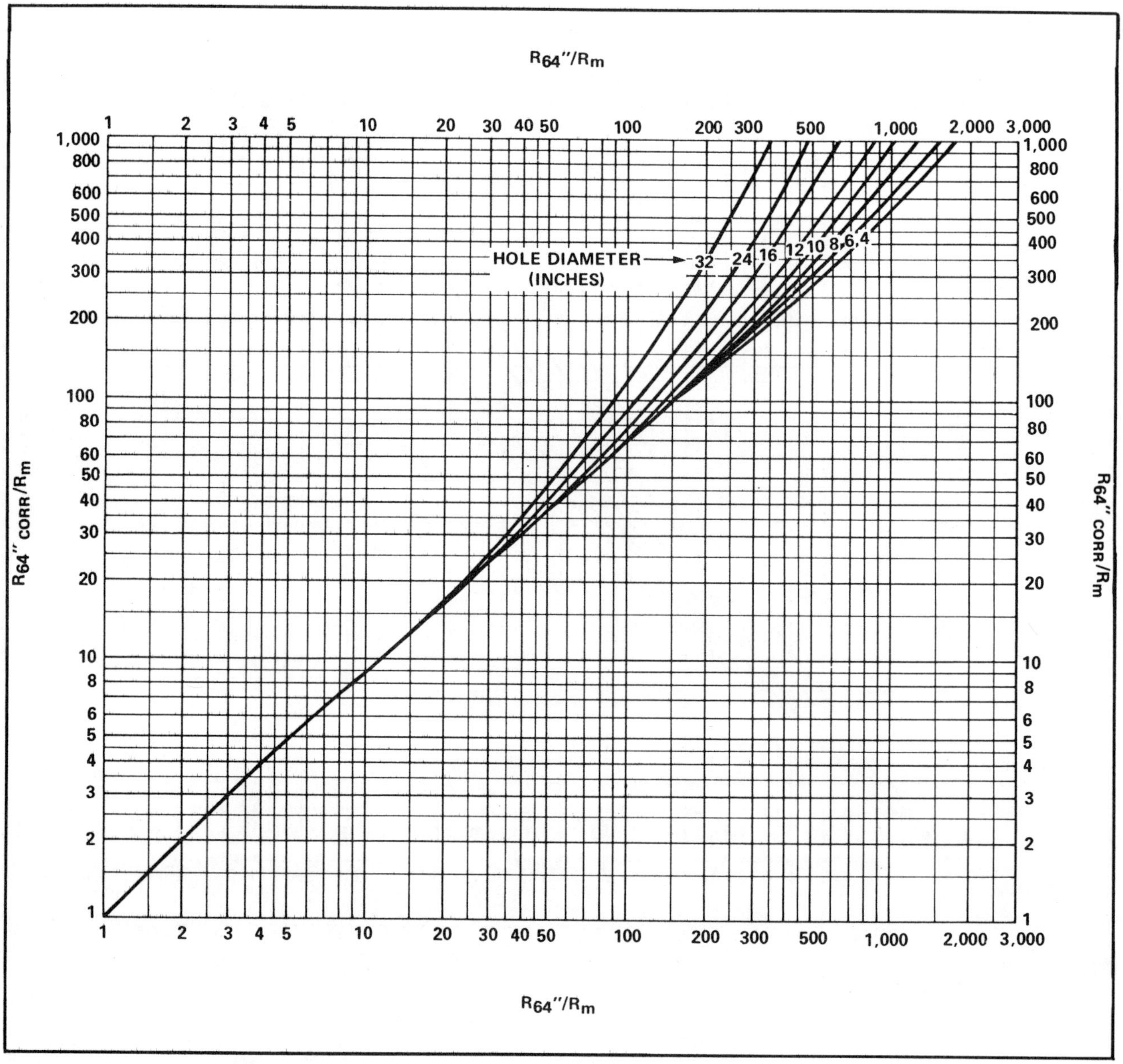

FIG. 4–27

Borehole correction for 64-in. normal readings, thick beds, no invasion or full invasion (courtesy Gearhart Industries Inc.)

moves out into and out of the formation surrounding the borehole. The magnetic field induces a current in the formation that flows around the borehole, or around the solenoid. This circular current sets up a second magnetic field that is 180° out of phase to the first field. A second solenoid on the axis of the borehole has a current induced in it from the second magnetic field. With suitable circuitry, the direct effect of the first or primary field can be eliminated (Fig. 4–38a).

The primary magnetic field strength is proportional to the primary current. The induced current is proportional to the magnetic field strength of the primary field and to the conductivity of the formation. The secondary field strength, and thus the signal, are proportional to the induced current in the formation. Therefore, with a constant coil current, the signal is a function of the formation conductivity.

By properly placing additional primary (transmitter) coils and secondary (receiver) coils and adjusting

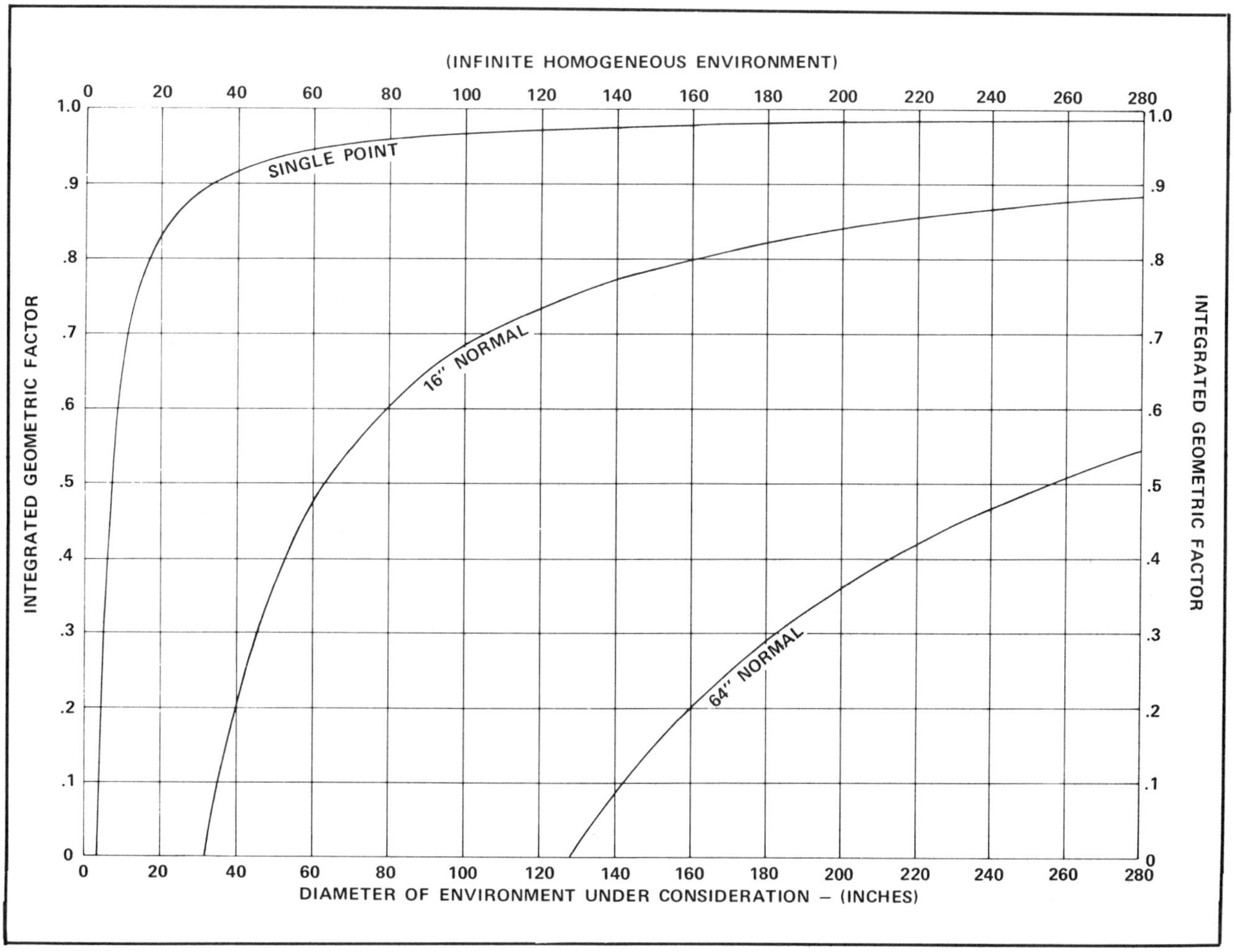

FIG. 4–28

Radial investigation characteristics of various resistivity tools, infinite homogenous environment
(courtesy Gearhart Industries Inc.)

the polarity and ampere turns, the primary magnetic field and the response of the tool can be shaped. The five- and six-coil induction log tools sense the conductivity of a torus of formation around the borehole. Thus, the induction log used in petroleum work is virtually independent of borehole and invaded zone characteristics. The oil-field induction log, such as the 6FF40, essentially reads R_t. Therefore, this tool has almost completely replaced the electric log—particularly the lateral curve—for deep resistivity determinations in petroleum work (Fig. 4–38b).

In most mineral work the induction log is not used because it is a large, complex system whose major value is to determine R_t and, thus, S_w. In mineral work the value of S_w is of little importance. Further, the induction log is a conductivity measuring system. In most mineral projects the conductivity is very low. Therefore, signal-to-noise problems plague the use of this tool for uranium and coal. An induction log instrument for mineral use could be designed but so far has not been.

The induction tool is important in oceanography. The resistivity and, thus, the salinity of ocean waters may be measured quickly and accurately with a simple two-coil device (nonfocused). Fig. 4–38c shows a comparison between a two-coil and a six-coil focused device. A two-coil laboratory device measures the conductivity of small fluid samples with lessened danger of contamination from electrode materials.

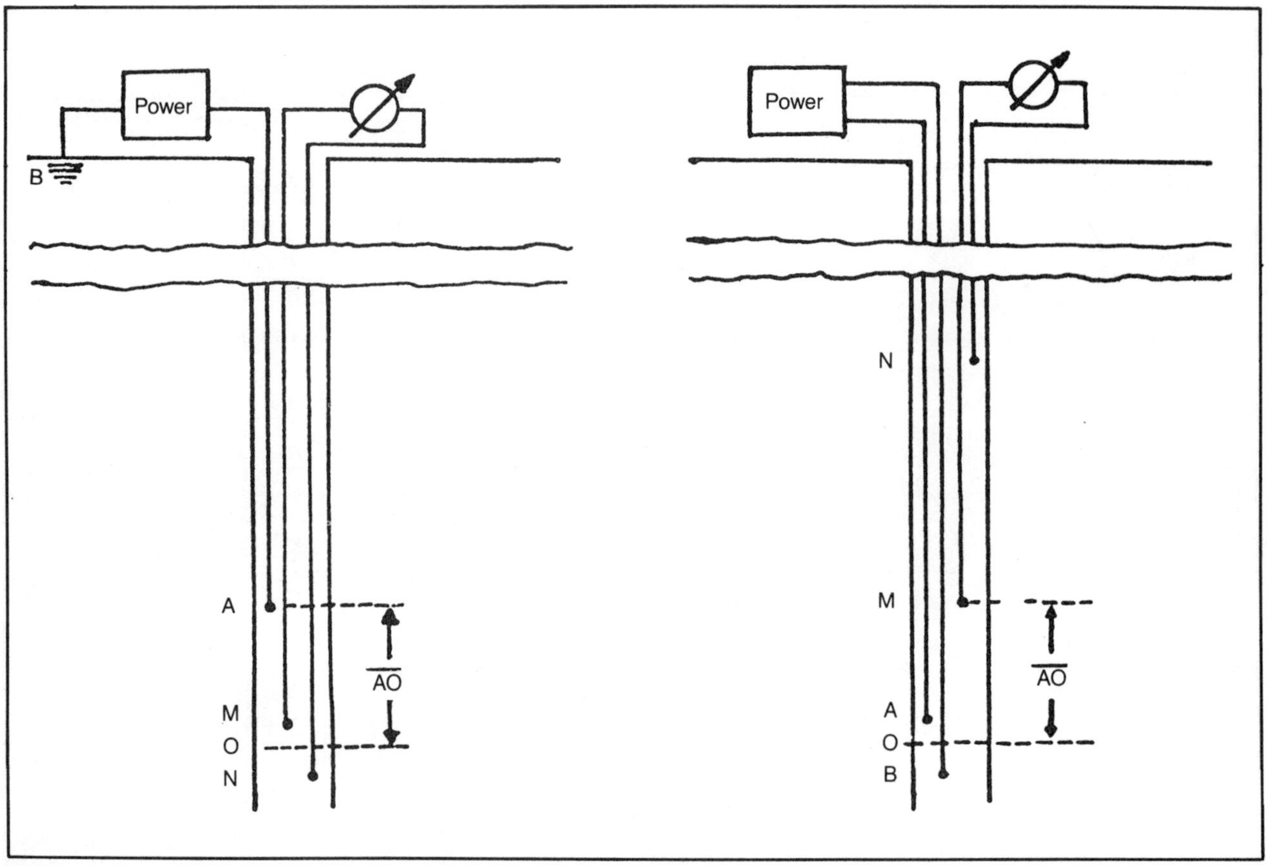

FIG. 4–29

Lateral array and inverse

■ Surface Systems

The principles outlined in the physical description section of this chapter apply equally well to surface resistivity measurements. There are three important differences, however. First, the arrays used and, thus, the sample volumes examined by surface methods are usually much larger than those in downhole methods. They are on the order of tens of meters to kilometers. Second, surface measurements examine a half volume rather than a full one. Third, resolution is poorer with surface methods because of the larger volumes. There are other differences, which we will examine, but in general, the similarities are greater than the differences.

Since the arrays used in surface resistivities are large, four-pole soundings or surveys are usually made: the Wenner, the Schlumberger, and the polar-dipole arrays. The four-pole system is used because the return electrodes, to be essentially at an infinite distance as in downhole arrays, would have to be many kilometers away. The types of common arrays are pictured in Fig. 4–39.

The Wenner and Schlumberger arrays are similar. They both use four colinear electrodes. The outer pair are the current electrodes (A and B), and the inner pair (M and N) are the measure or voltage electrodes. With the Wenner array all spacings, AM, MN, NB, are equal. With the Schlumberger array the MN spacing is much smaller than the other spacings. When a resistivity sounding is made, the Wenner array is expanded about the midpoint to increase the depth of investigation, keeping the spacings equal. Only the two outer or current electrodes are moved outward with the Schlumberger array.

With the polar-dipole system the current-array (A and B electrodes) distance and the measure-array (M and N electrodes) distance are both small compared to the distance between them. As any of the arrays are expanded, a deeper sample is examined—a sounding is made—and the sample becomes larger.

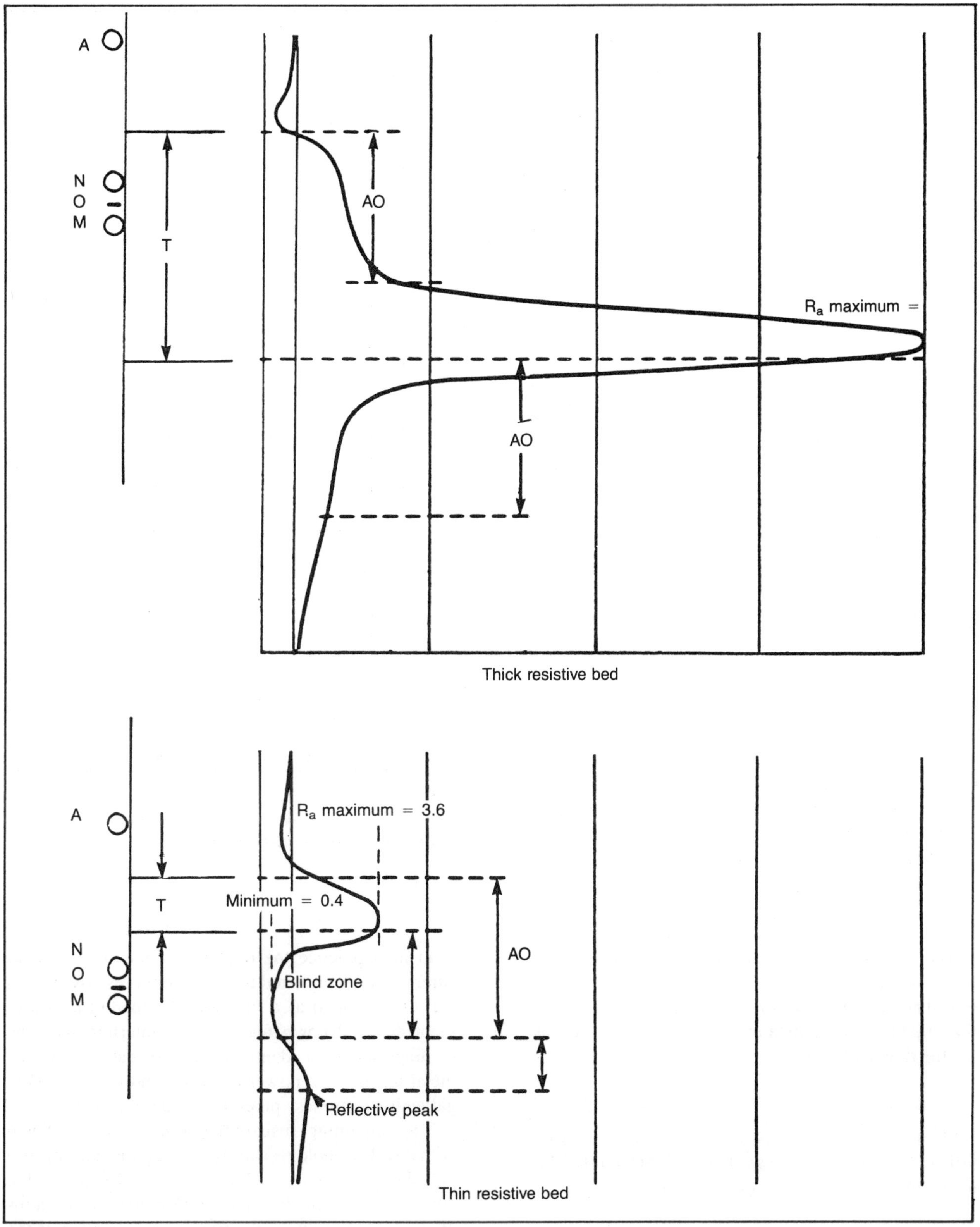

FIG. 4–30

Lateral resistivity response (courtesy Gearhart Industries Inc.)

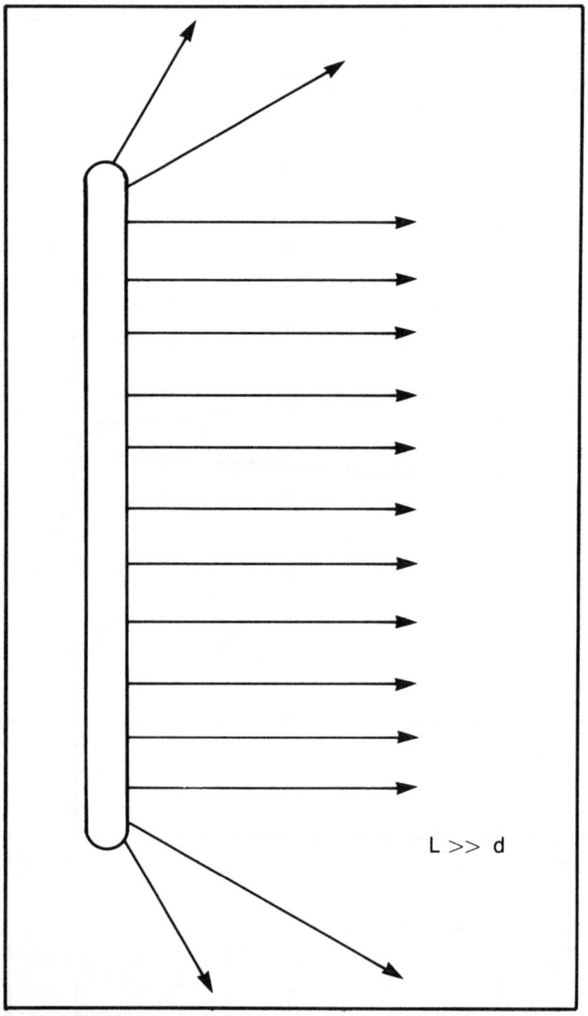

FIG. 4–31
Current flow from a single-point electrode that is long compared to its diameter

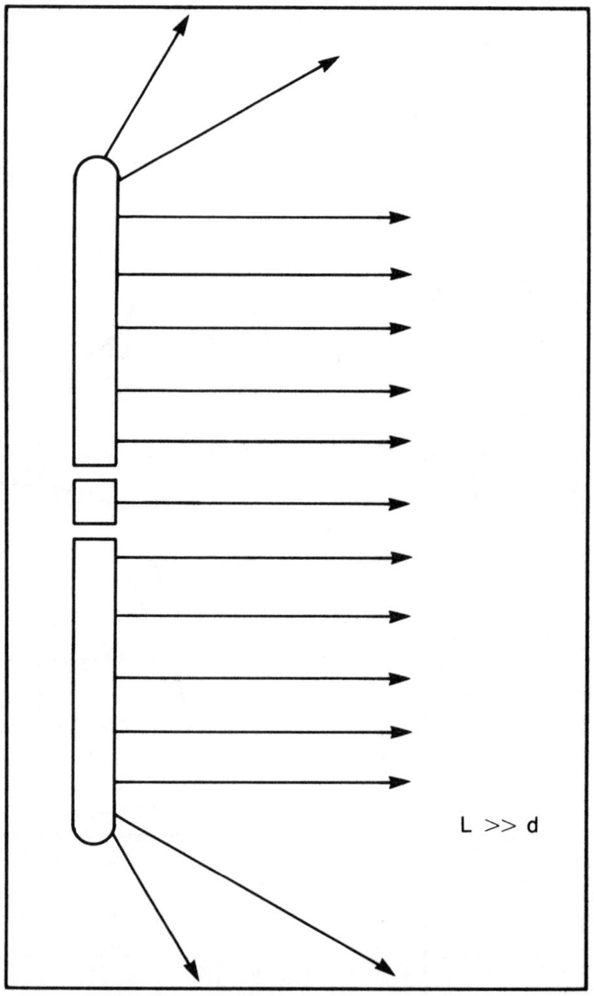

FIG. 4–32
Long electrode broken into three segments as a three-electrode guard system

The array data—current, voltage, and spacings—are converted to apparent resistivity with the assumption that the sampled volume is uniform. The relationship to calculate apparent resistivity is the same as for the downhole arrays:

$$R = K(E/I) \tag{4.5}$$

With surface arrays, however, the factor 2π is used instead of 4π, indicating a half volume limited by the earth's surface. Thus, the value of K is:

$$K = 2\pi \frac{1}{\frac{1}{\overline{AM}} - \frac{1}{\overline{BM}} - \frac{1}{\overline{AN}} + \frac{1}{\overline{BN}}} \tag{4.23}$$

In field practice the readings of the surface measurement are converted to an apparent resistivity with Eq. (4.5) plotted as a function of apparent measurement depth. Known features that interfere with the readings are then identified. These features include metal fence lines, power lines, grounding electrodes, galvanic electrodes, pipelines, and cased wells.

The remaining resistivities should be correlated with any downhole resistivity measurements. This is a valuable step that is seldom planned for and often not done. One of the major interpretation problems of surface resistivity is the proper identification of the subsequent layers or vertical anisotropy. These can be quickly identified and evaluated with a borehole

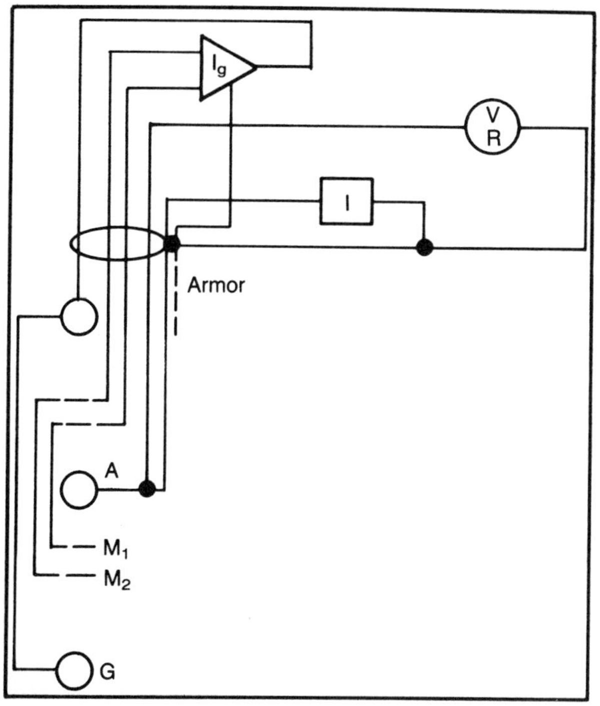

FIG. 4–33
Block diagram of Laterolog-7*
(*mark of Schlumberger)

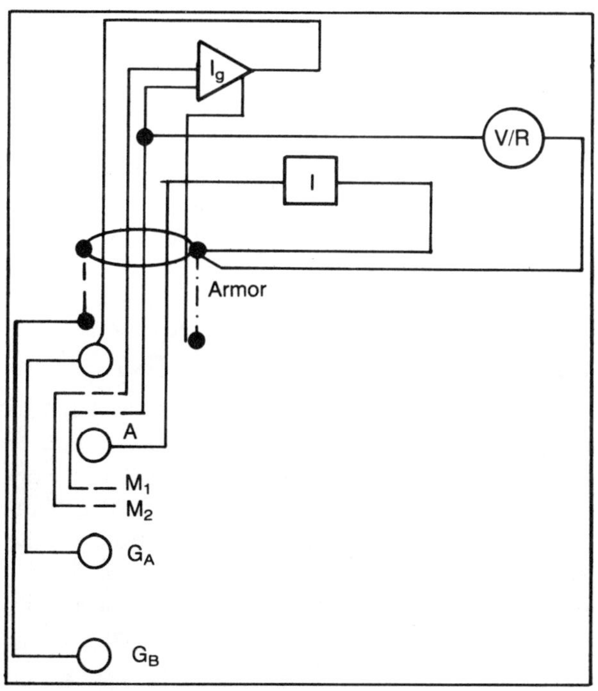

FIG. 4–34
Block diagram of Laterolog-8*
(*mark of Schlumberger)

resistivity curve. With this correlation a distribution of resistivities that fit known field data is picked from models or from a borehole resistivity log to match the calculated apparent resistivities. This last is best handled by a computer.

Surface methods are best suited to covering large areas in a short time because they have lower resolution than borehole resistivity measurements. The methods are particularly valuable when the terrain or landowner problems make access by drilling equipment difficult. This is especially true when they are combined with well-chosen borehole methods. The combination is ideal to map subsurface channeling and many other multilayered environments, rather than using it only to find specific mineral horizons. Surface resistivity methods combined with downhole measurements are valuable for engineering uses because of the complete picture they furnish. Surface resistivity surveys are often combined with induced polarization measurements. Only minor circuit modifications are required to allow one system to measure both parameters. The induced polarization measurements furnish further data about mineralization redox interface trends.

■ Microresistivity Devices

The microresistivity devices are known by various trade names. The MicroLog, the Microelectric Log, the Minilog ML, and the Contact Log are the unfocused devices. These devices are valuable in petroleum logging but are seldom used in nonpetroleum logging. The main value of microresistivity systems is to evaluate occurrences near and at the borehole wall. The nonfocused devices usually use a short normal (2 in.) and a short lateral or inverse (1.5 in.). The depth of investigation of these devices is extremely shallow: about 4 in. for the normal and about 2 in. for the lateral curves. They are sensitive to permeability by detecting the presence of mud cake. They also are used to determine R_{xo} because the limited depth of investigation restricts the measurement to the flushed zone of formation.

Since it has a shallower, more sharply bounded volume of investigation, the microinverse is more severely affected by mud cake than the micronormal. Therefore, when there is permeability and thus mud cake, the micronormal and microinverse curves separate. Otherwise, they coincide and read the resistivity of the zone immediately behind the hole wall.

The focused devices were developed to get around the great response to mud cake. These make use of

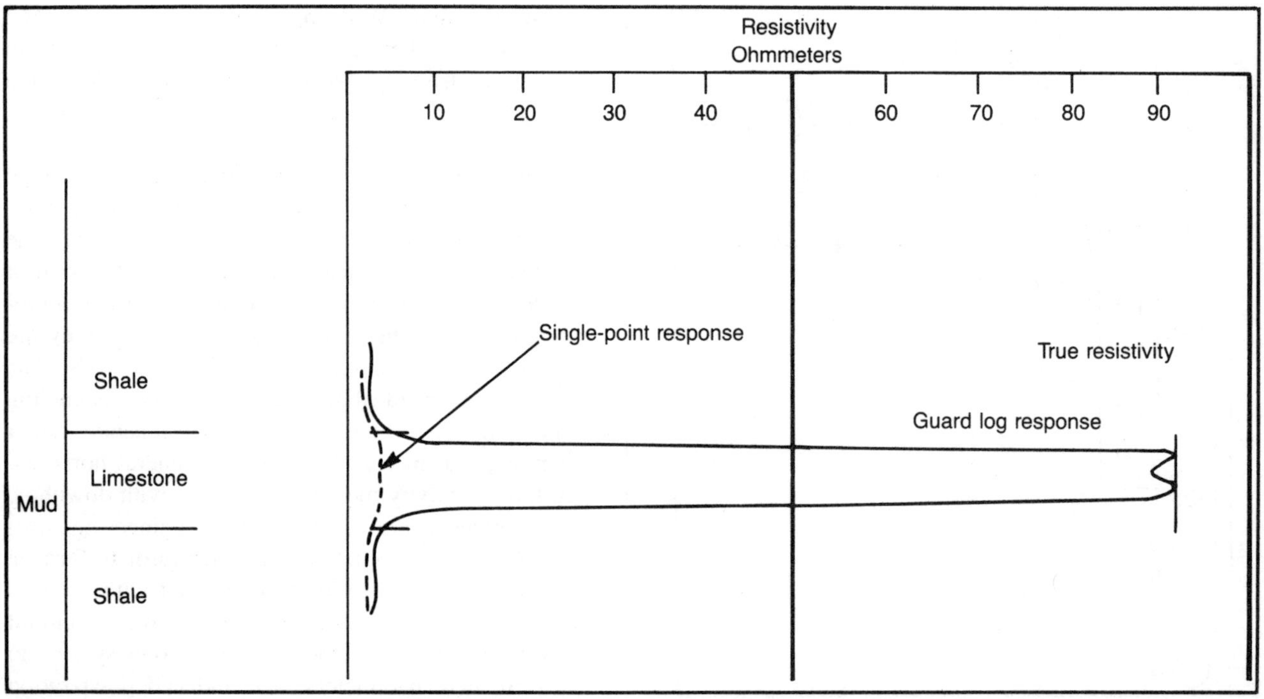

FIG. 4–35

Guard log response compared to single point

the guard log principle to minimize the effect of mud cake and to read more exactly the value of R_{xo}. These devices have depths of penetration of several inches to a foot.

The electrodes in all types of microdevices are imbedded in a soft, insulating pad that is larger than the radius of investigation of the devices. Therefore, they are all substantially independent of borehole effects. However, they must all be corrected for mud-cake thickness, especially at high formation resistivities (>10 ohmm). The microdevices are usually run in conjunction with a sensitive, one-armed caliper. This aids in assessing the mud-cake thickness. These devices are seldom run in nonpetroleum logging programs for several reasons. The main values of the devices are to assess the permeability and the porosity of the invaded zone. Neither of these two values has been considered important in uranium and coal logging. Also, the use of sidewall probes in poor hole conditions is risky.

The measurement or detection of permeability is perhaps a neglected tool in sedimentary uranium exploration. The transport of uranium compounds is directly controlled by permeability. Deposition often takes place at interfacies, which have sharp reductions in permeability. Porosity, too, plays an important part in sedimentary mineral exploration. The reservoir with which we are concerned in redox depositions is directly defined by the available porosity. Hydrothermal transport and deposition depend on porosity. In engineering and hydrological studies the amounts and movements of ground waters are directly linked to the porosity and permeability of the formations.

■ Calibrations

Calibrations of resistivity systems fall into two categories: the primary calibration used to verify or determine the array geometry (conversion factors and departures) and the secondary or field calibrations. The secondary calibrations are usually quite simple and often consist of only a precise resistor or network to verify or to set the current value. The value of the calibrating resistance is usually calculated from the resistivity expression using the conversion or K factor:

$$r = R/K$$

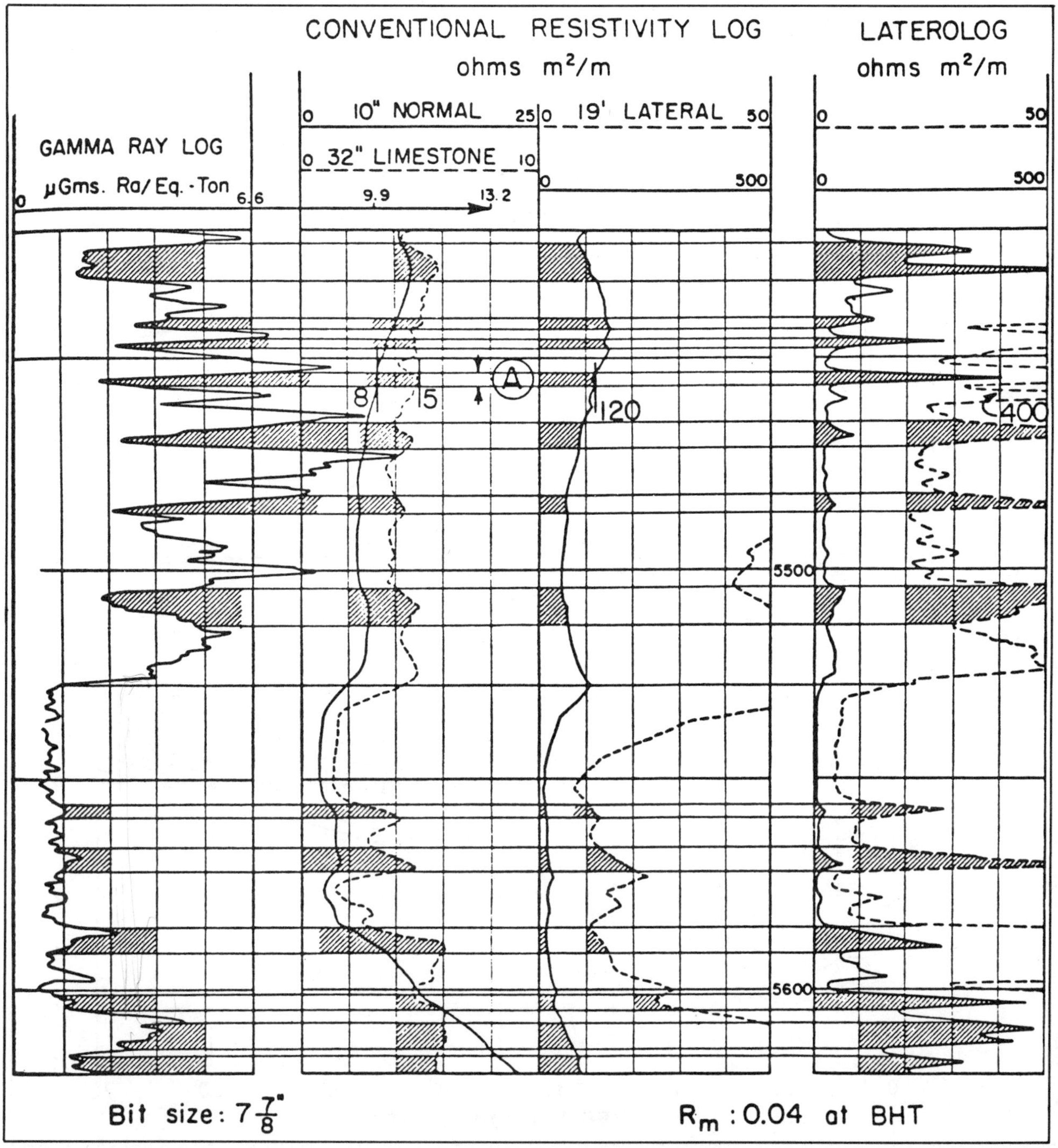

FIG. 4–36

Comparison of focused log with nonfocused set (courtesy Schlumberger Well Services Inc.)

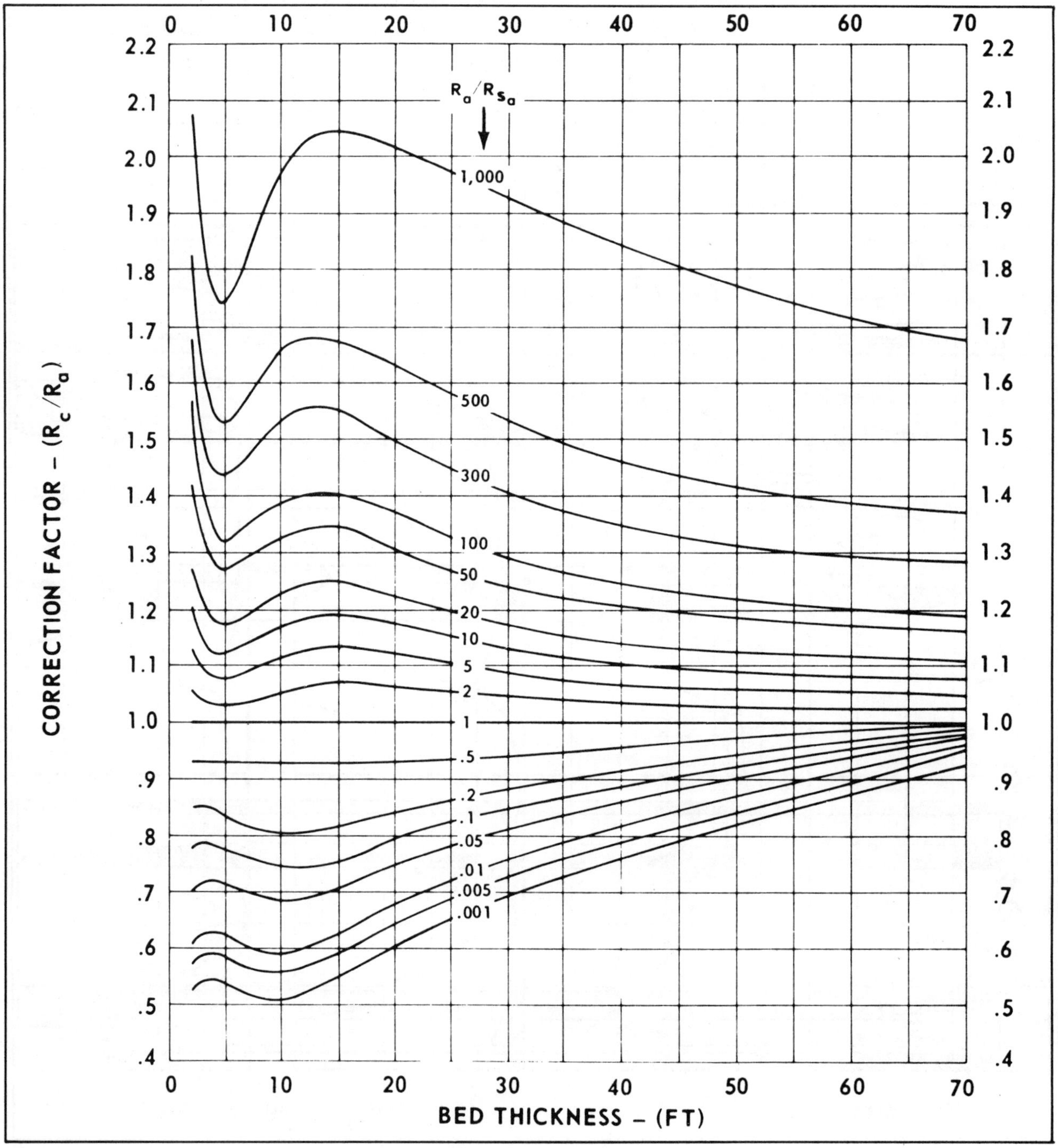

FIG. 4–37

Guard log correction for adjacent bed effects (courtesy Gearhart Industries Inc.)

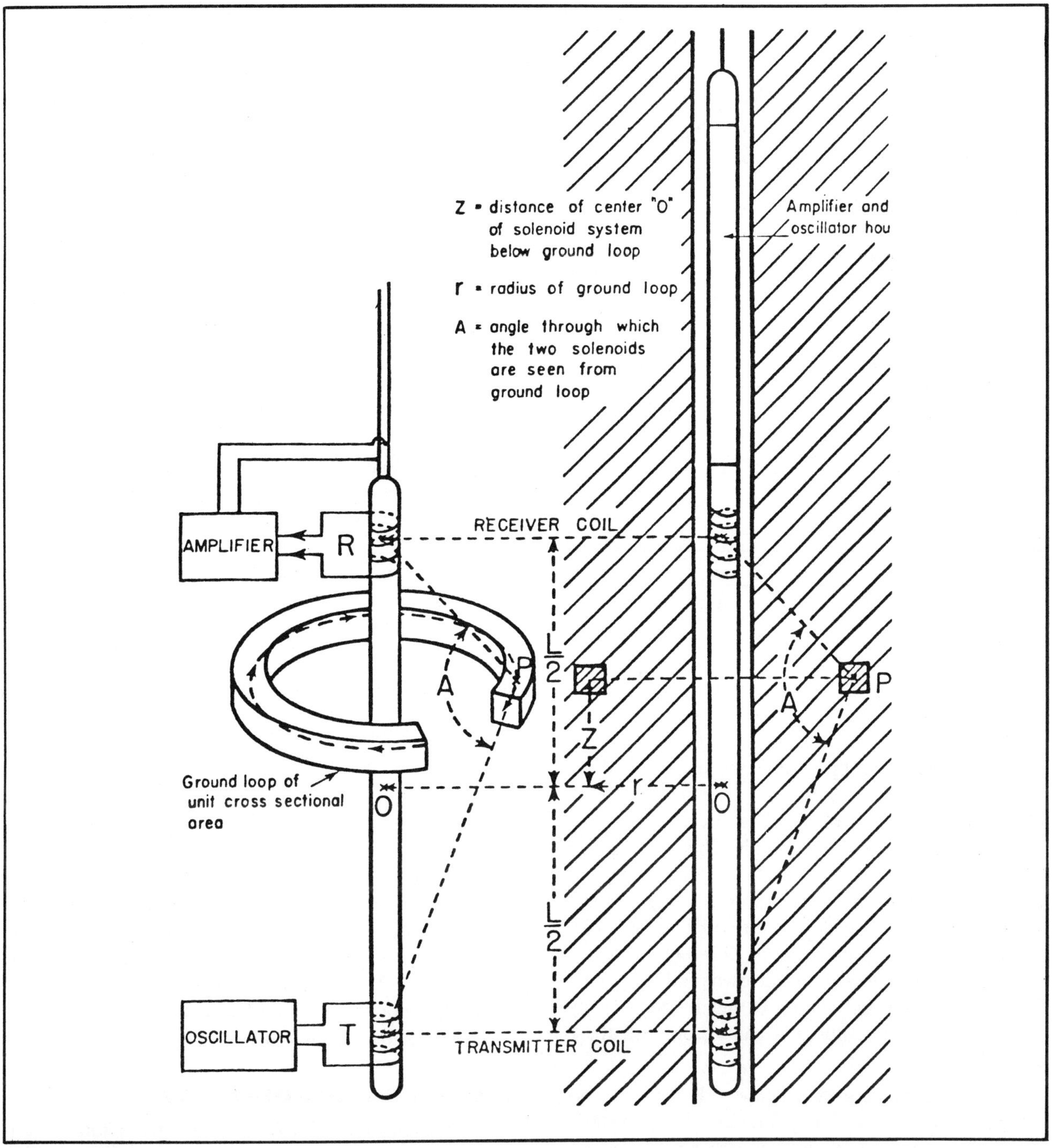

FIG. 4–38a

Diagram of the induction log principle (after Doll, *Introduction to the Induction Log and Application to Wells Drilled with Oil Base Mud*, courtesy Schlumberger Well Services Inc. and *Transactions*, volume 186, © SPE-AIME, 1949)

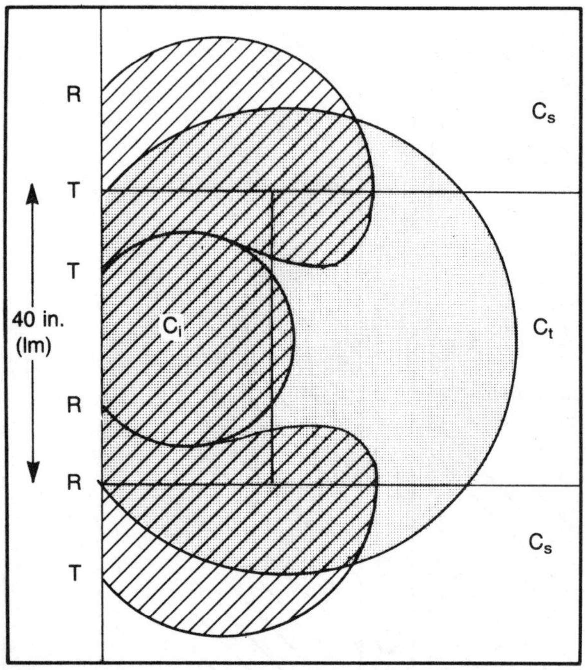

FIG. 4–38b

Response of the 6FF40 induction log, ILD

The resistor has a low value compared to the measure circuit impedance. Therefore, it is usually permanently placed in the measure circuit. The current circuit is switched to the calibration resistor, and the current required for a full-scale deflection is noted. This current is then set when the circuit is returned to the normal or logging position. This is shown in Fig. 4–40. With modern, stable circuitry and good current regulators, this step is probably superfluous. Almost any change to the circuit will be negligible.

The primary calibration is quite different, however. It is poorly understood by most users and designers. This step is long and complex, and involves many stages. A primary calibration can accomplish these calculations:

- It can determine if the calculated K factor is correct. This verifies the geometry of the system. This is usually done in a homogeneous medium of known resistivity. It is most easily done in a large tank of water of known salinity and temperature. The tank must be apparently infinite in size.
- It can determine the depth of investigation. This, too, verifies the calculated or assumed geometry. It is accomplished in the same type of tank as the one above. High-resistivity boundaries (plastic sheets) can be inserted at known distances, and the influence on the measurement can be noted.
- Primary calibrations can verify the geometry. This is a further verification of the first two steps and can be done in the same tank. A voltage probe returned to an infinite distance or a two-electrode gradient probe can be used to determine the shape of the equipotential surfaces in a homogeneous medium.
- The range of measurement must be determined. The same tank, again, may be used for the range measurements. Known amounts of sodium chloride should be verified with a resistivity tester and added to the tank water. The plotted results indicate the measurement range of the system.
- The calibration determines the departure from the ideal geometry. This is a difficult phase. Known resistivities, sizes of borehole, and formation and adjacent formations in many combinations must be used and the response of the system determined. This may be done in the tank with artificial formations, with a resistance network, or with computer modeling. The resulting departures from theoretical or homogeneous situations must be known for a realistic evaluation of the formation character. These departures occur at some point with any system.

As with other types of logging systems, the calibration phase is at least as lengthy and exacting as the design of the system itself. It is also a step or a phase that is usually neglected. The calibration of the resistivity and other systems is a phase that should be better documented and publicized by logging equipment manufacturers, designers, and operators. The usual assumption by the user is that the operator or contractor has made the measurements carefully, but this may not be true.

SPONTANEOUS POTENTIALS

Spontaneous potentials (SP) are the natural potentials that occur in the earth and are measured in a borehole. They occur because of imbalances in ion species and concentrations within the solutions in the formations and in the borehole. The SP principally involves four different mechanisms: diffusion, adsorption, redox reactions, and electrofiltration action.

The SP is a lithology log sensitive to sands and shales. It is used to calculate the formation water resistivity, R_w. The redox component of the SP

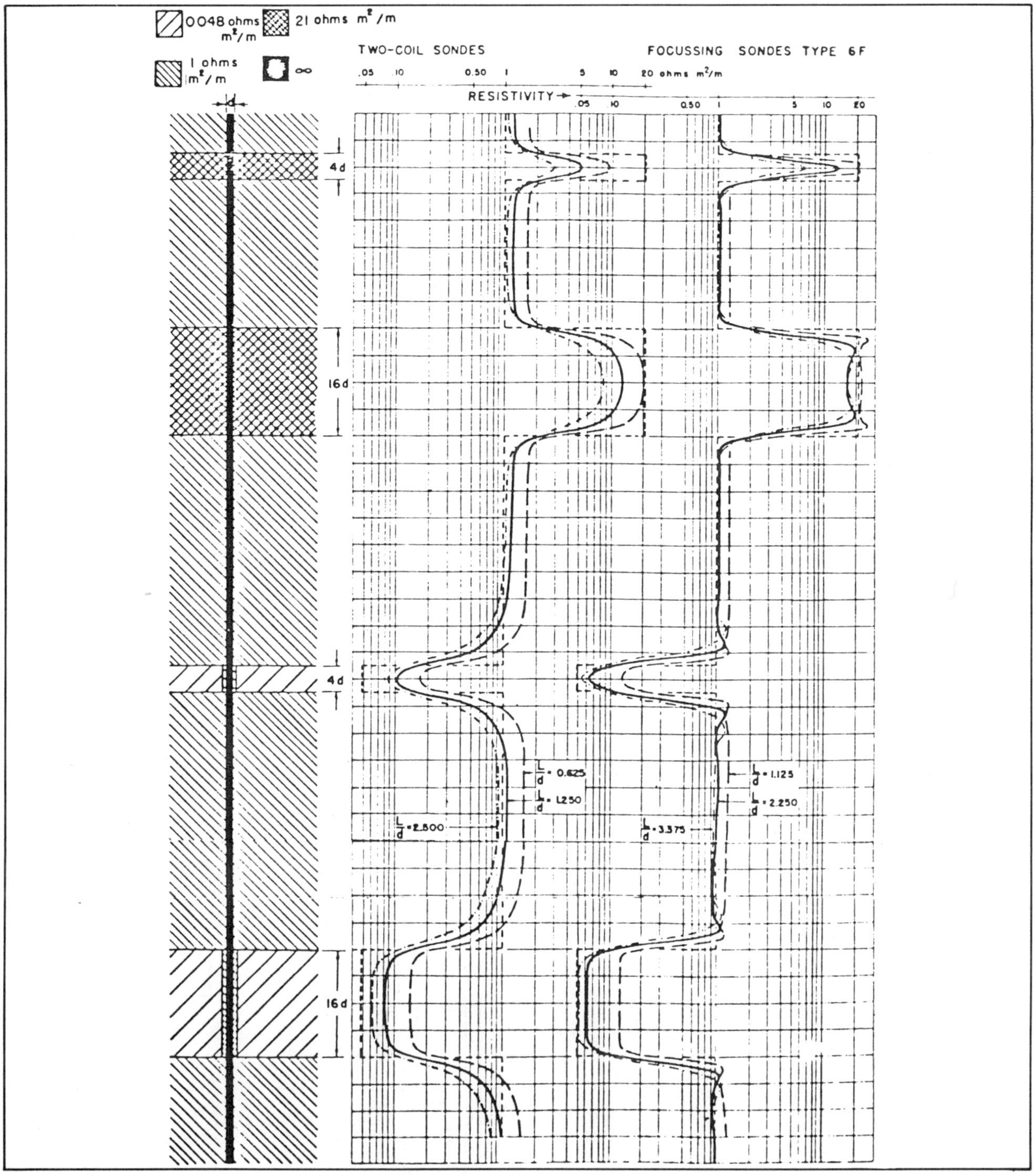

FIG. 4–38c

Response of a 6-coil and a 2-coil induction log system (after Doll, *Introduction to the Induction Log and Application to Wells Drilled with Oil Base Mud,* courtesy Schlumberger Well Services Inc. and *Transactions,* volume 186, © SPE-AIME, 1949)

curve can be applied in a mapping technique for mineral exploration. The electrofiltration component is sensitive to the movement of fluids into or out of the borehole. Fig. 4–41 shows an SP circuit.

The SP curve is used in petroleum logging to differentiate between porous, permeable beds and shales. The diffusion adsorption component of the SP curve is used almost exclusively. It is used to calculate the value of R_w to facilitate resistivity porosity determinations. The SP curve can show oil/gas/water contacts.

In nonpetroleum logging the SP curve is used as a lithology log to distinguish between sands, shales, and shaly sands and permeable and impermeable zones. It is used to calculate R_w for legal purposes, using the diffusion-adsorption component. It is also possible to use the redox component of the SP curve in a trend-mapping technique. The electrofiltration component is largely an interfering potential, but it can be used to spot fluid entry or loss. There are techniques for determining the amount of the electrofiltration potential. Other forms of potential measurements are the surface-potential measurements, redox measurements, the selective SP, the static SP, differential SP, and pH measurements.

In a borehole, the SP can only be measured in an open, uncased hole. Both metallic and plastic casing prevent the detection of an effective SP. SP measurements are difficult to make correctly. There are a number of disturbing phenomena that must be taken into account.

The SP measurement reflects:

- The shale or Nernst potential
- The diffusion potential
- The electrofiltration potential
- The relative redox potential
- Ion activity
- Ion mobility

It is affected by and must be corrected for:

- Borehole size
- Borehole mud resistivity
- Formation resistivity
- Ion type

It is adversely affected by:

- Variations in R_m point to point and time to time
- Cuttings in the mud column
- Mechanical damage to the electrode

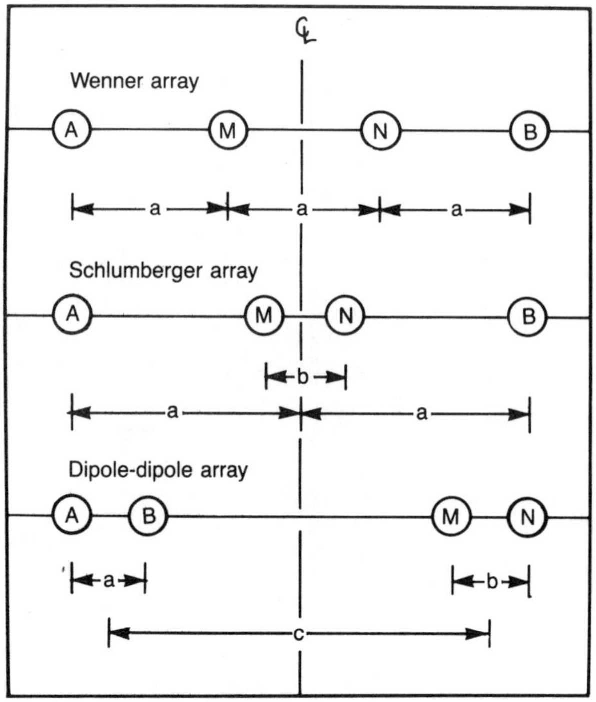

FIG. 4–39
Surface resistivity arrays

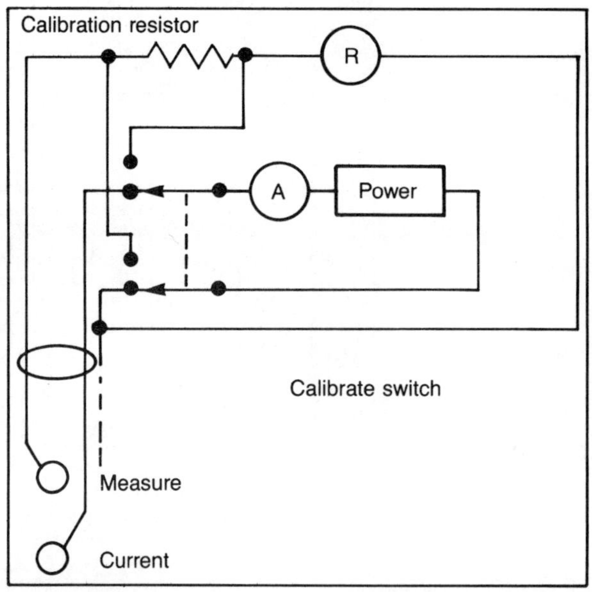

FIG. 4–40
Normal calibration

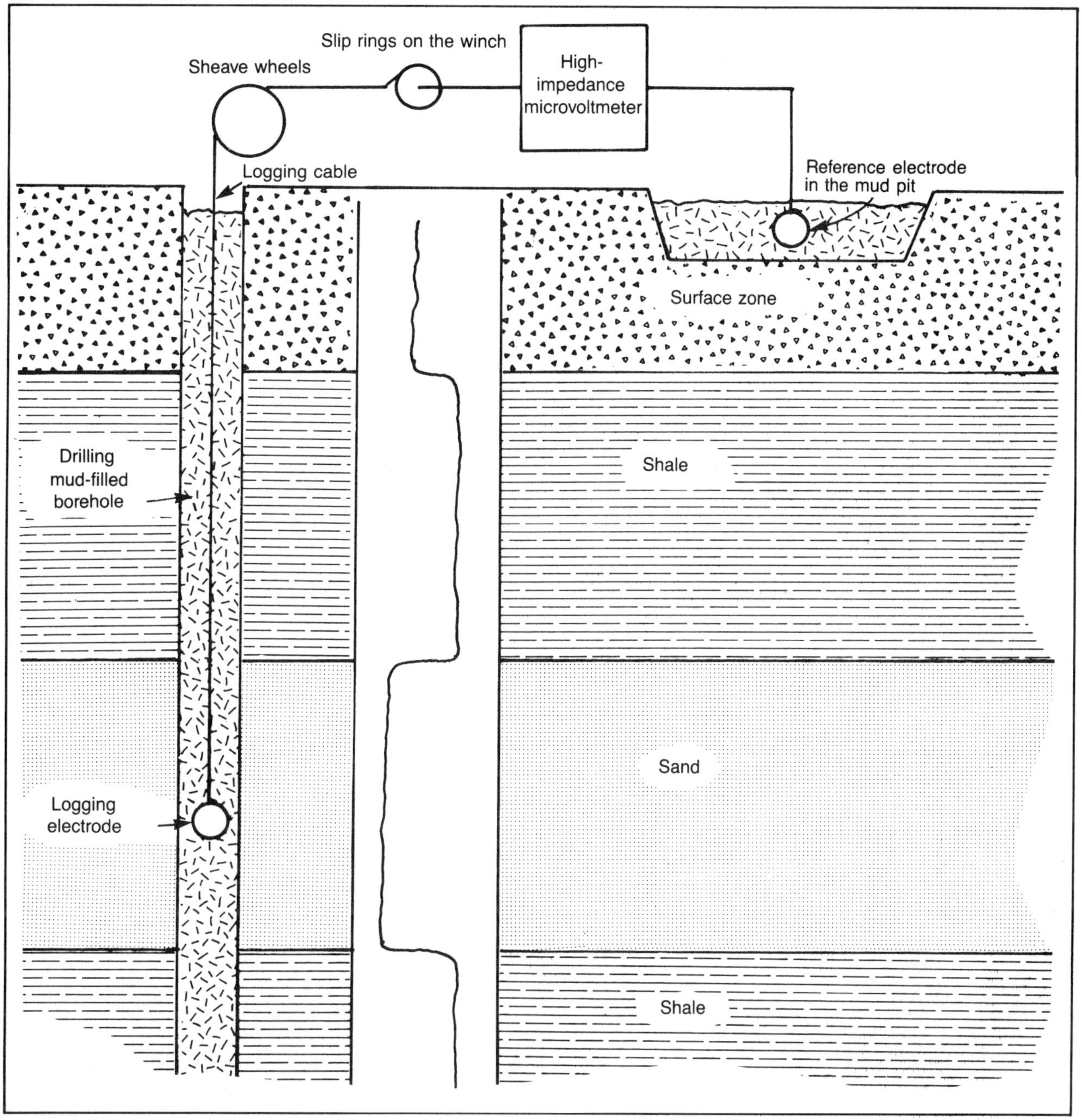

FIG. 4–41

Spontaneous potential circuit

- Exposed metal within the measurement volume
- Electrode metal type
- Electrode motion
- Fluid motion
- Stray currents
- Improper return electrode environment
- Temperature variations
- Drilling mud composition

Physical Description

Spontaneous potentials arise from an imbalanced distribution of ions caused by one or more mechanisms. The primary mechanisms appear to be diffusion, adsorption, oxidation-reduction, and electrofiltration. Any or all of these processes can act to unbalance the ion concentrations and cause potential differences and resulting current flows. There are other effects that may contribute, but mostly they are interfering or noise effects.

In borehole logging the combinations of the potentials in the formations cause electrical currents to flow in the borehole mud. The SP electrode assumes a voltage as a result of the potential drop in the mud column caused by this flow of current. This voltage is measured with respect to a distant, constant, identical electrode, illustrated in Fig. 4–42a, Fig. 4–42b, and Fig. 4–43.

In surface methods, the potentials themselves are detected or the potential drop through the formation may be detected. The mechanisms giving rise to the spontaneous potentials operate to separate anions from cations, deplete or enrich solutions of anions or cations, deplete or enrich solid portions of the formation with one of the ion types, or change concentrations that may cause any of these situations.

The potential differences set up within the formations can usually be considered to be across an interface. These potential differences result from various combinations of the four major mechanisms. The resulting currents flow through the mud column causing a potential drop along the mud column, as shown in Fig. 4–42b. The sum of the potentials is E_{SSP}:

$$E_{SSP} = I(r_s + r_{sh} + r_m) \quad (4.24)$$

The potential change as the electrode moves from a sand to a shale is SP:

$$E_{SP} = Ir_m \quad (4.25)$$

Where:

I = Current in the mud column
r_s, r_{sh}, and r_m = Resistances of the sand, shale, and mud, respectively

Eq. 4.24 and 4.25 can be combined:

$$E_{SSP} = E_{SP} \frac{(r_s + r_{sh} + r_m)}{r_m} \quad (4.26)$$

If the borehole could be plugged so no current could flow in it, the value of r_m would approach infinity:

$$\lim_{r_m \to \infty} E_{SP} = E_{SSP} \quad (4.27)$$

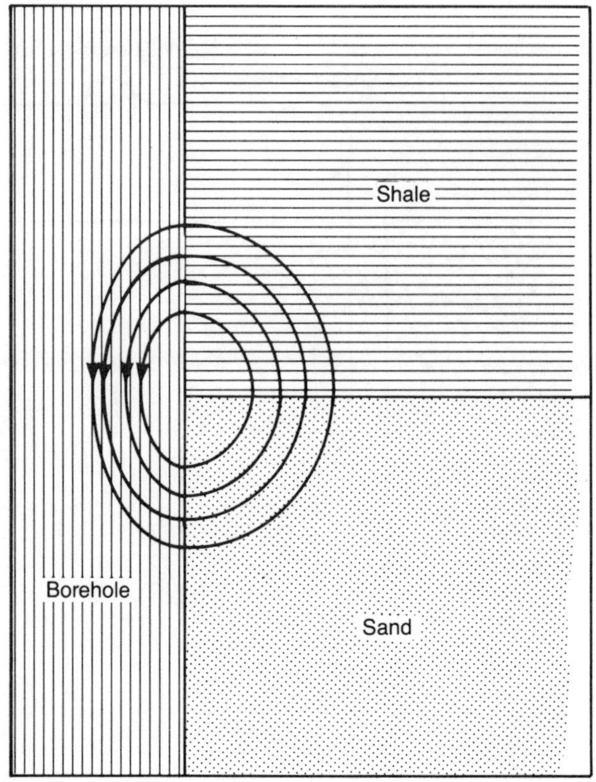

FIG. 4–42a
Electrochemical component of SP

This principle was used of in an early static SP method and the selective SP method.

If the beds are thick and the resistances of the sand and shale are low, E_{SP} and E_{SSP} are nearly equal because r_s and r_{sh} are small compared to r_m. However, in resistive beds the values of E_{SP} and E_{SSP} can be different because r_s and r_{sh} are no longer negligible.

The distribution of current in the shale/borehole interface and at the sand/borehole interface depends on the ratios r_{sh}/r_m and r_s/r_m. The higher these ratios are, the lower the current density is, and thus, the longer the SP slope from the shale potential reading to the sand peak. This is illustrated in Fig. 4–43.

The position of the bed boundary with relation to the slope depends on the ratio r_s/r_{sh}. This is actually the same effect as explained in the previous paragraph. The bed boundary is at a point halfway from the sand peak to the shale background if $r_s/r_{sh} = 1$. It is at a lower point, closer to the shale potential, if the ratio is high and at a high point if

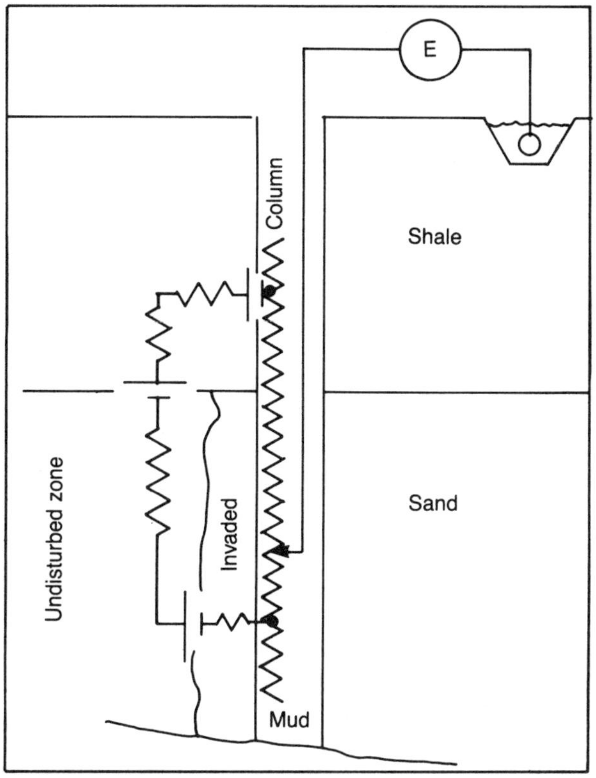

FIG. 4–42b

Formation equivalent circuit for the SP

the ratio is low. This is also shown in Fig. 4–43. In any event the bed boundary always coincides with the inflection point of the curve. Several examples of this effect are shown in Fig. 4–44a and Fig. 4–44b.

If the bed thickness, T, is considered in units of hole diameter, the potential E_{SP} is:

$$E_{SP} = T(4T^2 + 1)^{-1/2} (E_{SSP}) \qquad (4.28)$$

$$E_{SP}/E_{SP\,max} = (T^2 + 1)^{1/2} (4T^2 + 1)^{-1/2} \qquad (4.29)$$

Since the SP system measures the ohmic drop in the mud column, if the formation contributes nothing to the potential series as with a massive, impermeable limestone, the SP reading is merely the potential gradient between the two nearest contributing formations. It is usually a straight line through the limestone. This is illustrated in Fig. 4–45, Fig. 4–46a, and Fig. 4–46b.

■ Instrumentation

In principle the instrumentation of the SP logging system is quite simple. It should consist of a measure and a reference electrode that are identical, chemically inert, and stable; a reference position that is representative of the formation conditions; connecting wires; and a high impedance voltmeter. When differential or surface SP measurements are made, the reference

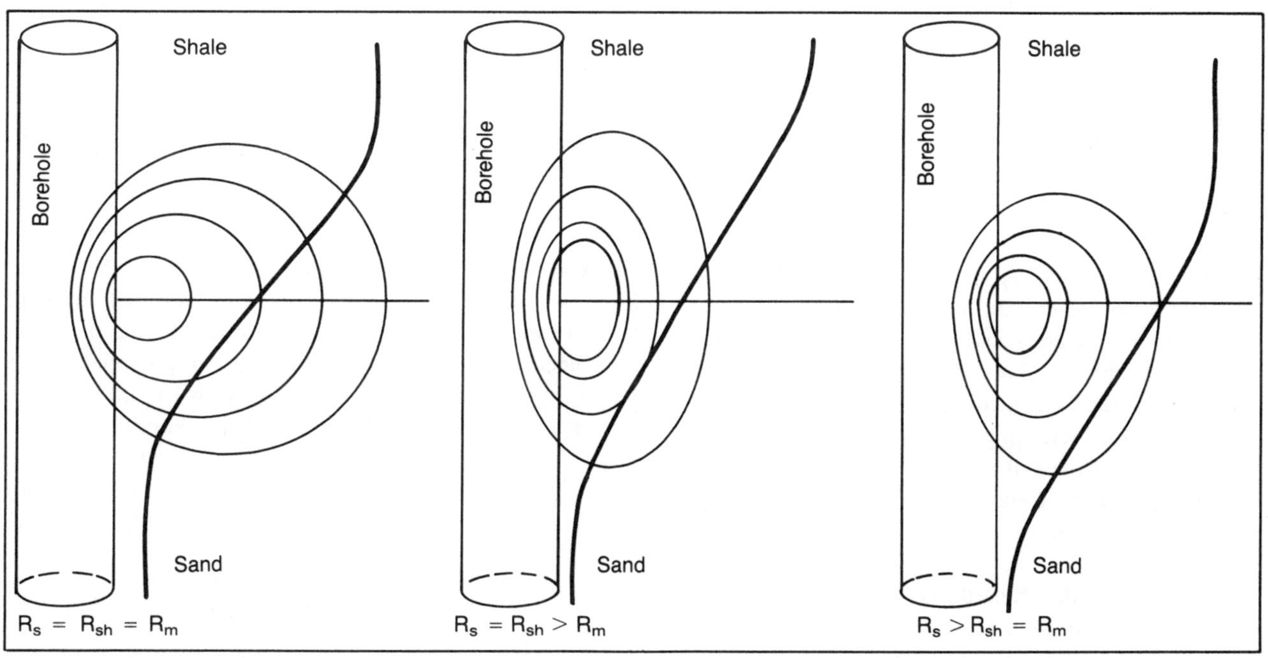

FIG. 4–43

The influence of the mud and formation resistivity on the SP curve

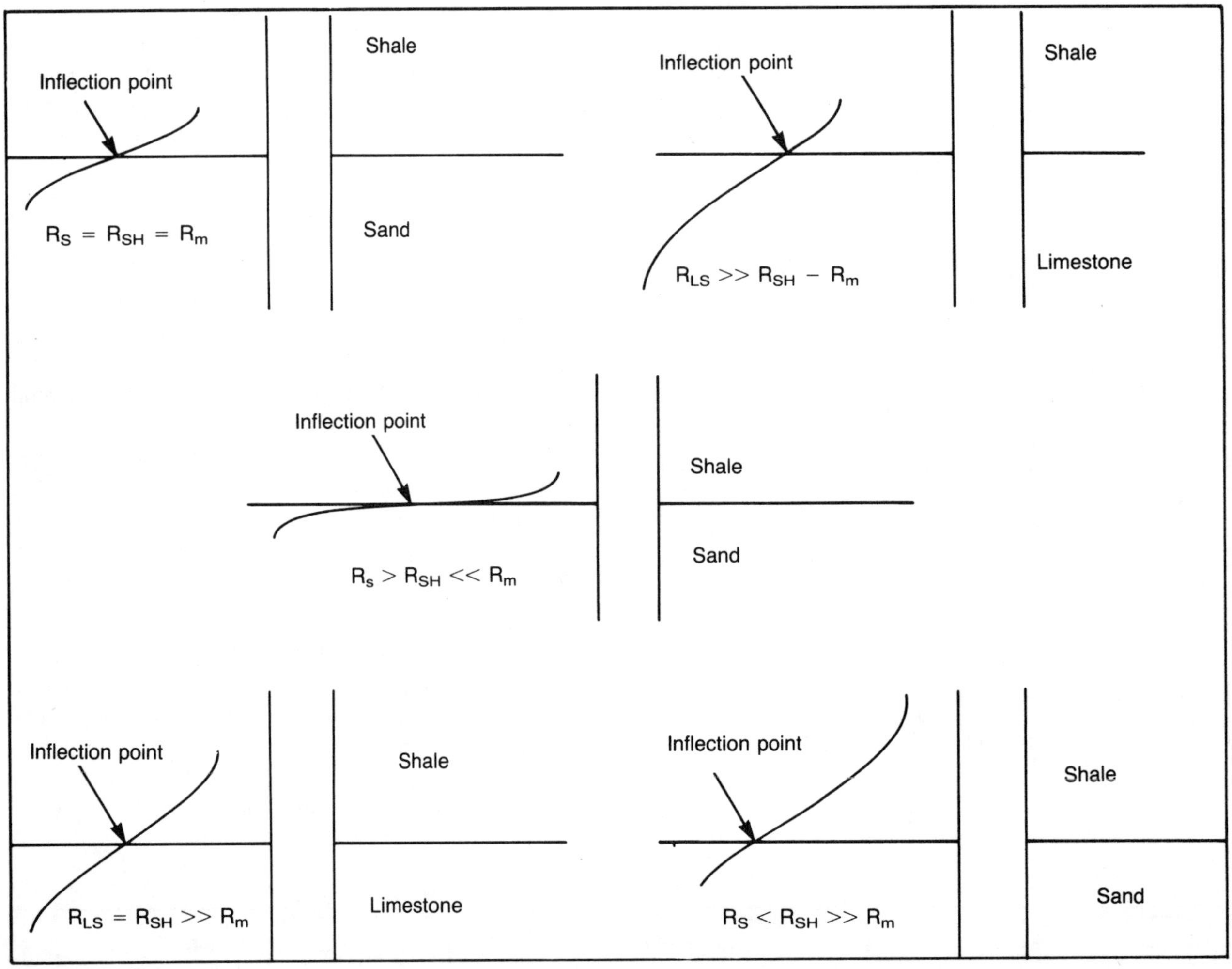

FIG. 4–44a

SP bed boundary effects (after Pirson, *Handbook of Well Log Analysis,* Prentice-Hall, 1963)

electrode is replaced by a second measure electrode.

Modern voltmeters are of high enough impedance (10^8 ohm or greater) and stable enough that they give few problems. This was not always so. Older texts show elaborate methods that were usually unsatisfactory for increasing the impedance and stability of the system. These older methods are not discussed except to discourage their use. When using old logs (pre-1955), keep the possibility of an SP error in mind.

The SP electrodes, regardless of whether the system is a conventional SP log, a differential SP, or any other, must be identical, inert, and stable. Since we do not know how to build electrodes that have no potential of their own, the measure and reference electrodes must have equal potentials. These potentials buck each other out, resulting in a measurement of only the changes of potential of the measure electrode as a result of formation potentials. The simplified diagram of the SP measure circuit is shown in Fig. 4–41. The SP potential measured is:

$$E_{SP} = E_{SSP} \frac{r_m}{(r_s + r_{sh} + r_m)} + (E_e - E_{ref.}) \quad (4.30)$$

Where:

E_e = Measure electrode potential
E_{ref} = Reference electrode potential

High-impedance measurements must be made for two reasons. First, the environment must be disturbed as little as possible. Some current must be drawn

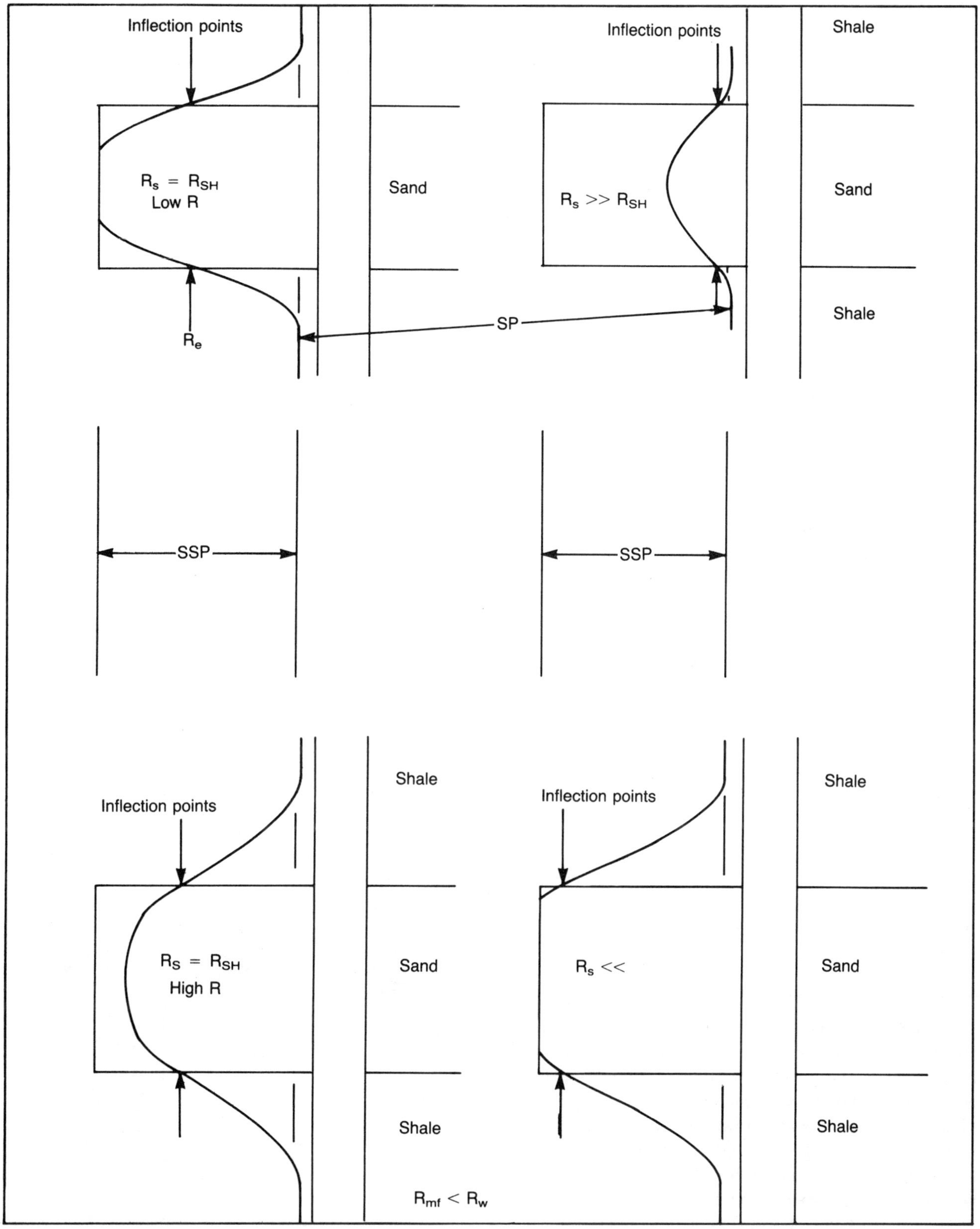

FIG. 4—44b

SP bed boundary effects (after Pirson, *Handbook of Well Log Analysis,* Prentice-Hall, 1963)

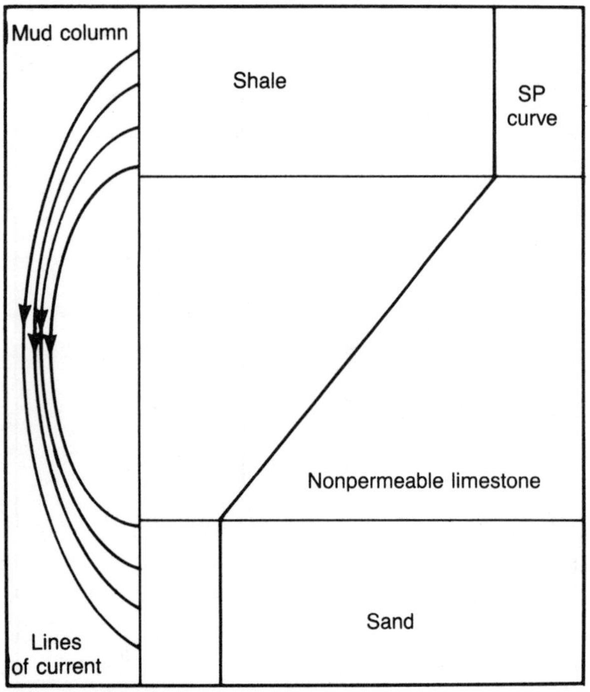

FIG. 4–45

SP response across a nonpermeable limestone

from a system or source on which we make a measurement of potential. This results in a lowering of voltage depending on the amount of current drawn and the impedance of the formation/electrode:

$$E_{measured} = E_{SP} - I_{meter} \times r_{formation} \quad (4.31)$$

A high-impedance measuring system draws negligible current, making the second term in Eq. 4.31 near zero. Second, the high-measurement impedance makes any other circuit impedance, such as cable resistance, negligible. Thus, these other circuit impedances can change greatly without appreciably affecting the measurement.

■ Electrochemical Potentials

The electrochemical component of the spontaneous potentials is the portion most used in petroleum logging. It is the portion of the SP curve used to determine lithology and to calculate the value of R_w. This component arises from two mechanisms: a separation of ionic charges through diffusion processes, and a separation of charges by adsorption of one ion type on a surface, such as on a clay.

The electrochemical or diffusion absorption potential is evaluated as the deflection opposite a permeable formation, compared to the potential of the adjacent shale potential. It is the sum of the voltages or potential differences developed across the liquid junction between the liquid in a sand and the mud filtrate in the invaded zone, the interface between the sand and shale, and the shale-mud interface. It depends on the following factors:

- The chemical and mineralogical composition of the rock composing the sand. These factors affect the apparent salinity of the sand and may vary in both amplitude and polarity.
- The composition and concentration of salts in solution in the electrolyte filling the pore spaces. This effect decreases with increased concentration and may change polarity with multivalent ions.
- The degree of saturation of the rock with electrolyte.
- The density of the rock.
- The grain size of the rock. Increases of the last three result in departures from a simple situation.
- The cation-exchange capacity (CEC) of the shale.
- The salinity of the mud.

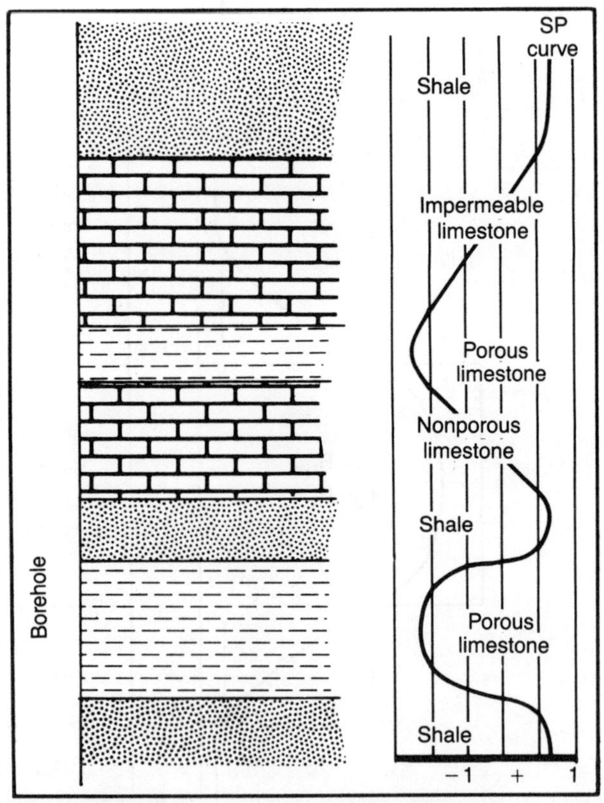

FIG. 4–46a

SP in limestone

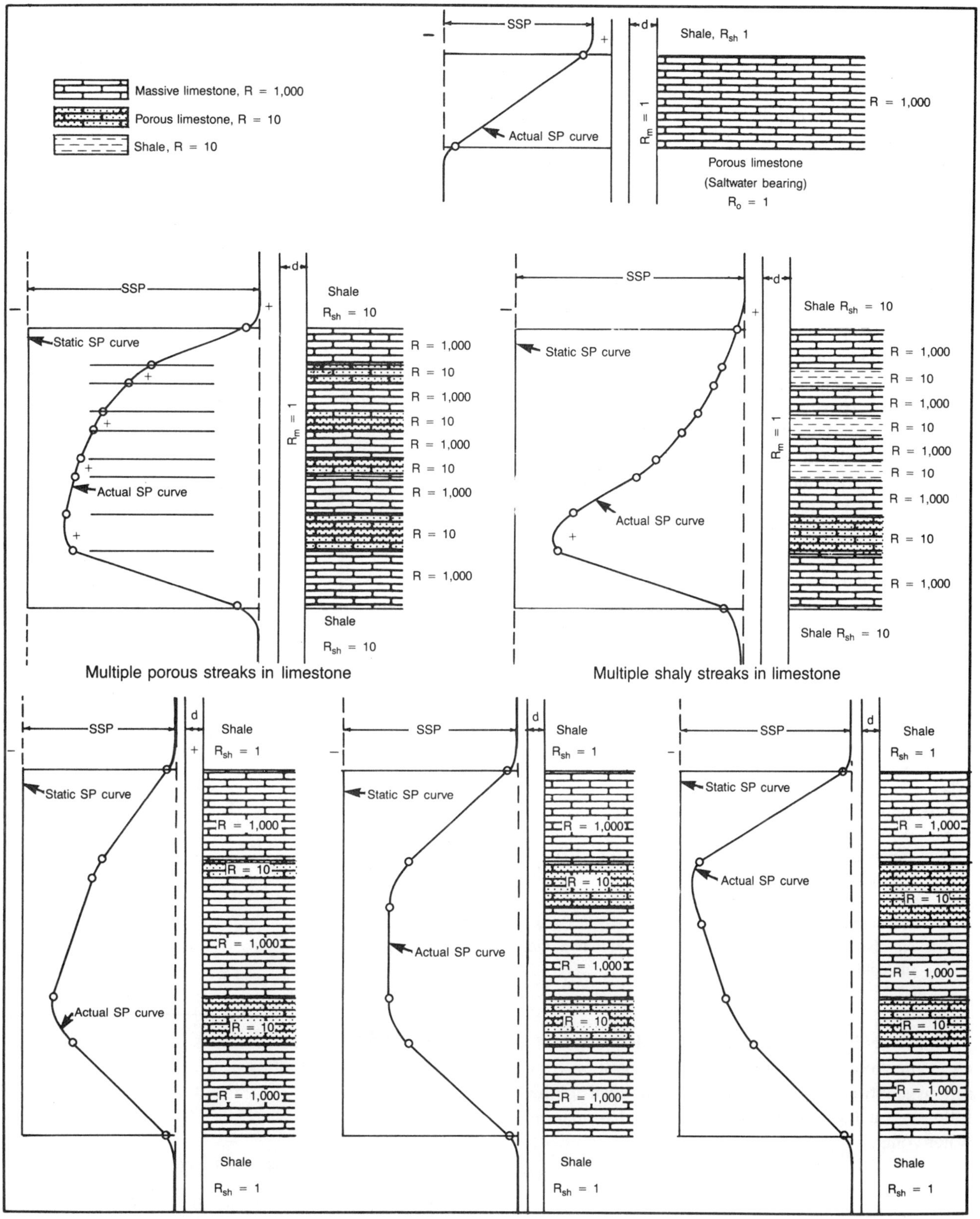

FIG. 4–46b

SP curve shapes in limestones (after Pirson, *Handbook of Well Log Analysis,* Prentice-Hall, 1963)

Diffusion Potentials

If a body of water is divided into two portions separated by a semipermeable membrane and if the two portions have the same concentration of sodium chloride dissolved in them, a pair of electrodes, placed one in each portion, shows no difference in potential (Fig. 4–47a). If the two portions have different concentrations, however, there eventually is a difference in potential between the two sides. This difference occurs because the chloride (−) ions have a higher mobility than the sodium ions (+). Therefore, the chloride ions penetrate the semipermeable membrane more often and more easily than the sodium ions. Thus, the chloride ions predominate, with their negative charge, on the side of the lower concentration. This leaves an excess of sodium ions (positive charges) on the higher concentration side. Therefore, a voltmeter connected to an electrode in each side of the cell shows a voltage difference across the membrane with the positive indication on the side of the higher concentration.

This type of diffusion action takes place across the interface between the invaded zone and the formation water in a permeable formation. It requires a water solution in the formation and mud and an interconnected or permeable porosity.

The relationship for the diffusion potential, E_D is:

$$E_D = \frac{RT}{F} \frac{\ell_c n_c - \ell_a n_a}{\ell_c n_c Z_c + \ell_a n_a Z_a} \ell n \frac{A_1 C_1}{A_2 C_2} \qquad (4.32)$$

Where:
- R = Universal gas constant
- T = Absolute temperature
- Z = Valence
- F = Faraday's number
- ℓ_a and ℓ_c = Mobilities of the anions and cations, respectively
- n = Number of anions and cations formed when the molecule is ionized
- A_1 and A_2 = Activities at concentrations C_1 and C_2*

Since the conductivity and, thus, the resistivity are a function of the activity and concentration of a salt in solution, Eq. 4.32 may be simplified to the following at the interface of the invaded/formation water zone:

$$E_D = K_D \log_{10}(R_2/R_1) \qquad (4.33)$$

*Activity is defined as the actual ion concentration divided by the concentration of that ion type in a standard reference state.

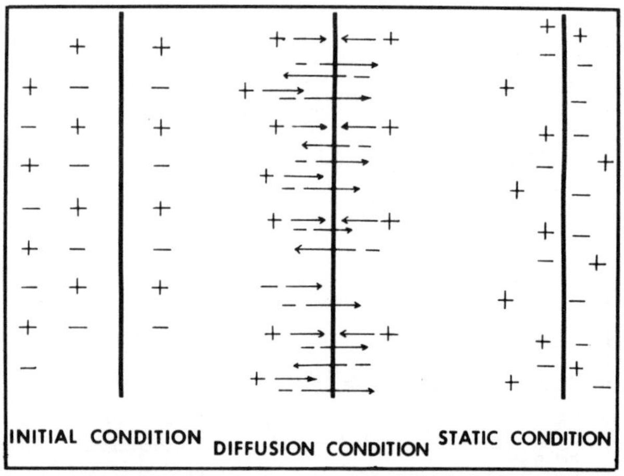

FIG. 4–47a

Diagram of the diffusion component of the SP curve

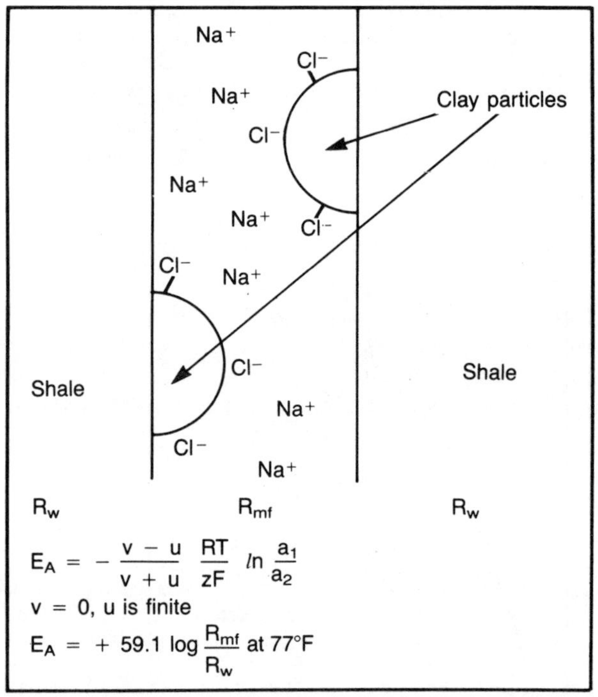

FIG. 4–47b

Shale or Nernst potential

Where:
- R_1 = Formation water resistivity, R_w, of the undisturbed formation
- R_2 = Mud filtrate resistivity, R_{mf}

The foregoing situation is valid for simple liquid junctions. In a clean sand or limestone the departure from Eq. 4.33 may be negligible. However, in a shaly sand the deviation may be substantial (Fig. 4–48a and Fig. 4–48b).

Adsorption Potentials

Many materials, such as clay and extremely fine quartz, have electrical properties because of their large surface area. These properties bind ions of one charge—usually the negative charge or anion—to the material surface. This results in a depletion of the solution of that charge and an excess of that charge associated with the solid material. This process is the Nernst diffusion layer process, and the resulting potential is the shale or Nernst potential (Fig. 4–47b). It is most prevalent in the clays of shales but can occur in fine-grain quartz materials in silts and mudstones.

The adsorption function can be combined with the diffusion function and simplified (Fig. 4–48a). The combined potential, the diffusion potential, E_D, plus the adsorption potential, E_A, is the SP deflection opposite a sand, E_{DA}. Thus, from a shale to a sand, the change of potential, E_{DA}, is for dilute solutions of NaCl:

$$E_{DA} = K_{DA}\, b \log (R_{mfe}/R_{we}) \qquad (4.34)$$

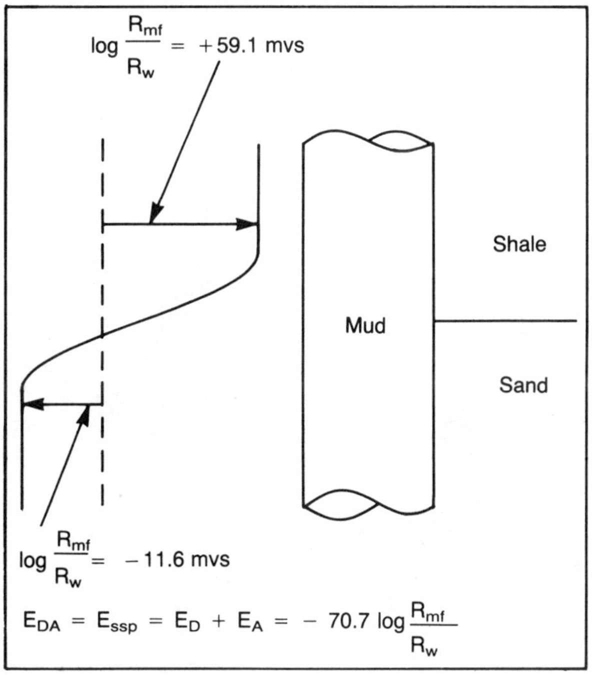

FIG. 4–48a
SSP as the combined potential

The resulting conversion constant, K_{DA} is a function of the temperature. Its value is:

$$K = -(0.133F + 60) \qquad (4.35)$$

Where:
- b = A parameter that depends on concentration (for dilute solutions b = 1)
- R_{mfe} and R_{we} = Effective resistivities of the mud filtrate and the connate water, respectively

The effective resistivity depends on the ion types present, since Eq. 4.34 assumes only Na^+ and Cl^- ion types.

The notation of effective resistivity is necessary because the assumption is made in all of the charts and relationships that the solution is only NaCl in water unless otherwise specified. The effective resistivity is what the resistivity would be if the ion types were NaCl. Obviously, this is not always true. Therefore, a correction must be made. Fig. 4–49 shows the relationships between various solutions of salts. Table 4–2 shows how the various ion types compare to a solution of NaCl. Fig. 4–50 shows the relationship between water resistivity and effective water resistivity.

The family of curves in Fig. 4–51 arises because of the temperature effect on the mobilities of the ion species. Below 0.1 ohmm the spread is caused by decreased relative activity due to concentration. Above 0.1 ohmm the mobilities of the calcium and magnesium ions, Ca^{++} and Mg^{++}, dominate and cause a spread as a result of temperature.

After the effective value of R_w has been calculated (R_{we}), it must be converted to the proper value for the particular mixture of ion types. Fig. 4–51 is representative of the ground waters found in the Gulf Coast, Midcontinent, and Rocky Mountain areas. It can be used with some degree of confidence elsewhere. If the ion type is known, Table 4–2 may be used.

Table 4–2
Conversion Factors to NaCl when Salinity Is Given in Ionic ppm

Ions	Factor
Na^+	1.00
K^+	1.00
Ca^{++}	0.95
Mg^{++}	2.00
SO_4^{--}	0.50
Cl^-	1.00
HCO_3^-	0.27
CO_3^{--}	1.26

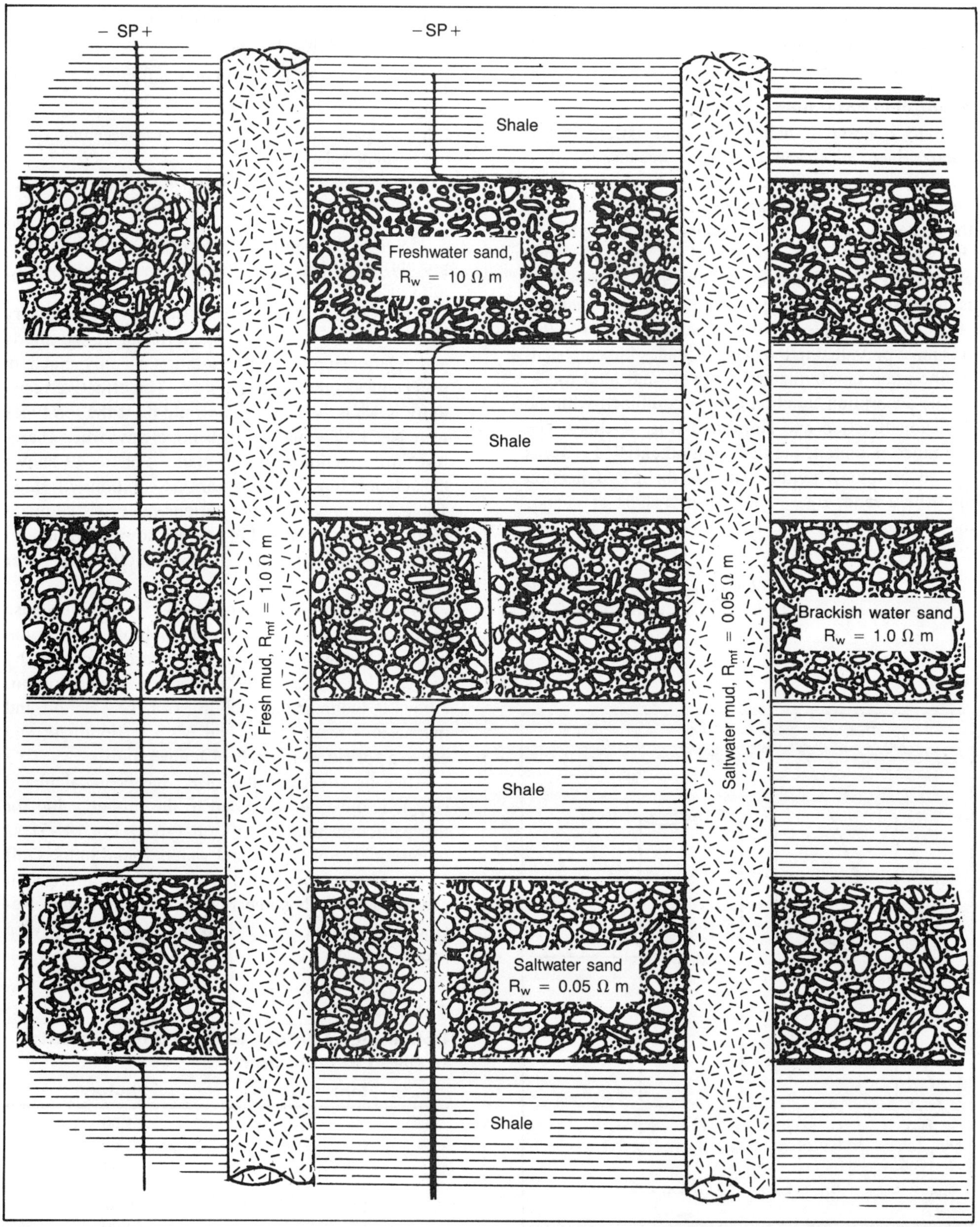

FIG. 4–48b

Effect of R_{mf}/R_w on the SP curve

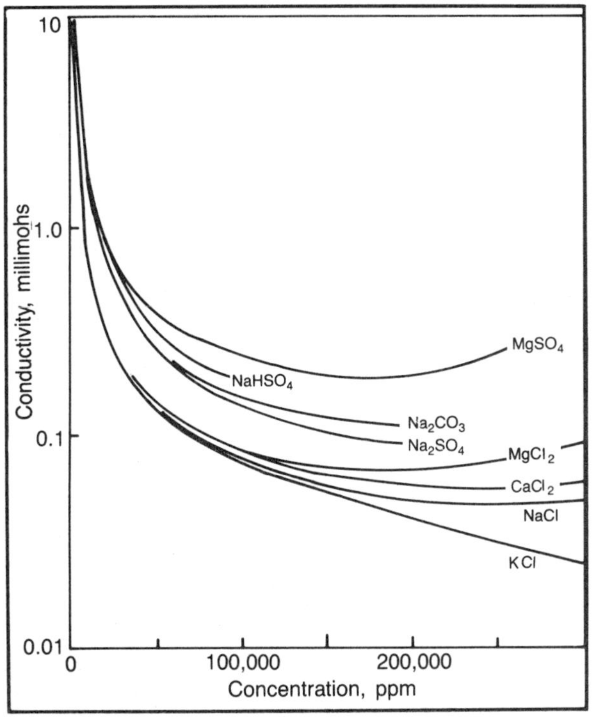

FIG. 4–49

Conductivity vs. concentration for several common salts

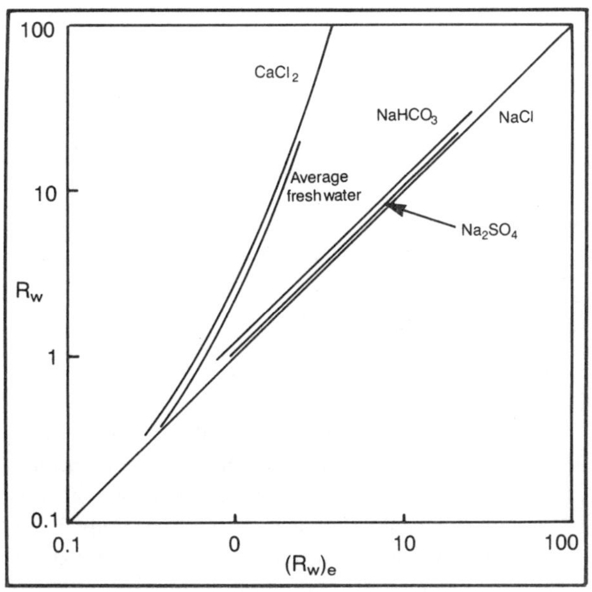

FIG. 4–50

R_w vs. $(R_w)_e$ for solutions of several salts

The mobility of an ion is a function of the absolute temperature. Therefore, one must correct all measurements to the same temperature for use and comparisons. The charts in Fig. 4–52a and Fig. 4–52b can be used to show the variation of a sodium chloride solution resistivity as a function of temperature at constant concentration and SP correction for bed thickness, respectively.

The value of K_{DA} in Eq. 4.35 for a clean sand and a shale for NaCl solutions is $K_{DA} = -70.7$ millivolts at 75°F. Note that the value of K is negative. Thus, if the resistivity of the mud filtrate, R_{mfe}, is higher than R_{we}, the value of the logarithm function in Eq. 4.35 is positive. Thus E_{DA} opposite a sand is negative.

If, on the other hand, the mud filtrate is more saline than the connate water, the logarithm function is that of a ratio less than one and is negative. This times the negative value of K_{DA} indicates a positive SP deflection opposite a sand when saline mud is used. Similarly, no SP deflection indicates $R_{mfe} = R_{we}$. These last two situations are prevalent in non-petroleum logging where low salinities are common and where formation waters are often surface run-off. At 50 ppm, which is not uncommon for Rocky Mountain surface water, the amount of salt (NaCl) is only 1 lb in 2,500 gal of water. Thus, it takes only a small additional amount of salt to change the SP curve dramatically. Literally, positive, negative, and zero SP deflections can be encountered in one hole and frequently are in the spring.

The value of R_w may be calculated from the SP curve in the following manner. Refer to Fig. 4–53, which is a graphical presentation of Eq. 4.35:

1. Draw a line from shale deflection to shale deflection across the sand deflection on the SP log
2. Measure the distance, in millivolts, from the shale line to the deflection in the cleanest portion of the sand
3. Solve Eq. 4.34 for R_{mfe}/R_{we}
4. Find the value of R_{mf} (Fig. 4–54)
   a. If a mud tester and mud press are available, measure R_{mf} and correct for formation temperature
   b. If no press is available, find the average value of R_{mf} from the R_m value, using Fig. 4–54 and correct for formation temperature
   c. Change the value of R_{mf} to R_{mfe} using Fig. 4–51
5. Divide R_{mfe} by the ratio found in step 3 to obtain R_{we}

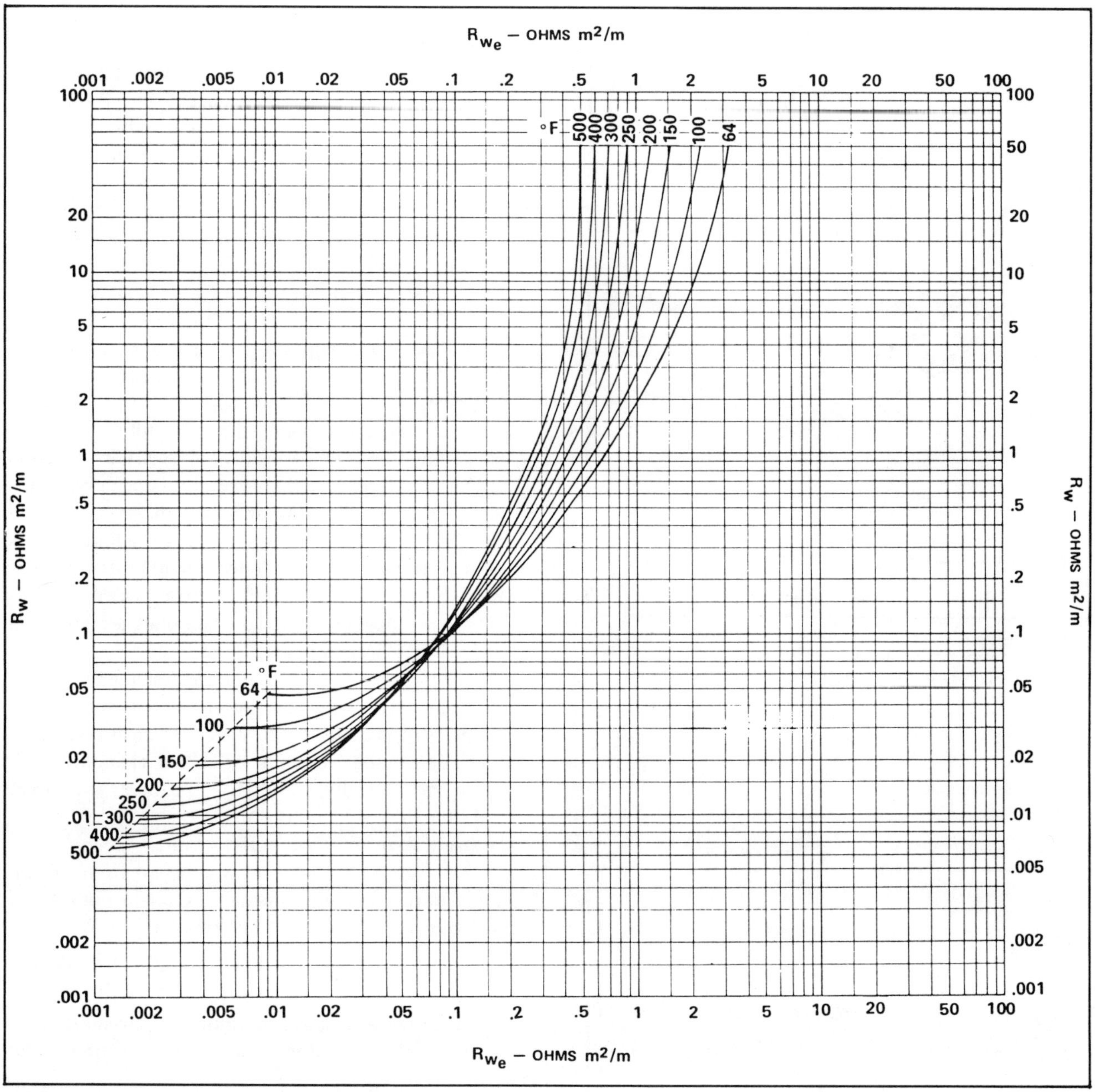

FIG. 4–51

R_w vs. R_{we} at various temperatures (courtesy Gearhart Industries Inc.)

6. Return to Fig. 4–51 to obtain a corrected value of R_w at formation temperature

7. The value of the total dissolved solids in terms of eNaCl can be obtained from Fig. 4–52a

Some of the additional mechanism that may change ion populations by adsorption are shown in Fig. 4–55a, b, and c.

Shaly Sands

When a sand contains a shale, the SP is reduced. The amount of reduction is a function of the amount of shale in the sand and the cation exchange capacity (CEC) of the shale. As the amount of shale is increased, the potential approaches that of a pure shale.

The value of α, the reduction factor, is:

$$\alpha = \frac{\text{pseudostatic SP in shaly sand}}{\text{static SP in clean sand}} \tag{4.36}$$

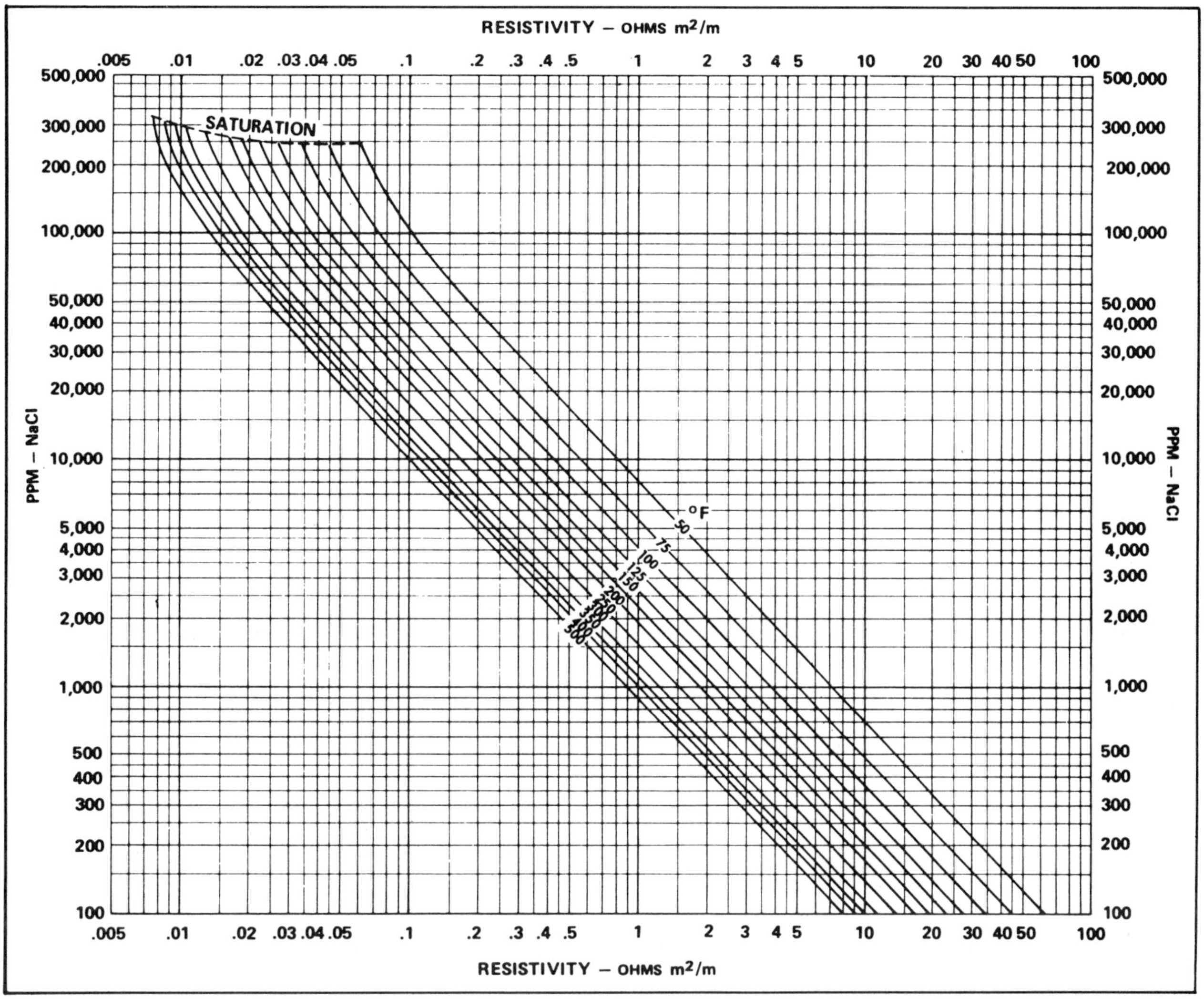

FIG. 4–52a

Resistivity vs. temperature—NaCl solutions (courtesy Gearhart Industries Inc.)

The value of α may be obtained from Fig. 4–56 or may be calculated:

$$\alpha = \frac{\log_{10} \dfrac{u + q}{1 + q}}{\log_{10}(u)} \tag{4.37}$$

Where:

$u = (R_t/R_i)$

$$q = (R_t/R_{sh}) \frac{V_{sh}}{(1 - V_{sh})}$$

V_{sh} = Fractional volume of shale

V_{sh} may be determined from the gamma-ray curve or from a crossplot. With the gamma-ray curve, the shale content, V_{sh}, by volume of total formation may be estimated by:

$$V_{sh} = \frac{\gamma_{sh} - \gamma}{\gamma_{sh} - \gamma_s} \tag{4.38}$$

Where:

γ_{sh} and γ_s = Gamma-ray readings, in any units, of the closest pure shale and clean sand, respectively

γ = Gamma-ray reading of the point in question

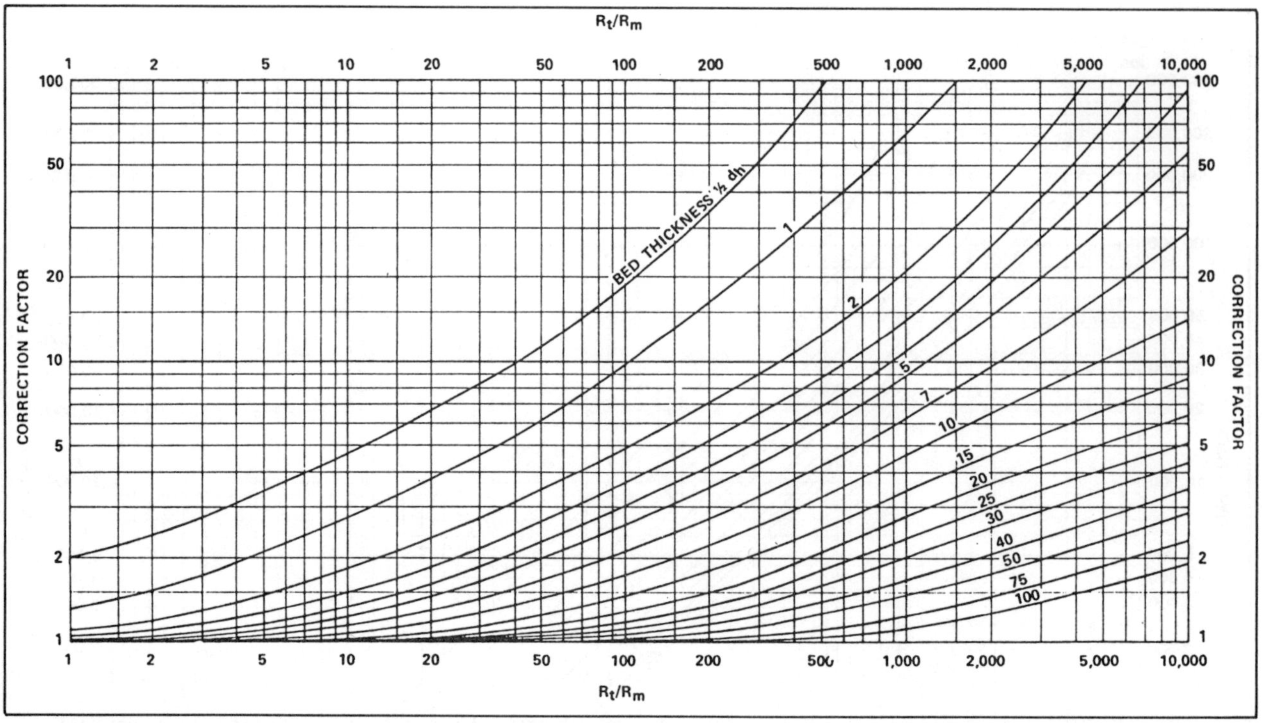

FIG. 4–52b

SP correction for bed thickness (courtesy Gearhart Industries Inc.)

By combining relationships, the value of R_w in a shaly sand may be found from:

$$\log \frac{R_{wa}}{R_w} = (1 - \alpha) \log \frac{R_{mf}}{R_w} \qquad (4.38a)$$

Where:

R_{wa} = Apparent water resistivity from a shaly sand deflection uncorrected

■ Redox Potentials

The redox component of the SP curve is seldom used in the petroleum industry. Occasionally, it is used to indicate the presence of pyrites but little beyond this. In mineral logging it promises to be quite helpful. Many important types of mineral depositions are accompanied by or caused by redox changes and reactions. These include some depositional mechanisms involving uranium, copper, silver, iron, and petroleum. Many compounds change solubility when undergoing a redox change. Some compounds change gross electrical and magnetic character with redox changes.

A redox reaction is one where an ion, such as a metal ion, changes valence. It is said to be oxidized if it loses electrons and becomes more positive. It is said to be reduced if it gains electrons and becomes more negative. The changes and reactions are called reduction-oxidation (redox) reactions. An oxidation of a material is always accompanied by something being reduced. The reactions can be abbreviated in equation form with the so-called half reaction. Thus, iron is oxidized from the ferrous state to the ferric state, and the reaction is shown thusly:

$$Fe^{++} \rightleftharpoons Fe^{+++} + e^- \qquad (4.39)$$

The double arrow indicates the reaction is reversible; e^- is the resulting electron. It represents energy, and it can do work. There are, of course, other parts to the reaction, but they are not needed to illustrate the process. It must particularly be remembered that something else is being reduced by gaining the electron. The reduction process requires energy, and work must be done to accomplish it.

Since there are changes of electrical charges involved in redox reactions, there are electrical phe-

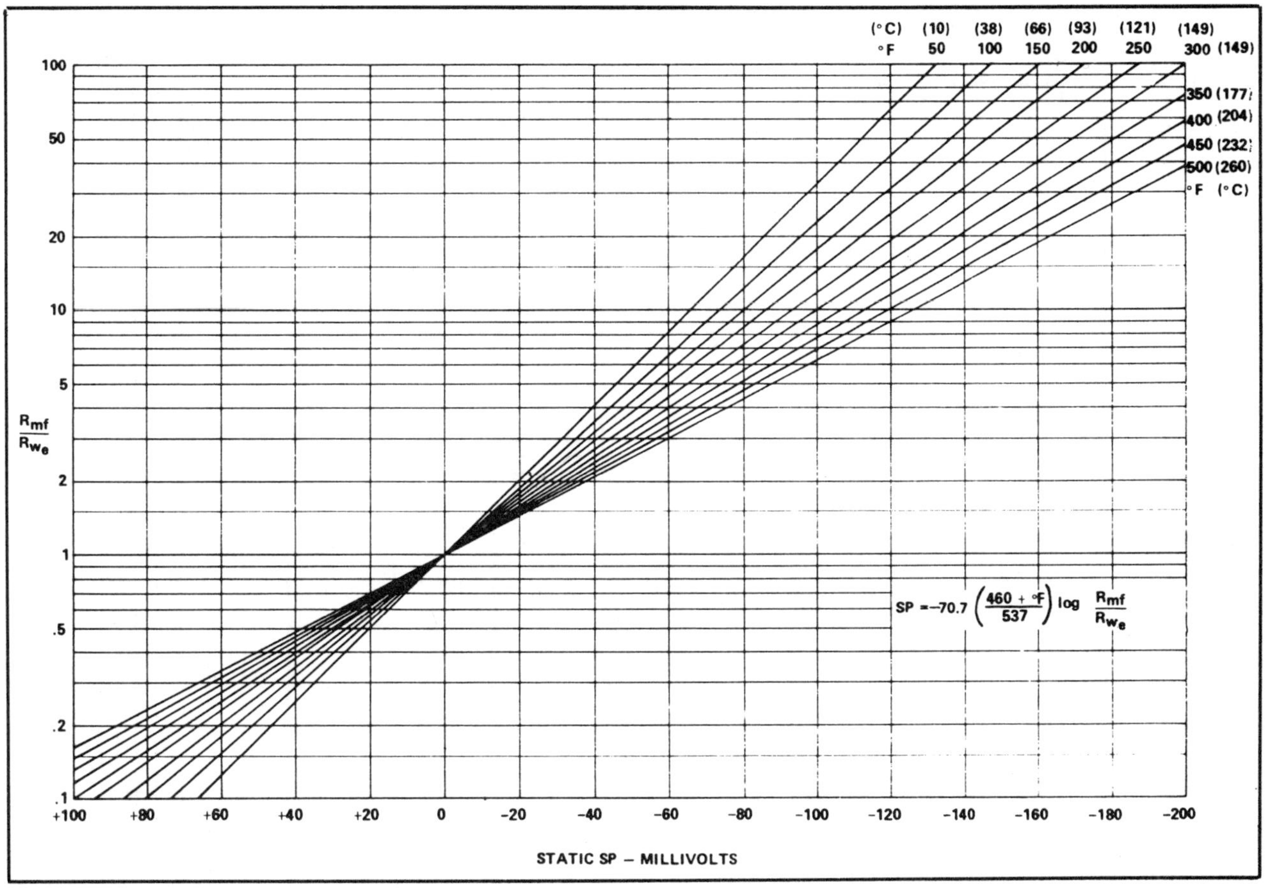

FIG. 4–53

Equivalent formation water resistivity for SSP (courtesy Gearhart Industries Inc.)

nomena accompanying them. Materials that are more oxidized, for example, tend to have more positive potentials than those that are reduced.

These redox/electrical interactions occur and are made use of in many common places. Ferric iron (hematite) is reduced to metallic iron in a reduction furnace. Carbon supplies electrons to the ferric iron and is oxidized in the process. A lead acid battery starts an automobile by oxidizing plumbous oxide to plumbic oxide. Chromium is plated on an automobile bumper by reducing chromic salts with electrons (electrical current). Steel is kept from rusting by galvanically protecting it with zinc, cadmium, magnesium, or aluminum, which oxidize before the steel does.

In the earth, redox reactions may take place through several mechanisms. One of the explanations of the action of a uranium roll front is shown in Fig. 4–57. Meteoric water containing dissolved oxygen percolates into a permeable outcrop. Any uranium compounds in dilute form are oxidized by the water-oxygen solution. In general, uranium compounds are more soluble when in an oxidized state than when in a reduced state. The difference can be many thousands of orders of magnitude. The oxidized uranium compounds remain dissolved until they encounter a reducing medium, such as a reduced sandstone. The uranium compounds are reduced, decreased in solubility, and precipitated out in that zone. The sandstone is altered by oxidation in the process.

More solution of uranium (oxidizing) dissolves the compounds at the back part of the deposit and carries them farther. They soon encounter more reducing material and precipitate. The process occurs repeatedly or continuously, concentrating the mineral as the cell moves slowly downstream.

Occasionally, a highly reducing zone is reached, such as a sulfide zone containing organic trash,

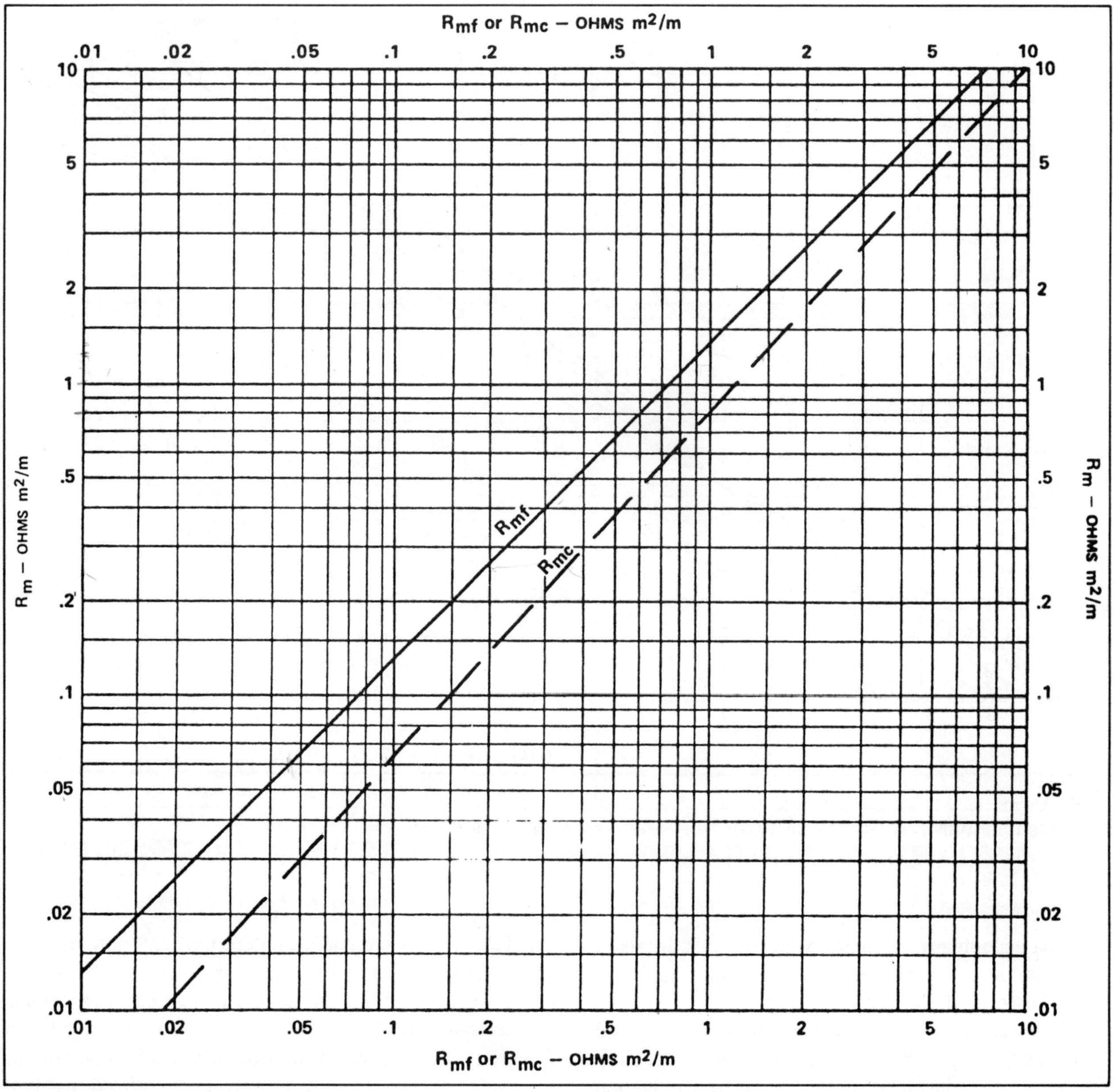

FIG. 4–54

R_m-R_{mf}-R_{mc} relationships, empirical chart (courtesy Gearhart Industries Inc.)

humates, organic or sulphurous gases, and probably reducing bacteria. Then the uranium compounds are concentrated in that zone. A change of permeability appears to be associated with this zone, also.

This process is only one of many variations of the redox process causing mineral deposition, leaching, and often concentration. Similar mechanisms occur near petroleum deposits, resulting in halos of minerals, many radioactive. Rabbit Lake and Beaver Lodge uranium deposits in Saskatchewan, Canada, seem to require redox mechanisms to hold the uranium mineral in place.

Surface measurements of spontaneous potentials for the location of sulfide ore bodies have been known since before the beginning of the twentieth century. Practice was to use two stable electrodes,

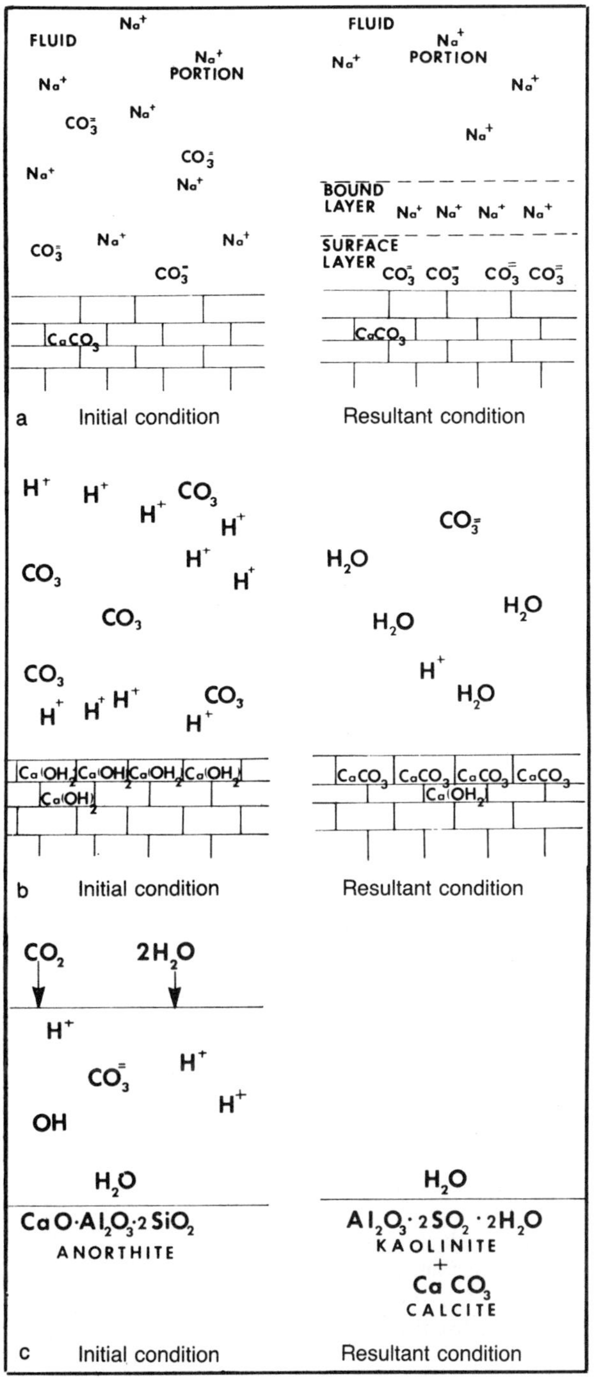

FIG. 4–55

Adsorption mechanisms that may change ion populations

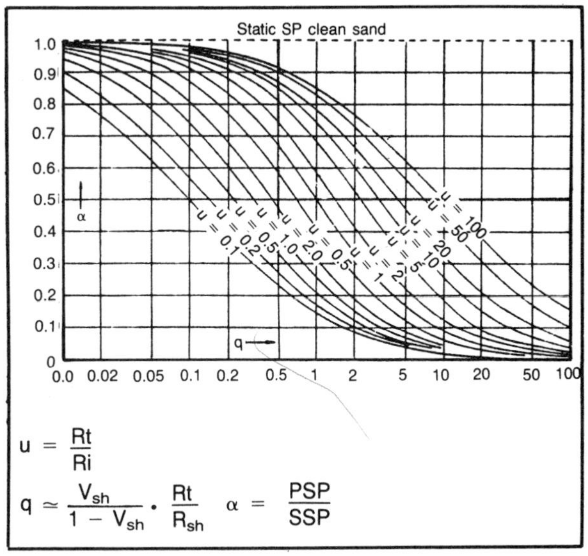

FIG. 4–56

The value of alpha

(after Doll, "The SP Log in Shaly Sands," *Transactions,* volume 189, © SPE-AIME, 1950)

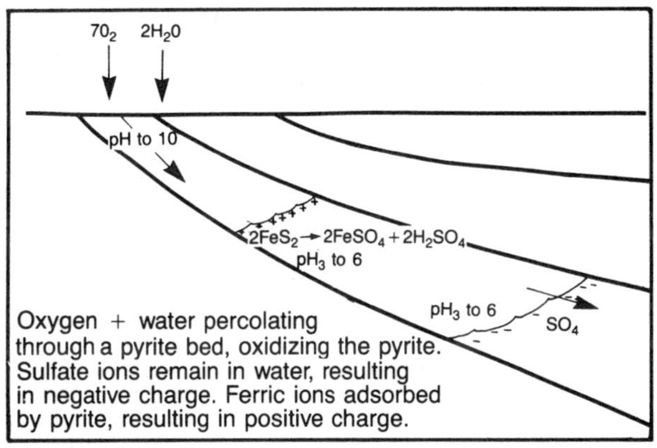

FIG. 4–57

Redox mechanism

such as copper in saturated copper chloride solution, in porous pots connected through a sensitive voltage indicator (galvanometer). A sulfide ore body lying across the redox interface of the surface oxidation and the reduced formations below has strong currents associated with it. The sulfide oxidizes at the top, and the highly conductive sulfide concentrates the resulting current flow. While there is disagreement as to the exact process, the mechanism appears to be redox reactions. As a result, voltage drops (ohmic drops) can be read on the surface (Fig. 4–58).

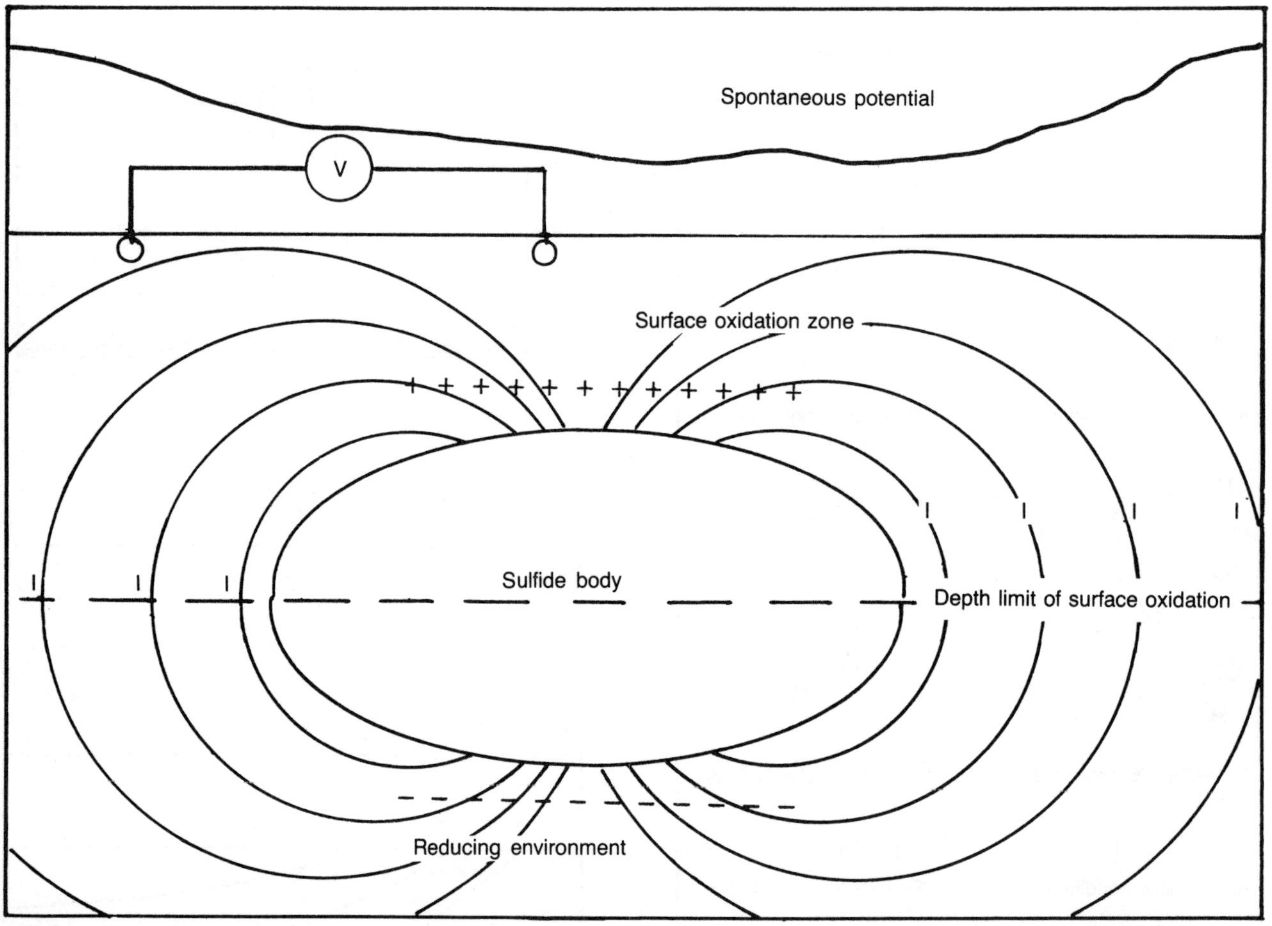

FIG. 4–58
Surface SP of a mineral body

Potential differences are associated with redox deposits because of imbalances of ion species. These potential differences seem to appear on the SP log as base-line shifts. The base-line shifts can be compared and mapped. They correlate well with the redox interface trends, as mapped from hematite staining in the drill cuttings and cores.

A base-line shift to the negative or left indicates the SP electrode is approaching a more oxidized zone. A shift to the positive or right shows the zone is less oxidized or more reduced. With the SP system the downhole potential is compared to the potential of a surface, which is usually in a highly oxidized environment.

The contribution of the redox component is the potential, E_{ox}:

$$E_{ox} = (RT/ZF)\,\ell n\,(k\,C_{ox}/C_{rd}) \qquad (4.40)$$

Where:
R = Gas constant
T = Absolute temperature
Z = Valence
F = Faraday's number
k = Reaction rate constant
C_{ox} and C_{rd} = Concentrations of the more and less oxidized solutions, respectively

More work must be done on this type of reaction. It is quite possible that many of the SP perturbations that have been blamed on other things actually originate with the redox potentials. This certainly appears to be true of the base-line drift of the SP curve as it approaches the oxidized surface formations. Examples of these base-line drifts that have been correlated with redox situations have been observed from a few millivolts to more than 90 millivolts. The mapping

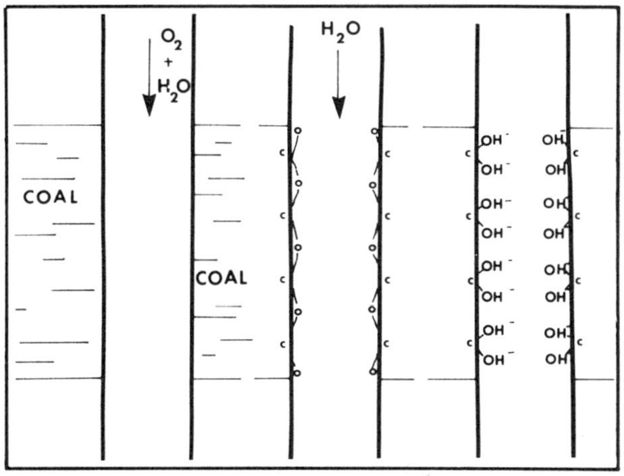

FIG. 4–59
Oxidation of coal by drilling mud

technique does appear to be a valid, useful one, even though the explanation is incomplete.

Redox reactions also appear in coal and in sulfides caused by the dissolved oxygen in the drilling mud. This is shown in Fig. 4–59 for coal. One patent has been issued describing a method for deliberately causing this type of oxidation by introducing highly oxidizing material to the mud and then logging with the SP.

Electrofiltration Potentials

Electrofiltration, electrokinetic, Helmholz, or streaming potentials are caused by the mechanical flow of a solution across a restricting or semipermeable barrier. Helmholz' Law states this potential to be E_H:

$$E_H = \frac{\xi R \zeta}{4\pi\mu} \Delta P \qquad (4.41)$$

Where:
- ξ = Dielectric constant of the liquid
- R = Electrical resistivity
- ζ = Zeta potential, the potential between the fixed and mobile layers of fluid
- μ = Viscosity
- ΔP = Pressure drop or differential pressure (Fig. 4–60)

Wyllie found the electrokinetic potential or streaming potential, E_s, to be:

$$E_s = kP^y \qquad (4.42)$$

Where:
- P = Differential pressure
- k and y = Constants depending on the system and temperature

This agrees for special cases with the Helmholz relation, Eq. 4.41.

It was believed by the Schlumbergers at first that this component was the major contributor to the SP curve. It has since been demonstrated that the streaming potential is usually negligible in oil-field logging. The electrofiltration may not be negligible in non-petroleum logging. In this case we log in freshly drilled holes filled with rather poor mud. Invasion rates are probably high, resistivities are high, and viscosities are low. Referring to Helmholz' Law, Eq. 4.41, these make ζ and R high and μ small, resulting in a possible high value of E_H.

One can detect the presence of streaming potentials by repeatedly logging the hole. These potentials decrease as the rate of invasion slows and as the differential pressure decreases (Fig. 4–61).

The streaming or electrofiltration potentials cause erroneous results in the calculation of R_w. If the SP curve is used quantitatively, the effects of the filtration potential must be considered. Also, if the ratio of R_{xo}/R_t is plotted against the value of the SP, the trend of the points intercepts a value of zero for the SP at a ratio of $R_{xo}/R_t = 1$ if there is no electrofiltration potential. If an electrofiltration potential exists, the intercept SP value is equal to it. Thus:

$$E_{SP} = E_{SSP} + E_H$$

Electrode and Probe Effects

So far, not much has been said about the potentials that can arise from the electrode materials and the probe body and their interactions with the fluids in the borehole and formation. It has been assumed,

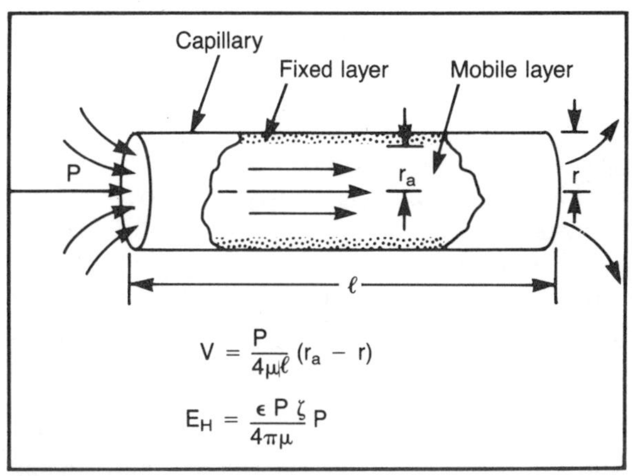

FIG. 4–60
Electrofiltration mechanism

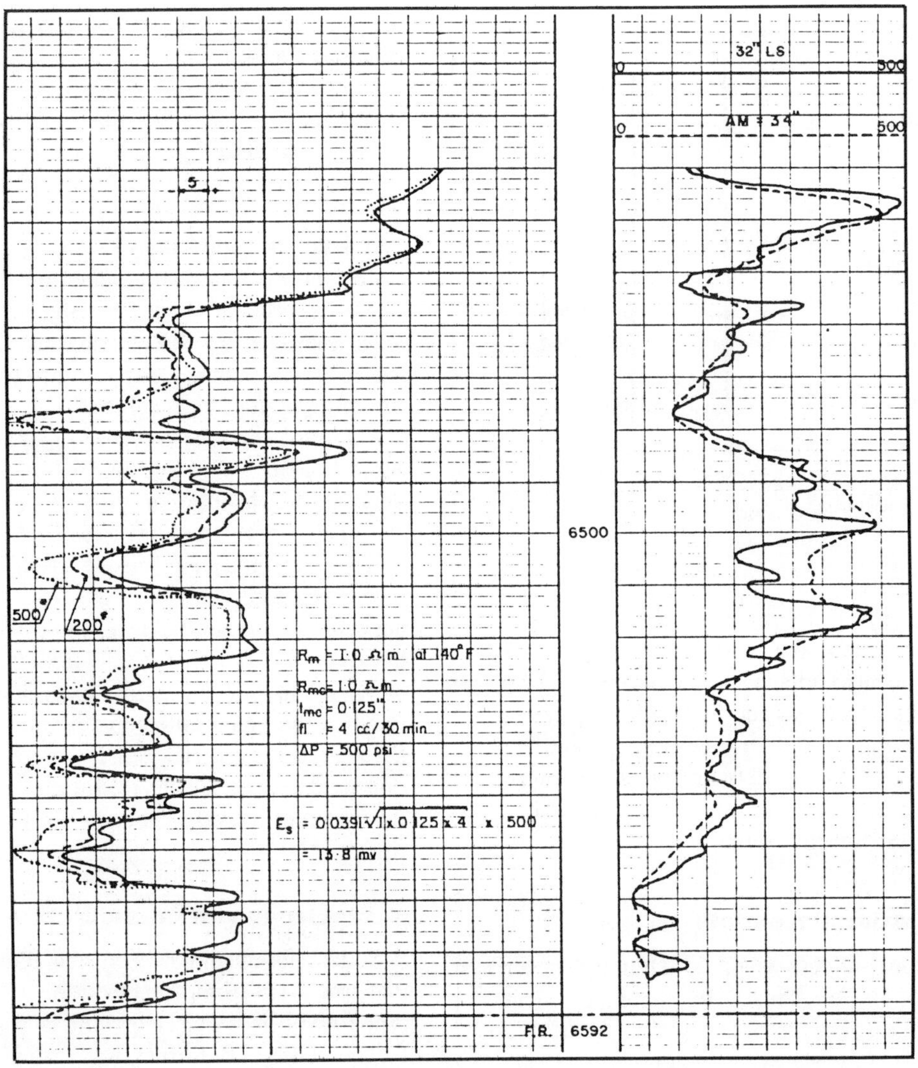

FIG. 4–61

A demonstration of the electrofiltration effect, caused by changing the wellhead pressure (courtesy Schlumberger Well Services Inc.)

generally, that the electrode and probe are inactive electrically. This may not be true. Electrode material is important. Obviously, one must use a material or an electrode type that can be used in the field. Hydrogen electrodes, for example, are excellent in the laboratory but would be quite difficult to use in the field.

Metals of several kinds are attractive. They usually lend themselves to fabrication; some are quite inexpensive. Most metals react with the surrounding solutions. While they are reacting, metal ions are going into solution, transporting charges, and leaving the electrode with a negative potential. As the surrounding solution gains more metal ions, the rate of reaction decreases. When the atmosphere around the electrode becomes saturated, the reaction stops. Thus, when a piece of metal is put into a water solution of NaCl for example, the potential of the metal being measured changes with respect to a stable reference electrode. It starts high and drops rapidly or slowly, depending on the metal.

Some metals do not react for two reasons. The noble metals (gold, platinum, etc.) are relatively inert chemically. They react little or not at all. Other metals, such as stainless steel, aluminum, or tantalum, form relatively inert, electrically insulting coatings that protect the surface. Some coatings, such as copper alloys and compounds, have undesirable electrical properties. Some metals react rapidly and continuously. Sodium is an extreme example. It reacts violently. Some metals change their mode of reaction depending on their thermodynamic state. Nickel alloys fall into this category. Some metals are mechanically not suitable. Mercury is an example. Others, like platinum, are too expensive for general use. Thus, if an electrode is to be made of a bare metal, it must be chosen carefully. It must have one major thermodynamic equilibrium state; it must have reaction products that are electrically symmetrical; it must have a limited reactivity and solubility; it must be mechanically competent; and it must not be costly. There are few metals or alloys that come close to these criteria. Table 4–3 lists the common electrode metals and probe materials and the problems associated with them.

Lead is close to being an ideal material. It is cheap, easily worked, and mechanically reasonable. It has two states of thermodynamic equilibrium, but one is not likely to change to the other under usual conditions. The solubilities of its reaction products are usually low. Its reaction products are bisymmetrical, and they cling to its surface. Therefore, lead is used with reasonable success as an SP electrode material.

Silver is another possibility for an electrode material, but it is not often used. It has all of the desirable properties of lead, and it is stronger mechanically and more stable thermodynamically. Other electrodes can be designed that are stable and should be investigated. These are the impolarizable or saturated electrodes. The electrical model of an electrode in an electrolyte, such as water, is shown in Fig. 4–62.

A lead SP electrode is presumed to be identical electrically and chemically to the surface reference electrode. In general, this is approximately true. The reference electrode should be in the mud pit, which contains the same fluid as the borehole. Both electrodes are made of lead.

As soon as the electrode, whether measure or reference, is put into the mud, reactions begin to put lead cations into solution around the electrode. This environment saturates very quickly (several seconds) because of the low solubility of lead compounds. Since the mud is viscous, it remains saturated. The reaction or solution stops, and the electrode stabilizes electrically. Therefore, the two electrodes quickly reach approximately the same potential. This is shown in Fig. 4–63a. However, as soon as the measure electrode is moved, it leaves behind the reaction products, the saturated atmosphere, and the reaction-rate increases. The potential of the electrode becomes more negative as a result of the motion through the borehole mud (Fig. 4–63b). If the electrode is moved at a constant rate, the reaction attains a constant rate and the electrode potential becomes pseudostable. However, it is at a different potential than the reference electrode.

If the electrode bumps the wall of the hole, the clinging reaction products are scraped off, leaving a bare, reactive metal surface. This results in a sharp potential change that dies away over several seconds.

Table 4–3
Materials Commonly Found as R/SP Electrodes

Materials	Remarks
Lead	Good material, clinging reaction products
Solder	Fair, tin content increases reactivity
Iron, rusty	Fair, large surface, reactive
Stainless steel	Poor, insulating surface products
Copper	Poor, oxide is rectifying
Brass	Poor, rectifying oxide, reactive
Aluminum	Poor, reactive, insulating surface layer
Silver	Excellent, poorly reactive
Gold plating	Excellent, almost nonreactive
Ag/AgCl	Excellent, stable, nonpolarizing
$Hg/HgCl_2$	Excellent, stable, nonpolarizing
Chromium plating	Poor, insulating surface layer
Nickel plating	Poor, insulating surface products
Cadmium plating	Poor, reactive, sacrificial
Zinc plating	Poor, reactive, sacrificial

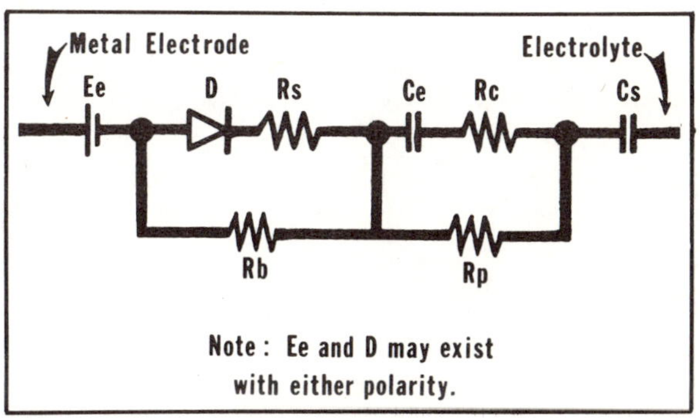

FIG. 4–62
Electrical model of electrode in electrolyte

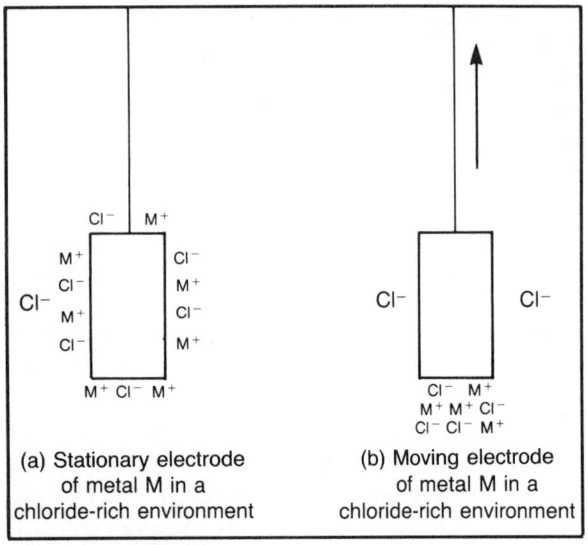

FIG. 4–63a and b

(a) Stationary electrode; (b) Moving electrode

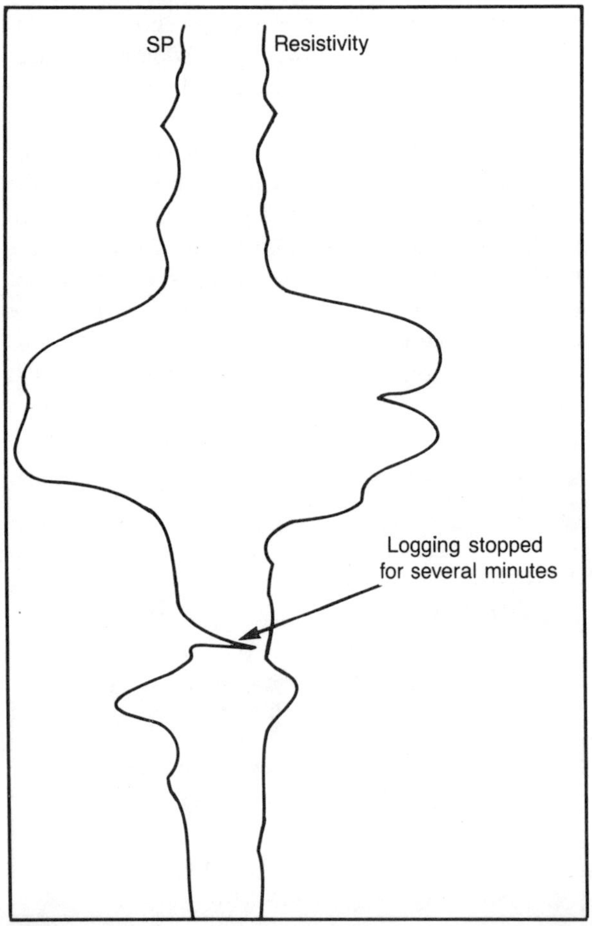

FIG. 4–63c

Effect of stopping a moving SP electrode

Alloy solder is often used to repair an electrode. It causes the electrode to exhibit a different potential because other metals are present. If the SP circuit is high impedance, this difference is not serious, except when the probe rotates. Then periodic changes are seen on the SP curve.

If the SP circuit is a low impedance circuit, any difference in potential between the measure and reference electrodes causes a current flow through the circuit and formation. This superimposes a single-point resistance component on the SP curve (Fig. 4–64). All of these effects are usually small with lead electrodes, high-impedance circuits, and a little care.

A more serious effect is bimetallism. If part or all of the probe insulation is removed—i.e., when bare metal is exposed at a tool shoulder through wear—the steel, brass, or aluminum housing has a much different potential than the zinc of the cable armor. This causes large currents (several amperes) to flow through the mud and the metal of the probe. A dc lateral curve is then superimposed on the SP curve. This lateral (Fig. 4–65) may be several times the SP curve in amplitude and may cause the SP curve to be nearly a mirror image of the resistivity curve. Since the SP that is measured is the ohmic drop in the mud

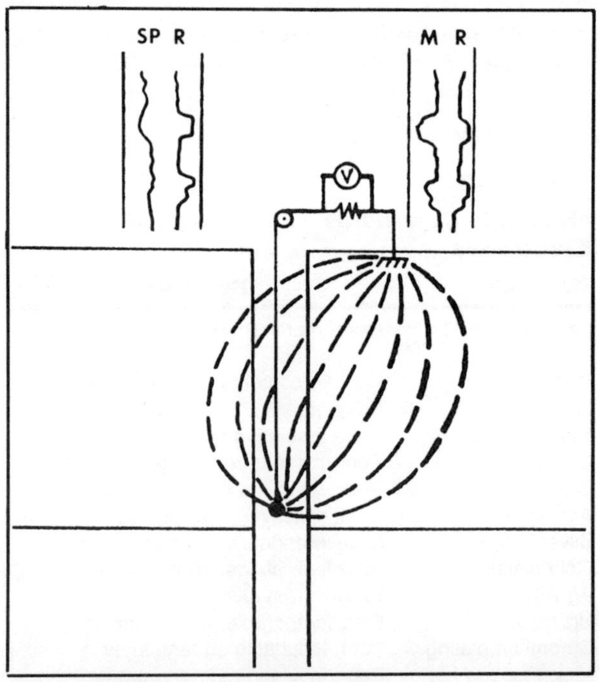

FIG. 4–64

Current in an unbalanced SP circuit

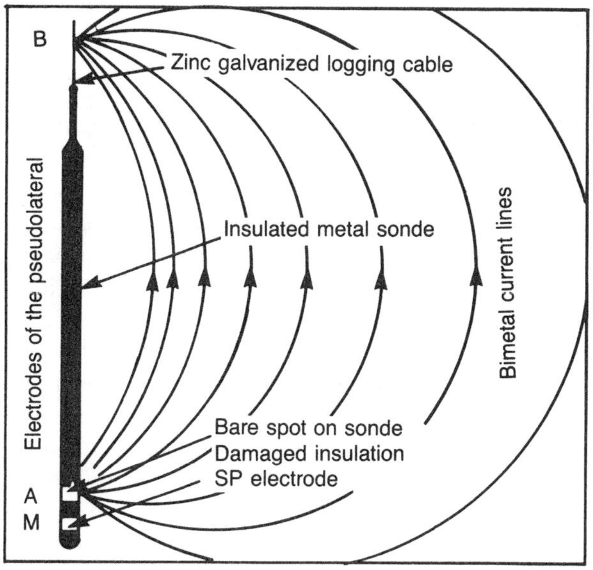

FIG. 4–65
Bimetallism pseudolateral resistivity

column, any bare housing section bypasses the mud column. It results in a reduction and distortion of the SP curve.

The SP curve is affected by many sources of noise, especially when logging in oil-well type environments, e.g., big drilling rig, deep-hole logging truck. The effect illustrated in Fig. 4–63 can give rise to road noise if solid formation material touches the surface of the electrode. This will result in a sharp deflection, usually in a positive direction, and a slow return to a normal reading. Stray voltages can arise from power lines, cathodic protection devices, and electric rig machinery. The cable on the winch constitutes a multiturn electrical transformer or generator. Any magnetic or electrical fields at the winch will affect the SP. Stray cable armor currents, which may be hundreds of amperes in a long cable, can affect the SP.

Surface Spontaneous Potential

The surface spontaneous or self potentials that are used arise mainly because of the oxidation of portions of a conducting metallic mineral body. The resulting potentials cause a current to flow through the mineral body and in the earth surrounding it. There can also be local concentration cells resulting in diffusion potentials. Changing magnetic fields, electrical storms, fluid flow, and biological activity can all give rise to potentials. Roll fronts and other geochemical anomalies have potential differences associated with them.

Surface SP measurements are usually made with a stable, sensitive, high-impedance voltmeter and a pair of stable electrodes connected through well-insulated cables. Stable surface SP electrodes are often made in porous ceramic pots. The pots contain copper electrodes in a saturated $CuSO_4$ solution. Presumably, other stable electrodes such as Ag-AgCl or Calomel electrodes could be used. It is not clear why copper-copper sulfate electrodes should be preferred, except perhaps by custom. The pots are placed in shallow holes in the ground where they can make electrical contact with the ground water. The spacings between electrodes are from 3–30 m. The electrode potentials should be checked often to ensure that they are within 3 millivolts of each other. This is usually done by placing them side by side in the same hole and checking the potential difference between them.

Isolated metallic bodies, such as a sulfide mineral body, may be detected by mapping equipotential lines (Fig. 4–58). Since a geochemical cell of any kind has potential differences associated with it, the SP should make a good method for mapping roll front trends in uranium exploration. However, uranium exploration is usually done in arid country. Common practice is to fill the holes for the pots with water to make good electrical contact. Electrofiltration potentials caused by the water used to fill the electrode holes may interfere severely. The signals measured in surface work may be quite small and difficult to detect. This method has not been popular because of the obvious difficulties. However, modern circuitry and a combination with borehole methods make it promising.

Special Spontaneous Potential Systems

There have been a number of special SP systems developed in attempts to circumvent the various problems of the conventional SP. These are the differential SP, the static SP, and the selective SP.

The differential SP (DSP) is attractive for two reasons. The differential SP makes use of two downhole electrodes at a fixed distance apart. It is the first derivative of the SP with respect to depth. The DSP was originally used to overcome the drifting base-line problem of the SP curve. Unfortunately, the base-line change—if it is real and not an instrument problem—is not steady, and it does convey information. The second reason is possibly a better one. The conventional SP must be referred to a surface reference electrode. The problems involved with having a stable, representative reference electrode on the surface have been discussed. With modern digital transmission systems all signals are encoded for transmission downhole. Therefore, the surface electrode must have one wire dedicated to its use. If it were not for that, a monocable could be used. If a DSP is recorded, the surface computer may be able to integrate the signal and obtain the SP signal. Because of drifts and errors, however, it may be necessary to check a reference potential periodically.

The static SP was an attempt to block the flow of current in the borehole and record the static SP (SSP) directly. Fig. 4–66 shows the system. The SP signal is monitored by an electronic circuit, and the potential of a guard electrode system is kept at that of the measure electrode. With no potential gradient there can be no current flow in the hole. Thus, the measure electrode assumes the static potential of the formation.

The selective SP system was designed to be somewhat similar, except the guard potential is swung periodically from the shale potential to the sand potential. The full-oscillating signal is recorded. The system indicates a zero current in a shale by one envelope side and in a sand by the other. Both the static and selective SP systems have been abandoned as being too complex for the improvement gained. Again, with modern circuitry the situation may be different.

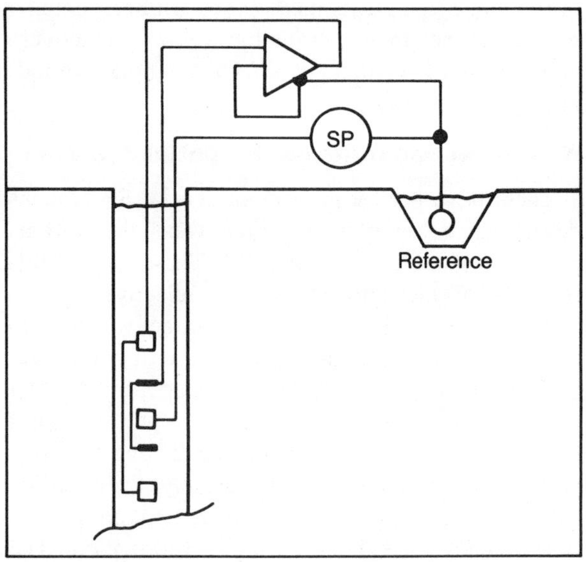

FIG. 4–66
Static SP setup

Calibrations

SP circuit calibrations are rather simple in principle. Dependence is placed on stable, high-impedance circuitry. A calibration circuit is sometimes used, but a poorly designed and maintained one can be more misleading than none at all. This circuit is a voltage divider supplied by a constant voltage source. The divider is designed to have a relatively low impedance so the loading by the SP circuit will not affect it. It usually has a switch to select several standard voltages.

Unfortunately, the assumptions made about the voltage source are faulty. The circuit is generally supplied by a battery for convenience and portability. A D-size mercury cell is usually used with the assumption that the potential of a mercury cell is constant throughout its life. This is definitely not true. The potential is constant throughout the cell life only at one particular load (about 200 ohm) and at one particular temperature. Like any such reaction, the mercury cell potential changes as a function of the square of the absolute temperature. This means that the potential, even with ideal loading, can change by a factor of about two in year-round usage because of temperature changes. Further, when a mercury cell is depleted, it is tempting to replace it with a more easily obtained, cheaper carbon-zinc cell (flashlight battery). Unfortunately, the potential of a flashlight battery under light load is 28% higher than a mercury cell. Any calibration box should be stabilized with an electronic regulator or at least by zero temperature coefficient zener diodes.

Most other care consists of good inspection and maintenance. It is imperative that all bare metal on the probe within 6–7 ft of the electrode, except for other electrodes, be completely insulated. Electrodes must also be protected from contact with the wall of the hole by recessing them in the sonde body. Tape bumpers can be built up and are a big help.

DIPMETER

Dipmeters are high-resolution devices that measure resistivity variations usually on three or four sides of a borehole (120° or 90° apart). The differences of the logged depths between the three or four simultaneous recordings of an anomaly are used to calculate the direction of the slope or dip of the plane of the anomaly with respect to the borehole. Formation dip can then be calculated from this, plus hole deviation measurements. High-resolution dipmeters (HRD) can also be used to locate fractures and fracture types.

There appears to be a good application of the HRD in nonpetroleum logging. Two obstacles—small hole size and poor hole quality—have interfered with its use so far. This is changing. Computation of dipmeter logs is complex, and contractors usually perform this as a service.

Dipmeter logs may use any high-resolution measuring system that is directional and that responds to lithological changes. They have been developed using the spontaneous potential curves, short normals, laterals, calipers, hole rugosity measurements, and (the most successful) the focused resistivity. Since the high resolution focused resistivity device is the major one in use now, this discussion is confined to it. The same principles apply to all others.

Dipmeter measurements are made at several places around a circle on a plane normal to the axis of the borehole. Typically, these are spaced at 120° (3 arm) 90° (4 arm). The position of the arms in the borehole—the direction of the number 1 arm—is known. The assumption is made that the same anomaly appears all around the circle, which coincides with the borehole wall. Thus, the relative depths of the anomalies of the three or four readings can be used to calculate the dip of the formation with respect to the borehole axis.

A hole deviation device is run simultaneously with the dipmeter. The calculation of the hole deviation defines the position of the borehole axis. Thus, the true dip angle and direction of the formation can be determined with respect to true north and to a horizontal plane.

The high resolution dipmeter (HRD) is valuable for its original purpose—determining regional dip. It is more valuable of late because of the patterns formed by the high-resolution presentation. It shows many structural and depositional features quite vividly.

■ Computation

Computation methods can be manual and usually involve the use of a graphical set or net. The reason is that the mathematics, while relatively simple, is rather long and involved. Typical methods can be found in the referenced literature.

Older computers were analog devices. Fig. 4–67 shows a typical one. These methods involved a large amount of manual handling, since the logged information was not digitized. It is still common practice to do much of the correlation of deflections manually, especially in high dip-angle zones.

FIG. 4–67

Early model of diplog computer
(courtesy Dresser Atlas Division, Dresser Industries)

Most modern methods use computer correlation and computation, and usually require digitized data on magnetic tape. Computations are made every 2–10 ft (optional). Occasionally, computations may be made every ½ ft. This is called the step length. The number of feet or meters of logged hole used for correlation of an anomaly is called the window length. The number of inches or centimeters that the computer is programmed to search for correlation is the search length. Typically an 8-ft window length, a 2-ft step length, and an 8–10-in. search length are used. Large dip angles require longer window lengths, as do large borehole deviations.

■ Presentation

Dipmeter data, like borehole deviation information, must be reduced and presented in a comprehensive form before they can be utilized. There are several popular formats. These are tabular printouts or vector (tadpole), cylindrical, linear polar, and azimuth frequency plots. Each has its use, and one should specify the plot or plots best suited to the needs. Of course, with the data on magnetic tape, there is usually no problem replotting the information in another format.

Usually, tabular data list the depth and thickness—or top and bottom depths—of a formation, the formation dip angle and direction, and often the borehole angle and direction. A tabular listing should also indicate the degree of confidence, or correlation coefficient, of the computation.

The vector plot presents a grid containing a vertical depth scale and a nonlinear horizontal degree of dip scale. Small arrows, usually headless, point in the direction or azimuth of dip. A small ball at the base of the arrow indicates the degree of dip. This presentation has the advantage of emphasizing patterns. Figs. 4–68a, b, and c illustrate this feature. This is the standard method of presenting dipmeter data.

The cylindrical, linear polar, polar, stick, and Schmidt plots are shown in Figs. 4–69a and 4–69b. These are not often used and will not be covered further.

The azimuth frequency polar plot can be quite useful. It is shown in Fig. 4–70. These are constructed using a polar plot and plotting the number of dips contained in a particular segment.

■ Uses

There are many uses for the dipmeter, including detection of traps, determination of fluid movement, estimation of drilling target depth, location and type of fractures, and many features of structural geology and sedimentary geology. The obvious use, determining formation dip, is the original and most frequent use of the dipmeter. In mineral exploration formation dip information is needed to calculate how deep a mineral-bearing horizon is at the next hole and the direction to it. The target depth, d, is:

$$d = e + (D \tan \theta) - s$$

Where:

D = Distance to the next hole
θ = Angle of dip in the direction to the next hole
e = Elevation of the horizon of interest in the first hole
s = Surface elevation of the next hole.

Formation dip indicates the present direction of fluid flow in the formation. The amount and the change of amount of dip give indications of the

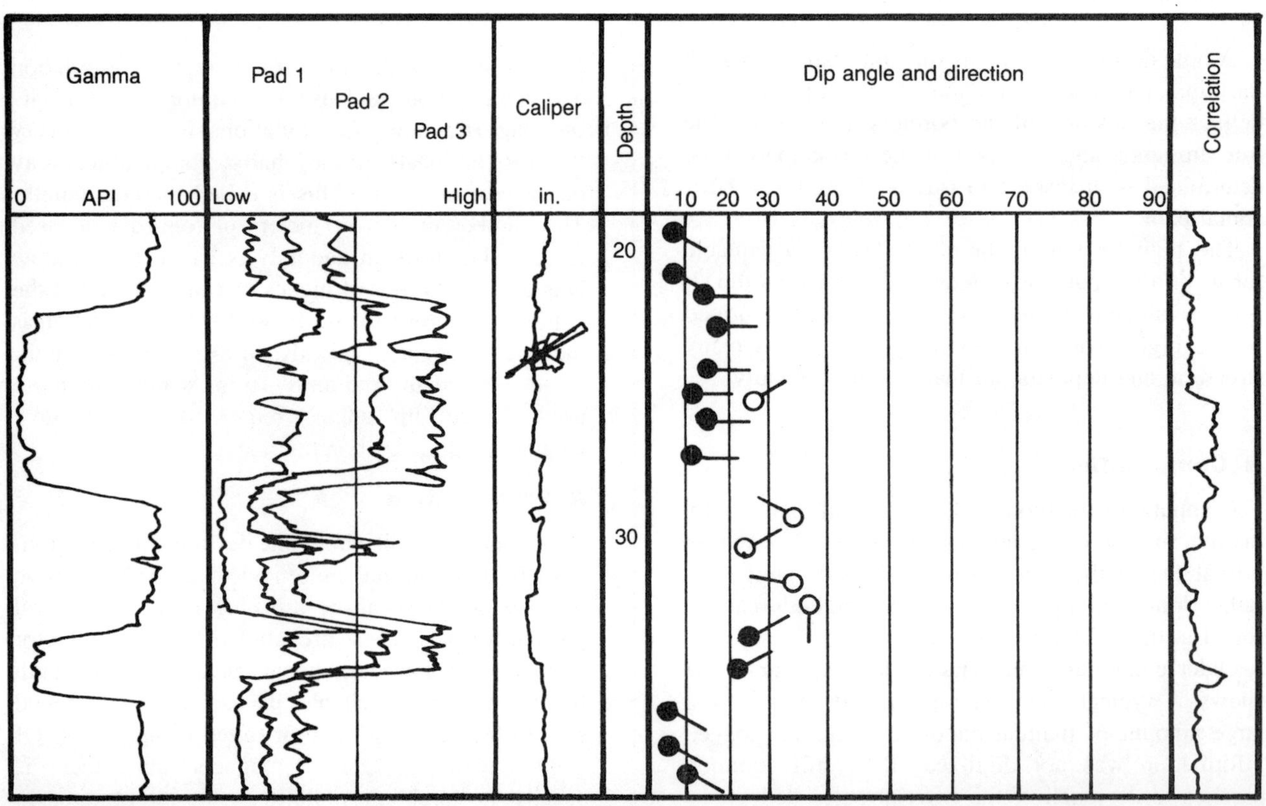

FIG. 4–68a

Sample dipmeter log with vector plot

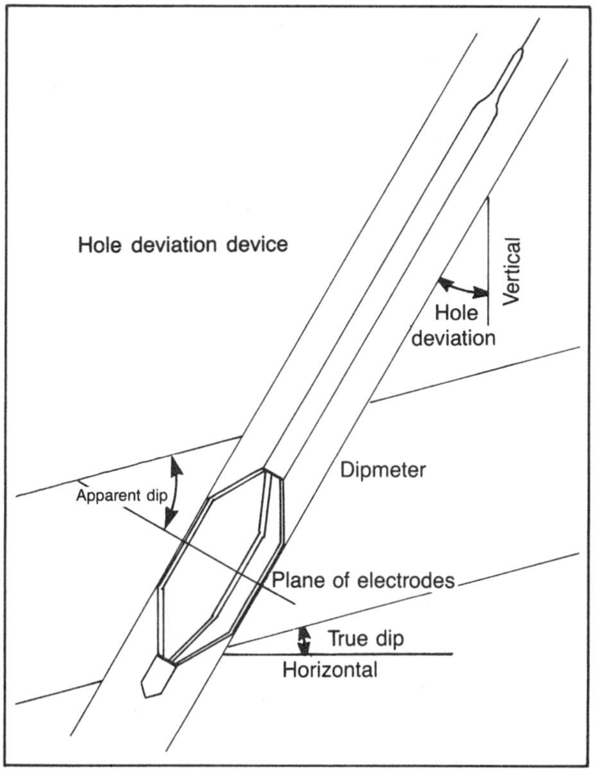

FIG. 4–68b

Geometry of the measurement of dip

FIG. 4–68c

Dipmeter tool
(courtesy Gearhart Industries Inc.)

proximity of basin axes and edges. In both petroleum and mineral logging dip indicates the direction to go for the next hole: updip for oil or gas, downdip for water, updip for altered zones, down for unaltered, and along the strike for trends of sedimentary minerals. This is a simplified generalization, of course.

In combination with other logs, such as the resistivity and spontaneous potential curves, the dipmeter can give excellent pictures of the occurrences, locations, and even orientation of faults, fractures, and stratigraphic features. This is covered further in the section on patterns.

At present, the dipmeter is seldom used in nonpetroleum work. It is, however, a very promising tool and probably will be used to a great extent in the future. At least two contractors are planning a small diameter, high-resolution dipmeter service. Fig. 4–71 through Fig. 4–78 illustrate the various uses for the dipmeter.

INDUCED POLARIZATION

Induced polarization methods are mainly surface methods. However, downhole induced polarization methods have shown usefulness in locating sulfide mineral deposits, in correcting surface methods, and in responding to alteration in clays and shales. Their greatest usefulness is when they are combined with other methods, surface and downhole. Induced polarization measurements are generally useful in combination with resistivity measurements.

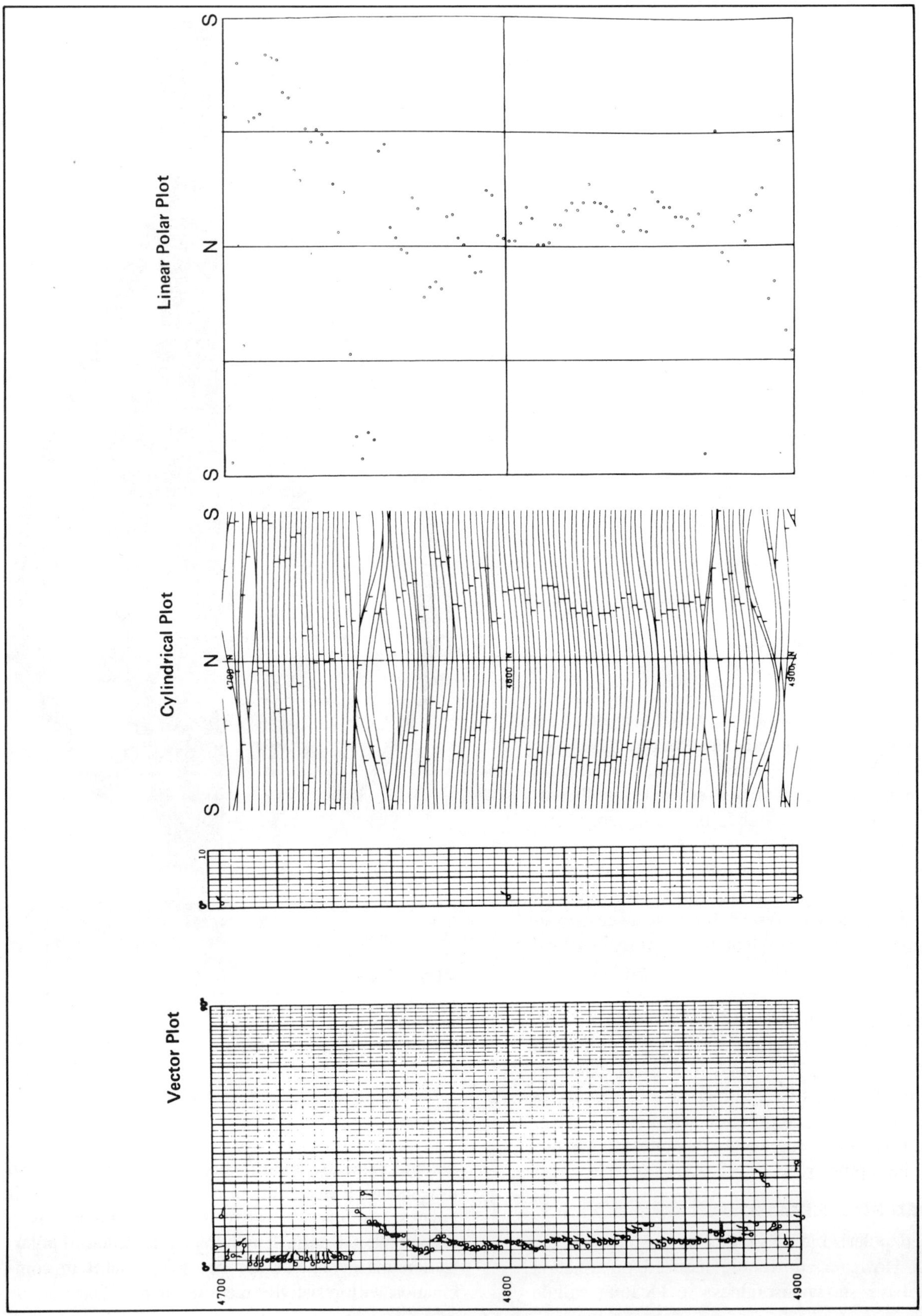

FIG. 4–69a
Dipmeter data presentation formats (courtesy Dresser Atlas Division, Dresser Industries)

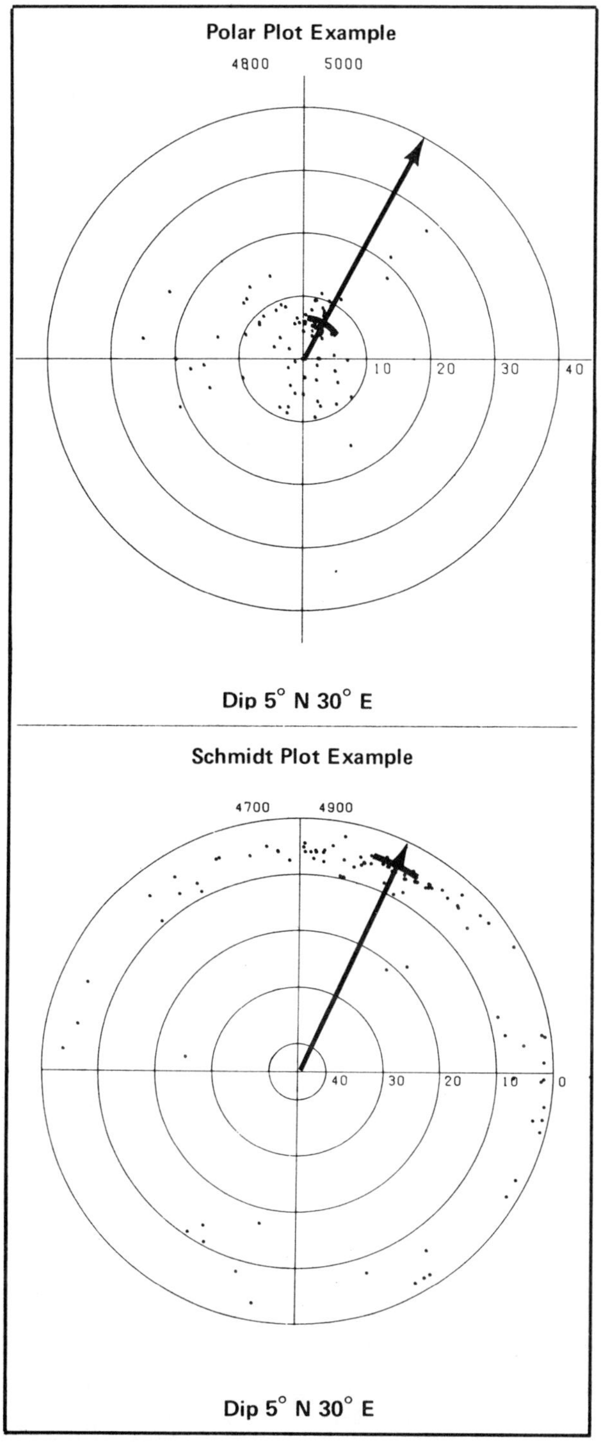

FIG. 4–69b

Dipmeter data presentation formats
(courtesy Dresser Atlas Division, Dresser Industries)

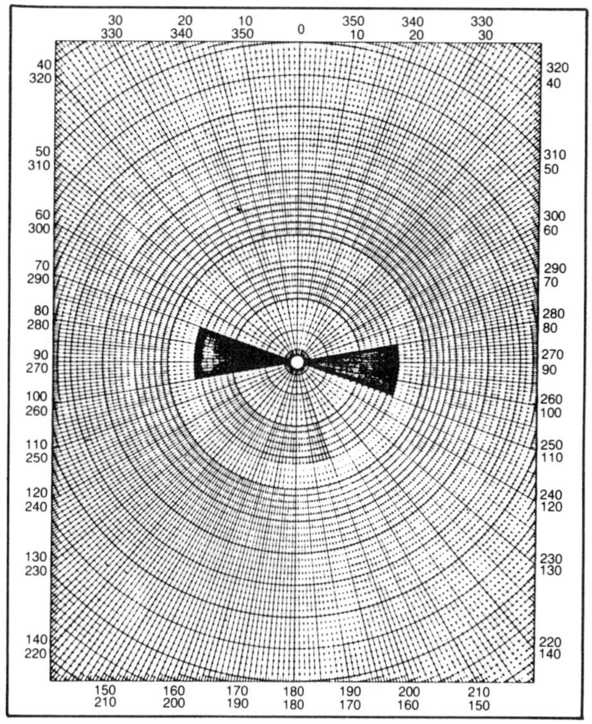

FIG. 4–70

Trimodal distribution of patterns. This pattern indicates probable deposition in a stream channel near the trough and with the current within the channel flowing to the south.

■ Principle

It is often assumed that the electrical resistivity measurements of the earth do not depend on frequency. That is, the impedance measurement contains only resistivity components. However, it is a complex measurement and contains imaginary or reactive components. The reactive portion is normally small, but measurements can be designed to enhance the detection of the electrical properties of the formations, which change with frequency. These are called induced polarization methods (IP) or complex formation resistivity measurements.

The IP methods are based on the electrochemical phenomenon of overvoltage. The technique establishes and detects double layers of electronic conducting materials when an electrical current is caused to pass across the interface (Fig. 4–79).

If a direct current is caused to flow through a formation, a voltage drop across the formation can be measured. The ratio of this voltage to the current is the resistance of the volume under investigation.

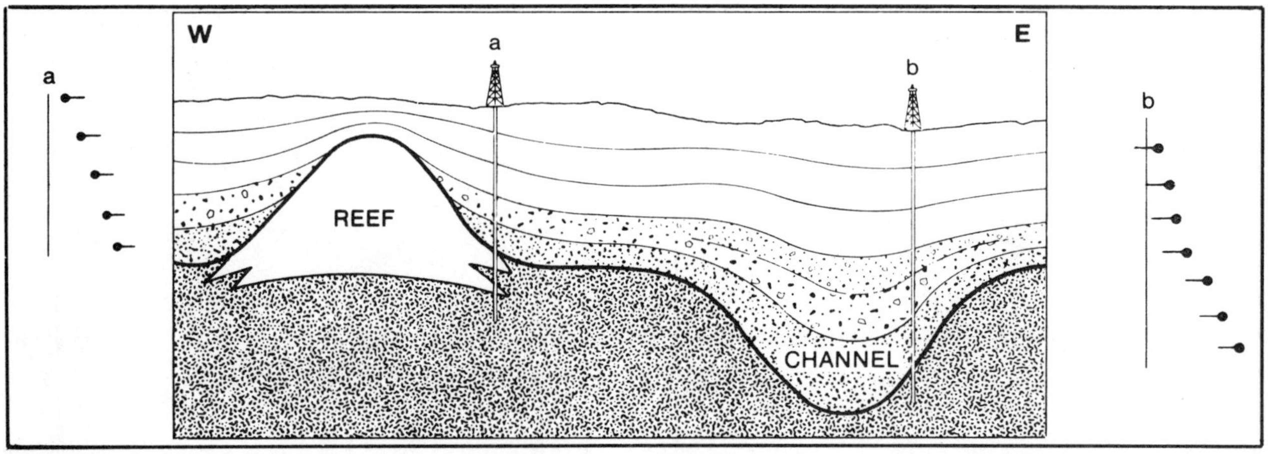

FIG. 4–71

Dip increasing with depth, drape, and channel filling

(courtesy Schlumberger Well Services Inc., *Dipmeter Interpretations*, volume 1, 1981)

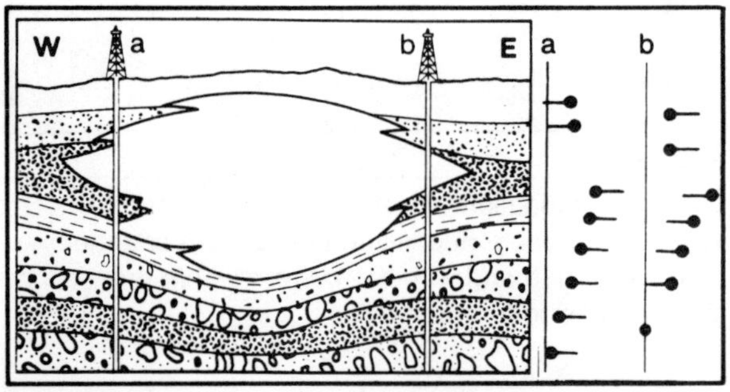

FIG. 4–72

Dip decreasing with depth, differential compaction beneath a sand lens

(courtesy Schlumberger Well Services Inc., *Dipmeter Interpretations*, volume 1, 1981)

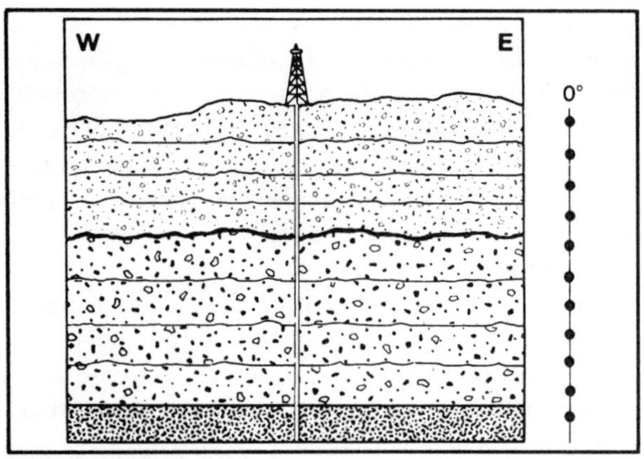

FIG. 4–73

Disconformity

(courtesy Schlumberger Well Services Inc., *Dipmeter Interpretations*, volume 1, 1981)

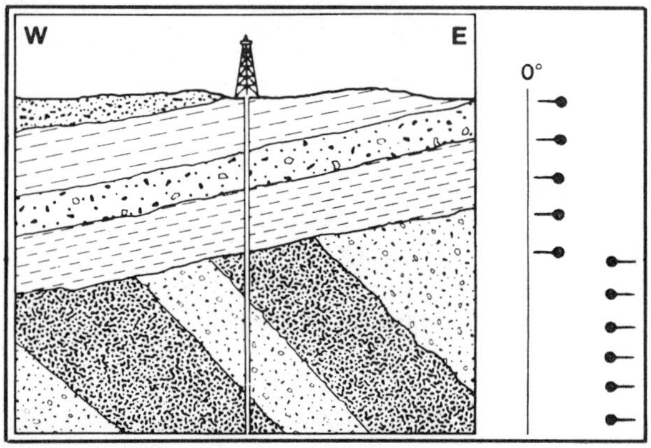

FIG. 4–74

Angular unconformity

(courtesy Schlumberger Well Services Inc., *Dipmeter Interpretations*, volume 1, 1981)

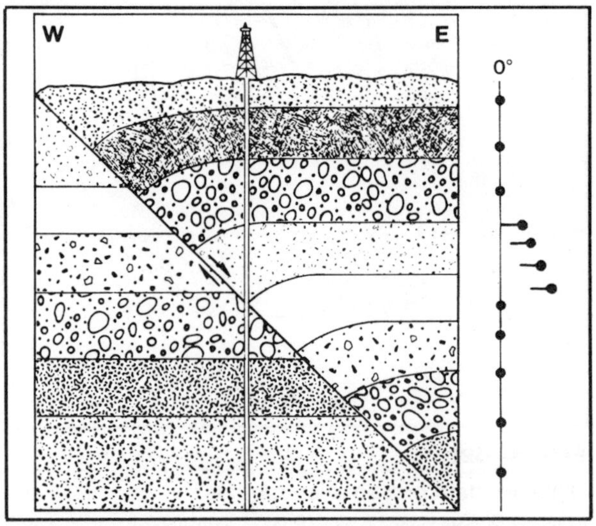

FIG. 4–75

Rollover in upper block of a growth fault

(courtesy Schlumberger Well Services Inc., *Dipmeter Interpretations*, volume 1, 1981)

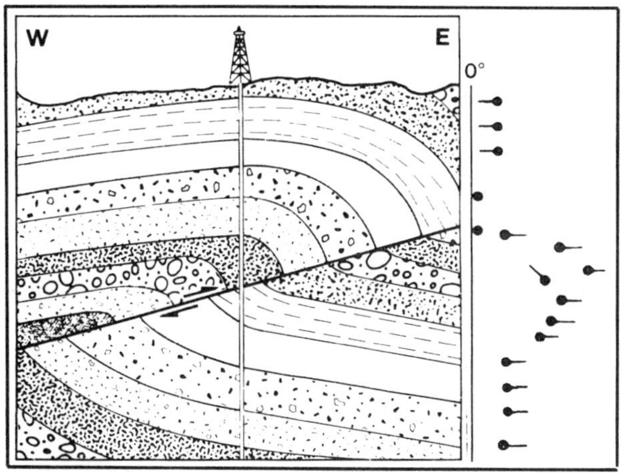

FIG. 4–76

Beds overturned by drag in the upper block of a thrust fault

(courtesy Schlumberger Well Services Inc., *Dipmeter Interpretations*, volume 1, 1981)

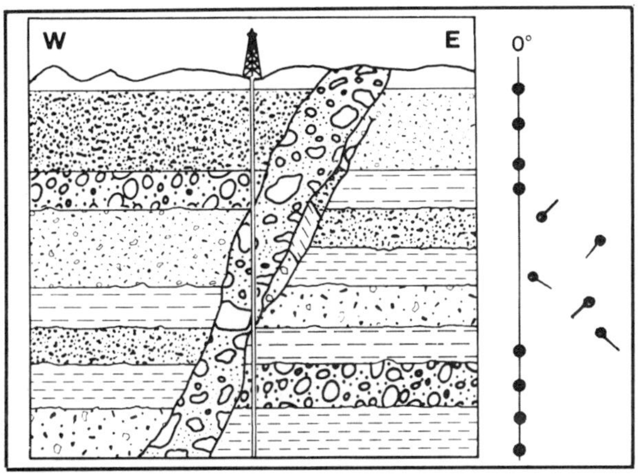

FIG. 4–77

Erratic dips, breccia zone of a fault

(courtesy Schlumberger Well Services Inc., *Dipmeter Interpretations*, volume 1, 1981)

If the current is interrupted, the voltage drops rapidly to a lower value and then decays slowly over a period of time from milliseconds to hours. If the current is an alternating one, the earth appears to have a reactive component whose reactance varies with frequency.

All naturally occurring sulfides of metallic luster, some oxides, and graphite give marked IP responses when present in sufficient volume. This response can occur even at low concentrations of these materials and when they are in the form of discrete, noninterconnected particles (Table 4–4).

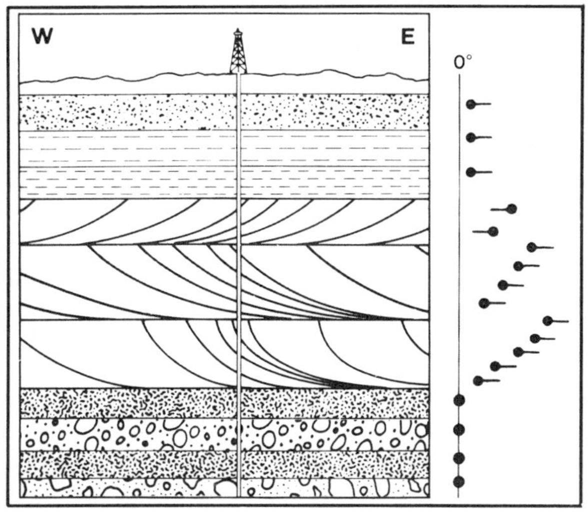

FIG. 4–78

Dip decreasing with depth, crossbedded strata

(courtesy Schlumberger Well Services Inc., *Dipmeter Interpretations*, volume 1, 1981)

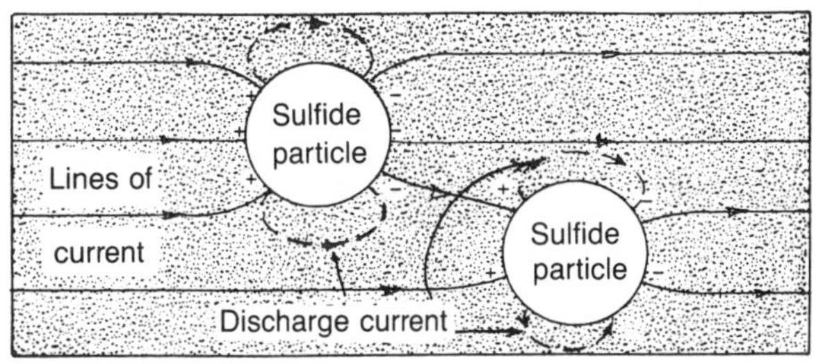

FIG. 4–79

Induced polarization effect

Table 4–4

Rock	Frequency Effect, %/Decade	Metal Factor mho/m
Rock with concentrated sulfides	>10	$>10^3$
High-magnetic iron ores		20–10,000
Porphyry copper ores (2%–10% sulfides)	5–10	30–3,000
Rocks with a trace of sulfide mineralization	2–5	1–100
Volcanic tuff	2–4	1–300
Sandstone	1–3	1–30
Basalt	1–2	½–7
Granite	0.1 to 0.5	<½

Induced polarization is the only method presently available that can be used to detect directly disseminated sulfides, such as the porphory type or bedded copper deposits, and bedded lead-zinc deposits in carbonate rocks. Each rock and soil type exhibits an IP response, usually of a relatively low amplitude, that is characteristic of the specific rock or soil (Fig. 4–80). Certain clays and platy minerals, such as serpentine, sericite, and chlorite, sometimes exhibit so-called membrane polarizations. This phenomenon can be used as an indicator of alteration.

Most IP methods measure voltages from overvoltage effects at the surface or in boreholes and those caused by currents flowing through the formation. This is the electrical IP (or EIP). Sintrex also has a technique that measures the magnetic fields associated with these currents and calls it the magnetic IP (MIP) method. This latter has the advantage of not requiring electrical contact with the ground for the measuring system, thus opening the way for moving vehicular measurements.

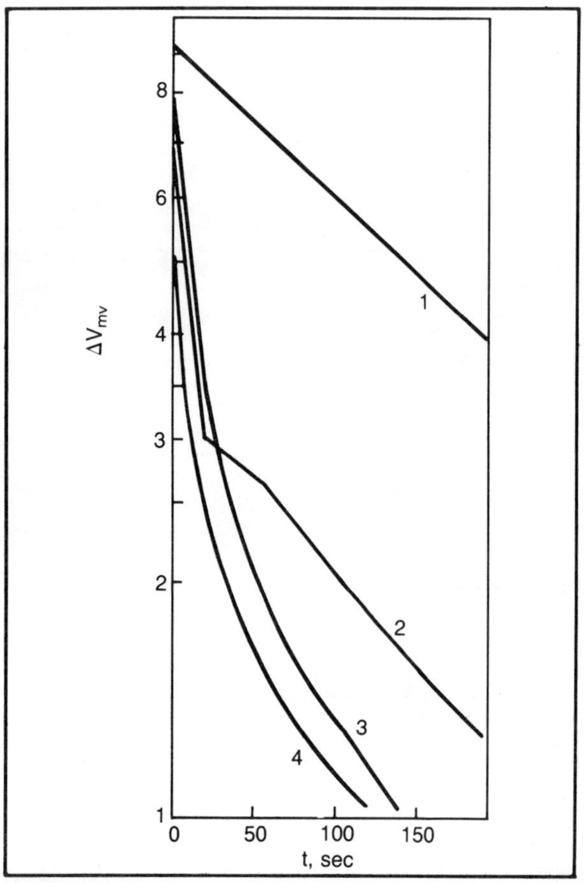

FIG. 4–80

Induced polarization decay curves for (1) plastic, (2) limestone, (3) dolomite, and (4) sandstone

There are three different techniques that are used in general field practice for both the EIP and MIP. These are the time-domain, frequency-domain, and phase-shift techniques. These techniques usually yield equivalent results if currents are less than 10 $\mu a/cm^2$. Frequencies of less than 0.125 Hz are usually used to eliminate electromagnetic coupling effects caused by conductive formations.

Time Domain

With the time-domain technique, an interrupted, alternating square wave current of low frequency is passed through the formation (Fig. 4–81). The IP voltage measurement is made during the off period between current pulses.

The relationship between the overvoltage, E_c, which is the extra voltage needed to charge the particle, and the current density, J, is:

$$E_c = a - b \log_{10} J \tag{4.43}$$

Where:

a and b = Experimentally determined constants

The IP response, or chargeability, M, is:

$$M_{(t)} = \frac{V_{s(t)}}{V_p} \tag{4.44}$$

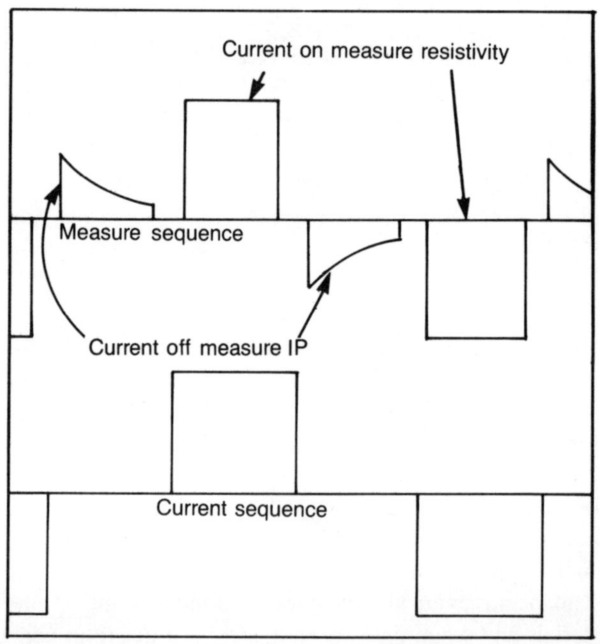

FIG. 4–81

IP current wave form

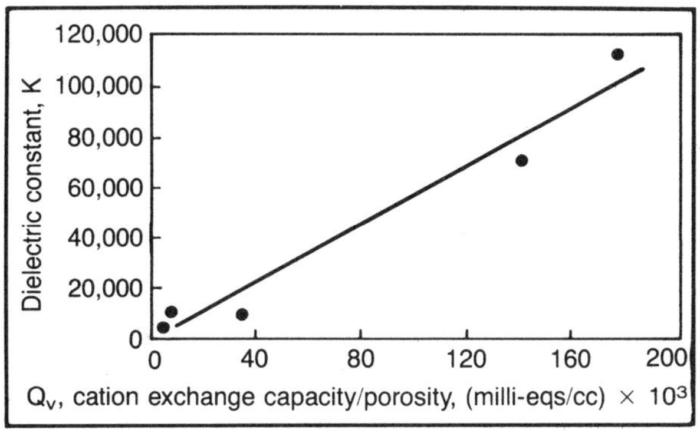

FIG. 4–82

Crossplot of specific cation exchange capacity (Q_v) as a function of the dielectric constant

(after Hoyer and Rumble, "Dielectric Constant of Rocks as a Petrophysical Parameter," SPWLA 17th Annual Logging Symposium, 1976)

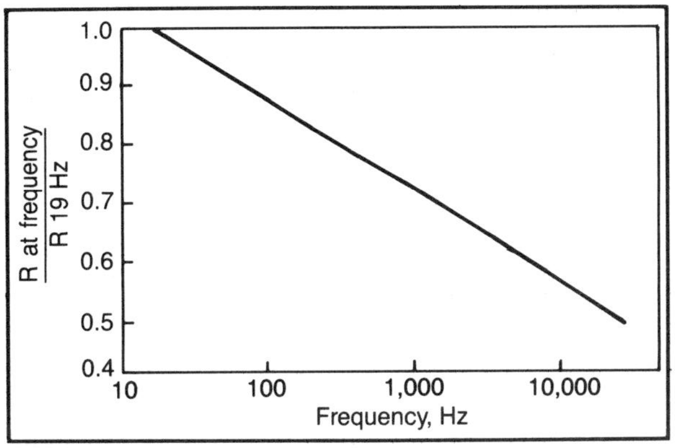

FIG. 4–83

Frequency effect of a pyritic core

(after Clavier, Heim, and Scala, "Effect of Pyrite on Resistivity and Other Logging Measurements," SPWLA 17th Annual Logging Symposium, 1976)

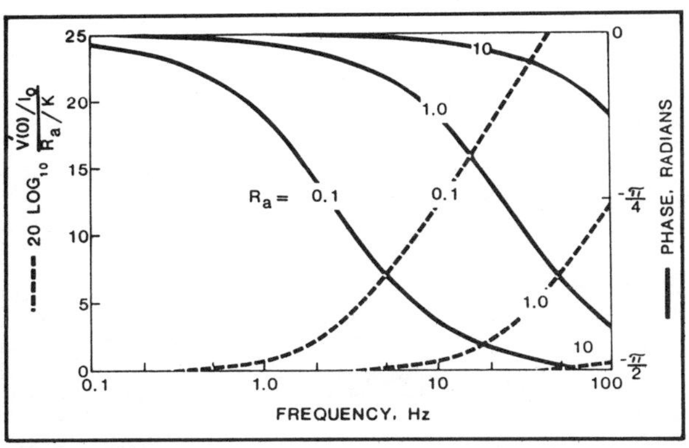

FIG. 4–84

The frequency dependence of amplitude and phase for the coupled normal array

(after Snyder, Merkel, and Williams, "Complex Formation Resistivity," SPWLA 18th Annual Logging Symposium, 1977)

This is the ratio of the average secondary voltage, V_s, or overvoltage effect, to the primary voltage, V_p, immediately before the current is cut off.

Frequency Domain

The frequency-domain technique usually uses a low-frequency, square-wave current but without the off-time between current pulses. The apparent resistivities are determined at several different frequencies (Fig. 4–84). Frequencies of 0.1–100 H_z are used.

When variable frequency measurements are made, results are usually given in terms of metal factor (MF) or percent change in resistivity, P, per decade of frequency change:

$$P = \frac{R_1 - R_2}{\sqrt{R_1 \times R_2}} \times 100\% \tag{4.45}$$

$$MF = \frac{P \times 10^5}{\sqrt{R_1 R_2}} \tag{4.46}$$

Where:

R_1 and R_2 = Apparent resistivity measurements made at frequency 1 and frequency 2

The IP also can be represented by the percent frequency effect, PFE:

$$PFE = \frac{R_f - R_{10f}}{R_f} \tag{4.46a}$$

Phase-Shift Method

Another method, which is really equivalent to the chargeability (M) and the frequency effect (PFE), is the phase-shift method. It is expressed in milliradians, ϕ:

$$\phi \cong \frac{V_q}{V_I} \tag{4.47}$$

Where:

V_q = Quadrature voltage

V_I = Resistive component

This is often called the metal conduction factor, MCF.

$$V_f = V_I + j\,V_q \tag{4.48}$$

Scintrex has a method that measures the relative phase shift of the first and third harmonics, instead of the shift of the fundamental, claiming an accuracy of ±0.15%.

Operation

To make EIP measurements, a receiver is connected to two impolarizable electrodes in contact with the formation. On a typical downhole system, the current electrode is at the bottom of the array and the cable armor is the current return. The measurement or potential electrodes are installed at intervals of 10, 40, 80, or 160 ft on the sonde or cable. These may be one or two electrodes, depending on the system and spacing desired. The depth of investigation is a function of the measure electrode spacings. A log is shown in Fig. 4–85.

For directional logging, the current electrodes may be placed on the surface. They are usually deployed on a north-south line, then on an east-west line. The resulting downhole measurements can determine the quadrant in which the target lies.

Since frequency effects are being measured, the IP measurements can be affected by the cable length in an analog system and by the spacing of the electrodes.

Surface Methods

Surface methods are similar to downhole methods and evaluations. Spacings are much greater, and thus, the currents are higher. Surface induced polarization/resistivity measurements are excellent in combination with downhole measurements for the location and evaluation of sulfide mineralization and redox trends. The surface-downhole combination is one the explorationist should seriously investigate.

Applications

The primary use of IP methods has been for the detection of sulfides. These semimetallic conductors exhibit strong IP effects, which are a function of the current, spacing, and sulfide quantity present within the measured volume (Fig. 4–86). Most of the work has been performed in igneous and metamorphic environments. Some interest has been evident in sedimentary environments for uranium and coal evaluation.

The IP technique appears to be an excellent one for uranium because of the alteration products and associated sulfides. In a sedimentary environment the altered feldspathic minerals on the barren side of the geochemical cell can result in low level but anomalous IP responses. On the reduced side of the cell, sulfide minerals are common.

In hydrothermal and other metamorphic environments, the uranium can be indicated by the often associated sulfide minerals and graphites. In coal the IP appears to give a response that is a function of the sulfide mineral quantity and the carbon content.

The volume percentage of conductive solids, S, within the investigated volume is:

$$S = K\,(\sigma)^{1/c} \qquad (4.49)$$

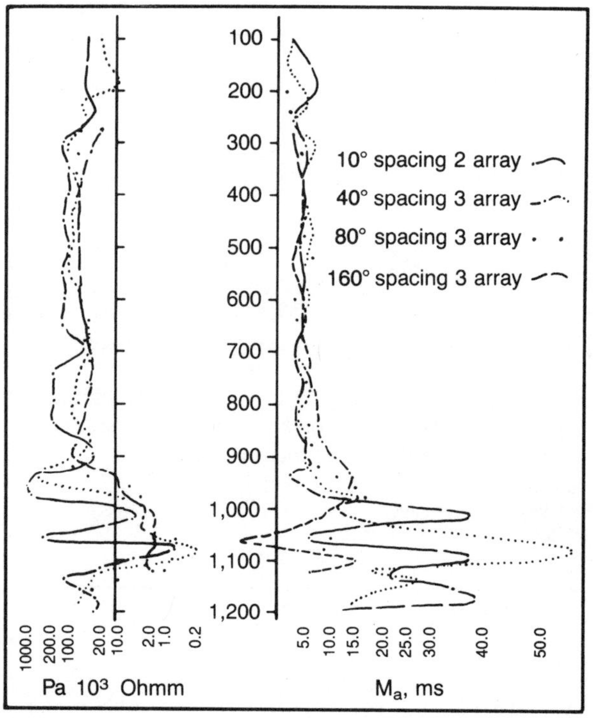

FIG. 4–85

Resistivity and time domain IP drill hole logs. A sulfide ore zone was discovered 15–30 m from the hole. Only a zone of slight alteration is seen in the drill hole

(courtesy Scintrex Limited)

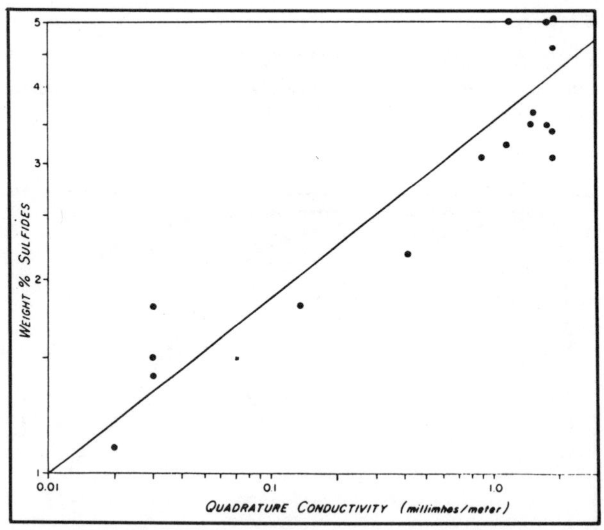

FIG. 4–86

Sulfide content vs. average quadrature conductivity crossplot

(after Snyder, Merkel, and Williams, "Complex Formation Resistivity," SPWLA Annual Logging Symposium, 1977)

Where:
K and c = Empirically derived constants
σ = Quadrature conductivity

If the phase shift is small compared to the resistivity, then:

$$\sigma = \frac{\theta}{R} \tag{4.50}$$

Several logs from Snyder, Merket, and Williams are included as Fig. 4–87 and Fig. 4–88.

A recent paper by Olhoeft and Scott indicates that electrical properties become frequency dependent when charges accumulate near energy barriers. As long as they are small relative to the driving force, the current density is proportional to the field strength. If the energy barriers are large, however, the linear proportionality breaks down. In rocks the two principle mechanisms of charge transport across these large barriers are the oxidation-reduction reaction and the ion exchange process, both of which occur at rock-fluid interfaces. From the specific potentials at which reactions are observed to occur, the types of reactions, the reaction products, and the reaction chemistries may be inferred.

Calibration to determine the specific potentials is obtained by driving the system to the point of breakdown of water into oxygen at one end and hydrogen at the other. Since these occur at known potentials, they may be used to determine the potentials of the other reactions and estimate the ground water Eh and pH. This appears to be a significant contribution to induced polarization and complex resistivity logging. Table 4–5 outlines induced polarization relations.

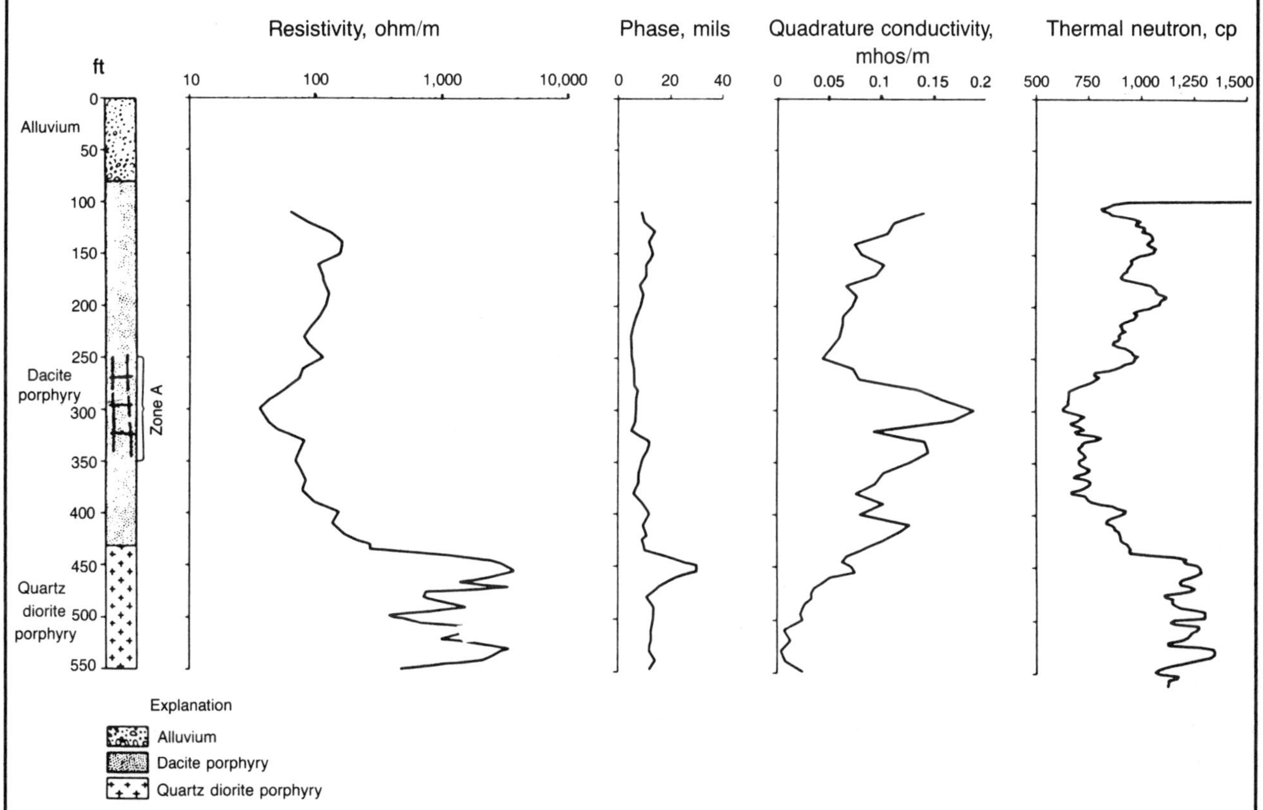

FIG. 4–87

A resistivity/IP log run in the southwestern U.S. to test an IP anomaly (courtesy Exploration Data Consultants Inc., after Snyder, Merkel, and Williams, 1976)

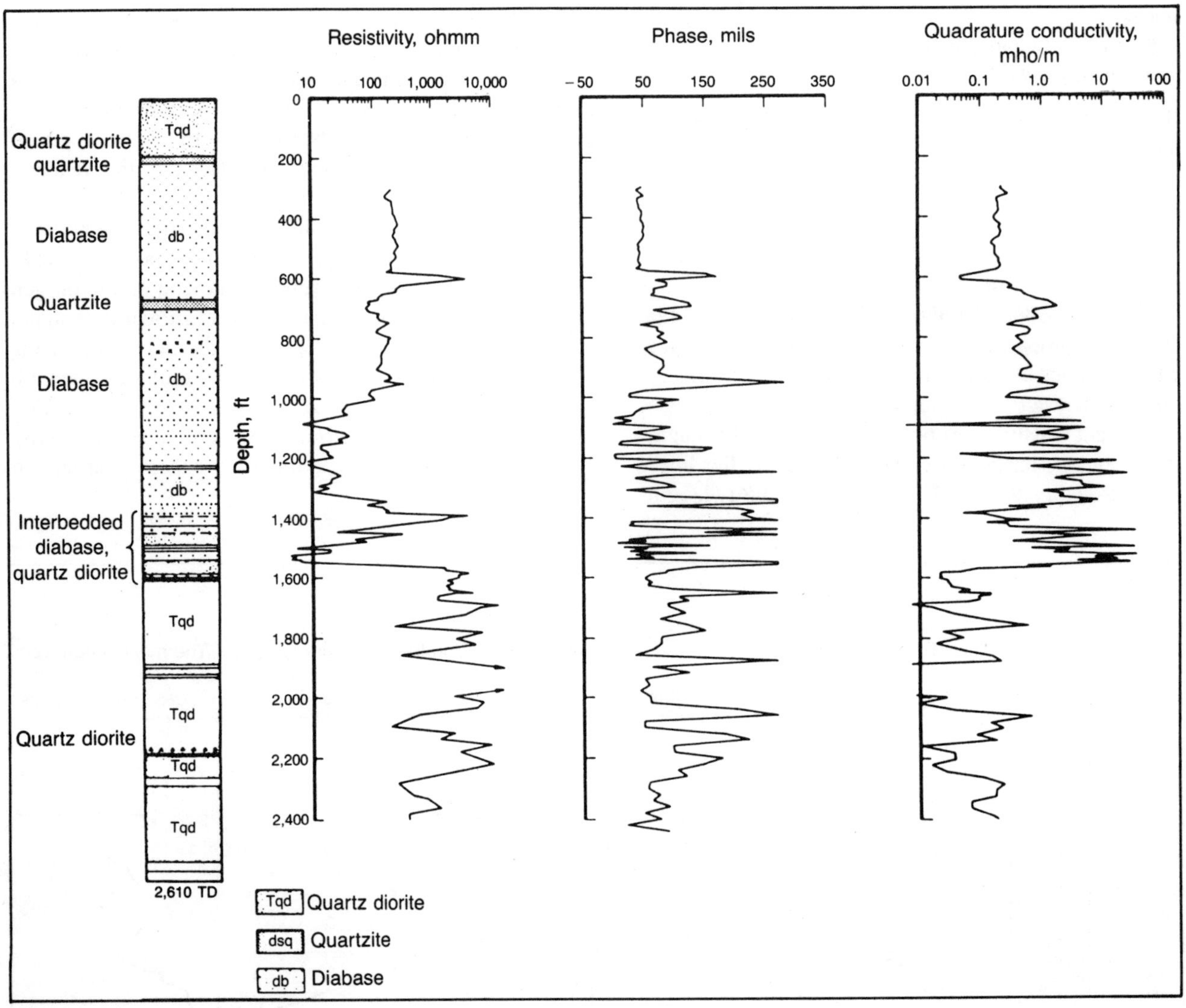

FIG. 4–88

IP log from a porphyry copper deposit in the southwestern U.S. (courtesy Exploration Data Consultants Inc., after Snyder, Merkel, and Williams, 1976)

Table 4–5

Induced Polarization Relations

Time domain:

$E_c = a - b \log_{10} J$

$M_c = V_s/V_p$

Frequency domain:

$P\% = (R_1 - R_2)/(R_1 - R_2)^{1/2} \times 100\%$

$M.F. = (P \times 10^5)(R_1 R_2)^{1/2}$

$P.F.E. = ((R_f - R_{10f})/R_f) \times 100\%$

Phase-shift method:

$\phi = V_q/V_i$

$V_f = V_i + jV_q$

Volume percentage of conductive solids, S:

$S = K(\sigma)^{1/c}$

$\sigma = \theta/R$ when the phase shift is small

REFERENCES

Alger et al. "Formation Density Log Applications in Liquid-Filled Holes." *Trans.*, AIME, 1962.

Allaud, Louis A., and Martin, Maurice H. *Schlumberger. The History of a Technique*. New York: John Wiley and Sons, 1977.

"Catalogue of Geophysical and Geochemical Instrumentation and Services." Concord, Ontario: Scintrex Limited Catalogue, 1980.

Cow, J.W. "The High Resolution Dipmeter Reveals Dip-Related Borehole and Formation Characteristics." *SPWLA Transactions*, 1970-D.

Dakhnov, V.N. "Geophysical Well Logging," *Colorado School of Mines Quarterly*, volume 57, No. 2, Golden: Colorado School of Mines, 1962.

Dakhnov, V.N.; Latishova, M.G.; and Ryapolov, V.A. "Well Logging by Means of Induced Polarization,"

Promislovaya Geofizika, 1952, translation by G.V. Keller, *Log Analyst,* volume VIII, No. 3, November–December 1967.

Doll, H.G. "The S.P. Log: Theoretical Analysis and Principles of Interpretation." AIME Technical Publication 2463, 1948.

Doll, H.G. "The S.P. Log in Shaley Sands." AIME Technical Publication 2912, 1949.

Doll H.G. "Introduction to Induction Logging and Application to Logging of Wells Drilled with Oil-Base Mud." *Trans.,* AIME, Technical Paper 2641, June 1949.

Doll, H.G. "Selective SP Logging." *Journal of Petroleum Technology,* Technical Publication 2850, volume 189, 1950.

Duncan, D.W., and Bruynesteyn, A. "Microbiological Leaching of Uranium." Socorro: Uranium Symposium, 1970.

Evers, John F., and Iyer, Babu G. "A Statistical Study of the SP Log in Fresh Water Formations of Northern Wyoming." SPWLA 16th Annual Logging Symposium, June 1975.

Evseeva, L.S.; Ivanov, K.E.; and Kochetkov, V.I. "Some Laws of the Formation of Epigenetic Uranium Ores in Sandstones, Derived from Experimental and Radiochemical Data." *Atomnaya Energiya,* volume 14, No. 5, May 1963, 474–481.

Fertl, W.H., and Hammack, G.W. "Resistivity Gradients Observed on the Dual Laterolog and the Laterolog-7." *Log Analyst,* volume XIII, No. 4, July–August 1967, 29.

Fertl, Walter. "Shaley Sand Analysis in Development Wells." *16th Annual SPWLA Symposium Transactions,* 1975.

"Fundamentals of Dipmeter Interpretation." Schlumberger Well Services document, 1970.

Garrels and Christ. *Solutions, Minerals, and Disequilibria.* San Francisco: Freeman Cooper & Company, 1965.

Gondouin, M.; Tixier, M.P.; and Seinard, G.L. "An Experimental Study of the Influence of the Chemical Composition of Electrolytes on the S.P. Curve. *Journal of Petroleum Technology,* February 1957.

Gondouin, M., and Scala, C. "Streaming Potential and the SP Log." SPE paper 864-G.

Guyod, Hubert. "Interpretation of Electric and Gamma Ray Logs in Water Wells." *Log Analyst,* volume VI, No. 5, January–March 1966.

Hallenburg, J.K. "Use of the Spontaneous Potential Curve in a Mineral Mapping Technique." *SPWLA 19th Annual Logging Symposium Transactions,* 1978.

Hilchie, Douglas W. "Old Electric Log Interpretation." Golden: Douglas W. Hilchie, 1979.

Hill; Shirley; and Klein. "Bound Water in Shaley Sands." *Log Analyst,* volume XX, No. 3, May–June 1979.

Holt, Olin R. "Relating Diplogs to Practical Geology." Dresser-Atlas document.

Ives, David J.G., and Janz, George J., editors. *Reference Electrodes, Theory and Practice.* New York: Academic Press, 1969.

Keller, George V., and Frischknecht, Frank C. *Electrical Methods in Geophysical Prospecting.* London: Pergamon Press, 1966.

LeRoy, D.O., and LeRoy, L.W. "Subsurface Geology." *Colorado School of Mines Quarterly,* Golden: Colorado School of Mines, 1977.

Log Interpretation. volume I. "Principles," Schlumberger Limited.

Lynch, Edward J. *Formation Evaluation.* New York: Harper and Row, 1962.

Okitsu, Fiunio. "Quick Interpretation of the High Resolution Dipmeter (HDT)." *SPWLA Transactions,* 1976-I.

Olhoeft, G.R., and Scott, J.H. "Nonlinear Complex Resistivity Logging." *21st SPWLA Symposium Transactions,* 1980.

Rodriques, A.R., and Pirson, S.J. "The Continuous Dipmeter as a Tool for Studies in Directional Sedimentation and Directional Tectonics." *SPWLA Transactions,* 1968-G.

Sato, Motoaki, and Mooney, Harold M. "The Electrochemical Mechanism of Sulfide Self-Potentials." *Geophysics,* volume XXV, No. 1, February 1960.

Schlumberger Interpretation Techniques B-2. "A Method for Interpretation of Shaley Sands." 1964.

Schlumberger Interpretation Techniques B-18. "More on Shaley Sands." 1966.

Sheriff, R.E. *Encyclopedia Dictionary of Exploration Geophysics.* Tulsa: SEG, 1973.

Sienke, Mitchell J., and Plane, Robert A. *Chemistry: Principles and Properties.* New York: McGraw-Hill, 1966.

Snyder, D.D.; Merkel, R.H.; and Williams, J.T. "Complex Formation Resistivity—The Forgotten Half of the Resistivity Log." *SPWLA 18th Annual Logging Symposium Transactions,* June 1977.

SPWLA, Houston Chapter. "The Art of Ancient Log Analysis." 1979.

Telford, W.M., et al. *Applied Geophysics.* Cambridge: Cambridge University Press, 1976.

Wyllie, M.R.J. "A Quantitative Analysis of the Electrochemical Component of S.P. Curve." AIME publication.

Wyllie, M.R.J. "An Investigation of the Electrokinetic Component of the Self Potential Curve." *Trans.,* AIME, Technical Publication 2940, volume 192, 1951.

CHAPTER

5 Radioactivity Methods

Radioactivity logging methods are important to nonpetroleum techniques. They gained initial importance because of the use of gamma-ray systems for the detection of uranium minerals. Since then, a number of important techniques have evolved. These include such techniques as gamma-ray spectrometry, the prompt fission neutron system, the delayed fission neutron systems, the neutron activation systems, and the filtered gamma-ray system.

In the petroleum field the radioactive methods started with the gross-count gamma-ray curve as a correlation curve in cased holes. This curve is still important in nonpetroleum work as one of the best open-and-cased hole correlation curves. The radioactivity logging methods are now a major tool in all logging efforts. They continue to grow in importance.

RADIOACTIVITY

Before entering into an explanation of radioactivity, there are a number of terms and units that should be defined.

A *curie* (Ci) is a unit of activity. It is the measure of the number of atomic disintegrations taking place per second regardless of the type of emission:

$$1 \text{ Ci} = 3.7 \times 10^{10} \text{ disintegrations/sec} \qquad (5.1)$$

A *roentgen unit* (R) is a unit of intensity of radioactivity. It is defined as the quantity of radiation that will produce one unit of ions in 1 cc of air under standard conditions of pressure and temperature:

$$1 \text{ R} = 2.083 \times 10^{9} \text{ ion pairs/cc of air} \qquad (5.2)$$

The roentgen unit can be related to the electron volt (energy):

$$1 \text{ R} = 6.76 \times 10^{10} \text{eV in 1 cc of air} \qquad (5.3)$$

The *Inverse Square Law* states that the intensity of radiation per unit area varies inversely to the square of the distance:

$$(d_1^2/d_2^2) = (I_2/I_1) \quad (5.4)$$

This assumes a spherical distribution in a homogeneous medium.

The number of atoms, N, that will disintegrate over a period of time, t, is:

$$N = N_o e^{-\lambda t} \quad (5.5)$$

Where:

N_o = Number of atoms present at the beginning of the time period

λ = Decay or disintegration constant or activity

Half-life ($T_{1/2}$) is a measure of the rate of emission of a radioactive element. The rate of decay of an element is a function of the relative amount of that element present and its activity. Thus, half-life is the length of time required for the amount of radioactive material to decay to one half of its original amount:

$$\text{Half-life} = \frac{0.693}{\lambda} \text{ sec} \quad (5.6)$$

Where:

λ = Disintegration constant or the relative number of particles disintegrating per second

Particles can have an apparent wavelength, ν:

$$E = eV = \frac{h\,c}{\nu} \quad (5.7)$$

Where:

E = Energy
e = Electronic charge
V = Voltage gradient (eV in electron volts)
h = Planck's constant
c = Velocity of light

The tendency of neutrons (or other particles) to interact is given in terms of the apparent target size presented to a neutron. It is a barn and is equal to 10^{-24} sq cm.

An electron volt (eV) is a measure of energy. It is defined as the amount of work done when one electron is accelerated by a potential difference of one volt. One thousand and one million electron volts are abbreviated keV and MeV, respectively:

$$1 \text{ eV} = 1.6 \times 10^{-12} \text{erg} = 3.8 \times 10^{-20} \text{ cal} \quad (5.8)$$

$$1 \text{ mass unit (mu)} = 931 \text{ MeV} \quad (5.9)$$

In geophysics the electron volt is the common energy term used for radioactive phenomena.

Neutrons are emitted by accelerators with an energy of about 14 MeV. Most isotopic sources are alpha emitters using the reaction:

$$^{4}_{2}\alpha + {}^{9}_{4}\text{Be} \longrightarrow {}^{12}_{6}\text{C} + {}^{1}_{0}\text{n} + 5.3 \text{ MeV} \quad (5.10)$$

Californium, Cf^{252}, spontaneously emits neutrons at about 2.5 MeV. Thermal neutrons have energies of 0.03 eV or lower.

In X-ray analysis, however, the energy term commonly used is the Angstrom unit (Å). It is a measure of wavelength:

$$1 \text{ Å} = 10^{-8} \text{ cm} \quad (5.11)$$

An Angstrom unit may be related to an electron volt:

$$1 \text{ eV} = \frac{1}{8.058 \times 10^3 \text{ Å}} \quad (5.12)$$

Basic Radioactivity

Since a number of the logging systems in use today involve radioactivity, a general discussion to cover all of the processes is included. Radioactivity logging is an important category, and an understanding of radioactivity itself is vital to understanding and using the equipment and logs. In many cases an understanding is necessary for safety reasons.

Radioactivity is the emission of radiant energy by the disintegration of nuclei of atoms. The emission occurs when the nucleus disintegrates spontaneously or because of a reaction with another particle, atom, photon, or electric or magnetic field. Radiant energy from radioactive processes may be one or more of several particles and photons. The more common particles are the alpha particles, beta particles, positrons, protons, and neutrons. There are others, but they normally are not encountered. The energy of the particles is manifested in their velocity, mass, and their ability to react, with part of the mass showing up as electromagnetic radiant energy, such as light, X-rays, and gamma rays.

The nucleus of an atom is made of protons and neutrons. A proton has a mass of approximately one atomic mass unit and a charge of plus one. A neutron has a mass of approximately one unit (1.008937 mu) and no charge. One mass unit (mu) is 1.66031 × 10^{-24} g or 931 million electron volts (MeV). Free protons and neutrons are seldom encountered naturally.

The mass of the atom, which is essentially the mass of the nucleus, is the sum of the masses of the protons and neutrons in the nucleus, plus the sum of the masses of the orbital electrons. The electrons lie, for the most part, outside the nucleus and determine the chemical characteristics of the atom. An electron has approximately 1/1,800th the mass of a proton or a neutron and has a negative charge of one. The electron mass is 9.1055×10^{-28} g or 5.4862×10^{-4} mass unit. A nucleus that has had one or more of its electrons stripped from it is said to be ionized. It has a positive charge whose strength depends on the deficiency of electrons; there is one charge unit for each missing electron.

When enough energy is expended on the orbital electrons to remove one or more from its normal position but not enough to free it, it is said to be in an excited state. The electron returns to its normal position or ground state by emitting its excess energy as electromagnetic radiation. A neutral atom has one electron in an orbit or shell around the nucleus for each proton in the nucleus. The number of protons is the atomic number, Z.

On the average, there is approximately the same number of neutrons in the nucleus as protons. The sum of the two is the atomic weight or mass, A. However, there may be more or fewer neutrons in the nucleus. These variations are called *isotopes* or *nuclides*. Isotopes of an element have similar chemical characteristics and atomic numbers, Z, but they have different masses, A. The published atomic mass of a natural element is the average mass of all of the naturally occurring isotopes of that element. The natural atomic mass usually is a fraction for a naturally occurring element. The atomic mass of a single isotope is an integral number.

An isotope is designated by its chemical symbol, atomic number, and atomic mass. Thus, the naturally occurring isotope of phosphorus, element number 15, has an atomic mass of 31. It has 15 protons and 16 neutrons in the nucleus. Its symbol may be written in any of the following ways: ${}_{15}P^{31}$, ${}^{31}_{15}P$, ${}^{31}P_{15}$, or P-15.

If the nucleus disintegrates, it emits particles. If it is an alpha particle, the element changes to a new element of lower mass and number because the alpha has a mass of four and a charge of two. A radium atom, ${}^{226}_{88}Ra$, number 88, with a mass of 226, becomes radon, ${}^{222}_{86}Rn$, with a mass of 222 and a number 86. The alpha particle is a helium nucleus. Because it is a charged (ionized) particle, its velocity can be changed by an electric or magnetic field.

If the particle is a beta particle, the mass of the nucleus does not change appreciably since the beta particle is an electron. The atomic number, however, increases one because one nuclear neutron has become a proton. Thus, the thorium isotope, ${}^{234}_{90}Th$, with a number 90, emits a beta particle and becomes protactinium, ${}^{234}_{91}Pa$, with an atomic number 91. Beta particle velocities can be changed with an electric or magnetic field because they have an electric charge.

A neutron is a neutral particle. That is, it has no electric charge. Therefore, it is penetrating, it is nonionizing, and it is unaffected by an electric or magnetic field. It can be considered to be the combination of an electron and a proton.

If a neutron is emitted or absorbed, the mass decreases or increases one mass unit, but the atomic number remains the same because it has the same number of protons—it becomes another isotope. The sulfur isotope, ${}^{34}_{16}S$, becomes ${}^{35}_{16}S$ on absorbing a neutron. When one of these reactions occurs with any of the particle interchanges, part of the mass of the nucleus is changed to energy. This usually is in the form of photons. Sometimes, especially when a nucleus absorbs a thermal neutron, the reaction product is unstable and, thus, radioactive.

The photons are particles or quanta of electromagnetic energy. They may be treated as discrete particles that travel at the speed of light. Their energy is manifested in the wavelength of the photon. Most of the photons we will deal with are gamma photons. There are, in addition, photons of X-rays, visible light, heat, and cosmic rays. X-rays will be investigated when we look at laboratory methods. X-rays are similar in nature to gamma rays but are generated by the sudden stopping or changing velocity of an energetic particle or mass.

Except for X-rays and alpha particles this text is restricted to gamma photons and neutrons. All are treated as particles. We are restricted to considering gamma rays and neutrons because of the shielding by the pressure housing of the downhole tool. The housing effectively eliminates everything else. Surface and airborne methods do not have these restrictions.

The penetration of radioactive particles and photons depends on their energy and the interactions with the atoms of the material being traversed. Alpha particles have high energies, but are not very penetrating. They have a large charge and cross section. Therefore, they interact and expend their energy very quickly. A few centimeters of air or one or two thousandths of an inch of aluminum will effectively stop alpha particles from radium.

Beta particles are likewise not very penetrating. They have little mass and a charge of one. Fairly energetic beta particles (2 MeV) can penetrate about 0.3 cm of aluminum. A beta particle is a negative electron. A positron is a positive electron.

Alpha and beta particles lose energy by interacting with the orbital electrons of atoms. The particle energy is expended removing electrons from their normal orbit. They are said to be ionizing particles.

Neutrons are extremely penetrating. Because they have no charge, they do not interact with the electrons. They are nonionizing. Neutrons interact and lose energy by collisions with the nuclei of atoms. As in any collision reaction, the most efficient energy transfer occurs between particles of the same mass. Since neutrons have almost the same mass as hydrogen atoms, these are the most efficient in slowing (moderating) neutrons. Fast neutrons lose energy through collision with atoms until they slow down to an epithermal range. This is an energy of about 100 eV to about 0.03 eV. Below 0.03 eV the neutrons are in thermal energy equilibrium with their environment. At thermal energies neutrons react easily and can be captured. They increase the atomic weight of the nucleus by one and usually make the atom unstable. A capture gamma ray is emitted when a neutron is captured. The energy of the gamma ray is characteristic of the reaction. The unstable nucleus then usually decays, giving off a gamma ray whose energy is characteristic of the decaying isotope.

Absorption

A thin beam of gamma rays of intensity, I_o, passing through any material of thickness x arrives at a detector at a reduced intensity I (Fig. 5–1). Attenuation of the gamma-ray beam occurs in three ways. If the gamma radiation is of low enough energy (long wavelength) or the nucleus is large enough (high Z), a gamma photon interacting with an orbital electron ejects the electron from its orbit and is completely absorbed in doing so (Fig. 5–2). This reaction is the photoelectric (PE) absorption, and the ejected electron is a photoelectron. Photoelectric absorption varies approximately as $(Z^{3.6}/E^3)$, where Z is the atomic mass and E is the photon energy. In sedimentary materials the photoelectric effect becomes important at energies below about 100 keV. The second form of attenuation to the gamma-ray beam is elastic or Compton scattering. In this reaction the incident photon ejects an electron from orbit but retains part of its energy. It scatters away from the beam at a lower energy, as if it were a particle (Fig. 5–3). This type of scattering is approximately proportional to the atomic number A, the number of orbital

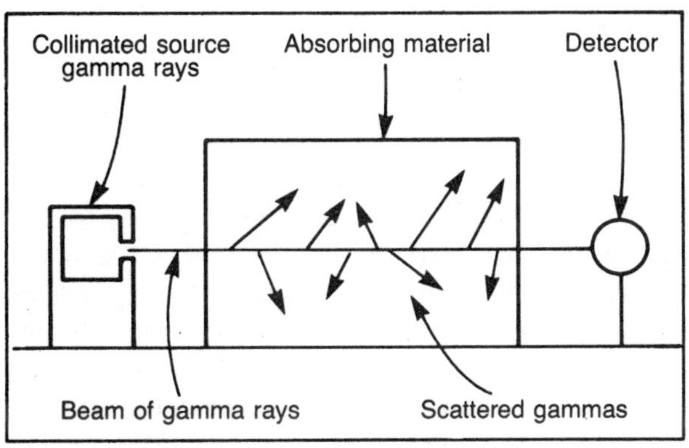

FIG. 5–1
Scattering and absorption of gamma rays

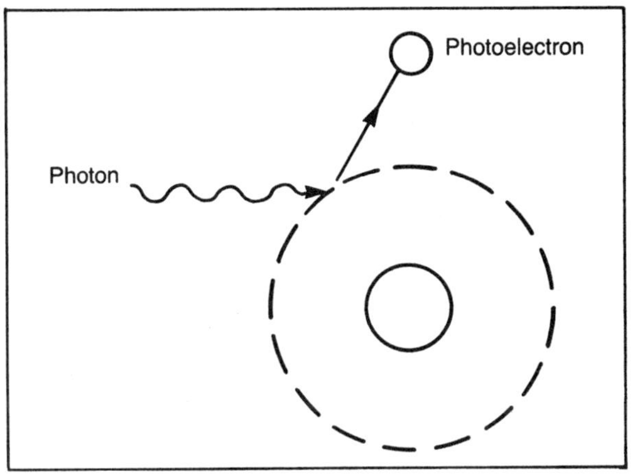

FIG. 5–2
Photoelectric absorption

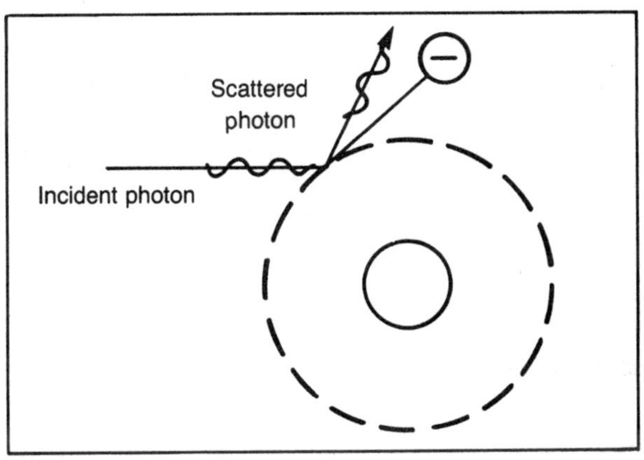

FIG. 5–3
Compton scattering

electrons, of the nucleus. The photon has an apparent mass, m, which is equal to:

$$m = \frac{h\nu}{c^2} \qquad (5.13)$$

Where:
h = Planck's constant
ν = Photon frequency
c = Velocity of light

The relation between PE and Compton scattering is shown in Fig. 5–4.

The third reaction is pair production. In this reaction, which has a threshold energy of 0.51 MeV/particle or 1.02 MeV, the gamma ray has sufficient energy to give rise to an electron and a positron (Fig. 5–5). The positron, however, has a short half-life and recombines quickly with the production of annihilation gamma radiation. Pair production is proportional to the atomic mass, A, of the target element. The reaction is:

$$h\nu = e^- + e^+ + E \qquad (5.14)$$

Where:
E = Kinetic energy of the particles

The overall effect of these processes is to reduce the intensity of the gamma radiation beam d, I:

$$I = I_o e^{-\mu \rho x} = I_o \exp(-\mu \rho x) \qquad (5.15)$$

Where:
μ = Constant of proportionality, called the mass attenuation coefficient, which is energy sensitive
x = Distance through material
ρ = Density

The relationship to energy is given in Fig. 5–6.

Many of the gamma-ray phenomena depend on this attenuation process. These phenomena are thick sample effects, and the sensitive volumes of the natural gamma neutron and other systems are large enough that they are always thick samples. Some of the detection methods make use of the ejection of photoelectrons. Density logging uses the Compton scattering of gamma rays. The base-line effects in spectroscopy are caused by Compton scattering. Therefore, these are important phenomena to understand.

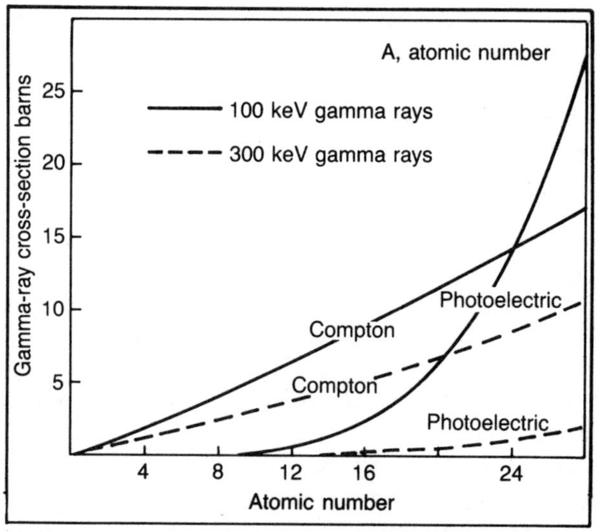

FIG. 5–4

Atomic attenuation coefficient for gamma radiation vs. atomic number
(courtesy Gearhart Industries Inc.)

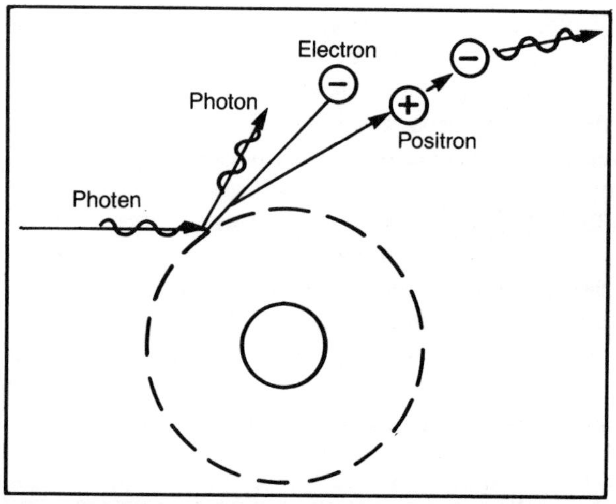

FIG. 5–5

Pair production

■ Health Physics

Radioactivity, as photons or as particles, is energy. This energy is neither more nor less dangerous than other forms of energy with which we are familiar. It must be treated with respect and handled properly. We do not worry about the tank of gasoline in our automobile, but it is a much more dangerous energy source than any radioactive logging source. We follow certain rules handling it and seldom have problems of safety. The electromagnetic radiation we call sunshine can burn and kill one just as surely as exposure to gamma rays, so we follow certain rules using it.

Exposure to gamma radiation and neutron emission can produce damage in the cells of the body. The damage is caused by ionization of the cell materials

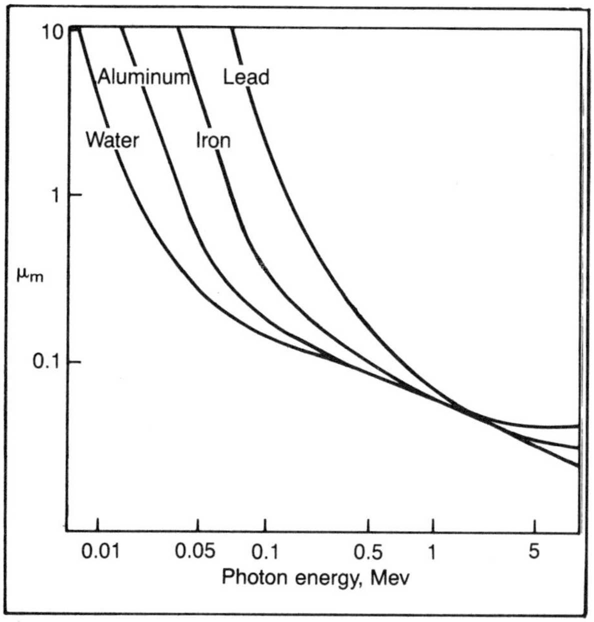

FIG. 5–6

Mass attenuation coefficient, μ

and actual disruption or breaking up of the cell molecules by the photon or particle. This damage, like a sunburn, is self-repairing most of the time. There are two important qualifications to this self-repairing, however. As with any burn if the damage is severe enough or if it is continuous at a high rate, the body mechanisms cannot handle the radiation products. The body is poisoned as with a burn. The other damage is an actual mechanical damage severe enough that the organ concerned is unable to function properly. This latter type of damage is most evident in the more rapidly growing tissues, such as mucous membranes, bone marrow, and reproductive organs. Therefore, the exposure allowable depends on the part of the body exposed.

We also must follow proper handling rules when using radioactive sources. Like sunlight, one should not be exposed too long to gamma rays. Just as any other source of radiation, the intensity of gamma rays decreases with the square of the distance from the source. If the distance is doubled, the exposure or intensity is reduced to one-fourth. This is true of neutrons, also. This is the Inverse Square Law. It is good practice to stay as far from a source as practical. Handle it with tongs or a handling tool, always. Never pick it up with bare hands. Never carry it in a pocket of your clothing. Hold it at arm's length away from your body and away from others. Warn others to remain at a distance. Adopt the philosophy that the minimum exposure is best. A dangerous source of 1 R/hr 6 in. away is safe for long exposure 8 ft away. Fig. 5–7 shows this relation.

In addition, there are materials from which storage, handling, and logging containers can be made that reduce the amount of radiation or neutron flux. Storage containers should always be used except when the source is actually in use. If nothing else, a source container makes a small source more visible. It warns people away. Logging tools should be designed to reduce exposure.

■ Gamma Rays

A gamma photon is electromagnetic energy and is produced from the conversion of part of the nuclear mass to energy. Its energy is proportional to its frequency. It normally interacts with the orbital electrons. Gamma rays are reduced by the number of electrons present. This is proportional to the density—mass of the material per unit volume—of the material. In other words the shield material should be as massive as possible.

The intensity, I, of gamma rays is:

$$I = I_o e^{-\mu \rho x} \qquad (5.16)$$

Where:

- I = Source intensity
- e = Constant 2.718
- μ = Mass attenuation coefficient
- ρ = Density of the material through which the gamma rays are passing
- x = Its thickness, cm

Refer to Fig. 5–6 for the value of μ.

Gamma rays are absorbed by being scattered. Each time a gamma is scattered, it loses some of its energy until finally it undergoes photoelectric absorption and the energy becomes heat, X-ray, light, or infrared energy. The process is similar to shining a flashlight through a fog.

Often, we see the expression half value thickness or HVT. This is the thickness of the material needed to reduce the intensity of that particular energy of gamma rays to one-half of its original intensity. Two HVT's reduce it to one-fourth, and so on. Table 5–1 shows some HVTs of common materials.

Commonly used gamma-ray source materials are shown in Table 5–2. With the exception of radium 226, these are all reactor products and one must have a license for possession and use of them. All gamma-ray source materials used downhole are in sealed capsules.

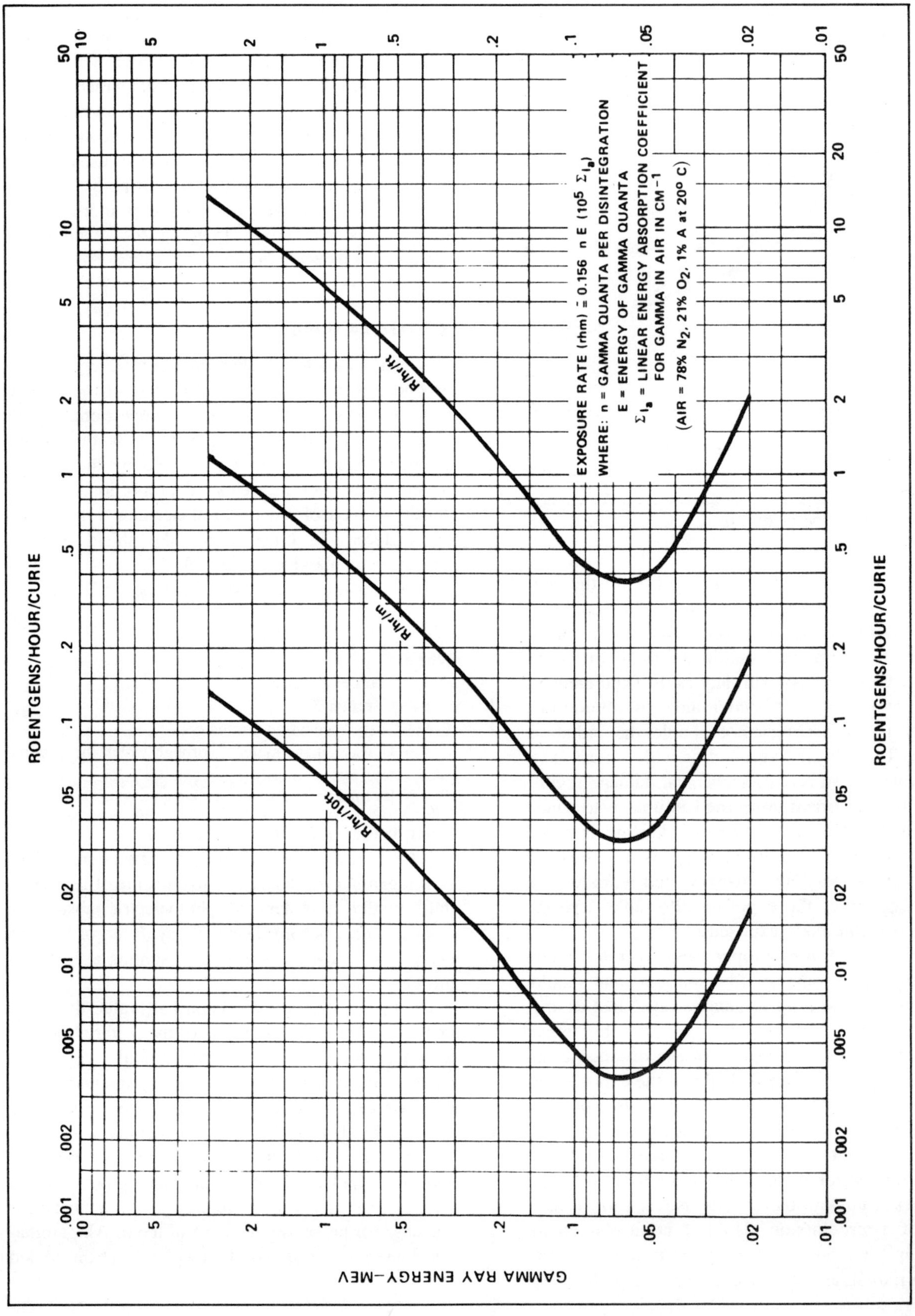

FIG. 5–7
Approximate gamma emitter dose in air assuming a point source and one photon per disintegration (courtesy Gearhart Industries Inc.)

X-Rays

X-rays are similar to gamma rays in character, but their method of production is different. They are also electromagnetic radiation. X-rays are produced by the change of velocity of a high-energy particle, such as an electron, beta particle, or any mass. X-rays can also be produced by fluorescent emission when excited electrons drop back to a ground state in the k, l, or m shells. These electrons absorb electromagnetic energy of the same wavelengths when in the process of being excited. If we do not know how they were produced, we cannot tell the difference between X-rays and gamma rays, except that X-rays are seldom as high energy as gamma rays.

The fluorescent emissions are characteristic of the element involved, and identification by spectrographic analysis is an excellent elemental analysis tool. X-rays can also be produced in neutron-shielding material from the capture gamma rays. This is a factor to be considered in shield design.

Filters for gamma-ray detectors must also take the production of secondary or fluorescent X-rays into consideration. Therefore, one usually sees a lead

Table 5–1

Half-Value Thicknesses, X(½), for Various Materials*

Photon Energy, MeV	Water, cm	Aluminum, cm	Iron, cm	Lead, cm
0.20	5.1	2.1	0.66	0.14
0.50	7.8	3.0	1.11	0.40
1.0	10.2	4.5	1.56	0.87
1.5	12.0	5.1	1.74	1.25
2.0	14.4	5.7	2.05	1.39
2.5	16.5	6.9	2.22	1.47
3.0	18.3	7.8	2.31	1.44
4.0	21.0	8.4	2.55	1.47
5.0	23.1	9.9	2.88	1.47

*$X(½) = 0.693 \div (\mu_m \times d)$

Table 5–2

Common Source Materials

Source Material	Half-Life	Neutrons/ Ci/Sec	Neutron Energy Average	Neutron Energy Maximum	Gammas/ Ci/Sec	Gamma Energy, MeV	Gamma Radiation, r/h/m/c	Other Emissions Type	Other Emissions MeV
Americium-beryllium, ^{241}Am	433Y	2.2×10^6	4.0	11	13.7×10^9	0.017 (48.2%)		α	5.44 (13.8%)
					10.0×10^8	0.026 (3.5%)		α	5.48 (85%)
					22.2×10^6	0.043 (0.08%)			
					13.7×10^9	0.060 (48.2%)			
					7.4×10^6	0.099 (0.03%)			
					2.8×10^{10}		0.001		
Californium, ^{252}Cf	2.6Y	4.4×10^9	2.3	6	2.4×10^{10}	0.04–0.16	0.15	α	6.11
Polonium-beryllium, ^{210}Po	138d	2.5×10^6	4.2	10.9	1.54×10^6	0.8	10^{-4}	α	5.3
Radium-beryllium, ^{226}Ra	1,600Y	1.5×10^7	3.9	13.1	1.5×10^{11}	0.19–0.61	0.974	α	4.8 (94%)
								α	4.6 (6%)
Plutonium-beryllium, ^{239}Pu	24,400Y	2.2×10^6	4.5	10.7	11×10^8	0.017 (99.4%)	0.011	α	4.9
Cesium, ^{137}Cs	30.1Y					0.662	0.32	β	0.51 (92%)
Barium, ^{137}Ba	2.6m							β	1.17 (8%)
Cobalt, ^{60}Co	5.27Y					1.17 (100%)	1.24	β	0.318
						1.33 (100%)			
Radium, ^{226}Ra	1,600Y					0.19	0.825	α	4.8 (94%)
						0.26–0.61		α	4.6 (6%)
Iridium, ^{192}Ir	74.3d					0.135–1.38	0.50	β	0.67
								β	0.54

gamma-ray energy filter incorporating copper and cadmium layers as X-ray filters. Industrial X-ray sources are always used on the surface. They are small, low-energy sources and usually are sealed plastic and metal wafers. They are easily shielded and handled.

Neutrons

Neutrons are neutral particles with a mass of about one mass unit and no electrical charge. They can be considered the combination of a proton and an electron. They are produced in several ways for logging purposes. Alpha particles are reacted with beryllium-9 (or boron) in most isotopic sources:

$$^{4}_{2}\alpha + ^{9}_{4}\mathrm{Be} \rightarrow ^{12}_{6}\mathrm{C} + ^{1}_{0}\mathrm{n} \qquad (5.17)$$

The result is carbon-12 and a neutron with an average energy of 4 MeV. The source half-life depends on the half-life of the alpha emitter. However, californium 252 (Cf 252) decays spontaneously with the emission of neutrons with an average energy of 2.3 MeV, with a half-life of 2.63 years.

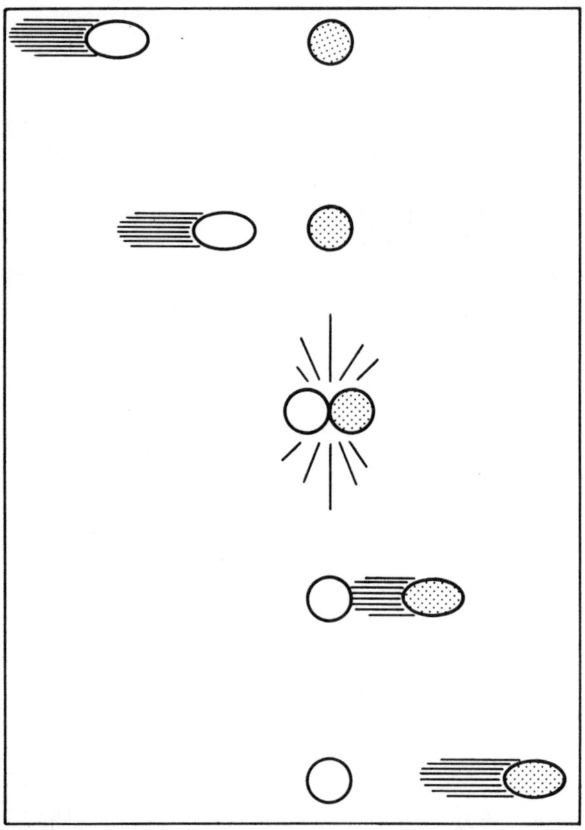

FIG. 5–8

Neutron striking a target nucleus of about equal mass, giving up all of its energy

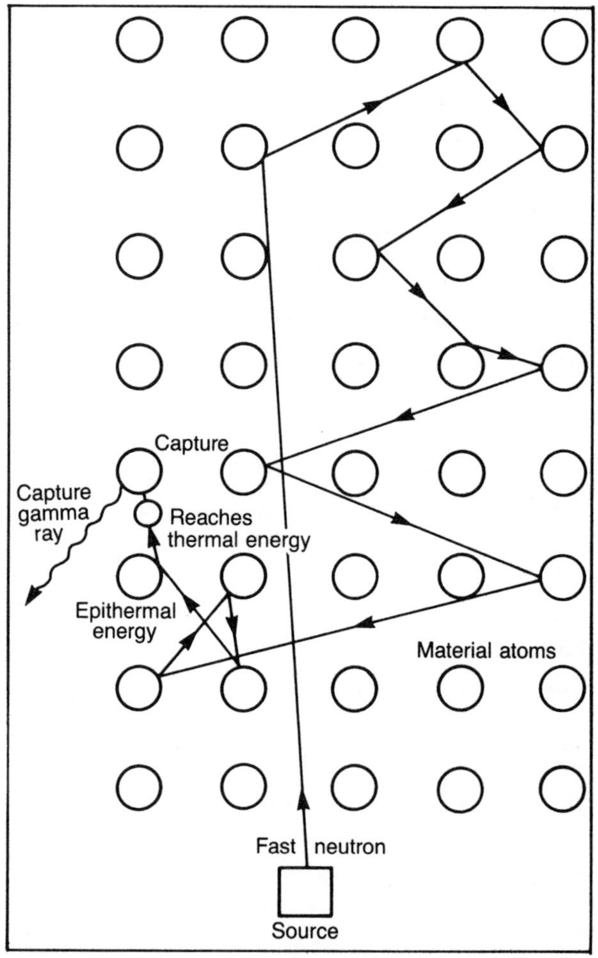

FIG. 5–9

Typical life history of a neutron

Deuterium ($^{2}_{1}$H) can be reacted with tritium ($^{3}_{1}$H) to produce 14 MeV neutrons:

$$^{2}_{1}\mathrm{H} + ^{3}_{1}\mathrm{H} \rightarrow ^{4}_{2}\mathrm{He} + ^{1}_{0}\mathrm{n} \qquad (5.18)$$

Deuterium can also be reacted with deuterium:

$$^{2}_{1}\mathrm{H} + ^{2}_{1}\mathrm{H} \rightarrow ^{3}_{2}\mathrm{He} + ^{1}_{0}\mathrm{n} \qquad (5.19)$$

This neutron has an average energy of 10 MeV. Neutron source materials are shown in Table 5–2.

Neutrons lose their energy mostly by elastic collision with the nuclei of atoms. As with any elastic collision, the most efficient energy transfer takes place when the two particles have the same mass. Therefore, hydrogen is an efficient moderator of neutrons. For the same reason hydrogeneous material is the best neutron shield. The energy transfer, ΔE, for a head-on collision is:

$$\Delta \mathrm{E} = \mathrm{E}\left[1 - \left(\frac{\mathrm{A} - 1}{\mathrm{A} + 1}\right)^{2}\right] \qquad (5.20)$$

Where:
E = Neutron energy
A = Atomic mass of the atom

The number of collisions for a 1 MeV neutron to be moderated to thermal velocity is about:

Mass Number	*Number of Collisions*
1	18
4	40
9	50
12	110

A number of reactions may take place as the neutron moderates. However, the neutron is most likely to react with an atom when it has reached thermal velocity, or less than 0.03 eV. Fast neutrons have kinetic energies greater than 0.1 MeV. Intermediate neutrons have energies from 100 keV to 100 eV, epithermal neutrons from 100–0.03 eV, and thermal neutrons less than 0.03 eV. The lowest kinetic energy a neutron may have is determined by the ambient temperature. The minimum kinetic energy, E_{min}, is given by:

$$E_{min} = 1/2\ mv^2 = 2/3\ KT \qquad (5.21)$$

Where:
m = Mass of the neutron, 1.6751×10^{-28} g
v = Velocity of the neutron
K = Boltzman's constant, 1.38032×10^{16} erg/degree
T = Absolute temperature

A neutron has a half-life of about 30 min. It will decay into an electron, a proton, and a neutrino:

$$^{1}_{0}n \longrightarrow {}^{1}_{1}p + {}^{0}_{1}e + {}^{0}_{0}n + Q \qquad (5.22)$$

Where:
Q = Mass difference energy of about 0.75 MeV

Fig. 5–10 indicates neutron exposure time in air as a function of distance from the source.

Detectors

The detector is the primary component that differentiates one radioactivity logging system from another. The type of detector also determines many of the interpretation details and methods. Therefore, a fair amount of study must be devoted to them. There are many excellent texts and papers written on the subject of radiation detectors. Some of these have been listed in the references for this chapter. Since the general characteristics of radiation detectors are similar for all of the radioactivity logging methods, these have been combined in this general section rather than in each specialized section.

All detectors used in logging make use of ionization or excitation produced in a detection medium as a result of the absorption of all or part of the energy of the photon or nuclear particle. An X-ray and gamma-ray particle or photon must undergo a photoelectric or Compton interaction, which transfers all or part of its energy to an electron. The electron, in turn, produces a track of ionization or excitation. Neutrons react negligibly with electrons since neutrons have no electromagnetic field. However, they interact by collision with a nucleus that will produce a charged particle, which then produces the ionization or excitation.

Neutrons are usually reacted with a material that has a reasonably large cross section. Ionizing recoil particles or gamma radiation are detected. These reactions are commonly used:

$$^{3}_{2}He + {}^{1}_{0}n \longrightarrow {}^{3}_{1}H + {}^{1}_{1}p + 765\ keV$$

$$^{10}_{5}B + {}^{1}_{0}n \longrightarrow {}^{7}_{3}Li + {}^{4}_{2}\alpha + 2.78\ MeV$$

$$^{6}_{3}Li + {}^{1}_{0}n \longrightarrow {}^{3}_{1}H + {}^{4}_{2}\alpha + 4.78\ MeV$$

$$^{235}_{92}U + {}^{1}_{0}n \longrightarrow \text{fission fragments} + {}^{1}_{0}n + 160\ MeV$$

Ionization Detectors

Many of the early gamma-ray logs were made with ionization chambers. They are little used now. Some of the older logs have been run with ionization chambers.

An ionizing particle loses energy in a gas. The amount of energy lost, ΔE, produces N pairs of charge carriers. An electrostatic field between two electrodes in the gas causes the charge carriers (ion pairs) to move, resulting in a current in the circuit. The total charge, $\overline{Q}$, is proportional to the energy loss from the particle and the energy amount, W, required to generate one ion pair (about 35 eV):

$$\overline{Q} = Ne = \pm \frac{\Delta E}{w} e \qquad (5.23)$$

Where:
e = Charge of an electron

Thus, individual pulse amplitude is proportional to the energy of the incident particle.

Radiation is detected by measuring the current in the chamber circuit. The circuit is shown in Fig. 5–11. The ionization chamber is slow and inefficient because of the low field strengths and low gas density used. They do not use gas amplification, and they are relatively large. Their signal to noise ratio is low, so external amplification is of only limited value.

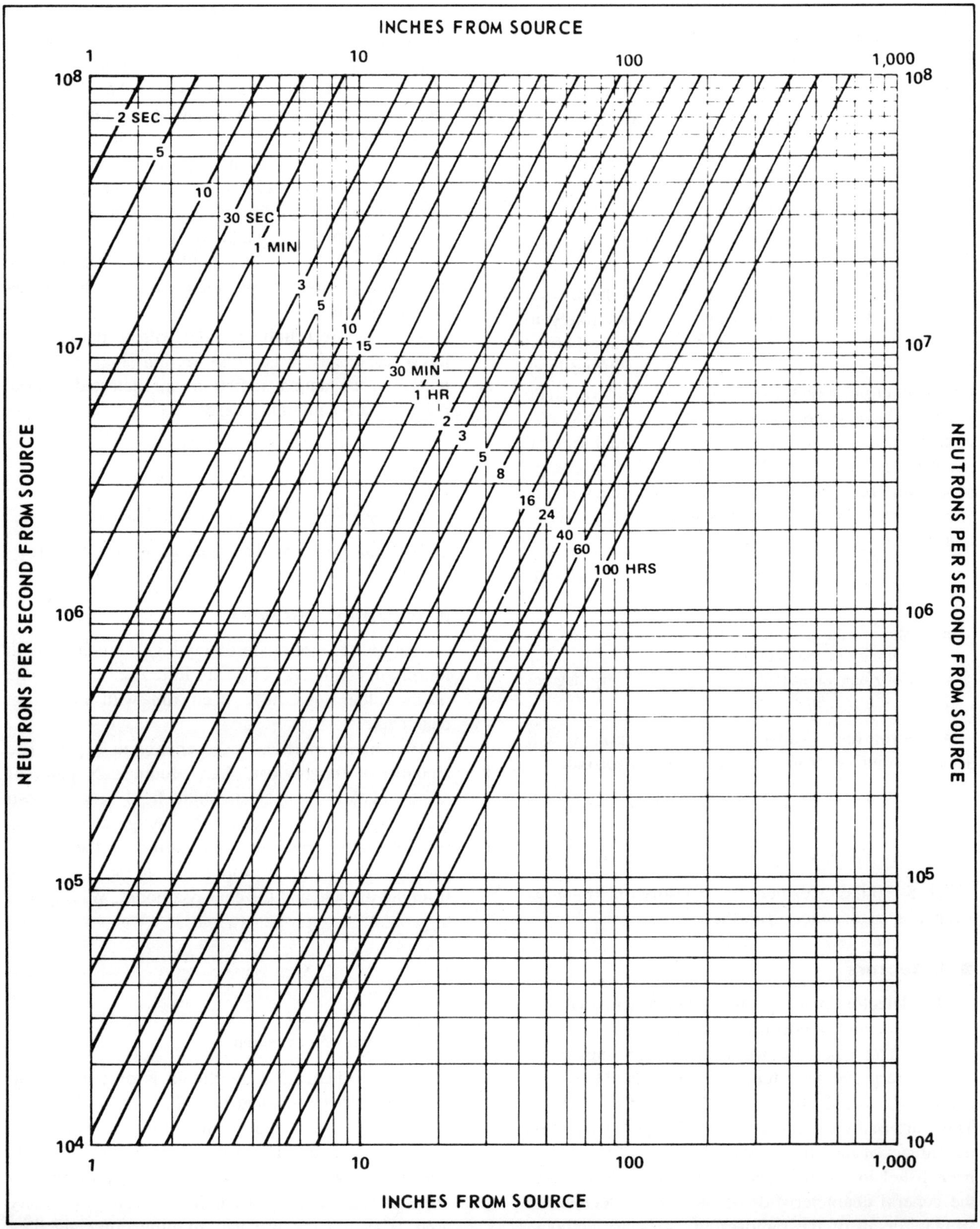

FIG. 5–10

Neutron exposure time in air vs. distance from source to obtain a 0.3-rem dose (courtesy Gearhart Industries Inc.)

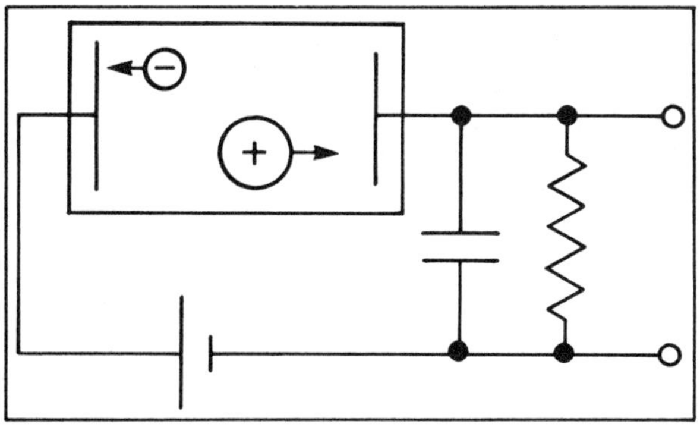

FIG. 5–11
Ion chamber

■ Proportional Counters

If the field strength of the ion chamber is increased sufficiently, the electrons can accumulate enough energy after an ionizing collision to ionize another impacted gas molecule. The number of carriers is increased by a factor of two with every collision. Therefore, the chamber current is higher by a factor M, where:

$$M = 2^n \tag{5.24}$$

n = Number of collisions

The total charge remains proportional to the energy loss:

$$Q = \pm M \frac{\Delta E}{W} e \tag{5.25}$$

Thus, the pulse amplitude of a proportional counter is proportional to the energy of the incident particle, as in the ionization chamber, but is increased by the factor M.

The proportional counter was used in early gamma-ray spectroscopy, but it was large and inefficient. It is seldom used for this now.

The greatest use of proportional counters in logging equipment is in neutron detection. One popular form is the helium-3 detector. The helium-3 detector is a proportional detector filled with $^3_2$He gas. The pressure of the gas is made as high as possible to increase the density and to make collision by a neutron more likely. Pressures of 4–10 atm are common. Also, helium-3 has a large cross section to thermal and epithermal neutrons. Sometimes, the detector is lined with uranium-235 to increase the sensitivity further to slow neutrons. Uranium-235 has a large cross section to slow neutrons and also is fissile with radioactive fragments.

The pulses from a helium-3 detector are small and require external amplification, but the detectors are extremely efficient and reasonably compact. The helium-3 detector can be made into an epithermal-only detector by surrounding it with sheet cadmium. Cadmium has a tremendous cross section to thermal neutrons. Thus, only the epithermal neutrons pass the cadmium for detection. Fig. 5–12 shows the principle of the proportional counter.

■ Geiger-Mueller Detectors

If the gradient of the electric field in the chamber is increased, the probability increases that light photons will be emitted. These photons may produce further electrons through photoionization, causing avalanche effects throughout the chamber. In this case the charge and, therefore, the pulse height are constant. It is independent of the energy of the primary ionizing particle.

The avalanche effect is responsible for many of the desirable features of G-M detectors. Pulses are of high amplitude, as much as 100 v, and of uniform amplitude. This eliminates the need for some of the circuitry needed for other types of detectors.

When the slower moving positive ions reach the cathode shell, their impact will cause emission of light and a repetitive avalanche causing the tube to destroy itself. Therefore, external and internal quenching mechanisms are designed in the process. These are current-limiting, voltage-dropping methods and gases for filling that combine rapidly with the ions created by the process.

One result of the avalanche process is that the G-M tube is insensitive to another pulse for about 300 μsec.

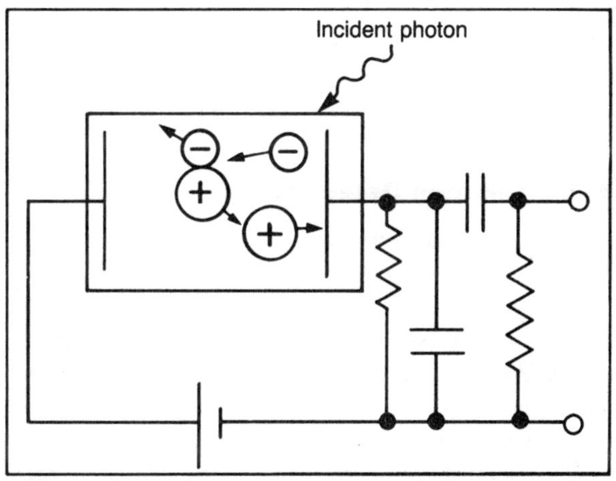

FIG. 5–12
Proportional counter

Thus, the counting rate is severely limited (see the dead-time effect explanation at the beginning of the next section). This effect can be reduced by using a bundle of tubes. However, the dead-time is still long. G-M tubes are often operated in bundles that can, if the circuit is designed properly, divide the dead-time by the number of tubes. The tubes must be electrically isolated from each other, however, by resistors. If not, a pulse in one tube triggers a pulse in the others.

G-M detectors were a great improvement over earlier detectors. Their sensitivity is reasonable, about 1%. The required circuitry is simple; no ultra-stable power supply or amplifiers are needed, and the signal-to-noise ratio is excellent. These detectors are still very much a part of our geophysical instrumentation. Fig. 5–13 shows the operating principle of the G-M detector.

If the voltage across a G-M tube is varied (Fig. 5–14), the counting rate increases. This is the proportional range. With a further increase of voltage, there is a level above which the counting rate will not increase with voltage. This is the G-M range and is commonly called the plateau. The avalanche action is taking place within the tube and occurs regardless of the voltage within this range. If the voltage is increased above the plateau range, the positive ions are accelerated to the shell rapidly and the avalanche process is repeated continuously.

The operating voltage of the tube should be set a small amount below the center voltage of the plateau. This allows fairly substantial power supply changes without affecting the counting-rate sensitivity. Setting the voltage slightly below the center point allows a greater range of voltage variation with changes as a result of higher temperature downhole. In surface equipment the operating voltage should be set at midplateau. The only change due to power supply voltage changes is in the pulse amplitude.

■ Scintillation Detectors

Scintillation detectors were developed during World War II. They are an outgrowth of a device called a spinthariscope used by the English physicist, Lord Rutherford. It is a zinc sulfide screen and microscope. When the screen is exposed to alpha radiation, the events result in tiny flashes of light or scintillations. They once were counted by observation. A modern scintillation detector uses one of many types of scintillator material, and the flashes of light are detected by a photosensitive device.

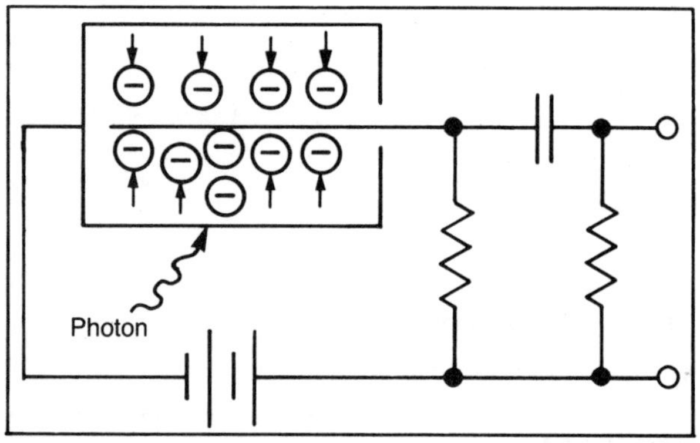

FIG. 5–13

Geiger-Mueller detector

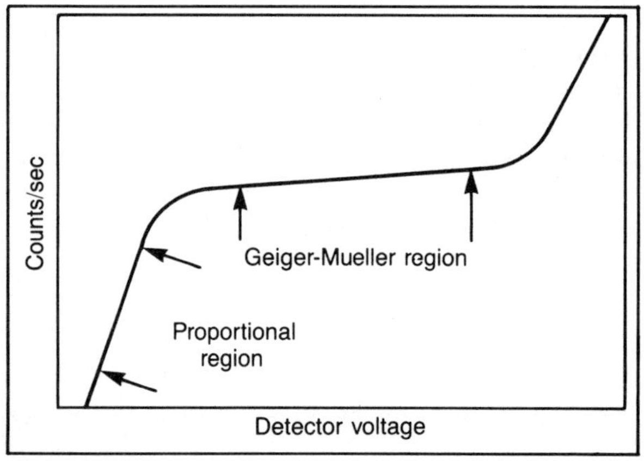

FIG. 5–14

Plateau effect

Most geophysical gamma-ray detectors use a sodium iodide, thallium-activated crystal as a scintillator. Occasionally, neutron detection is done with a lithium-6 compound. Radon sniffing (alpha detection) uses zinc sulfide.

The mechanism of scintillators is not well understood. Apparently, the individual molecules are loosely bound and retain much of their individual identity, electronic structure, and luminescence. The molecules are excited by the passage of a charged particle. They then transfer the excitation to the impurity centers (thallium in the sodium iodide). The impurity centers then de-excite with a photo emission that is characteristic of the impurity. This process takes place in about 230 nanoseconds in a NaI(T1) crystal.

The photoemission from the scintillator is usually chosen to be in the visible range. This is detected by an amplifying photosensitive device—a photomultiplier tube or PM tube.

The PM tube has a photosensitive cathode (one that gives up photoelectrons on impingement of light photons), a series of dynodes (a secondary electron emission surface) of successively higher potential, and a collection anode. Fig. 5–15 shows a PM tube diagram. Each photoelectron is accelerated through a field of 100–200 volts to the surface of the first dynode. There, it has enough energy to produce five or more secondary electrons. These are, in turn, accelerated to the next dynode where a further multiplication takes place. Typical PM tubes in logging equipment have 9–11 dynodes and produce an amplification of 10^5–10^7.

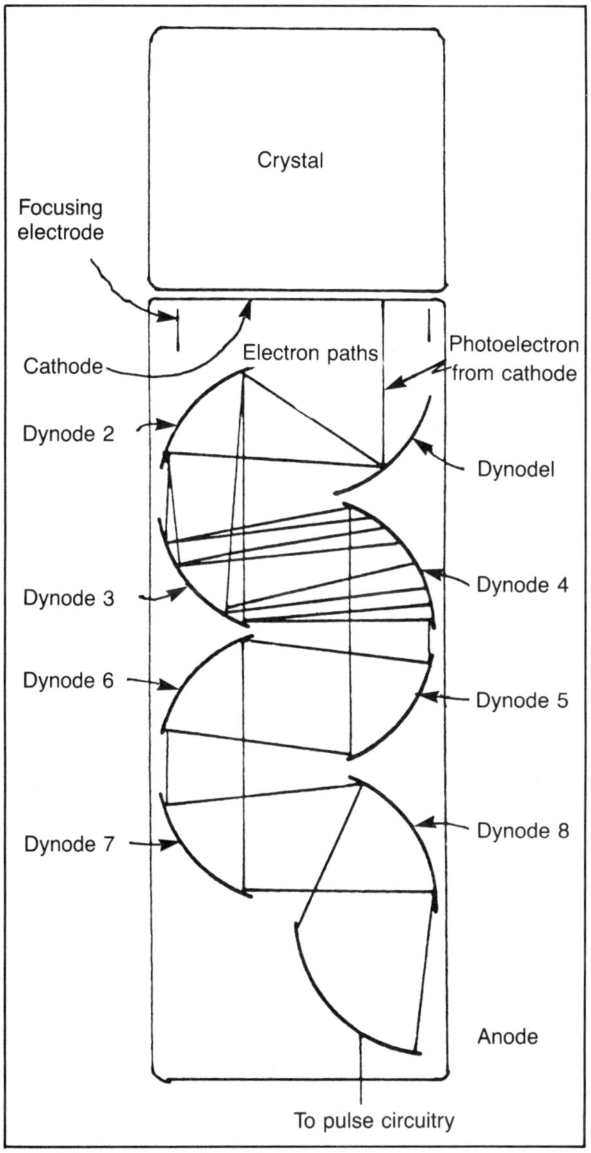

FIG. 5–15

Scintillation detector

The path lengths of individual electrons are not all the same, so the spread of transit time lengthens the original pulse slightly. The pulse widths at the anode of the PM tube can be from 250–500 nanoseconds long. They are, however, substantially shorter than those of even a bundle of G-M tubes. Thus, scintillation devices are capable of much greater counting rates than G-M tube systems.

Because of the higher counting rates of scintillation detectors, the statistical variation of the rate, or the probable error, is lower. The size of the detector, especially the active length, is much smaller than G-M detectors. Because of the high density of the scintillator material, the efficiency is greater than for G-M detectors (10–15%). On the other hand, scintillation detectors are mechanically fragile, they have severe temperature limitations, they require stable power supplies, and the pulse outputs require amplification. In general, though, they are an improvement over most other detectors.

The light output of a scintillator is a function of the path length of the particle and the degree of ionizing capability. Thus, the scintillator is sensitive to the particle energy. The energizing particle must be completely absorbed, however, for the measurement to be meaningful. Therefore, the scintillator must be as large and as dense as possible.

There are limits to the size and density of the scintillator. Since no material is perfectly transparent, there are losses of light transmission through the crystal. There is a practical limit to the length. The downhole tool design puts a limit on the scintillator diameter. It must be small enough to get down the borehole. The material must be as transparent as possible to the emitted light. Thus, the choice of materials is limited.

In logging equipment scintillation detectors exhibit a pseudoplateau. If the exciting voltage of a photomultiplier is increased with the detector in the field of a monoenergetic source, such as ${}^{137}_{55}Cs$, the counting rate at first increases. This is because the primary gamma rays and their small amplitude variations are increasingly detected. When the voltage has been increased enough to bring the main body of pulses above the discriminator level, further voltage increases result in only moderate counting rate increases. These additional counts result from the Compton scattering within the formation producing gamma photons of lower energy. As the energy of the scattered gammas decreases, more are absorbed by the steel housing of the logging tool and eventually none get through to the detector. Further voltage increases do not result in substantial counting rate increases. This is the plateau of a scintillation detector. Further increases of PM voltage bring the circuit noise above the discriminator level, and there is a sharp counting rate increase. This latter is independent of ambient radioactivity. The PM voltage should be set about one-third of the way up the plateau.

The length of the pseudoplateau is a function of the system noise level. Thus, a good, low noise circuit and PM tube have a long plateau. A marginal PM tube results in little or no plateau.

■ Semiconductor Detectors

Free-charge carriers can be generated in solids by ionizing radiation in a manner analogous to gas ionization. Ionizing particles, such as gamma-ray photons, create hole/electron pairs in solids. An electric field across the solid causes the pair to migrate in the electrode directions. Semiconductive materials that do not trap the electrons or holes quickly are used. Reversed biased p-n junctions of semiconductors are used in this manner. The semiconductor detectors are, in effect, solid-state ionization chambers. The semiconductor detectors have extremely good energy resolution for gamma rays. Therefore, they are excellent for spectrographic work.

The resolution of semiconductor detectors is dominated by the noise of the semiconductor. The efficiency depends on the leakage current. Both of these factors are temperature dependent. Therefore, these detectors must be operated at cryogenic temperatures. Some germanium detectors must be stored at cryogenic temperatures, also. The commonly used solid-state detectors are silicon and two types of germanium—lithium drifted and intrinsic. The lithium-drifted germanium detector must be stored at cryogenic temperatures.

Their use has been hindered by the need for low operating temperatures, however. Liquid nitrogen, when used as a coolant, must evaporate in order to cool. To evaporate at a constant temperature, it must be at a constant pressure, and it must be exhausted into the borehole. Whether it is stored or exhausted, the changing pressure on the liquid nitrogen results in increased boiling temperatures, which cannot be tolerated. This problem has been overcome using a heat sink of melting propane. This involves little or no change in volume and makes a viable downhole spectrographic detector of gamma rays.

Germanium detectors are being operated experimentally in neutron activation tools and commercially as a direct uranium tool. Recent investigations in the U.S. by Stromswold and in Canada by the Geological Survey of Canada with CsI (Na) and bismuth germanate ($Bi_4Ge_3O_{12}$) crystals indicate counting rate advantages over NaI(Tl). These advantages may be enough to overcome some disadvantages, such as lower light amplitude and poorer energy resolution. These crystals should be considered in any tool design or redesign.

■ Instrumentation and Instrumentation Effects

Radioactivity circuitry is more complex in general than that of the electric logging systems because it handles pulses of current rather than steady-state signals. Pulse rates and amplitudes are usually determined. Circuit frequency response and bandwidth must be very great. The response of the detector and the circuit determines the method of interpretation to a much greater extent than with other kinds of logging. Table 5–3 compares several types of systems.

Radioactivity instrumentation is generally similar for most types of systems whether they are natural gamma ray, scattered gamma ray, neutron, or spectrograph. The major difference of usage is centered in the detector. Therefore, a general discussion of radioactivity instrumentation is covered, rather than repeat it for each type of use.

The detectors play a bigger role than the circuitry in determining the type, use, and interpretation methods of the various radioactivity systems. Therefore, a substantial portion is devoted to detectors.

The geometry of the radioactivity measurement is as important as in other types of logging. Again, this is a region not generally well understood. An instrument, such as a hand-held scintillometer, has been calibrated for use above a flat plane of emitting

Table 5–3

Radioactive Circuitry Requirements

	Uranium Gamma Ray	Coal Gamma Ray	Petroleum Gamma Ray
Bandwidth	1 kHz to 10 MHz	1 kHz to 1 MHz	1 kHz to 0.1 MHz
Pulse length	$< 1\ \mu S$	$< 10\ \mu S$	$< 50\ \mu S$
Dead time	$< 1\ \mu S$	$< 20\ \mu S$	$< 100\ \mu S$
Counting Rate range/sec	1–100 k	1–10 k	100–1 k
Cable length maximum	< 3,000 m	< 1,000 m	5 k to 10 km

material. To use it in another environment, such as downhole, requires a different calibration. Fig. 5–16 illustrates detector geometry for several uses.

Detectors have already been examined. They usually have a pulse output whether the rate or the amplitude is a measured. Therefore, the associated circuitry is designed to handle pulses rather than steady-state signals.

Radioactivity circuitry falls into two major categories: analog and digital. Each type of system has its advantages and disadvantages. In spite of the newness of the full digital systems, they are extremely popular and their use is growing rapidly. Most units in use today are analog systems. Therefore, both types are covered.

Analog and digital systems are not interchangeable. In order to handle the data properly, one must know which system or combination of systems is used. This information should be required on the log heading since it is not always apparent from the appearance of the log or from the description of the operator.

Further divisions of circuit types into gross count, single channel, or multichannel pulse height analyzer, are variations of the two major classes and will be described as we come to them. These types are available in both analog and digital form.

Preamplifier

The first stage of either the analog or the digital system may be a preamplifier. Not all detectors (i.e., G-M detectors) require preamplification. The purpose of this is to provide a low-noise linear amplification for low-level signals. It is usually found as close to the detector as possible. It may have any amplification factor from one to more than 1,000, depending on the need. It may be considered part of the detector.

There may be other circuits for special purposes. There may be further amplifiers, base-line restorers, differentiators, pulse stretchers, and others. Their purposes depend on the individual problem, and they may or may not be included. They will not be covered in this text but are well covered in the references.

Discriminator

A discriminator is included in both types of systems. The main purpose of the discriminator is to eliminate circuit noise. In spectrographic circuitry, the discriminator is used in multiplicity to separate channels of different pulse heights. The type of discriminator may differ from analog to digital circuitry and between gross count and spectrographic systems.

In all radioactive systems the setting of the discriminator level is critical. A discriminator operates by passing pulses of amplitude above a preset level in gross-count systems and between preset levels (two discriminators) in most spectrographic systems. The level may be adjustable, or it may be fixed by design. The operation of the discriminator is illustrated schematically in Fig. 5–17.

When a discriminator is used with a G-M detector, there is little problem setting it. One sets the operating voltage of the detector within the plateau and then sets the discriminator level below the uniform pulse

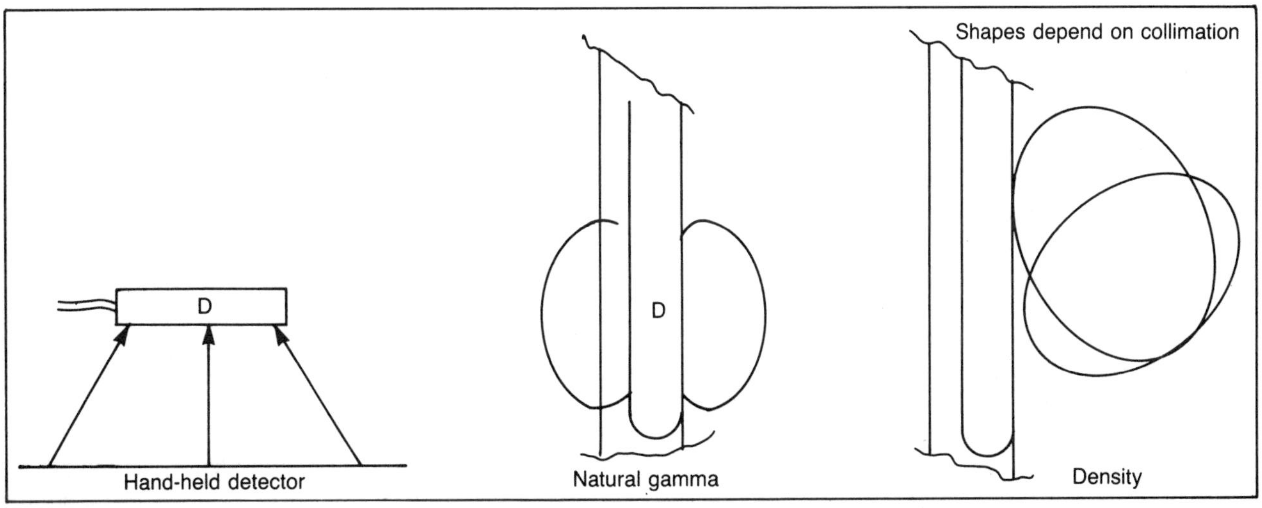

FIG. 5–16
Detector geometry

height. This can be judged by varying the discriminator level until changing it does not change the counting rate. The level should be set as high as possible without affecting the counting rate. Fixed-level discriminators are often found in gross-count radiation-measuring equipment. With this type of discriminator, the voltage level for the G-M tube determines the pulse height. The circuits are designed so that setting the plateau at the proper point with the G-M-tube power supply results in a pulse high enough to exceed the discriminator level.

When the circuit uses a scintillation detector or a proportional counter, the setting of the discriminator level in a gross-count system can be very important. If a pulse height analyzer is not available, the pseudo-plateau can be used. The discriminator level should be lowered until there is negligible change in the counting rate. Then it should be raised up to the point before the counting rate is affected. This should be performed with a moderate counting rate, not a low or background rate, and with a wide range of gamma energies, not a monoenergetic source. A range of energies can be obtained by scattering the gamma rays in a bulk of material (Fig. 5–18).

If the discriminator level is set too high in a gross-count system, the level is apt to be at a point where the counting rate can change rapidly with small power-supply changes. This invariably results in erratic sensitivity of the system. If it is set too low, noise is admitted. In a density system the level must be well below the peak of the single, scattered gamma rays. With spectrographic systems of any kind, the initial discriminator setting is a design problem and should not be changed. From the output of the discriminator, the components depend on whether the data handling is analog or digital.

The type of detector, its size, the discriminator setting, and the operating voltage determine the counting rate for a given amount of radioactivity. Therefore, these things should be determined and set initially for any design of probe. Once they are determined, the probe or system must be calibrated. Then they should be kept unchanged. If one setting is changed, a calibration must be performed afterward. A pulse height with a specified monoenergetic source at a fixed distance should be noted at the discriminator input. Amplification should be adjusted to obtain the correct, specified pulse height when the probe is checked. It should be checked regularly. The discriminator level must then be checked and, if necessary, reset after the input pulse height has been set. These steps ensure a minimum change of sensitivity.

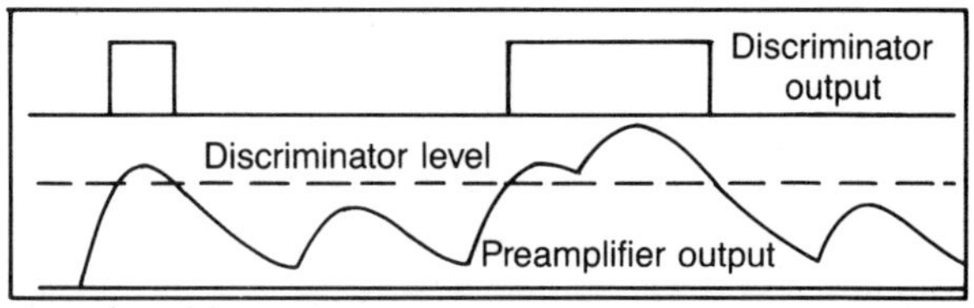

FIG. 5–17

Signal into (lower) and out of (upper) discriminator

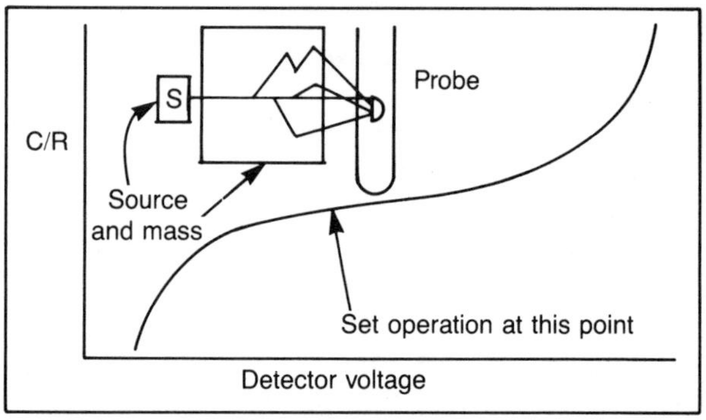

FIG. 5–18

Pseudo plateau of a scintillation detector

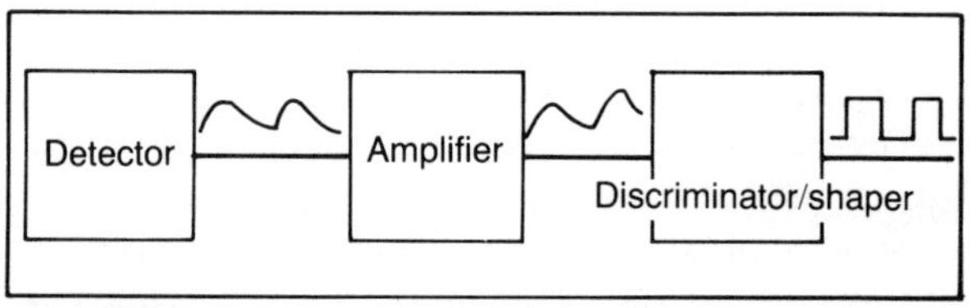

FIG. 5–19

Block diagram of a gross-count gamma ray to the discriminator

Fig. 5–19 is a block diagram of a typical circuit to this point.

Pulse Shaper, Analog

Following the discriminator in an analog system or incorporated in it, there usually is a pulse-shaping system. If the tool is a gross-count tool, the pulses form to a uniform shape. If the system is a spectrographic one, the pulses vary in amplitude but have a uniform shape.

Output Amplifier, Analog

A power amplifier is also needed to put the pulses on the cable to the surface so there is minimum distortion in the cable. This stage is usually the tool output in a simple analog circuit. The output amplifier increases the voltage and available current of the pulses so the attenuation of the cable can be tolerated.

The characteristics of the power amplifier are important because an output impedance that is poorly matched to the cable can result in substantial increases in system dead time. This is a design problem and should be considered before buying a new system.

One class of problems with radioactivity measuring systems is caused by the randomness of the occurrence of the radioactive events and, thus, the electrical signal pulses. These problems are the resulting probable error and the dead-time losses. The cable is a component that can increase the dead time of the system substantially. A short pulse (5 μsec or less long) put on 3,000 ft of logging cable has a dead time of about 20 μsec by the time it is taken off at the top. Fig. 5–20 shows this problem. Fig. 5–21 is a block diagram of a simple analog circuit.

One way to alleviate dead-time problems is to derandomize the signal pulses and make them periodic, especially before putting them on the cable. Then there is no dead-time losses until the pulse rate reaches an easily determined rate of overlap. At this overlap point the system ceases to transmit pulses completely. This is also shown in Fig. 5–20.

There are several methods of accomplishing derandomizing. One of the better ones is to use a buffer memory or storage register to store the random pulses. Then they may be read out at a periodic rate. This type of system usually is inserted before the final amplifier. It can result in real dead times as short as the design of the downhole equipment permits.

Some systems, such as count-rate divider circuits, only offer a small improvement in dead-time effects. They are, however, offered as improving or shortening the apparent dead time and are somewhat effective, particularly when high counting rates are encountered.

Cable

The logging cable is by far the most detrimental component in an analog system. Unless there is a full derandomizer, an analog system must be recalibrated any time a change is made to the cable. For example, if a cable is 3,000 ft long and 100 ft are cut off during reheading, there is about a 5% decrease in system dead time at the surface.

A logging cable is the result of a compromise of many conflicting demands. The cable is the main support for the downhole probe and the major way to retrieve it if it becomes stuck. Therefore, it must be strong. A 3/16-in. logging cable has a new breaking strength of 2,500–3,500 lb. The cable is the depth measurer. Therefore, it must have a low, predictable elasticity. It must go over sheave wheels, so it must be flexible. Finally, it must carry electrical signals, including power. All of these demands conflict. The result is that, electrically, the logging cable is only a fair device. It has a substantial distributed resistance and capacitance and, therefore, has a high time constant that stretches and distorts the pulses.

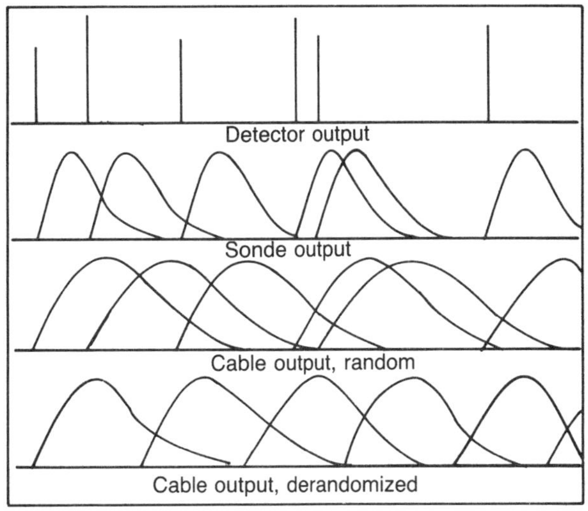

FIG. 5–20

Comparison of random and derandomized pulses

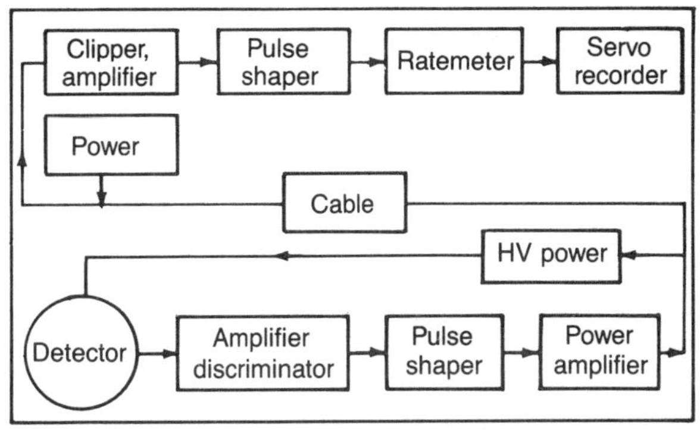

FIG. 5–21

Simple analog system block diagram

With an analog system, especially one without a downhole derandomizer, the log analyst should know how much cable and what type are used. Furthermore, any time a tool is changed, a cable is altered, or any surface change is made, a calibration should be performed. In a simple analog system the dead-time measurement must include the whole system. This is not true of a fully derandomized or a digital system.

Cable Termination

Some types of cable termination improve the dead-time characteristics of the cable. They are not covered except to point out that the use of one should be noted on the log heading. They should be used with random-type analog systems.

Amplifier

The cable attenuates the analog pulses severely. Therefore, it is customary to amplify the pulses to their former height at the output of the cable and the input of the surface equipment.

Pulse Shaper and Ratemeter

At the output of the amplifier is another option point. A straightforward, simple analog system has a pulse shaper and ratemeter at this point. The pulse shaper is to ensure that all of the pulses are of uniform width and amplitude—that is, they have a uniform charge, Q. The pulse shaper is usually a trigger circuit.

The pulse shaper is usually one of the major problem points with commercial equipment. The pulse width and sometimes the amplitude are usually adjustable. They frequently are not set properly, resulting in nonlinearities of the following ratemeter. Pulse widths out of the shaper must be less than the incoming pulse width and uniform. If the pulse width from the shaper is longer than the incoming pulse, pulses may be lost because of the dead time of the trigger circuit. This circuit must be checked on any new equipment and periodically during the life of the circuit.

The ratemeter circuit uses the uniform pulses from the pulse shaper and charges a capacitor, C, through a diode pump. Pulses arriving at the capacitor charge it up. Thus, the voltage across the capacitor increases to the maximum set by the pulse amplitude, V. The capacitor is discharged by a parallel resistor, R, at a time constant:

$$\tau = CR \text{ sec} \tag{5.26}$$

Where:
C = farads
R = ohms

The voltage across the capacitor is then a function of the rate at which pulses are put into the circuit. The voltage is read with a high-impedance voltmeter or recorder. A ratemeter circuit is shown in Fig. 5–22.

Digital ratemeters are becoming more popular and probably represent a significant advancement over the analog ratemeter. They usually are an accumulating register, which is read every tenth second or every

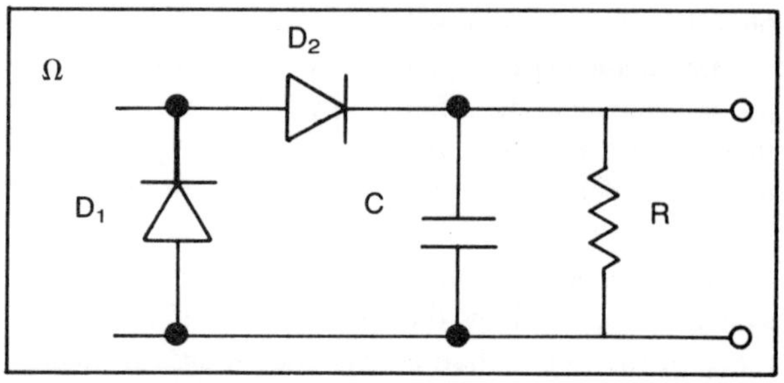

FIG. 5–22
Ratemeter circuit

second. This type of ratemeter should be buffered so as not to lose counts while it is reading out. A multiple buffer is preferable but not often available. Digital ratemeters usually have an analog output from a digital to analog converter (D/A) for operating a strip chart recorder and a digital output for operating a printer.

Note that a digital ratemeter or a digital counter, while often a big improvement over an analog system, does not make this a digital system. It is still subject to the problems of an analog system.

The smoothing of the irregular signal into an analog ratemeter is determined by the time constant, τ, of the ratemeter. The output voltage, $\overline{V}_o$ of the ratemeter (average) is:

$$\overline{V}_o = NQR \tag{5.27}$$

Where:
N = Number of pulses/unit time
Q = Charge of the capacitator
R = Resistance in series

The ratemeter is linear as long as the output voltage, $\overline{V}_o$ remains much less than the pulse amplitude, V:

$$\overline{V}_o << V \tag{5.28}$$

The root mean square value of the output voltage is V_{rms}:

$$V_{rms} = \left[\frac{nQ^2R}{2C}\right]^{1/2} \tag{5.29}$$

Thus, the standard deviation, if N is the counting rate and τ is the time constant, is:

$$\sigma = V/V_{rms} = (2RCN)^{1/2} = (2\tau N)^{1/2} \tag{5.30}$$

This is the same fractional standard deviation that would be given by a counter over a time, T = 2RC. Thus, 2RC is the effective counting period of an analog ratemeter. Increasing RC would reduce the standard deviation and make the recorded curve smoother, but the response would be more sluggish and important detail (resolution) would be lost.

The time constant, τ, puts severe restrictions on the use of an analog ratemeter circuit for logging purposes. The smoothness and, thus, the ease and accuracy of reading a log curve of a radioactivity measurement depend on the value of τ. The smoothness of the recorded curve is a function of the standard deviation; a smoother curve has a lower percentage standard deviation. The response time of the curve going across a bed boundary or through a deflection also depends on the value of τ. Logging speed must be kept slow enough that the bed boundaries, the anomaly itself, and the details within the anomaly are not skewed. Skewing takes place with an analog system only. It occurs in the direction of logging, and it is equal to L, ft:

$$L = \frac{\tau, \text{ sec} \times \text{logging speed, fpm}}{60 \text{ sec/min}} \qquad (5.31)$$

This, of course, places a severe restriction on logging speed. It has resulted in the practice of logging slowly through critical zones and speeding up through noncritical zones. For example, we must log at 3 fpm if we wish a depth accuracy of 1/10th ft with a 2-sec time constant. The normal 20 fpm results in a lag (or upward) displacement of 0.7 ft. The normal recommended 5 fpm speed and 1-sec time constant through an anomaly give about 0.1-ft displacement.

Standard practice with an analog system should be to use as short of a time-constant setting as possible; certainly no more than 2 sec and preferably 1 sec. Limit logging speed to no more than 20 fpm (6 m/min). When nearing an anomaly of any kind, positive or negative, slow down to 3–6 fpm (1–2 m/min). While making the rerun, reduce the time constant if the anomaly is positive or increase it if the anomaly is negative. At 1,000 counts/sec, a τ of 0.1 sec is quite satisfactory. At 10 counts/sec, the σ is $\pm 10\%$ even with a 5-sec τ. Reducing logging speed to less than 5 fpm (1.5 m/min), results in the optimum situation with respect to time, lag, and standard deviation.

Digital Readouts

At the output of the surface pulse amplifier, the pulse signal may be used to operate one or more digital systems. These systems use the analog pulses from downhole and present a digital readout, recording, or data processing. It should be remembered that these mixed systems are still analog systems, at least as far as data gathering and handling are concerned. They are subject to many of the disadvantages and limitations of analog systems. However, each has its definite advantages because of the digital system in use.

Digital Displays

Examples of digital displays in block form are shown in Fig. 5–23. In its simplest form the digital section takes the pulses from downhole and counts them with a pulse-counting system. The pulse counter is usually a commercial timer/scaler or some other form of counting circuit. Usually, the count is for 1 sec at a time, so the readout is in pulses per second or counts per second. This is then displayed on a digital display panel. This display closely resembles the strip chart recorded value, especially if the ratemeter is set at a 1/2-sec time constant.

Note that the 1-sec digital display is not the same thing as the ratemeter reading on a 1-sec time constant. The ratemeter responds 63% to the second just past; 24% to the one before that, 7% to the second one before, 4% to the third one before, 1% to the fourth one before, and so on. The digital counter, on the other hand, examines one separate second only. Everything else is ignored. Thus, the counter reading of the digital readout only corresponds to the reading that contributes 63% to the ratemeter reading. Therefore, the two readings, analog and counter, closely resemble each other but are seldom identical.

There is no lag or skewing of an anomaly introduced by the digital readout. It reads the total number of pulses counted during the second displayed. Therefore, a peak reading of the digital system may or may not coincide with the peak reading of the analog system, depending on the logging speed. Some digital displays take the output of the ratemeter,

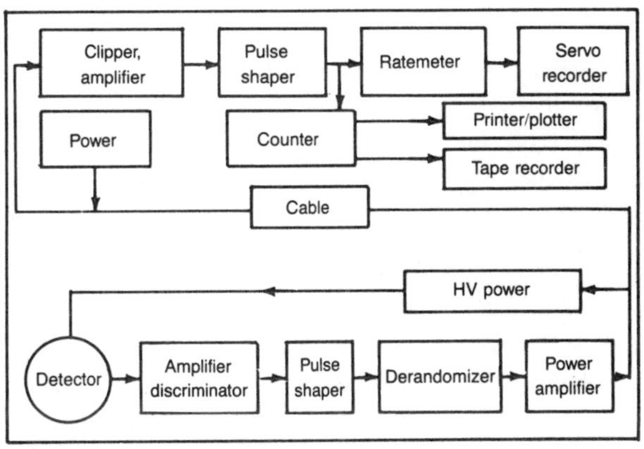

FIG. 5–23
Derandomizer analog with digital recording

digitize it, and process it. This type of system has the extreme disadvantage that all of the faults of the analog system are added to all of the faults of the digital system. If the digital readout or other digital processing takes data from the pulse amplifier and operates in parallel with the ratemeter, each is independent of the faults of the other.

The less expensive counting devices cannot count while displaying a count. All of the data available during display is lost. A better counter has a buffer memory, which allows the system to display and count at the same time. The extra cost is well spent.

A digital printer or a recorder magnetic or perforated tape can record the output of the counter. This machine records the count every second, along with the depth from a depth encoder. The printout and recording also can be triggered by the depth encoder. For uranium and coal logging this latter is the preferred system.

When the counter output is triggered by the depth encoder, some convenient interval, such as each half foot or 10 cm is picked for the readout interval. This allows mineral-grade calculations to be done more easily. With this type of system, most of the data are discarded, even when the counter has a buffer memory. This is because logging a half foot of hole at 5 fpm requires 6 sec. Thus, five out of six readings or 83% of the data are discarded.

Some commercial circuitry incorporates interrogation- or scanning-type digital systems. This allows the printer or other digital recorder to accept the outputs of a number of panels in sequence. These can be depth encoders, other counters, digital ratemeters, analog to digital converters, etc. Thus, the whole logged output may be recorded.

Recording data on a tape allows data playback at any time in the future. This has many advantages, which will be covered in chapter 8.

Microprocessors and Computers

In place of the entire analog and digital recording and display just described, a digital data-processing system can be used. This can be a microprocessor system or a computer system. Furthermore, some of the functions of these systems can be handled on-site, at the base, or both. Digital data processing allows recording of the data for later processing and for further processing. It also allows the presentation of log data in terms directly usable in the field, such as percent uranium, barrels per square foot, percent porosity, and any other useful parameter that can be put into equation or table form. We must ascertain, however, exactly what processing has been done to the data. Therefore, insist that the contractor explains exactly and note in the log heading what processing has been done.

The major differences between the microprocessor approach and the digital computer approach are the amount of memory available and the fact that the microprocessor system is usually dedicated to a single set of procedures. These differences are disappearing quite rapidly.

Digital Data Processing

A digital data-processing system is an extension of the digital readout and recording systems described. These systems use the information in real time or from the recorded tape to process the data to a more usable form. The most common system of this type is a microprocessor that accepts the gross-count gamma-ray counting rate for each half foot and the depth readout. Then it performs the following calculation for the percentage of uranium grade, G:

$$G = KNF \tag{5.32}$$

Where:

K = System conversion factor
N = Dead-time corrected counting rate
F = Borehole correction factor (water factor, casing factor, etc.)

One must know if this is really the relationship used to determine K, if N is actually corrected, what dead time has been used for N, what has been included in F, and how F has been determined.

Many valuable computations may be performed at the site where they are useful. However, one must be aware of the methods used and the capabilities of the system. Also, the full information is often not available for complete, accurate processing. Therefore, these field computations must be considered an interim device only.

It is also possible to use a computer at the output of the cable. The computer accepts the signals from downhole and records them in raw form on magnetic tape or some similar device. This is a permanent recording. Then the computer reads the recorded data and processes it to finished form such as a printout or a log. The amount of processing and the final form of the data are only limited by the capabilities of the computer and the programmer.

These are fine systems, but one must remember two things: (1) These systems are analog to the input of the digital system; and (2) the actual processing done is important and must be known. Fig. 5–24 shows a block diagram of one system.

A calculation of the thickness, T, the grade, G,

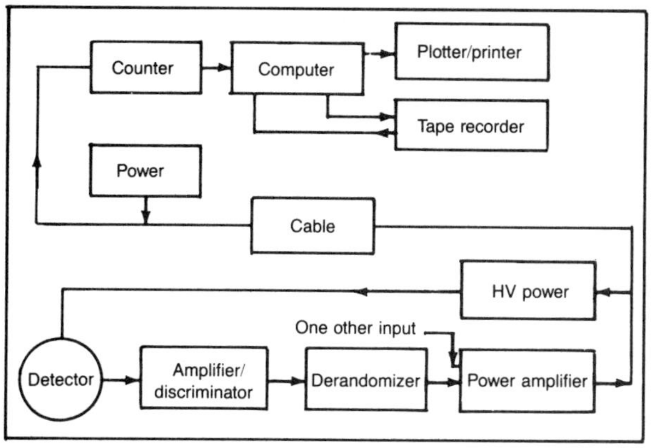

FIG. 5–24

Hybrid analog with digital recording

and the grade-thickness product can be performed by this type of system. The relationship is:

$$GT = KAF$$

Where:

K = Universal K factor
A = Area beneath the anomaly
F = Product of any correction factors needed

It must be remembered that this expression gives values for the average parameters over a whole anomaly. It will be in error if part of an anomaly or a thin anomaly is examined. This type of system can also handle some of the deconvolution-type programs, such as the GAMLOG program or one of the more advanced ones. These have many advantages and should be considered.

Full Digital Systems

Immediately at the output of the sensor or the sensor preamplifier, the signal can be digitized. In some cases sensors themselves have a digital output. The digital signals can be handled by digital circuitry downhole and encoded for transmission to the surface. This has many advantages. Digital signals are of the on-or-off form, with a large difference between the two states. A great amount of signal attenuation and distortion can be tolerated without destroying or modifying the information content of the signal. Thus, distorting components, such as a logging cable, have little or no effect on the signal. Sampling methods and encoded transmissions can increase the information density.

For example, an analog system with a 1 μsec dead time has a practical limit of 30,000–40,000 pulses/sec/cable conductor (four conductors). One digital transmission system in common field use now is using 11 channels, each with a limit of the equivalent of more than 600,000 pulses/sec/cable conductor. Fig. 5–25 shows a block diagram of a full digital system.

Digital systems also use other averaging systems than time averaging in real time. Time averaging in real time is always asymmetrical because we never have exact information, except for what is past. However, with digital systems it is easier not to operate in real time. Thus, besides the data for any point, we have a recording of data that precede it and data that follow the point. Therefore, we have available a whole range of symmetrical processes for averaging and otherwise handling data that do not distort the data and that are not easily available to analog processes.

Finally, digital systems handle data in a form that does not have to be converted for computer handling. Usually, when the data are faulty, it is obvious.

Again, as with the analog transmissions and computer handling, the data-processing capabilities are limited only by the computer and the programmer. These systems are excellent for any type of logging and are much to be preferred. They are free of many of the errors that appear on analog recordings because of operator errors. The information is available in a more usable form. One must know, though, what has been done to the data, and the data must be available in raw form.

Radioactivity Statistics

Radioactive events are completely random. Also, we detect a large number of individual events, whether we are counting gross amount or detecting

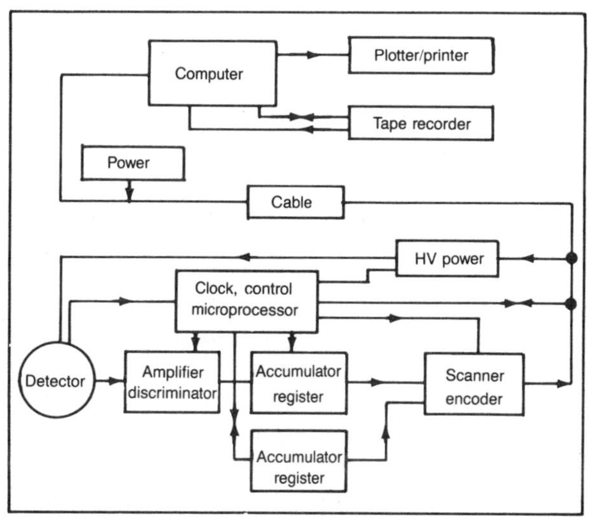

FIG. 5–25

Full digital system

the amplitudes of pulses. Thus, we must handle the data, the data reduction, and the analysis of radioactive logs with statistical mathematics.

Since the occurrence of events is random, we cannot arbitrarily pick a reading, N, and expect it to be the true, unchanging value representative of that steady-state sequence of events. The next one we pick, even though nothing has been changed except time, will be different. We may, however, sample a number of times, s, and take the mean, m. The more samples we take—the larger n is—the more confidence we can have in the value of m. The mean is:

$$m = \left[\sum_{i=o}^{n-1} N_i \right] \div s \qquad (5.33)$$

If we take one particular rate reading, N, it will not be equal to m because of the random variation of the signal. Approximately two-thirds of the time, the value of N falls between the limits.

$$m \pm (2m)^{1/2} \simeq m \pm \sigma \qquad (5.34)$$

Expressed as a percent:

$$m \pm \frac{2m^{1/2} \times 100}{m} \% \simeq m \pm \sigma \% \qquad (5.35)$$

If we are using an analog system, we must take the averaging effect of the time constant, τ, into account:

$$\sigma \simeq (2m\tau)^{1/2} \qquad (5.36)$$

Due to the random nature of the signal pulses, there is always a finite chance that one pulse may occur so close to the preceding one that they cannot be separated by the circuit. The length of time during which a succeeding pulse cannot be detected is called the system dead time, μ. The longer the dead time is, the more likely it will be that we will have such a loss of pulses. The dead time is most reliably determined by the 2-pit or multipit method (see reference for Crew and Berkhoff).

If the real or correct counting rate is N and the resultant or recorded counting rate is n, the corrected counting rate is:

$$N = n/(1 - \mu n) \text{ and } n = N/(1 + \mu N) \qquad (5.37)$$

It is evident from these relations that it is desirable to have as high a counting rate as possible to reduce the percent variance. Also, it is desirable to have μ as short as possible in order to make as small of a correction as possible. The dead-time correction is a probability correction. That is, within the limits the probability of N being correct is greater than it being incorrect. The upper limit on the expression is about 50% correction.

GAMMA-RAY LOGGING

Gamma-ray logging measures natural gamma ray and X-ray emissions. It includes several types of logging that detect, record, and otherwise use the natural gamma radiation from the formation rocks. These include gross-count gamma ray, total count, gamma-ray spectroscopy, KUT systems, and surface and airborne systems. All natural gamma-ray systems are passive systems.

Gamma-ray logging is important to nonpetroleum logging. It was the first and, for a long time the only, quantitative logging system used in nonpetroleum logging.

Gamma-Ray Systems

Gamma ray measurements reflect the following:

- The total radioactive, gamma-ray emitting content of the formation material
- The shale or clay content of the formation material
- The presence and amounts of any anomalous gamma-ray emitting materials in the formation

Gamma-ray measurements are influenced by and must be corrected for the following:

- Borehole diameter
- Borehole fluid
- Thin beds
- Formation density
- Casing
- Position within the borehole
- Crystal size
- Radioactive material in the borehole
- High alpha, beta, or neutron emission

The total or gross-count gamma-ray systems count all of the gamma and X radiation in the formation available to the detector. It was first used as a cased-hole substitute for the SP curve in the petroleum industry. It gives a sand-shale log that is easily correlated with the SP log because the shales have a much higher radioactivity than the sands. Shale content is also estimated with it. The penetrating action of the high-energy natural gammas allows them to penetrate casing materials easily where an SP log cannot be made.

Shale gamma radiation was originally thought to result from the activity of potassium-40. However, this only accounts for about 20% of the emission from a shale. The rest comes from uranium and thorium daughters in the shale.

Nonpetroleum logging uses the gross-count gamma ray for correlation and for determining shale content. In coal logging it is used to estimate the ash content. In the uranium business it is used to calculate the equivalent uranium content for mill design. In the potash industry it is used to determine amounts of potassium mineral in place.

Gamma-ray spectrographic measurements can be tremendously useful. They allow a determination of the origin material of the gamma rays. They measure narrow, specific bands of gamma energies quantitatively. This allows the determination of the mineral composition and the depositional history.

Spectrographic measurements were first made by surface methods. Then airborne methods were used because of the easier restrictions on equipment size. Now, however, especially with on-site computers, various gamma-ray spectrographic methods are beginning to be used downhole. Hand-held instruments are common for monitoring and exploration purposes. These systems include single-channel systems, KUT systems, multichannel analyzer systems, and the PGT system.

All gamma-ray systems lend themselves to quantitative use. The uranium and potash industries have used the calibrated gross-count systems to determine mill size for many years. Nonpetroleum calibrations result in an accuracy that can be better than $\pm 1\%$. Petroleum calibration is more of a standardization than a calibration. It results in being able to compare any logs, regardless of the equipment or contractor.

Origin of Gamma Rays

Natural gamma rays in the earth originate primarily from three sources: potassium-40 ($^{40}_{19}K$), uranium daughter products, and thorium daughters. Many of the common natural minerals are potassium compounds. Sylvite and potash contain substantial amounts. Most clays and shales contain appreciable amounts of potassium compounds, since they are the weathering and alteration products of potassium minerals such as feldspar and mica. About 20% of the gamma-ray emissions from shale are from potassium-40.

Many sands contain potassium compounds. On the average sands contain about 12% feldspar. Orthoquartzite sands contain about 10% feldspar, and the orthoclase or potash bearing varieties are common. Some sands may contain as much as 50% feldspar, and arkosic sandstones may contain 25% even though they are nonshaly. Graywacke sands are generally shaly and may contain more than one-third alterable materials that may contain potassium (Fig. 5–26). Both sands and shales contain trace amounts of uranium and thorium and their daughters.

Shales contain a total of about 5 μg radium equivalent/ton of shale. This amounts to an equivalent of about 0.0025% eU_3O_8 or about 16 g of uranium/metric ton.

In igneous environments the granites and extrusive rocks, such as rhyolitic volcanic flows and tuffaceous ash, are high in sodium and potassium. These rocks may also have the greatest uranium and thorium concentrations. They are the source rocks for many sands and shales. In such environments varying amounts of gamma radiation originate from the K^{40} and from the daughters of uranium and thorium.

When they are oxidized, uranium compounds usually become soluble in water. Thus, trace amounts are carried in ground waters to be precipitated by reducing materials, such as carbonaceous material or sulfide minerals or gas. Sea water contains about 3 ppb of dissolved uranium. This precipitates in the black, carbonaceous shales or may be absorbed by marine phosphorites.

Thus, most formations have some low degree of gamma emission. The relative amounts vary in a typical manner, allowing a separation by lithology on a gamma-ray log. In general, marine shales have the highest emission, aside from anomalously mineralized zones. Continental shales average about twice the emission as sands. Carbonates are generally low, and coals are extremely low in gamma-ray output. Fig. 5–27 shows the gamma emissions from some of the common minerals.

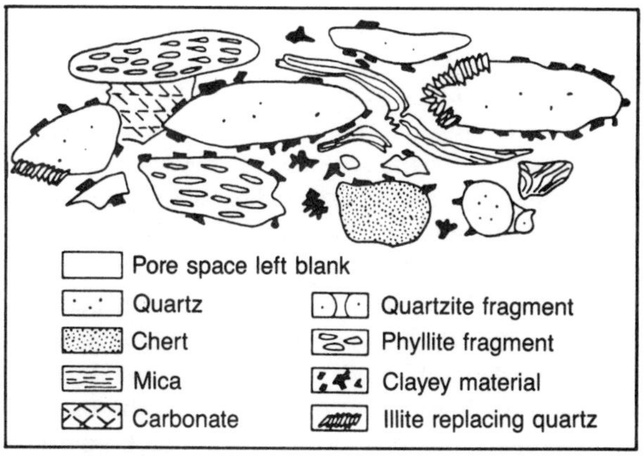

FIG. 5–26

Graywacke-type reservoir rock, Bradford sand, Pennsylvania

(after Krynine, "Sediments and the Search for Oil," *Mineral Industries*, volume 13, No. 3, 1943)

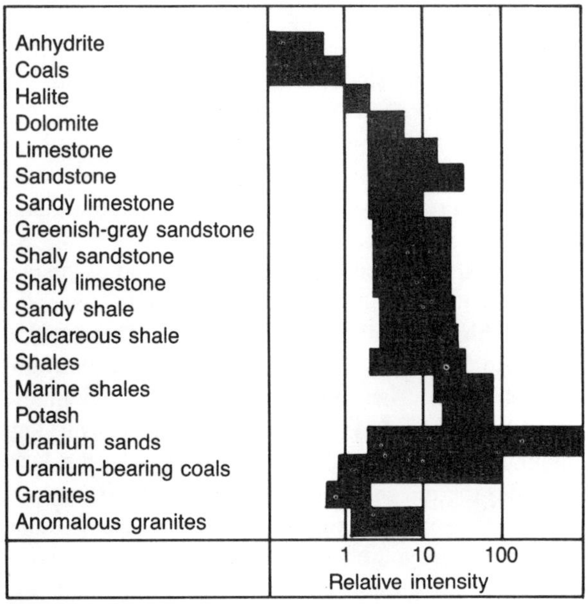

FIG. 5–27

Relative natural gamma-ray intensities

(after Russell, "Well Logging by Radioactivity, *Bulletin,* AAPG, volume 5, No. 9, 1941)

Table 5–4 shows the effects of formation potassium content on gamma-ray response. About 0.012% of all natural potassium is radioactive (i.e., K-40) giving off a 1.5 MeV gamma ray on disintegration. Approximately 20% of the gamma-ray emissions from shale are caused by the isotope K-40. The remaining radiation from shales is generally caused by uranium and thorium series elements.

Gross-Count Gamma-Ray Systems

Gross-count gamma-ray systems (GCGR) read all of the gamma-ray emitting events that can trigger the gamma detector. No attempt is made to discriminate against any legitimate pulse. The GCGR is used in petroleum and engineering logging as a cased-hole substitute for the spontaneous potential curve, a correlation curve, and as an indicator of sand/shale. Often, it distinguishes between continental and marine shales. In nonpetroleum logging it is used for all of the above calculations. In addition, it can be calibrated so it can be used to calculate radioactive mineral grade for uranium and potash. In coal logging it is often definitive for coal, and it can be used to estimate the coal-ash content. In hard-rock environments it is an excellent stratigraphic, lithographic curve.

The greatest use of the GCGR curve is probably as a correlation curve. Hole conditions—casing, hole size, and hole fluid—have only a minor effect on the curve and can usually be corrected for this effect. Thus, it can be used for correlation in sand/shale sequences over wide areas. It is especially good when used with a resistivity curve but is used as a control curve with most other types.

Table 5–4

Potassium Content of Various Formation Materials

Material	Potassium Content, wt% Average	Range
Sylvite	54	
Potash	44.9	
Langbeinite	20	
Microcline	16	
Kainite	15.1	
Carnallite	14.1	
Orthoclase	14	
Polyhalite	12.9	
Muscovite	9.8	
Biotite	8.7	
Illite	5.2	3.51–8.31
Arkose (sandstone)	4.6	4.4–5.1
Synite	4.53	
Glauconite	4.5	3.2–5.8
Granite	4.0	2.0–6.0
Norite	3.3	
Granodiorite	2.90	
Shale	2.7	1.6–9.0
Igneous rock	2.6	
Graywacke (sandstone)	1.8	1.2–2.1
Diorite	1.66	
Basalt	1.3	
Sandstone	1.1	0–5.1
Gabbro	0.87	
Diabase	0.75	
Kaolinite	0.63	0–1.49
Limestone	0.27	0–0.71
Montmorillonite	0.22	0–0.60
Orthoquartzite (sandstone)	0.08	0–0.12
Dolomite	0.07	0.03–0.1
Dunite	0.04	
Sea water	0.035	

Detector

Most GCGR systems use a sodium iodide, thallium-activated scintillation detector. The size of the NaI crystal may be about 0.5 in. long by 0.5 in. diameter in analog tools designed for high mineral concentrations to about 8.0 in. long by 3.0 in. diameter in oil-field stratigraphy/lithology probes. No attempt is made to absorb the gamma ray completely within the crystal nor to evaluate pulse amplitude (gamma-ray energy). Maximum and minimum counting rates are the major considerations.

In analog equipment the maximum counting-rate capability is usually low. Maximum rates are usually 30,000–90,000 cp with good analog equipment. Therefore, detectors are generally smaller than in full digital systems. A typical mineral analog ore probe has a crystal 0.5 in. long by 0.75 in. diameter. A lithology tool has a crystal 2.0 in. long by 0.75 in. diameter. A compromise tool has a crystal about 1.0 in. long by 0.75 in. diameter.

Digital mineral logging systems are capable of handling substantially higher counting rates than analog systems. Thus, mineral probes have crystals of 2.0 in. long by 1.0 in. diameter to 4.0 in. long by 1.25 in. diameter.

Oil-field equipment is designed for use entirely in low-radiation environments. Also, tool diameters are larger. Therefore, crystal sizes of 8.0 in. by 3.0 in. diameter are common. Dead times are longer than with mineral equipment.

The counting rate of a GCGR system varies approximately as the longitudinal cross section of the crystal increases. In reality, the variation is slightly greater. A 1.5-in. diameter crystal has approximately a 50% greater counting rate than a 1.0 in. diameter crystal for the same length. Detectors using bismuth germanate (BGO) crystals promise to increase the counting rate over the same size NaI (T1) detector. This is because the BGO has a greater density and does not need a protective can. BGO crystals are also less fragile mechanically.

Format

The GCGR log heading should include the detector type and size, the conversion or K factor if it is a mineral log, the relation of counting rate to API units if it is scaled in API, the system dead time, the date of the last calibration, the reading of the truck test source at calibration, the tool serial number, the test source number, the logging speed, the time constant (analog) or time base (digital), and the averaging factor (digital). Some analog equipment uses a constant probable error system in place of the time constant. This should be indicated. No information should be omitted.

The GCGR log is usually presented with the curve at the left side of the chart with counting rate or API units increasing to the right. If it is different, it should be well marked. The body of the log should have the scales marked unambiguously. The term full scale can have several meanings on a multitrack log, for example. Also, the practice of reducing log sizes for correlation sometimes makes such labels as 100 counts/sec/in. confusing, unless an inch is marked on the log. All reruns or offscale traces should be marked clearly as to scale as well as time constant for analog. All reruns on analog equipment must have enough overlap with the regular scale that the marked scale can be verified. Any scale or time-constant changes or shifts or changes of logging speed should be marked on the body of the log. The zero position and the truck calibrator reading should be recorded at the bottom of the log and should be clearly and completely labeled.

Uses

The GCGR responds to the differing amounts of radioactive, gamma-ray emitting materials in sands and shales or any other material in the formation. The continental shales have approximately twice the radioactivity that sands have. Marine shales have about 2.5 times the radiation of a sand. This amount of contrast allows the GCGR curve to be used effectively in sand/shale sequences for lithology and correlation, and it led to the initial use of the gamma-ray curve as a cased-hole substitute for the SP curve.

One should be familiar with typical counting rate responses to various lithologies of the detector size and type. This allows the use of the GCGR for lithology determinations. Often, shales have surprisingly uniform, indentifying emissions over wide areas. This can be used to identify and map members. Knowing typical counting rates is also valuable, especially with analog systems in the event of instrument problems.

The GCGR curve can also be effectively used to estimate the percentage by volume of shale, V_{sh}, in a shaly sand or a sandy shale (Fig. 5–28). The method

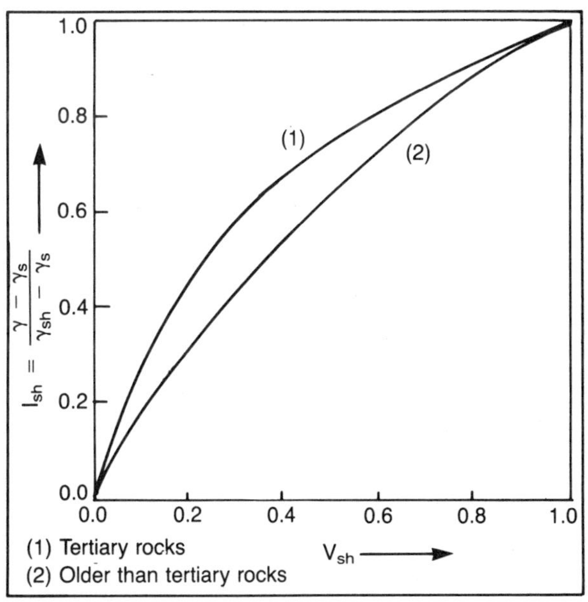

FIG. 5–28

Gamma-ray reading vs. shale index (coal ash determination)

(courtesy Dresser Atlas Division, Dresser Industries)

compares the gamma-ray reading in a shaly sand or sandy shale (γ) with the readings in a shale (γ_{sh}) and in a sand (γ_s):

$$V_{sh} = f(I_{sh}) = \frac{\gamma - \gamma_s}{\gamma_{sh} - \gamma_s} \tag{5.38}$$

The estimation of the ash content of coal uses a similar method. The assumption is made that the ash in the coal is or closely resembles the nearby shales. The reading in a pure shale immediately above or below a coal is assumed to indicate an index, I_{sh}, of 1.0. This is equivalent to a pure ash or 100 wt% ash. Initially, since it is difficult to determine the other end—the reading of an ash-free coal—it may be assumed that the coal contributes little or no radiation. Therefore, a zero or small gamma-ray deflection (5–10 API) is assumed to be 0% ash and, thus, an index of zero. A plot as shown in Fig. 5–29 can then be drawn to estimate the ash content in any portion of the hole. When core samples are analyzed, the actual slope and intercept of the curve can be corrected. The relationship is of the form:

$$A = a + b\gamma$$

Where:

- a = Gamma-ray reading in an ash-free coal on the y intercept
- γ = Gamma-ray reading at any point in the coal away from the edges
- b = Slope of the curve equal to ($V_{sh} - V_s$).

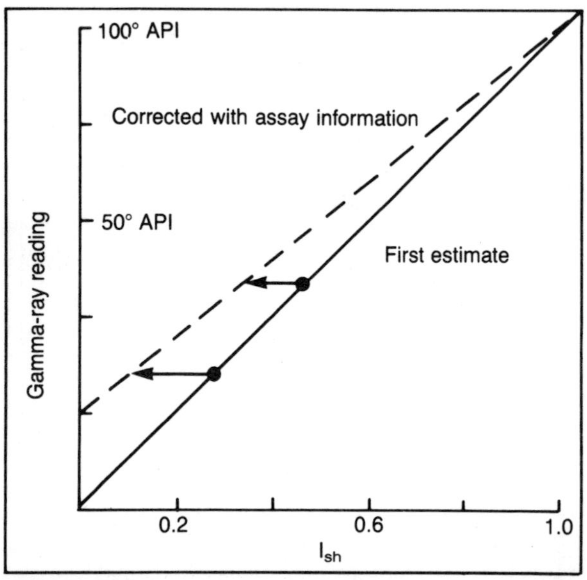

FIG. 5–29

Shale index vs. shale volume or ash volume

Since both the gamma-ray reading and the resistivity readings are functions of the clay content of a formation material, a crossplot of the two is informative even though we may not be able to formulate the relationship. If Fig. 5–28 is examined, the relationship between V_{sh} and I is of the form:

$$V_{sh}\ a + I^b$$

It is seen that the relationship is nonlinear. Therefore, when the gamma-ray readings are plotted, they have a nonlinear pattern.

Similarly, the relationship between the resistivity and the shale content is of the form:

$$\frac{1}{R} = \frac{V_{sh}}{R_{sh}} + \frac{\phi}{R_w} \tag{5.39}$$

The result is that the crossplot pattern has a rather sharp break in it at approximately the point where a shaly sand changes to a sandy shale.

If the zones within 2.5–3.0 ft (0.7–0.9 m) of the bed boundary or coal boundary are eliminated, this method gives a reasonably accurate estimate of the ash content of a coal. Bed boundaries give a particular problem with the GCGR curve because it has a relatively large volume of investigation. A detector, such as a gamma-ray detector, receives signals from a volume of material. In the case of gamma rays, the volume depends on the energy of radiation and the density of the formation material. Note that the volume does not depend on the instrument, only on the formation. The contribution of any point volume within the material contributing depends on the energy of radiation, the density of the material, and the distance from the detector. The intensity at the detector from any point volume of material is:

$$I = I_o e^{-\mu \rho x}$$

Where:

- I_o = Original intensity
- μ = Mass absorption coefficient
- ρ = Bulk density
- x = Distance from the detector

The result is that the detector sees a radioactive bed while it is still in the barren zone, before it arrives at the bed boundary (Fig. 5–30). When the detector is at the bed boundary, the sensitive volume is filled half by radioactive bed and half by barren formation. Therefore, if the beds are thick, the deflection of the GCGR is half of the maximum. The GCGR curve does not read the maximum or true value until the sensitive volume is completely within the radioactive bed. The distance of uncertainty, edge effect,

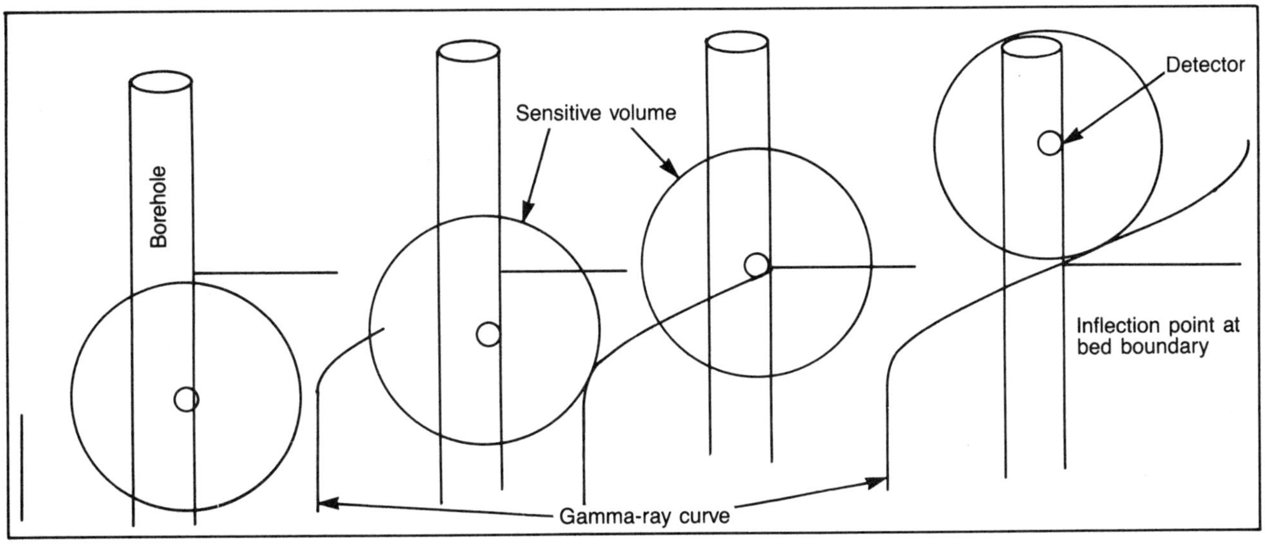

FIG. 5–30

Gamma ray—crossing a bed boundary

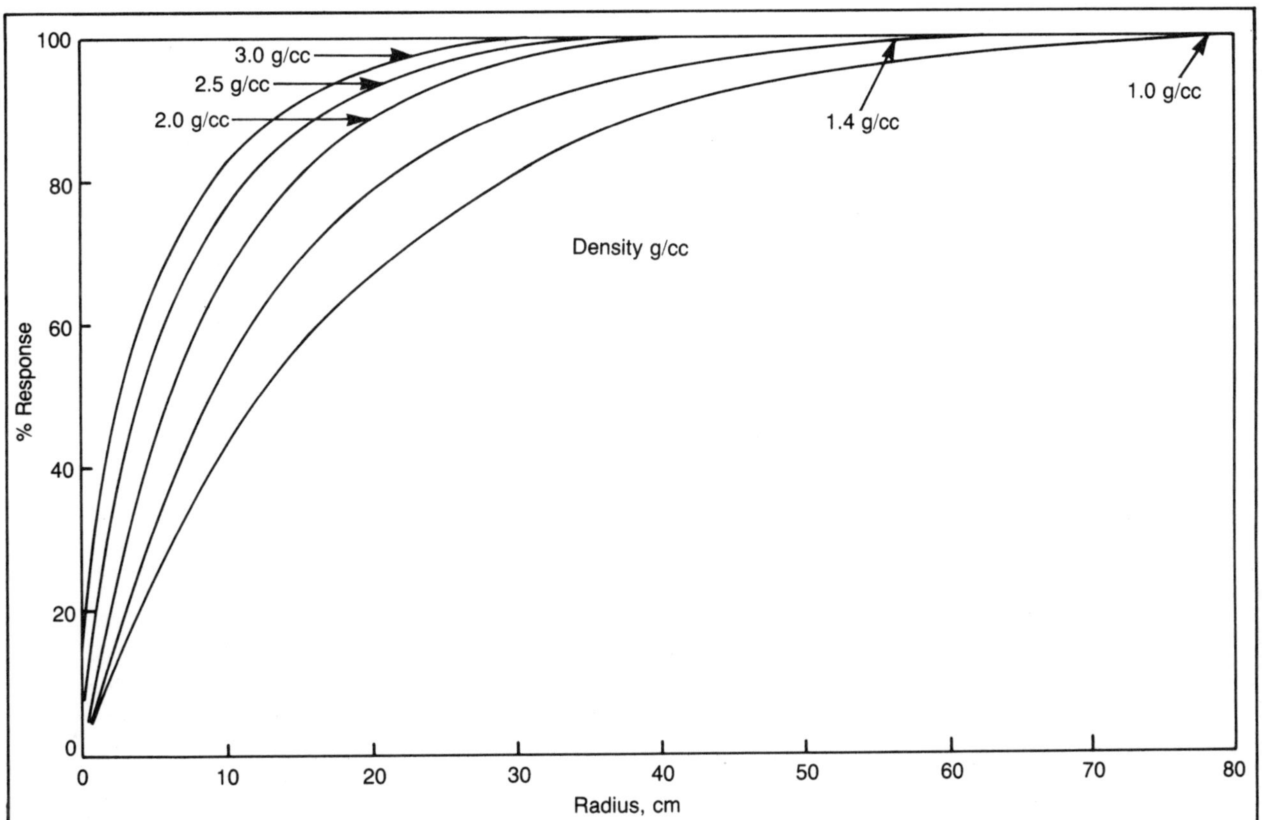

FIG. 5–31

Percentage of gamma-ray response as a function of distance into the formation

or tail is about 19 in. (0.5 m) each side of the boundary in a sand/shale contact. In a homogeneous medium with a density about that of a typical sand and shale, the sensitive volume is a sphere, neglecting the borehole and tool body, with a diameter of about 1 m. In higher densities the diameter decreases. A dolomite results in about a 2-ft diameter sphere. A coal or lignite may give a 5–6-ft diameter sphere. Fig. 5–31 shows the relationship between density and volume (diameter) of investigation.

When the gamma-ray characteristics are examined, the portion of the curve near the boundary must be eliminated or a correction must be made. Fig. 5–32 can be used to correct thin-bed response of the gamma-ray system. The gamma-ray values can be effectively crossplotted with the resistivity, neutron, or acoustic curves to determine clay mineral characteristics.

Hole-Size Factor

The borehole usually contributes a negligible amount of radiation to the GCGR system. However, it is a major portion of the measured volume, and it is close to the detector where it has the greatest influence. Therefore, a correction must be made for the reduction of counting rate caused by the inclusion of the borehole in the measured volume. This correction factor, F_w, is variously called the hole-size factor, the water factor, or the mud factor.

Tools such as the gamma-ray tool are expected to be eccentered in the borehole. However, it may be anywhere from centered to eccentered. If the probe is centered, it is usually accomplished with some device such as bowsprings or flexible fingers. These are designed to hold the tool in the center of the hole. When the hole is the same size as the probe, the borehole has no effect on the measurement and the factor

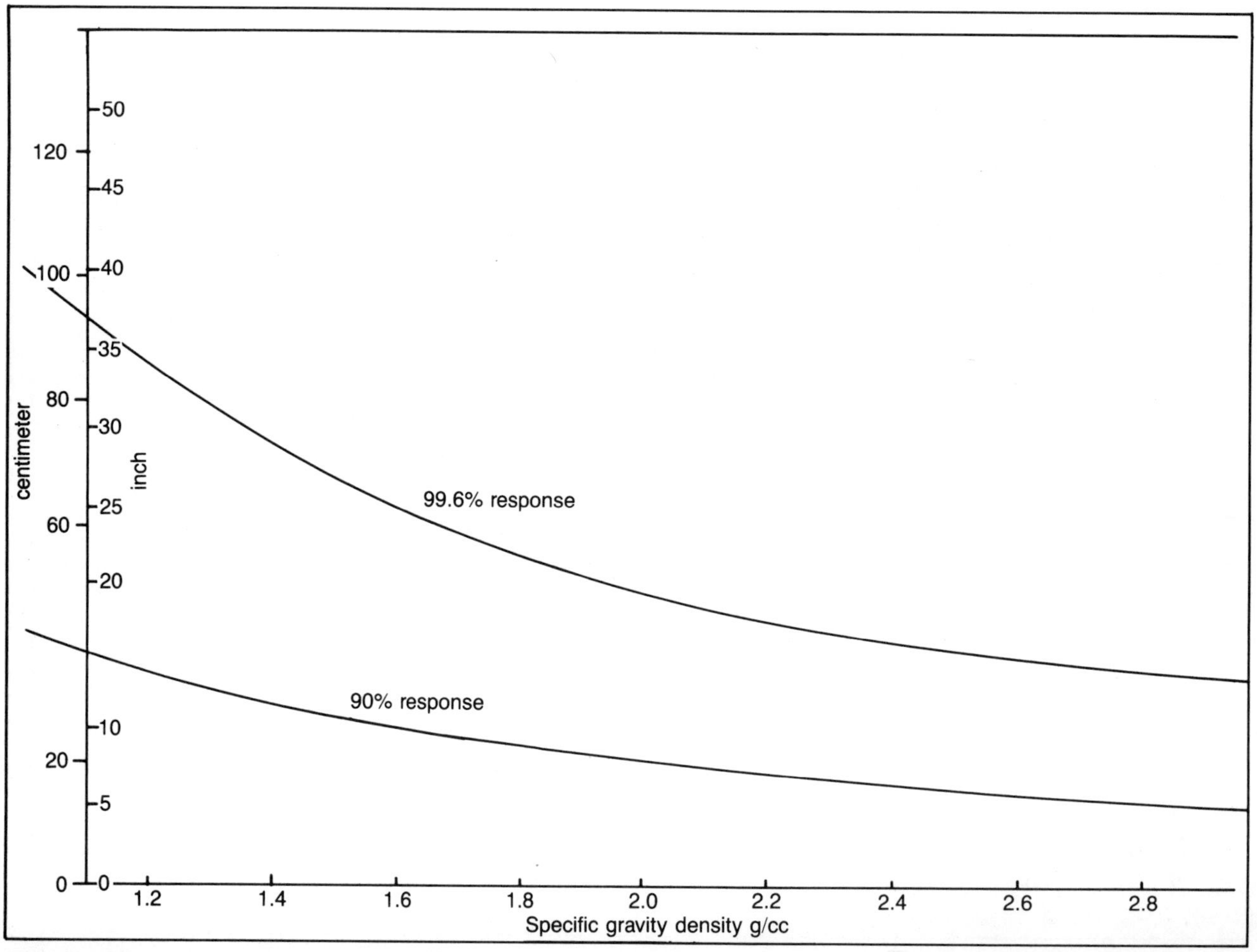

FIG. 5–32

Radius of investigation of the natural gamma-ray system as a function of formation specific gravity or bulk density

F_w is equal to 1.0. As the borehole becomes larger, the probe becomes farther from the formation, until in a very large hole it receives a negligible amount of radiation. Thus, the value of F_w for a centered situation varies from 1.0 to infinity as a function of borehole size. Fig. 5–33 shows a typical centered-tool response curve for a gamma-ray tool.

If the probe is against the wall of the hole—if it is eccentered—again the value of F_w is 1.0 if the hole is the same size as the tool. If the hole is extremely large, however, the shape of the wall approaches that of a flat surface. With the probe against the flat surface, approximately half of the measured volume is filled with borehole fluid and half with radioactive formation. Therefore, the value of F_w approaches approximately 2 as the hole diameter increases (F_w is asymptotic to 2). Fig. 5–34 is a typical hole-factor curve for an eccentered probe. Fig. 5–35 illustrates the two situations.

The error can be appreciable if it is assumed that the gamma tool is in one position when actually it is in the other. It becomes more serious as the hole size increases. In a 6-in. hole, for example, if it is assumed the probe is eccentered when it is actually centered, the error can be −14% of the correct reading. Fig. 5–36 shows curves evaluating the positional errors in the borehole.

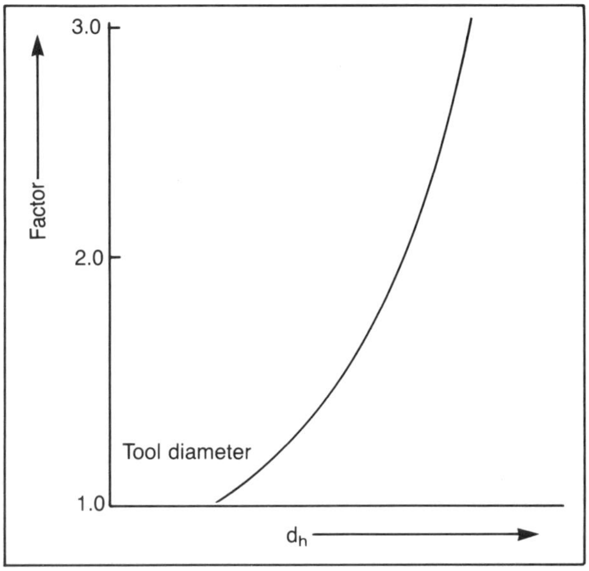

FIG. 5–33

Hole correction factor for centered tool

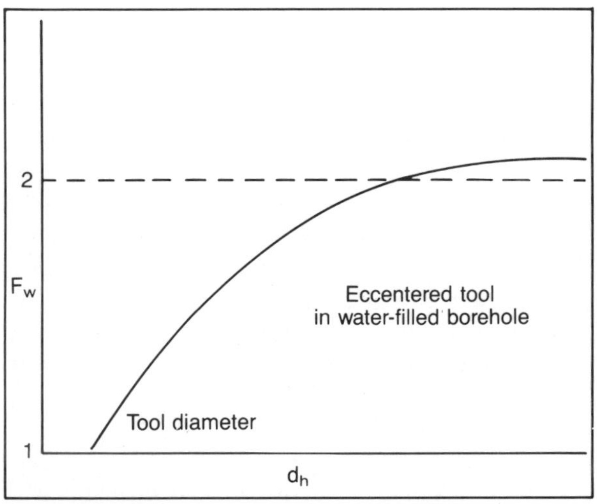

FIG. 5–34

Hole correction factor for an eccentered tool

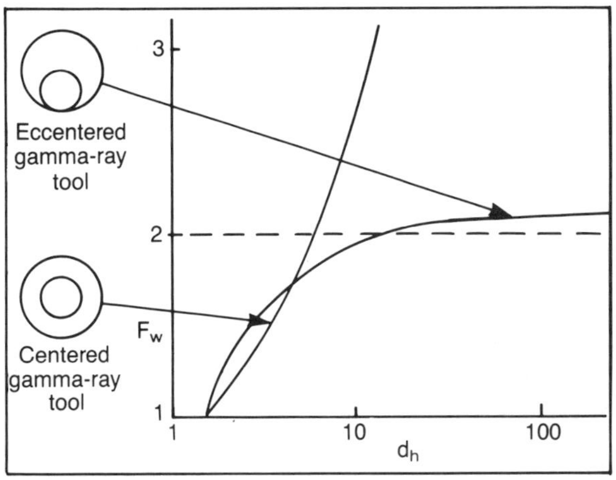

FIG. 5–35

Comparison of hole correction factors

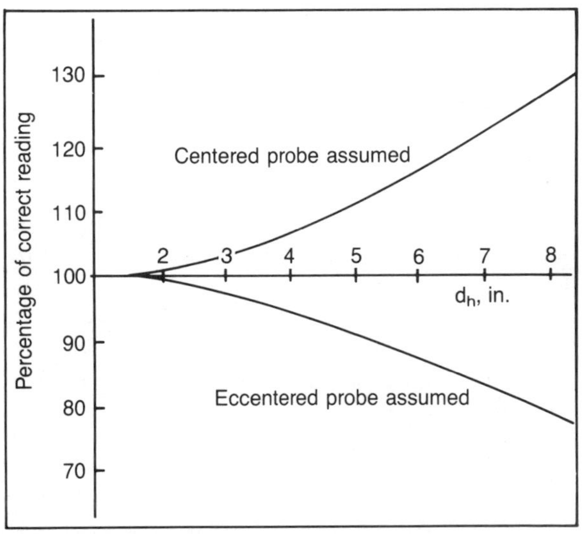

FIG. 5–36

Error incurred with the wrong assumption

There are two contributing systems to the borehole factor. As the borehole gets larger, the gamma rays must traverse more absorbing material to reach the detector. This decreases the amount of radiation reaching the detector for a given amount of radioactive material. The attenuation is an exponential function of hole diameter. If water or mud fills the hole, the effect is more severe than with a gas-filled hole.

Because the attenuation is low in gas and the radiation surrounds the detector, for ordinary hole sizes the distance attenuation in a gas-filled hole is negligible. On the other hand, even though a volume of formation a meter in diameter contributes to the reading, 50% of the reading originates from the first 9 cm (3.5 in.) of formation around the borehole. As the borehole becomes larger, more material is exposed to the detector. A 16.0-in. hole radiates twice as much into the borehole as a 5.0-in. hole. This tends to make the borehole factor increase with hole size. The net result is that the factor F_w slowly decreases with hole size in gas-filled holes. In water-filled holes it increases. A typical response curve set is shown in Fig. 5–37.

The determination of F_w is made in a block of concrete that has uranium ore distributed uniformly through it. It has four holes of different diameters in it (Fig. 5–38). The probe is put into each hole, and the response is noted. Then the hole is filled with water, and the response is noted again. The ratio of the air-filled to water-filled response is the hole-size factor for that hole. A measurement in each hole defines a response curve for the normal range of hole-size factors.

Whether the hole fluid is water or mud is usually ignored since a 10% change in the hole factor (i.e., from 1.14 to 1.25) would require a 300% change in mud density in a 4-in. hole for an eccentered sonde.

Casing Factor

One of the values of the GCGR system is that it can detect formation parameters through casing. With steel or aluminum casing, the attenuation caused by the metal in the path of the gamma rays must be corrected. This casing factor, F_c, can be large or small.

The value of F_c for any type of casing approaches 1.0 as the casing approaches a negligible thickness. The factor is equal to:

$$F_c = e^{\mu \rho t}$$

Where:

μ = Linear absorption coefficient for the average energy of the gamma rays, about 0.065

ρ = Density

t = Thickness of the casing

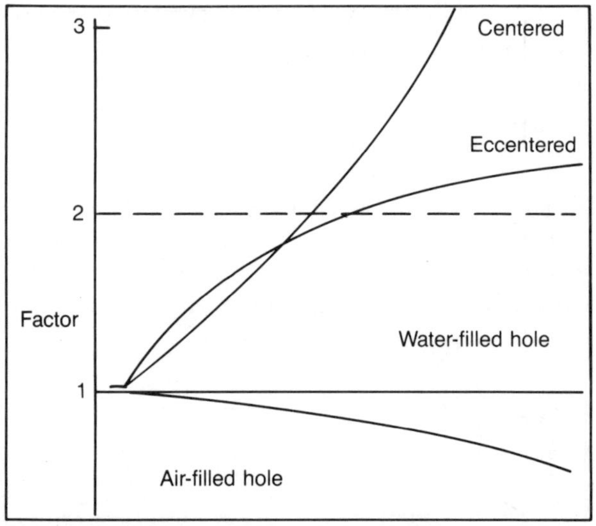

FIG. 5–37

Gamma-ray borehole correction factors

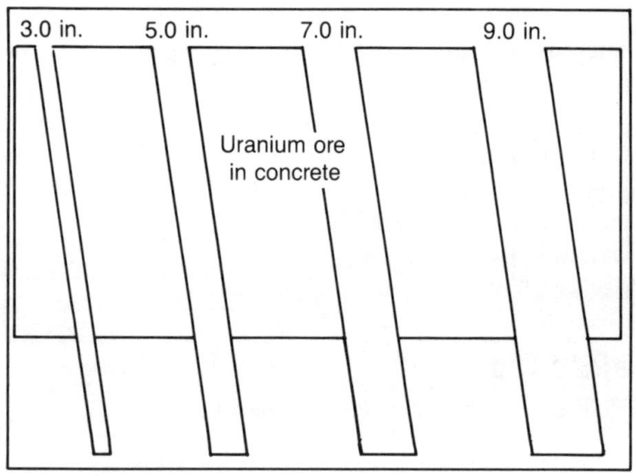

FIG. 5–38

Hole size model

Casing may be steel, aluminum, polyvinylchloride (PVC), fiberglass epoxy, fiberglass polyester, polypropylene (ABS), or polystyrene (plexiglass). Steel has a density of about 8.7 g/cc. Aluminum has a density of about 2.8 g/cc for commercial pipe. Fiberglass has a density of about 1.5 g/cc. The others have densities close to 1.0 g/cc. F_c values for different casing thicknesses and types are given in Fig. 5–39.

There are several things to keep in mind about casing corrections. First, drillpipe and drill-collar corrections are the same as casing corrections. Second, casing corrections should be applied in addition to borehole-size corrections. Third, drilling crews often do not know the thickness of drillpipe and drill collars. They are concerned with weight, not wall thickness.

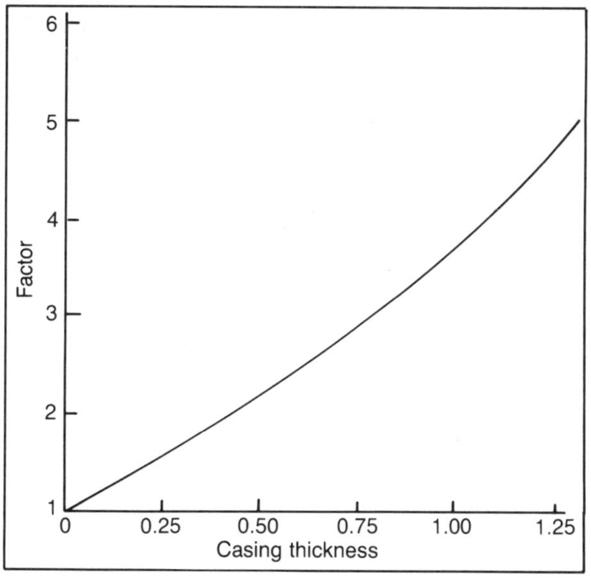

FIG. 5–39

Casing thickness factors for steel

They usually know the weight, so they use a casing table to obtain thickness. Fourth, oil-field casing is usually centered in the borehole by external centralizers. Therefore, the distance from the wall of the hole can be calculated. Nonpetroleum casing, on the other hand, is seldom centered. Furthermore, the weight makes the casing corkscrew and accordion in the hole. One literally does not know where the probe is within the borehole of a cased hole. This can cause serious errors in correction. However, it does not mean that the cased-hole logs should not be used. The larger uncertainty of the determinations should be recognized and noted in the calculations.

Grade Calculation

Since the GCGR curve is a function of the amount of radioactive material in the formation within the sensitive volume of the tool, the response can be used to calculate the equivalent grade of the mineral, provided a suitable and careful calibration has been made. The amount of radiation from a formation traversed by a borehole and, thus, the amount detected in the borehole, is a function of the amount of radioactive material in the formation. Thus, the mineral grade, G, is proportional to the area under the GCGR curve:

$$GT = KAF$$

Where:

T = Bed thickness
K = Constant of proportionality
A = Area under the curve
F = Product of the several correction factors (borehole, casing, etc.)

This method can be used with uranium, potassium, or any radioactive mineral with the appropriate value of K.

To determine the area under the anomalous deflection of the gamma-ray curve, first determine the bed boundaries and the bed thickness. The bed boundaries always are at the inflection points of the gamma-ray or any other symmetrical curve. The distance between inflection points is the bed thickness, T. Start 1.5 ft (0.5 m) before the bed boundary and divide the anomaly into equal 1/2-ft (10-cm) intervals to at least 1.5 ft (0.5 m) past the other bed boundary. At the center point of each interval, read the counting rate and record it in tabular form. This is, in effect, reading the average counting rate of each interval. Correct each counting rate for probable dead-time effects. The relationship is:

$$N = n/(1 - \mu n)$$

Where:

N = Corrected counting rate
n = Recorded counting rate
μ = Dead time, μsec

This equation must be applied to each value independently. The sum of the corrected values multiplied by the interval thickness (i.e., 0.5 ft or 0.10 m) is the value of the area, A. This is an areal estimation method that has been used in engineering for more than 100 years. It is closely related to integration in calculus.

Another method that is popular for calculating uranium grade is the GAMLOG program. Basically, the program determines the response caused by each individual half-foot layer of mineral-bearing formation. The area under the gamma-ray curve is the sum of the individual responses. The grade of each half-foot interval required to match the original curve is taken to be the average mineral grade for that interval.

Czubek has suggested several methods for more accurate calculation of mineral-grade determination and detection of uranium and other minerals. These methods have not all been developed to the point of general use. One method of Czubek deserves more than casual mention. Czubek has suggested that photoelectric absorption of natural gamma rays takes place primarily within the uranium content of a formation because of the high atomic number of uranium. Gamma rays are emitted by the daughters of uranium at high energy (>1 MeV) and are degraded by Compton scattering as they proceed to the detector. If the formation contains uranium compounds, fewer soft (low-energy) gamma rays arrive at the detector because of photoelectric absorption by the uranium. Therefore, a comparison of the low- and high-energy content of the gamma spectrum in a uranium anomaly is a measure of the uranium content of the formation. This method is usually done with a GCGR system that is run with and without a gamma-ray filter to control the low-energy detection. The method works reasonably well once a given environment is evaluated.

Czubek's filtering method has a further advantage. The presence of the uranium within the formation causes degradation of gamma rays and a resulting filtering of the low-energy gammas by the housing of the tool and detector. This introduces a nonlinearity in the relationship between uranium content and counting rate from a formation in equilibrium. This is called the Z effect. If a graded filter is used to eliminate the low-energy radiation, only the high or primary radiation is detected. The correspondence is then nearly linear. This method should be used more than it is.

A final method for determining equivalent uranium content has been proposed by Jonas. The Jonas method uses Fourier analysis in which a curve, such as a gamma-ray anomaly, is broken into its component sinusoidal parts (deconvolved) for easier evaluation of area. The system is applicable to digital processing and should be seriously studied.

When the grade is calculated correctly and is averaged for the whole interval when the interval is thick, there is a negligible difference in the values by any of the methods. Thus, one should use the simplest. The methods of enhancement must, however, be used when thin zones, less than 1 m thick, are being examined. These methods are equally valid with the potassium minerals. One should use only a different conversion constant.

■ Gamma-Ray Spectrographs

In contrast to the gross-count gamma-ray systems that detect as broad an energy spectrum as possible, there is a family of gamma-ray measurement systems that separates the various gamma-ray energies into distinct energy channels. These systems detect narrow bands of energy. The methods range from the mechanical filtering system described previously to sophisticated multichannel systems. Their purpose is invariably to extract more information than is possible by GCGR detection alone. They further lead to activation systems that are covered in later sections.

When an atomic nucleus is in an excited state (radioactivity), there is a finite chance that it will drop to a normal or ground state by the emission of gamma radiation. Excitation may occur in several ways, but the gamma radiation emitted is characteristic of the excited state and not dependent on the method of excitation. Analysis of detected gamma-ray energies furnishes information about the types of excited states or the energy levels that characterize and, thus, identify the emitting nuclei.

Single-Channel Methods

When a pulse-height discriminator is used with a scintillation or a solid-state detector, we are using energy discrimination. The pulse height or amplitude of such detectors is a function of the energy absorbed in the detector. In a GCGR system the discriminator level is set to eliminate noise only (if possible). However, it can also be used to eliminate pulses below a chosen amplitude. The signal that is passed is the remaining pulses or those from gamma rays of energy higher than the cutoff determined by the discriminator.

A gamma-ray system could be set with a discriminator level at about 1.5 MeV. Most of the radiation would be from bismuth-214, which is a daughter of uranium-238. Thus, assuming the sample is in equilibrium, the resulting curve would be a measure of the uranium content of the formation, independent of the potassium and thorium content. This method can be improved by setting the discriminator of a second system just above the range desired and subtracting its output from the first. There are a number of different circuits that will accomplish this, which results in a window of gamma-ray energies that can be set to center on the desired energy. This has the advantage of eliminating both the unwanted higher and lower energy contributions. The window-type system can be set to center on any peak energy anywhere in the spectrum. Thus, a log can be made of the bismuth-214, potassium-40, or the thallium-208.

Three of the single-channel systems, each on one of the above windows, operating in parallel constitute one form of the so called KUT system (KUT is an abbreviation of potassium [Kalium], uranium, and thorium). Usually a KUT system incorporates a total-count reading.

Many of the single-window systems have controls by means of which the lower threshold and window width or upper threshold are set. If the window position is motor controlled, the entire spectrum can be scanned and recorded for a given sample. The recording is a spectrum (as in Fig. 5–40) that includes the characteristic peaks superimposed on the Compton scattering base or continuum. The resolution depends on the window width.

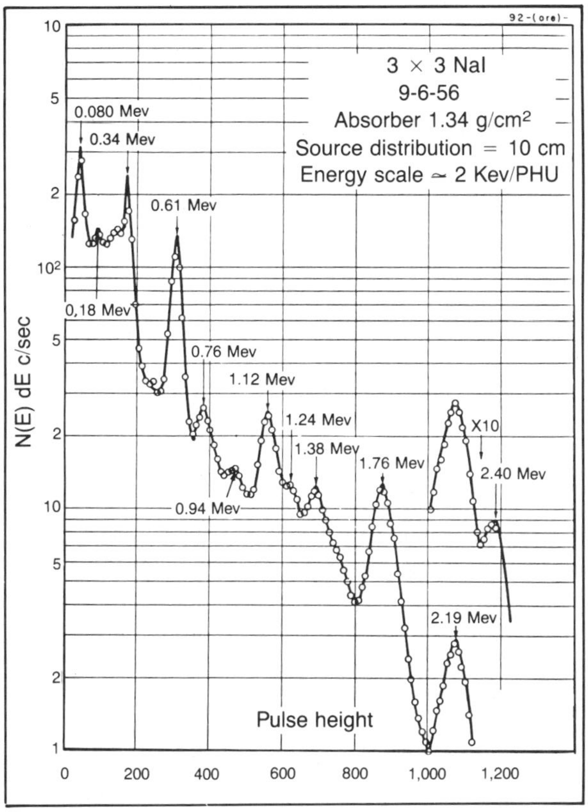

FIG. 5–40

Gamma rays from uranium ore

A large number of very narrow windows in a single instrument is possible, especially with modern solid-state circuitry. These are called multichannel analyzers (MCA). The number of channels available depends on the capabilities of the individual instrument and the setting chosen by the operator. Typically, 64–1,024 or more are used. MCA's can collect data for a full spectrum or a portion of it continuously for a preselected length of time. Typically, the spectrum is displayed on a cathode-ray tube (CRT) as it is accumulated. On completion of the collection time period, the resulting data are read out in one manner or another. Readouts may be individual spectra presented in graphical form or alphanumeric form. Portions of the spectrum can usually be expanded. Some units permit recording one or more channels or peaks in graphic form as a function of depth. The PGT system measures the protactinium-234 line and interprets it as a uranium log.

Spectrographic systems may be either analog or digital. They may use a scintillation or solid-state detector. Usually, a constant, known counting rate is used to stabilize the spectrographic circuit and reduce drift to a negligible amount. This may be from a detector crystal doped with a trace of a monoenergetic radioactive material or from an oscillator-driven light-emitting diode. The known signal controls the voltage to the detector. Occasionally, use can be made of a prominent, known energy, natural emission line of gamma radiation from the formation.

Spectrum

The spectrum obtained with a downhole spectrographic system of any kind is from a thick sample. During the transport from the point of emission to the detector, the gamma photon may undergo Compton scattering and arrive at the detector at a lower energy than original. The result is that the thick sample spectrum shows a progressively larger buildup of Compton-scattered gamma rays as lower energies are examined. Fig. 5–40 shows typical downhole, thick sample spectra. This Compton continuum must be taken into account before a realistic tally of the number of pulses for a given channel may be determined.

In a scintillation detector, the light output is proportional to the photon path length through the detector. High-energy photons have greater penetrating power and, thus, longer path lengths than lower energy photons. Crystal detectors, especially in borehole logging equipment, have a limited size. Therefore, it is quite possible for the photon to go completely through the crystal and still retain some energy upon emergence. Thus, this light output is not a function of the photon energy, which necessitates corrections. Compton scattering within a crystal detector, such as a sodium iodide detector, results in broadening the characteristic peak. Solid-state detectors are better, in this respect, by about an order of magnitude.

General practice at this time with a spectrographic recording for potassium, uranium, and thorium is to use the upper thallium-208 peak at 2.62 MeV for thorium indication, the bismuth-214 peak at 1.76 MeV for uranium indication, and the potassium-40 peak at 1.46 MeV for potassium indication. The height, N (number of events), of each peak is:

$$N_{(Th)}\text{corrected} = N_{(Th)} - Bkg_{(Th)} \quad (5.40)$$

$$N_{(U)}\text{corrected} = N_{(U)} - Bkg_{(U)} - \alpha N_{(Th)}\text{corrected} \quad (5.41)$$

$$N_{(K)}\text{corrected} = N_{(K)} - Bkg_{(K)} - \beta N_{(Th)}\text{corrected} - \gamma N_{(U)}\text{corrected} \quad (5.42)$$

Where:

Bkg = Background rate at each window

α = Contribution factor for thorium in the uranium window
β = Contribution factor for thorium in the potassium window
γ = Contribution factor for uranium in the potassium window

Absolute values of each of the three elements are important because they are functions of the amounts of these elements present.

Equally important are the ratios of one to another. The U/Th ratio gives an indication of the depositional history of the mineral because uranium and thorium are transported differently. The U/K and Th/K ratios are important in low-grade ore or mineral deposits because the gamma contribution of potassium may be a major part of a low-grade deposit.

Marine shales generally have high U/Th ratios. Because the uranium is more easily transported than the thorium, the thorium content in clays is relatively uniform. The potassium content of clay depends on the clay type. Therefore, the thorium content is probably a better indicator of clay or shale proportion than the gross-count gamma ray or the potassium content is. Uranium may be precipitated in fractures from solution. Therefore, a uranium curve can be used as a fracture log. Permeable streaks often have higher uranium content than nonpermeable ones. Relative potassium amounts can identify clay types. A high uranium to thorium ratio can indicate a redox solution deposition, such as a geochemical cell. High thorium to uranium and high potassium to uranium ratios can indicate a silty-clayey depositional environment.

Coals typically have low radioactivity because of the mode of deposition and the short half-life of carbon-14. However, some have high uranium contents, probably because of an anomalous radioactivity in the clay of the ash codeposited with the coal. Also, uraniferous solutions moving through fracture systems within the coal deposit uranium compounds because of the reducing environment of the coal. The best known of these are in Sweetwater County, Wyoming; the Paleocene lignites in Harding County, South Dakota; Eocene subbituminous coals at Rosa, Yugoslavia; subbituminous coals in Hungary; Wackersdorf, West Germany; Trimmelkamm, Austria; and Eocene lignites in Serres, Greece. Fertl describes coal as follows:

> The ash content of the coal is usually and mostly the clay and silt content of the partings. Low-potassium clays, such as kaolinite, and secondary enrichment by uranium mineral can be detected. Use of the thorium curve allows a more accurate calculation of the ash content than does the gross-count gamma ray.
>
> The retention of uranium from solution by low-rank coals amounts to about 98% of the solution content. This is in metallo-organic form. In contrast, clays and limestone only retain about 28% in solution form.
>
> Carnotite-stained coals are occasionally encountered. These contain radium or zircon.

PGT System

The PGT system was developed under contract to the U.S. Bureau of Mines and the U.S. Geological Survey. In its commercial form it uses an intrinsic germanium detector cooled with frozen propane. The isotope of protactinium, Pa-234, gamma emission is recorded. Since Pa-234 is the second daughter of U-238 and both Th-234 (the first daughter) and Pa-234 have short half-lives (24 days and 2.2 min, respectively), the presence of Pa-234 means U-238 was in place within the last few days or weeks. The gamma-ray output from Pa-234 is low (about 4% of the total), but it is energetic enough (up to 1.9 MeV) to be detected through an instrument housing with little attenuation.

The PGT system has a low-counting rate, so it must be logged extremely slowly (as low as 20 ft/hr in low grades) in order to obtain statistically reliable samples. The system uses a computer program to vary the logging speed to get a good sample within an anomaly. Then it speeds up and records Bi-214 outside the anomaly. Thus, the overall time used is only slightly longer than with a GCGR system.

Gamma-ray spectroscopic logging systems are a promising method for the immediate future. They will extensively increase the amount of data obtained from gamma-ray logs (Fig. 5–41a–d).

Airborne Systems and Surface Methods

Airborne and surface gamma-ray logging systems parallel downhole systems in types and usage. In spectrographic systems the airborne units were used earlier and have been more advanced than the downhole systems, probably because of the easier restrictions on size.

Airborne gamma-ray mapping is a widely used, powerful, large-area exploration technique for finding radioactive deposits. It can also be used with care for exploration for some other nonradioactive minerals that may have radioactive halos. Surface gamma-ray methods along with backpacked downhole systems have gained popularity in Canada and Australia, where mobilization problems are difficult with large equipment such as drilling rigs.

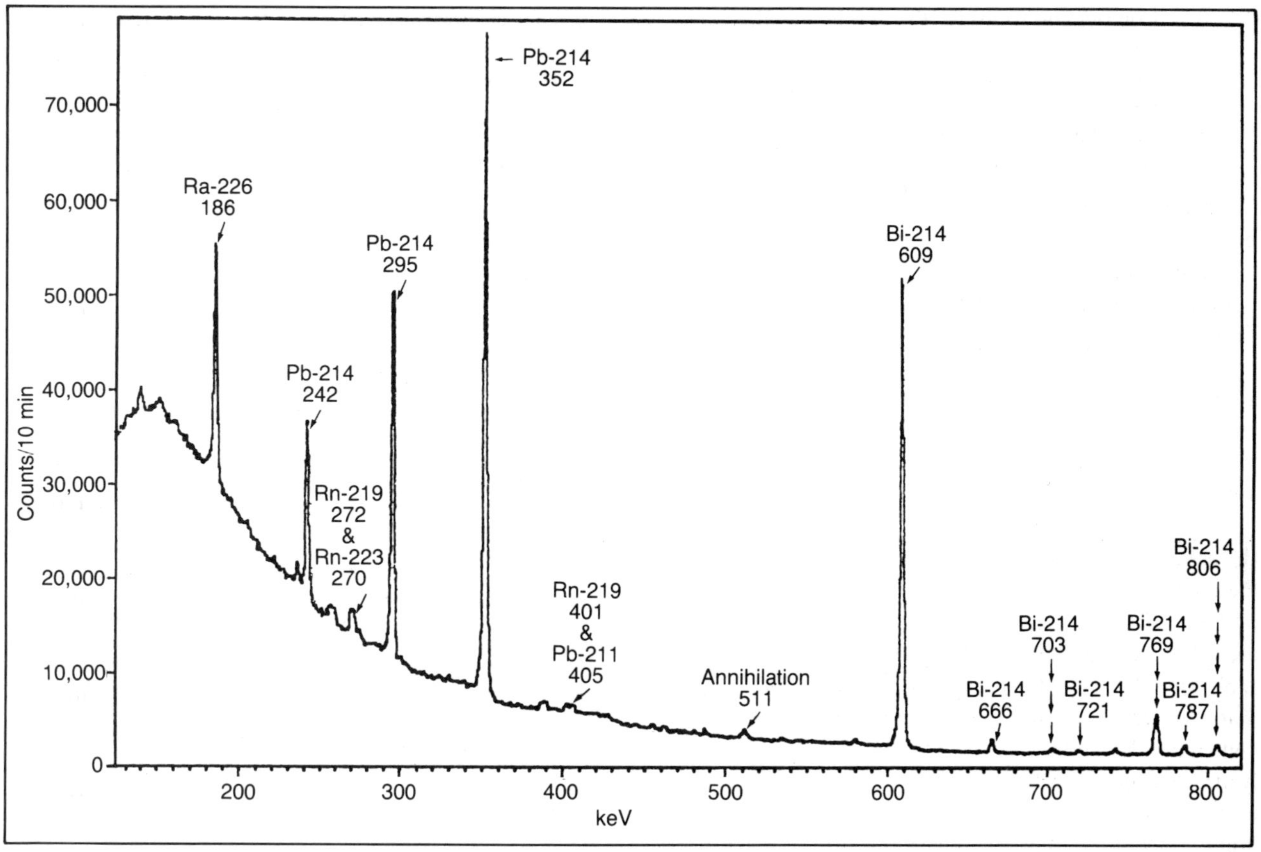

FIG. 5–41a

PGT system—inhole uranium mineralization gamma-ray spectrum made with an intrinsic germanium detector
(courtesy Princeton Gamma Tech.)

Instrumentation

The instrumentation of airborne and surface gamma-ray systems is identical in concept to downhole systems. Lack of weight and size restrictions allow the incorporation of many features that are difficult to include in downhole systems. These systems have been computerized for years.

Airborne gamma-ray systems have large volume detectors in general, some as large as 3,000 cu in. There are two reasons for this. Gamma-ray densities are low, so detection cross sections must be made large to obtain statistically satisfactory samples. A common small size in a gross-count airborne system is 3 x 3 in. Since counting rate or sample size is a primary consideration in gross-count airborne systems, crystals tend to be smaller than in spectrometric systems. Crystal shape has a major effect on the response geometry of the detector. Crystals whose diameter and length are about equal exhibit a minimum of directionality. As the length is increased, the relative sensitivity drops off with angular displacement from the detector axis.

Detector size becomes more important in airborne spectrometric detection. This is both because of the need for high counting rates and for the increased spectral resolution of a crystal large enough to absorb a photon.

Sodium iodide crystals are mechanically fragile. Large ones become increasingly sensitive to thermal shock as well as more expensive. Modern practice is to use clusters of smaller crystals, multiple photomultipliers, and combined signals.

■ Airborne

Virtually all of the useful signal in an airborne survey originates in the upper 10–20 cm of soil. Therefore, direct detection of radioactive mineral deposits deeper than a meter is virtually impossible. Apparently, the anomaly detected by an airborne gamma-ray survey results from a halo caused by small

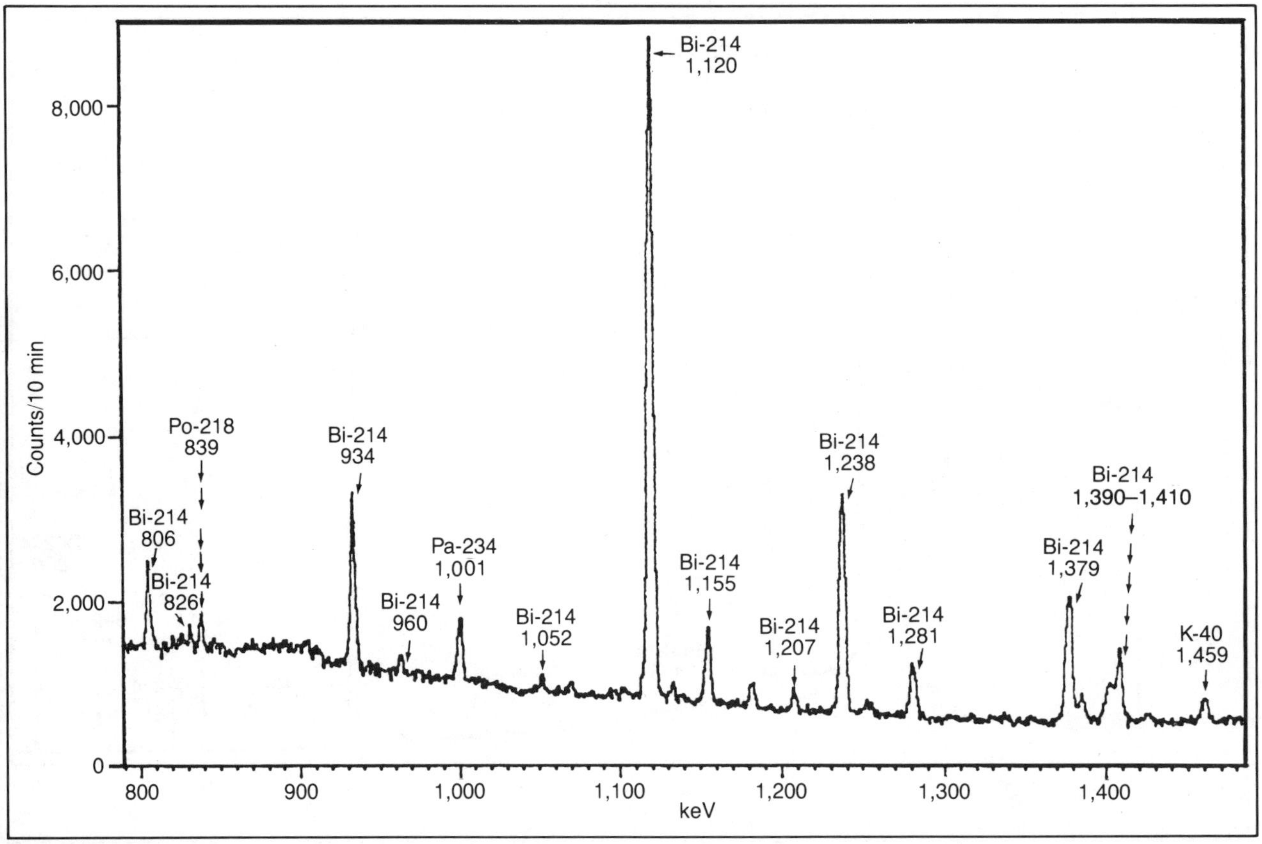

FIG. 5–41b

PGT system—inhole uranium mineralization gamma-ray spectrum made with an intrinsic germanium detector (courtesy Princeton Gamma Tech.)

amounts of solutions and radon gas migrating from the mineral body to the surface.

Since most rocks have some gamma-ray emission, it is necessary to subtract a background counting rate from the observed signal. This rate only comes from the direction of the earth. Also, the atmosphere supplies a component to the background radiation as a result of small amounts of radon gas. This latter comes from all directions. A cosmic-ray component from the upper direction must also be subtracted.

The counting rate, N_{corr}, must then be determined for each channel. The correction is the same as for the downhole data, but the constants α, β, and γ are different. Then the signals must be corrected for terrain clearance, using the relationship:

$$N = N_o e^{-\mu H} \tag{5.43}$$

Where:

μ = Attenuation factor
H = Height above the ground at the time of recording

Fig. 5–42 shows a typical airborne gamma-ray spectrum.

Airborne gamma-ray readings, in gross count, spectral channel count, or ratio form, are usually plotted on a map of the area in isorad form. Anomalies show up as changes in the radiation gradient.

■ Surface

Surface gamma ray measurements are made for a number of purposes in mineral exploration. It is common practice for a uranium geologist to carry a gross-count gamma-ray system with field vehicle. The output of the meter is connected to a buzzer, light, or some other warning device. That way, any anomalies are called to attention.

Uranium and other mineral exploration geologists often use a hand-held scintillometer to examine outcrops and drillhole cuttings. These readings become part of the field notes. Since the geometry of the measurement is uncertain, the readings are qualitative. In general, they are given as so many times the background.

Portable gamma-ray spectrometers are in common use in Canada, especially in areas where thorium is

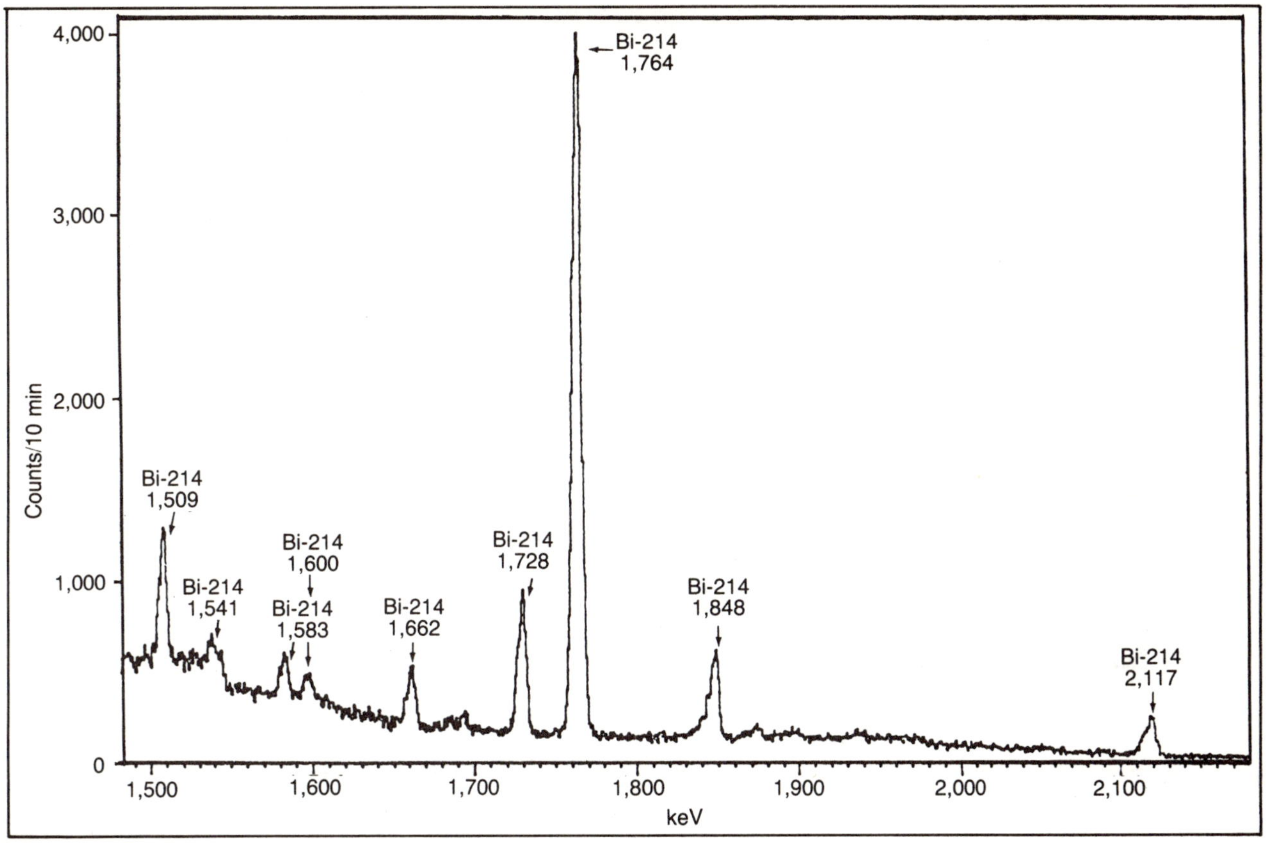

FIG. 5–41c

PGT system—inhole uranium mineralization gamma-ray spectrum made with an intrinsic germanium detector (courtesy Princeton Gamma Tech.)

a major component of the mineral. These may be single-channel, adjustable-channel, or three-channel analog systems. Recently, digital systems that record all signals on magnetic tape for later playback have made their appearance. Many of these systems also can be used in shallow boreholes.

Gamma-ray measurements are used extensively during uranium mining. Long holes in the face of a drift are often logged in underground mining to ensure following the ore trend. In open pit mining it is common practice to map the gamma radiation as a guide to stripping. This is usually done with a G-M detector in a pointed probe with a tee handle.

Calibrations

Calibration of the gamma-ray curve is more important in nonpetroleum logging, especially uranium logging, than in petroleum work. AEC (now the U.S. Department of Energy) facilities for calibration were installed in the mid-1960s and were in general use by 1970. However, even in petroleum work API standards define a calibration method. The API calibration facilities have been available since 1960 at the University of Houston.

Three purposes are served in calibrating a gamma-ray system. The first is the initial determination of sensitivity. This determines if the system meets design specifications. Second, the electronic circuitry, particularly analog circuitry, changes with time, and the response sensitivity and discriminator positions drift. This is a relatively long-term drift if the circuitry has been designed to be insensitive to temperature changes. It is usually much less than 5% over a 2-week period with good equipment. Third, there are settings (e.g., gain and discriminator settings) that may be incorrect and components that may fail. The surest way to detect some of these is by periodic calibration and checking. For example, if the discriminator is set improperly, the circuit is quite temperature sensitive. It may also show a sharp drop in sensitivity compared to the previous calibration, or it may have a high background counting rate. Any of these should make one suspicious of improper settings. If a photomultiplier fails, it may still detect pulses at a reduced

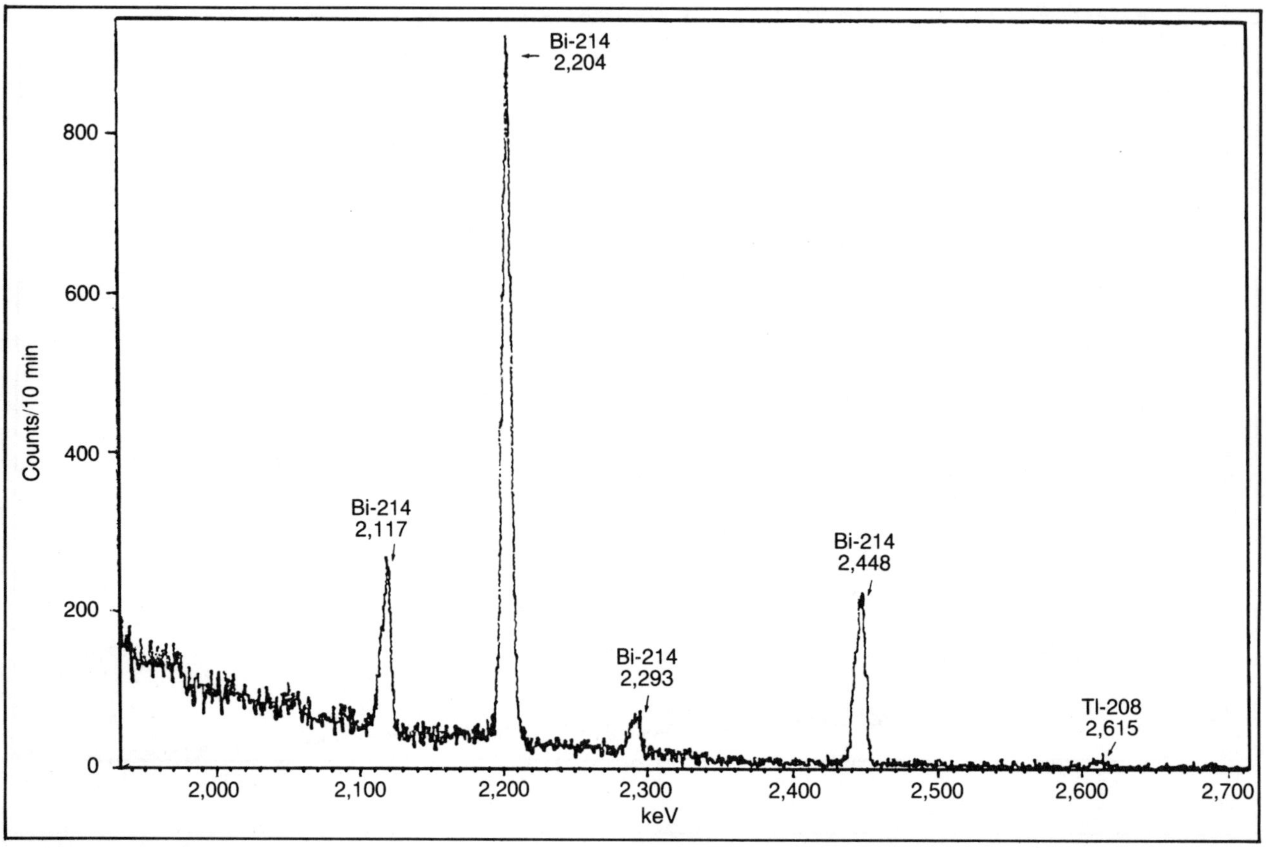

FIG. 5–41d

PGT system—inhole uranium mineralization gamma-ray spectrum made with an intrinsic germanium detector (courtesy Princeton Gamma Tech.)

rate. Not only does a periodic calibration procedure detect this, but it also provides a time frame for the failure, which aids in correcting irreplaceable data and salvaging them.

The primary purpose of calibration is to evaluate tool parameters so the formation characteristics can be considered independently. The two main methods in current use are the API system and the uranium or AEC system. One is fully convertible to the other. Probably, a uranium log should initially be scaled in API units to eliminate tool parameters from the log.

The API system makes use of a calibration model at the API facility at the University of Houston. Note that the API gamma-ray model is described as having twice the radioactivity of a typical midcontinent shale. Thus, a gamma-ray system response can be determined that is a function only of the formation parameters, hole size, and dead-time effects. One log then can be compared with another regardless of contractor, crystal size, or discriminator setting. It must, however, be corrected for hole diameter. This is usually not done on the log.

API calibrations are gaining favor in nonpetroleum logging because of the ease of comparing logs. It is quite simple to convert an API reading to equivalent U_3O_8. After a correction is made for dead-time effects, 100 API gamma units are equal to 0.0025 wt% eU_3O_8 of dry formation. This relation is approximate and must be verified.

The AEC method is the one generally in use in the uranium industry in the U.S. and other countries. This method records a raw gamma-ray counting rate curve and notes the conversion information in the log heading. This calibration should be performed at least every 2 weeks on uranium logging equipment.

The vital information in the log heading includes the conversion factor (K factor), the system dead time, the date of last calibration, the truck source reading at the time of the last calibration, the detector size, the detector type, and the instrument serial

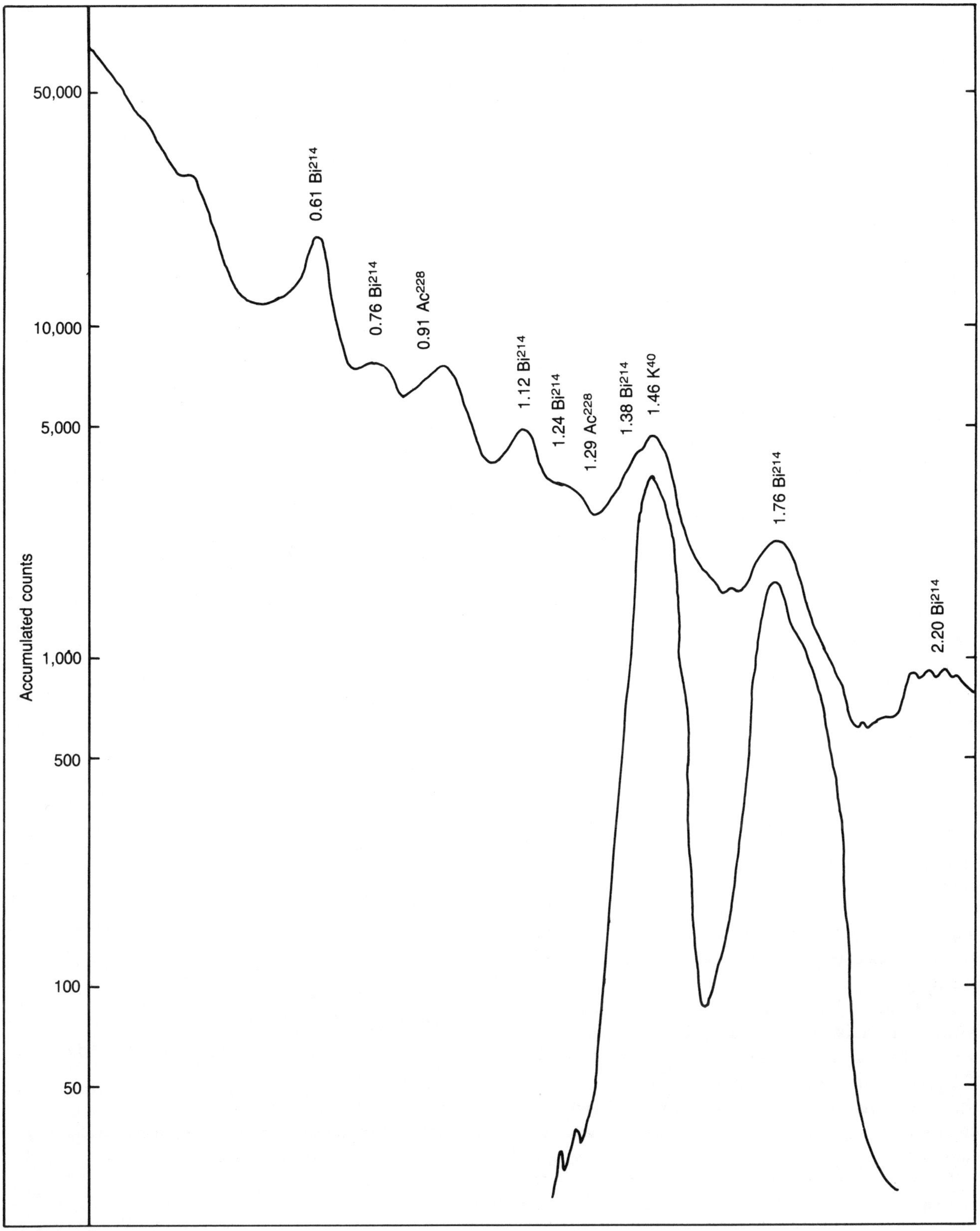

FIG. 5–42

Airborne gamma-ray spectrum NaI-Tl

number. All of these data are necessary.

There are several ways of calculating the conversion or K factor. Probably the best is the *multipit* method. This method can be operated by computer or manually. To use this method, two or more model values are used to solve for K with an assumed dead time in the expression:

$$GT = KA \qquad (5.44)$$

Where:
G = Uranium grade
T = Thickness
A = Area under the anomaly curve

The dead time is arbitrarily increased incrementally, and the resulting values of A are compared with the actual GT values by means of a linear least-squares fit. The values that give the best fit or minimum value of $\Sigma\ (Y - Y_1)^2$ are accepted as the correct conversion factor and dead time.

The second method, and at present the most popular, is the *2-pit* method. This method also calculates the apparent system dead time, μ, and the K factor:

$$\mu = \frac{N_L + N_H R}{N_L N_H (1 - R)} \qquad (5.45)$$

Where:
N_L and N_H = Measured counting rates in the low- and high grade models, respectively
R = Ratio of low to high-grade U_3O_8 concentration by dry weight and in equilibrium

$$K = \frac{GT}{A} \qquad (5.46)$$

Remember that A is the area under the anomaly. It is the sum of the products of the incremental widths times the dead-time-corrected counting rates at the midpoints of the incremental rectangles. N is in half-foot or other increments from one sensitive radius before the bed boundary to one past the other bed boundary.

The K factor illustrated in Eq. 6.46 is the universal K factor. Its units are percentage of eU_3O_8 per count per second. In contrast, most K factors used in the U.S. and listed on log headings are half-foot K factors, K_m. The value of K_m may contain some increment other than 0.5 ft. Thus, it can be confusing and can cause errors.

The relationship between the two, if the depth interval is i, is:

$$K = K_m/i$$

It would be simpler if the gamma-ray curve were calibrated in API units while recording. However, the relationship between counting rate and API units must be noted on the log heading. Also, in this case a dead-time correction should be made on the log.

Calibrations are required where a computer deconvolution of mineral grade, such as the GAMLOG program, is printed on the log. All logs, both analog and digital, must have the truck-source deflection on the log. While this is not a calibration, it is absolutely necessary to detect sensitivity changes. The deflection caused by the test source at the time of logging should be compared with the test source reading (deflection) at the time of calibration. If there is a greater difference than a set standard (usually less than ±5%), the source of the difference or error should be found.

SCATTERED GAMMA-RAY SYSTEMS

The physical principles of gamma-ray scattering density measurements have been known for many years. It has only been since about 1960, however, that the solution of the practical problems have allowed the successful use of the density tool downhole. These problems were the expression of the process in physical terms to allow an orderly approach, the advance of technology permitting the development of viable sources, detectors, solid-state circuits, and the determination of characteristic density ranges and composition of geological materials.

Density tools are the most common of the scattered gamma-ray systems. These systems measure the electron density of the formation, which can be used to derive the bulk density and the probable porosity rather accurately. The system is an excellent one for location and evaluation of coal and oil shale because of the large density contrasts between the fuel material and the surrounding rocks. In combination with the acoustic methods, it can be used to determine elastic parameters more reliably than from the cores because of the probability of mechanical damage to the cores. Density can be effectively plotted against several other systems to determine lithology, corrected porosity, gas content, water content, and several other parameters.

Density measurements reflect:

- Electron density of the material within the sensitive volume of the system
- Bulk density of the material within the sensitive volume
- Average porosity of the measured volume
- Average lithology of the measured volume

Density measurements are affected by and must be corrected for:

- Source strength
- Crystal size
- Mud-cake thickness and density
- Z/A ratios
- Borehole size and density

Density measurements are adversely affected by:

- Stand-off, the distance the tool is from the surface
- Mud cake
- Insufficient pad pressure
- Low source energy
- Low source strength

Density Logging

Density logging uses the same type of circuitry as the gross-count gamma systems. Density detectors are standard gamma-ray detectors. They may be G-M tubes but are usually sodium iodide (thallium-activated) crystals and photomultipliers.

Discriminator levels are set to respond preferentially to the single scattered energies and higher. However, often the discriminator is set in the same manner as the gross-count gamma system, that is, to respond to all Compton-scattered energies above the housing cutoff energy. Response to single scattered energies is probably to be preferred but requires more careful settings and more stable circuitry.

Studies of the depth of investigation indicate that about 90% of the signal of the density systems originates within an ovaloid volume whose major diameter is slightly larger than the source to detector spacing and whose minor radius is about one-third the source to detector spacing.

The maximum scattering angle, θ_M of this system is about 112°. The energy of a scattered gamma ray, E'_γ is:

$$E'_\gamma = \frac{E_\gamma}{1 + (E_\gamma/m_o c^2)(1 - \cos\theta)} \qquad (5.47)$$

Where:

m_o = Rest mass of the electron
c = Velocity of light
E_γ = Original gamma-ray energy

Thus, if a monoenergetic source, such as Cs-137 is used (663 keV), the single scattered gamma energies will cover the whole spectrum from the source energy down to the housing cutoff energy. However, the average energy will be near that of the 112° scattering energy. This energy is about 480 keV.

If the gamma-ray intensity, I, is examined in a direction outward, normal to the probe axis, and in the direction of greatest intensity, it decreases exponentially. The relationship is:

$$I = I_o e^{-\mu dx} \qquad (5.15)$$

With a Cs-137 source of 663 keV gamma rays, 90% of the scattering takes place within 6.5 in. (16 cm) of the source. Thus, the density system is a shallow investigation system.

All scattered gamma-ray systems carry a source of gamma rays. This is usually an isotopic source but can be an electrical source. Isotopic sources are usually used because of their high yield, compactness, ease of maintenance, and simplicity. Most isotopic sources have gamma-ray energies of less than 2 MeV.

Source type and strength must be chosen with care because they must meet several conflicting criteria. A gamma ray interacts with the orbital electrons through which the gamma quantum or photon passes. The photon acts much as a particle. It may be scattered or absorbed or it may disrupt the normal electron pattern. In any event of this type, the photon imparts some or all of its energy to the electron.

If the photon energy is low, the photon may eject an electron from its orbit, but in doing so it loses all of its energy to the electron and ceases to exist. The free electron is a photoelectron. It has a velocity that depends on the angle of the reaction with the photon and the photon energy. This reaction also depends on the atomic number, Z, of the atom. The higher Z atoms require less energy to lift the outer electrons from atomic orbit. Thus, the photoelectric reaction is a function of atomic number and a function of the reciprocal of the photon energy. Both are nonlinear. The photoelectric relationship for the absorption coefficient, μ_p is:

$$\mu_p \simeq (Z^{3.6}/E^3) \qquad (5.48)$$

The relationship of attenuation coefficient, μ, to the photon energy is shown in Fig. 5–43. If the photon energy is high and the atomic number is high, the gamma photon may cause a pair production of the electron. The electron is split into an electron and a positron at a threshold energy of 1.02 MeV. The positron rapidly recombines, with the emission of two gamma-ray photons of 0.51 MeV each (see Fig. 5–41b, annihilation peak).

In between these two types of reactions, the photon

may eject an orbital electron and lose part of its energy to the electron. However, it retains part of its energy. The amount retained depends on the energy of the photon, of course, but it also depends on the angle of scattering. A 90° scattering has a Gaussian distribution of energies centered about a value of approximately half that of the original photon. Note from Fig. 5–43 that on a logarithmic scale this reaction is nearly linear with respect to atomic number. This is Compton or elastic scattering.

When a source is chosen for a density system, it must be of a high enough energy not to cause an appreciable number of photoelectric reactions because of the extreme nonlinearity of these. On the other hand, the energy must not be high enough to cause a significant number of pair productions. Fig. 5–43 shows that a lower limit of the scattered energies for linear, quantitative results is about 300 keV. Further, the upper limit for primary energy is 1.04 MeV. Further considerations are a high enough specific activity that the source is small and essentially a point source and a long enough half-life that frequent corrections need not be made—this consideration is not as important in computer systems as in analog systems. Also, the energy must be low enough or the shielding good enough that the detector does not receive unscattered or direct gamma photons from the source.

The number of photons scattered within the formation from the source and back to the detector depend on the number of orbital electrons present and the source strength and energy. The number of the photons arriving at the detector, however, depend on the number scattered away from the path of the photons and away from the detector, whether the system is a backscattering one or a transmission one. Thus, if it is assumed that only Compton scattering takes place, the detected scattering or counting rate is inversely proportional to the number of electrons or electron density present for a given tool configuration. This is illustrated in Fig. 5–44.

The nucleus of an atom contains virtually all of the mass of the atom. This is the combined mass of the protons and neutrons in the nucleus. Approximately 0.1% of the mass of an atom is caused by the electrons. On the average for light elements (Z < 25), the number of neutrons in the nucleus is nearly equal to the number of protons. Since the atomic number of an element, Z, is the number of protons and the atomic mass, A, is the number of neutrons plus protons, the ratio of Z/A is, on the average, about 0.5. The major exception to this is hydrogen, which has one proton only in the nucleus. Thus, hydrogen has a Z/A ratio of 1.0.

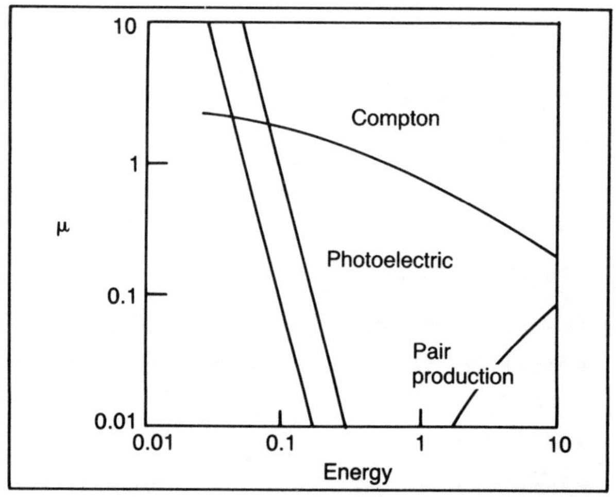

FIG. 5–43

Gamma-ray mass absorption coefficients
(courtesy Schlumberger Well Services Inc.)

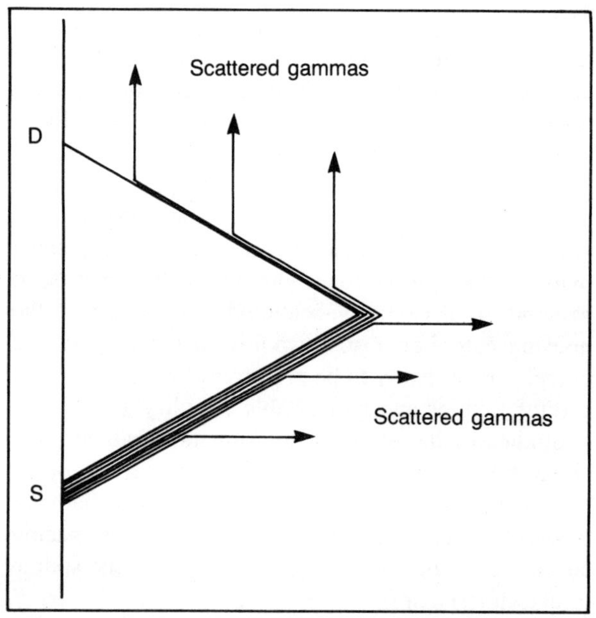

FIG. 5–44

Scattering of gammas from path, density

The number of orbital electrons of an atom is equal to the number of nuclear protons of that atom. Thus, if the Z/A ratio is constant, the amount of Compton scattering of gamma photons from orbital electrons is a simple function of the mass of the atom. Z/A ratios are not always 0.5, however. It is relatively easy to calculate the Z/A ratio of any compound whose composition is known. The actual Z/A ratio is equal to the sum of the partial Z/A ratios. Thus, the Z/A ratio is:

$$Z/A = \Sigma \text{ (relative mole volume) (Z/A of each element)} \quad (5.49)$$

In addition to being easily calculable, Z/A ratios have been determined for a great number of formation minerals. These are available in tabular form. Such a list is given in Table 5–5. For many purposes it is sufficiently accurate to use a Z/A ratio of 0.5, which is how the tool is calibrated.

If a Z/A ratio is assumed, the reading of the density tool will be in apparent density, which is really relative electron density. The apparent density differs from the real bulk density by an amount proportional to the real Z/A ratio compared to 0.5. Thus, apparent bulk densities, ρ_a, may be converted to real bulk density, ρ_b:

$$\rho_a = \frac{Z/A}{0.5} (\rho_b) \quad (5.50)$$

If water or petroleum is present in any quantity in the formation or borehole, the apparent densities of

Table 5–5
Formation and Mineral Densities, Z/A Ratios, and Thermal Neutron Capture Cross Sections*

Material	Z/A Ratio	Matrix Density	Apparent Density**	Σ_{ma}, 10^{21} barns/cc
Lead, Pb	0.3953	11.34	8.97	5.61
Uraninite, UO_2	0.4000	8.25 (6.5–10.8)	6.60	49.69
Cinnabar, H_gS	0.4143	8.1 (8.0–8.2)	6.71	7,981.16
Iron, Fe	0.4687	7.87	7.38	214.90
Galena, PbS	0.4093	7.5 (7.4–7.6)	6.14	12.47
Wulfenite, $PbMoO_4$	0.4187	6.9 (6.7–7.0)	5.78	32.50
Arsenopyrite, FeAsS	0.4605	6.1 (5.9–6.2)	5.62	165.22
Cobaltite, CoAsS	0.4517	6.1 (6.0–6.3)	5.51	936.73
Chalcocite, Cu_2S	0.4610	5.65 (5.5–5.8)	5.21	173.56
Hematite, Fe_2O_3	0.4787	5.26 (4.9–5.3)	5.04	100.47
Magnetite, Fe_3O_4	0.4774	5.18 (4.97–5.18)	4.95	112.10
Bornite, Cu_5FeS_4	0.4643	5.15 (4.8–5.4)	4.78	145.63
Pyrite, FeS_2	0.4850	5.06 (4.95–5.17)	4.91	89.06
Illmanite, $FeTiO_3$	0.4757	4.75 (4.5–5.0)	4.52	158.23
Zircon, $ZrSiO_4$	0.4691	4.69 (4.2–4.86)	4.40	5.42
Stibnite, Sb_2S_3	0.4436	4.57 (4.52–4.62)	4.05	17.82
Pyrrhotite, Fe_5S_6	0.4812	4.55 (4.58–4.64)	4.40	90.52
Barite, $BaSO_4$	0.4454	4.45 (4.30–4.60)	3.96	19.40
Chromite, $FeCr_2O_4$	0.4753	4.45 (4.30–4.60)	4.23	102.20
Rutile, TiO_2	0.4756	4.20 (4.15–4.25)	3.80	202.75
Chalcopyrite, $CuFeS_2$	0.4751	4.2 (4.1–4.3)	3.99	80.95
Corundum, Al_2O_3	0.4904	4.02 (3.95–4.10)	3.94	11.04
Carnotite, $K_2O \cdot 2UO_3 \cdot V_2O_5 \cdot 2H_2O$	0.4350	4+	3.48+	56.21
Rhodocrosite, $MnCO_3$	0.4793	4.0 (3.5–4.0)	3.84	278.93
Sphalerite, ZnS	0.4720	4.0 (3.9–4.1)	3.78	38.33
Siderite, Fe_2CO_3	0.4797	3.88 (3.0–3.88)	3.72	68.81
Limonite, $2Fe_2O_3 \cdot 3H_2O$	0.4897	3.8 (3.51–4.0)	3.72	74.10
Dunite, 4 samples	0.4978	3.3 (3.24–3.74)	3.29	17.03
Olivine, $(Mg, Fe)_2SiO_4$	0.4892	3.3 (3.27–3.37)	3.23	31.74
Magnesite, $MgCO_3$	0.4992	3.1 (3.0–3.2)	3.1	1.48
Norite, 11 samples	0.4970	2.984 (2.720–3.020)	2.97	12.88
Diabase, 6 samples	0.4954	2.98 (2.96–3.05)	2.95	17.12
Gabbro, 27 samples	0.4938	2.976 (2.850–3.120)	2.94	21.47
Anhydrite, $CaSO_4$	0.4995	2.95 (2.89–3.05)	2.95	12.30
Aragonite, $CaCO_3$	0.4995	2.94 (2.85–2.94)	2.94	8.12
Muscovite, $KAl_2(AlSi_3)O_{10}(OH)_2$	0.4966	2.93 (2.76–3.1)	2.91	17.30
Biotite, $H_2K(Mg, Fe)_3Al(SiO_4)_3$	0.4900	2.90 (2.65–3.1)	2.84	25.20
Dolomite, $CaMg(CO_3)_2$	0.4994	2.85 (2.80–2.99)	2.85	4.78

(Courtesy of Gearhart Industries Inc.)
*Density at room temperature and 1 atm pressure unless otherwise stated
**Based on tool calibration using a Z/A ratio of 0.5.

Table 5–5 *continued*

Material	Z/A Ratio	Matrix Density	Apparent Density**	Σ_{ma}, 10^{21} barns/cc
Illite, $KAl_5Si_7O_{20}(OH)_4$	0.4954	2.84 (2.60–3.0)	2.81	39.90
Diorite, 13 samples	0.4964	2.839 (2.721–2.960)	2.82	14.33
Langbeinite, $K_2Mg_2(SO_4)_3$	0.4961	2.83	2.61	78.87
Polyhalite, $2CaSO_4 \cdot MgSO_4 \cdot K_2SO_4 \cdot 2H_2O$	0.5013	2.78	2.79	21.00
Synite, 24 samples	0.4971	2.757 (2.630–2.899)	2.74	16.43
Granodiorite, 11 samples	0.4963	2.716 (2.668–2.785)	2.696	11.33
Chlorite, $(Mg, Al, Fe)_{12}(Si, Al)_8O_{20}(OH)_{16}$	0.5056	2.71 (2.60–3.22)	2.74	17.56
Calcite, $CaCO_3$	0.4996	2.71 (2.71–2.72)	2.71	7.48
Aluminum, Al	0.4818	2.70	2.60	13.99
Plagioclase feldspar, $xNaAlSi_2O_8$, $yCaAl_2Si_2O_8$	0.4925	2.69 (2.62–2.76)	2.65	6.99
Limestone, average of 345 samples	0.5000	2.69 (2.66–2.74)	2.69	8.72
Granite, 155 samples	0.4969	2.667 (2.516–2.809)	2.65	11.62
Quartz, SiO_2	0.4993	2.65 (2.65–2.66)	2.65	4.36
Sandstone, average of 12 samples	0.4990	2.655 (2.59–2.84)	2.655	8.66
Kaolinite, $(OH)_8Al_4Si_4O_{10}$	0.5103	2.63 (2.40–2.68)	2.68	13.06
Albite, $NaAlSi_3O_8$	0.4885	2.62 (2.61–2.65)	2.56	6.77
Orthoclase feldspar, $KAlSi_3O_8$	0.4958	2.57 (2.55–2.63)	2.55	16.00
Kieserite, $MgSO_4 \cdot H_2O$	0.4724	2.57	2.43	12.77
Concrete		2.35 (1.98–2.35)		
Montmorillonite, $(OH)_4Si_8Al_4O_{20} \cdot nH_2O$ (n = 1)	0.5009	2.35 (2.00–3.00)	2.35	8.10
Gypsum, $CaSO_4 \cdot 2H_2O$	0.5111	2.32 (2.30–2.35)	2.37	19.40
Glauconite, KMg (FeAl) $(SiO_3)_6 \cdot 3H_2O$	0.4998	2.30 (2.20–2.80)	2.30	16.80
Graphite, C	0.4995	2.22 (2.09–2.23)	2.22	0.38
Serpentine, $Mg_3Si_2O_5(OH)_4$	0.5062	2.20	2.23	8.80
Halite, NaCl	0.4799	2.16 (2.135–2.165)	2.07	752.36
Nahcolite, $NaHCO_3$	0.4905	2.20	2.16	
Kainite, $MgSO_4 \cdot KCl \cdot 3H_2O$	0.5140	2.13 (2.1–2.13)	2.19	196.13
Trona, $Na_2CO_3HNaCO_3 \cdot 2H_2O$	0.5043	2.125 (2.11–2.15)	2.14	16.21
Sulfur, orthorhombic, below 95.4°C	0.4990	2.07 (2.05–2.09)	2.07	19.06
Potash, $K_2CO_32H_2O$	0.5049	2.04	2.06	39.70
Sylvite, KCl	0.4829	1.99 (1.97–1.99)	1.92	570.68
Cement, (32 samples)		1.99		about 13
Sulfur, monoclinic, above 95.5°C at 1 atm; above 150°C at 1.304 atm S	0.4990	1.96	1.96	18.05
Kernite, $Na_2B_4O_7 \cdot 4H_2O$	0.5026	1.91	1.92	12,793.69
Carnallite, $KMgCl_3 \cdot 6H_2O$	0.5095	1.61 (1.60–1.61)	1.64	370.92
Anthracite coal, 0.9350(C) 0.0281(H) 0.0097(N) 0.0272(O)	0.5134	1.60 (1.32–1.80)	1.64	1.08
Bituminous coal, 0.8424(C) 0.0555(H) 0.0152(N) 0.0869(O)	0.5201	1.35 (1.15–1.7)	1.40	1.54
Lignite		1.10 (0.5–1.5)	1.16	
Water: 300,000 ppm NaCl	0.5325	1.219	1.298	146.22
250,000 ppm NaCl	0.5363	1.1825	1.268	122.55
200,000 ppm NaCl	0.5401	1.146	1.238	100.08
150,000 ppm NaCl	0.5438	1.109	1.206	78.75
100,000 ppm NaCl	0.5476	1.073	1.175	58.69
50,000 ppm NaCl	0.5513	1.0365	1.143	39.02
30,000 ppm NaCl	0.5528	1.022	1.130	32.56
pure water	0.5551	1.00	1.11	22.08
NaCl solution density at stp $\cong$ 1 + (0.00000073 × ppm)				
oil $n(CH_2)$, 10°API, stp	0.5703	1.00	1.14	28.02
30°API, stp		0.88	1.00	25.89

**Based on tool calibration using a Z/A ratio of 0.5.

Table 5–5 *continued*

Material	Z/A Ratio	Matrix Density	Apparent Density**	Σ_{ma}, 10^{21} barns/cc
40°API, stp		0.85	0.97	24.22
50°API, stp		0.78	0.85	22.23
(C_8H_{18}), 70°API, stp	0.5778	0.70	0.81	22.12
n-pentane, C_5H_{12}, stp	0.5823	0.626	0.733	20.80
200°F, 7,000 psi		0.603	0.702	20.02
n-hexane, C_6H_{14}, stp	0.5803	0.659	0.765	21.38
200°F, 7,000 psi		0.628	0.739	20.37
n-heptane, C_7H_{16}, stp	0.5778	0.684	0.790	21.80
200°F, 7,000 psi		0.657	0.759	20.84
n-octane, C_8H_{18}, stp	0.5778	0.703	0.812	22.12
200°F, 7,000 psi		0.673	0.778	21.12
n-nonane, C_9H_{20}, stp	0.5768	0.718	0.828	22.37
200°F, 7,000 psi		0.686	0.791	21.37
n-decane, $C_{10}H_{22}$, stp	0.5763	0.730	0.841	22.55
200°F, 7,000 psi		0.701	0.808	21.65
n-undecane, $C_{11}H_{24}$, stp	0.5759	0.740	0.852	22.71
200°F, 7,000 psi		0.713	0.821	21.87
Methane, CH_4, stp	0.5703	0.000677	0.00076	0.028
200°F, 7,000 psi		0.2189	0.2497	10.88
Ethane, C_2H_6, stp	0.5986	0.001269	0.00015	0.051
200°F, 7,000 psi		0.4104	0.4913	16.34
Propane, C_3H_8, stp	0.5896	0.00186	0.0022	0.067
n-butane, C_4H_{10}, stp	0.5850	0.00246	0.0029	0.085
Helium, He, stp	0.4997	0.00017	0.00017	0.0000
Carbon dioxide, CO_2, stp	0.4999	0.001858	0.001857	0.0001
Nitrogen, N_2, stp	0.4998	0.001182	0.001185	0.004
Oxygen, O_2, stp	0.5000	0.001350	0.001350	0.00001
Hydrogen sulfide, H_2S, stp	0.5281	0.001438	0.001519	0.029
Air, dry, N, 78%; O, 21%; A, 1%	0.4997	0.001224	0.001223	
Argon, A, stp	0.4859	0.001688	0.00144	0.017
Average natural gas, stp	0.5735	0.0007726	0.000886	
200°F, 7,000 psi		0.252	0.289	
Hydrogen, H	0.9921	0.00009		
Oxygen, O	0.5000	0.001429		
Nitrogen, N	0.4998	0.00125		
Carbon, C	0.4995	3.52		
Calcium, Ca	0.4990	1.5		
Sulfur, S	0.4990	2.07		
Silicon, Si	0.4985	2.4		
Magnesium, Mg	0.4975	1.74	1.73	
Potassium, K	0.4859	0.86		
Phosphorous, P	0.4845	1.83		
Aluminum, Al	0.4818	2.70	2.60	
Sodium, Na	0.4804	0.97		
Chlorine, Cl	0.4795	0.0032		
Nickel, Ni	0.4769	8.90		
Iron, Fe	0.4687	7.86		
Boron, B	0.4625	2.45		
Chromium, Cr	0.4614	7.1		
Titanium, Ti	0.4593	4.5		
Zinc, Zn	0.4589	7.14		
Manganese, Mn	0.4568	7.4		
Copper, Cu	0.4564	8.92		
Vanadium, V	0.4514	5.96		
Cobalt, Co	0.4432	8.9		
Arsenic, As	0.4405	5.7		
Zirconium, Zr	0.4385	6.4		

**Based on tool calibration using a Z/A ratio of 0.5.

Table 5–5 *continued*

Material	Z/A Ratio	Matrix Density	Apparent Density**	Σ_{ma}, 10^{21} barns/cc
Bromine, Br	0.4380	3.12		
Strontium, Sr	0.4336	2.6		
Tin, Sb	0.4212	7.2		
Molybdenum, Mo	0.4159	10.2		
Barium, Ba	0.4077	3.5		
Mercury, Hg	0.3988	13.56		
Lead, Pb	0.3953	11.34		
Uranium, U	0.3865	18.7		
Water, H_2O	0.5551	1.0		
Carbon dioxide, CO_2	0.4999			
Calcium carbonate, $CaCO_3$	0.4996			
Calcium sulfate, $CaSO_4$	0.4995			
Calcium oxide, CaO	0.4993			
Silicon oxide, SiO_2	0.4993			
Magnesium oxide, MgO	0.4985			
Aluminum oxide, Al_2O_3	0.4904			
Phosphorous oxide, P_2O_5	0.4935			
Boron oxide, B_2O_3	0.4884			
Potassium oxide, K_2O	0.4883			
Sulfur trioxide, SO_3	0.4874			
Sodium oxide, Na_2O	0.4855			
Hematite, Fe_2O_3	0.4787			
Magnetite, Fe_3O_4	0.4766			
Iron oxide, FeO	0.4757			
Titanium oxide, TiO_2	0.4756			
Titanium oxide, TiO	0.4695			
Manganese oxide, MnO	0.4652			
Zirconium oxide, ZrO_2	0.4555			
Strontium oxide, SrO	0.4439			
Barium oxide, BaO	0.4173			

the materials, especially the water or petroleum, must be used for any calculations. This is because of the large percentage of hydrogen in both of these compounds. Therefore, an apparent density of 1.11 g/cc must be used for water instead of the real density of 1.00 g/cc. As a general rule it is safer to use apparent densities for any calculation.

A scattered gamma-ray density system uses a gamma-ray detector. Therefore, the density system is sensitive to natural gamma radiation as well as to the radiation scattered from the source. It is imperative that a density tool be designed to minimize the effects of natural gamma radiation while, perhaps, enhancing the response to scattered gamma radiation. In general, a well-designed density tool employs shielding to block as much natural gamma radiation as possible. Then the detector is designed to have as low a sensitivity to natural gamma rays as possible. In order to have effective counting rates with an insensitive detector, it is necessary to use a large source (high output). With cesium 137, probably 100 millicuries is the lowest-strength reliable source that should be considered. In sidewall probes, one to three curies is preferable, even with a relatively short spacing.

Dead-time effect corrections are as important as they are with gross-count gamma-ray logging. They will correct some substantial nonlinearities and should be used. If a computed log is presented scaled in density units, the relationship between counting rate and density must be stated on the log heading or scale.

Single Spacing

The single-spacing density systems have two major variations. The omnidirectional systems are the original and simplest type. The sidewall directional systems are more complex but can be used more effectively.

The omnidirectional tools are usually standard gamma-ray tools with a spacer and a gamma-ray source a fixed distance from the detector. Gamma rays are radiated in all directions and the detector is sensitive in all directions. Fig. 5–45 shows a tool of this type. Because of the shallow depth of investigation, the omnidirectional systems must have long source-to-detector spacings and large source strengths in order to minimize borehole effects. Typical spacings are 12–20 in. They are so sensitive to borehole effects, however, that they should never be used to calculate density. Also, they usually are standard GCGR tools that have been converted. The result is a rather poor resolution that has the borehole response superimposed on it. These systems should only be used in small diameter holes and should only be used quantitatively. The possibility of substantial natural gamma-ray and borehole effects must always be kept in mind. Omnidirectional systems are useful where the sidewall system cannot be used. For example, when the tool must be lowered through small diameter drillpipe, there usually is not room for a sidewall system. These systems are excellent for rapidly locating coal depths.

The sidewall density tools usually employ an eccentering device, such as one or more bowsprings or spring-loaded arms causing the tool to ride against the hole wall. The gamma-ray source is collimated in the direction of the resultant force of the eccentering device, so most of the radiation leaving the tool enters the formation. The gamma-ray detector is also

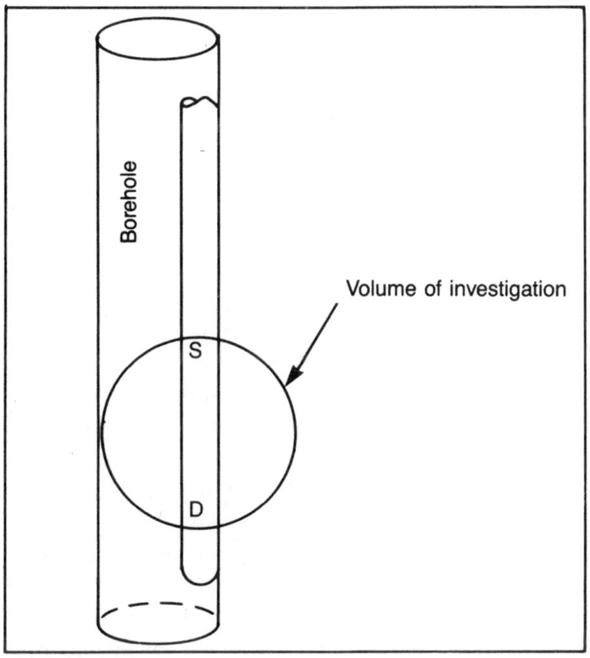

FIG. 5–45
Omnidirectional density

highly collimated in the same direction. Fig. 5–46 and Fig. 5–47 show tools of this type. Notice that these are not focused systems. The gamma rays and sensitivities are collimated by shielding.

This type of tool responds primarily to formation density. The ratio of formation response to borehole response is the front-to-back ratio and is ideally infinite. The actual ratio depends on the tool design, the tool diameter, the type of shielding material, and the source energy. A well-designed, 2-in. diameter tool can easily have a ratio of 100 to 200. Neglecting the small borehole effects, the response of a sidewall tool is ρ_a:

$$\rho_a = f + e^{-b(c - \gamma_n)} \tag{5.51}$$

Where:
γ_n = Counting rate caused by natural gamma radiation
c = Counting rate caused by radiation scattered from the source, both corrected for dead-time effects
b = Constant of proportionality and depends on tool design and source characteristics
f = Constant that is a function of the fixed background radiation because of the presence of the borehole.

In actual practice the bulk density determination deviates from a pure exponential at the high-density end because of the increased likelihood of photoelectric reactions. At the low-density end, the deviation from exponential is caused by increased amounts of hydrogen in the sedimentary formations. At both ends the effects of γ_n and f (Eq. 5.51) are apparent. Because of these departures from a logarithmic relationship, the response of density tools is best represented by a section of parabolic arc rather than an exponential relation. This takes the form for apparent density, ρ_a:

$$\rho_a = a + bN + dN^2 \tag{5.51a}$$

Where:
N = Counting rate
a, b, and d = Constants that depend on the tool designs and settings

This relation is best determined by a least squares fit and is valid only for a limited range.

One occasionally sees density tools using 10–100 microcuries of radium. This is done to circumvent license problems. However, such a system is generally useless quantitatively because of natural gamma sensitivity.

One special form of the single-spacing sidewall tool is built for locating coal-bed boundaries. This tool uses an extremely short spacing—2 in. or less—and is usually of small diameter. In order to obtain

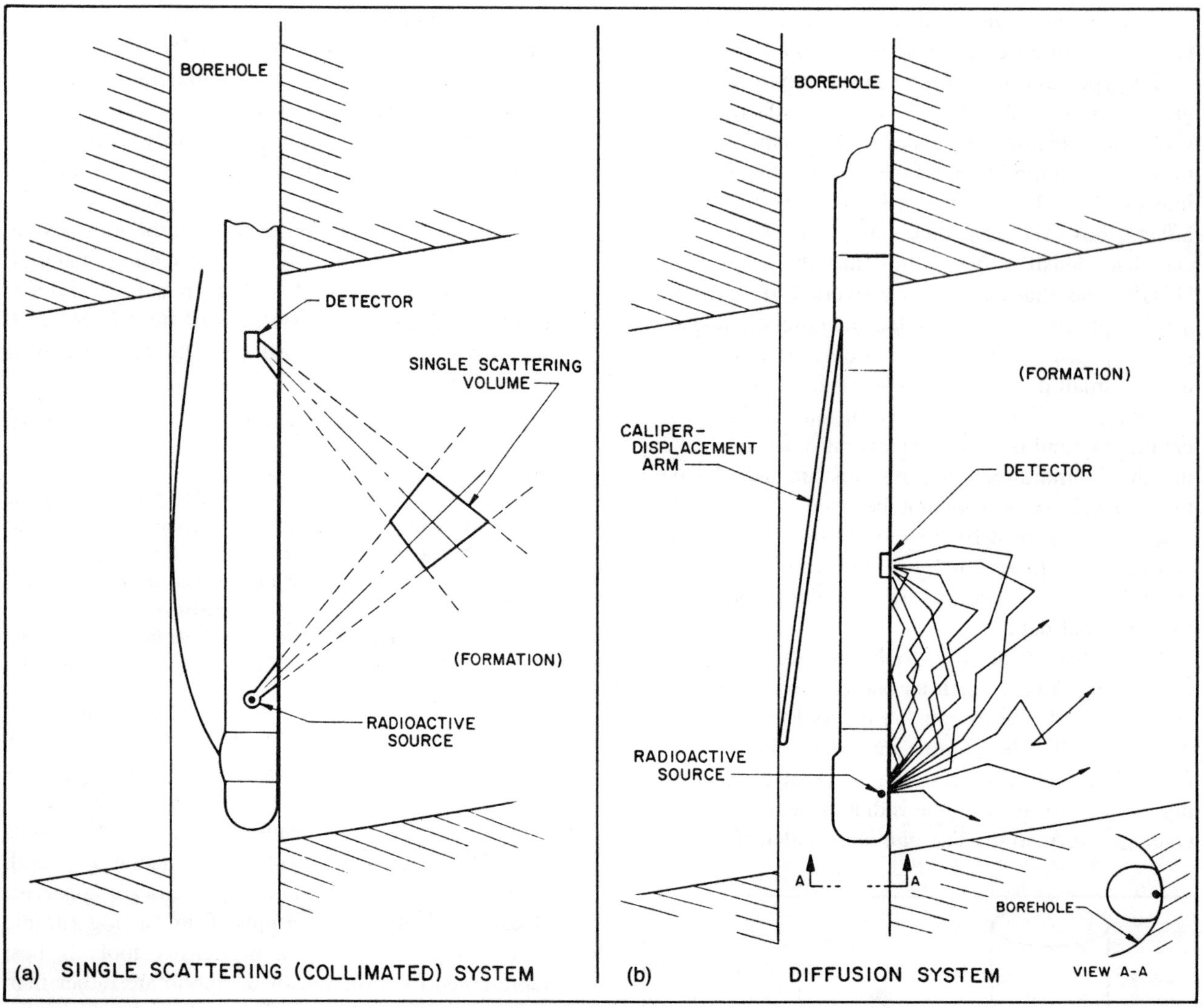

FIG. 5–46

(a) Single-Spacing (collimated) and (b) diffusion systems (courtesy Mt. Sopris Inc.)

enough shielding between the source and the detector and between them and the borehole, a low-energy gamma-ray source, such as americium-241 (60 keV) is used. As a result of the short spacing (shallow investigation) and the high amount of photoelectric absorption in the sand or shale, the bed-boundary response is excellent. However, these systems are only suitable for locating bed boundaries in smooth-wall, good borehole conditions. No attempt should be made to use this system quantitatively.

Compensated or Dual Spacing Density

By using various combinations of spacings and sometimes gating, it is feasible to eliminate many of the problems inherent in the single-spacing tools. These are the compensated systems.

A long spacing measurement has a deeper investigation, and it measures the response of a larger volume than does a short spacing. The differences in response of a short and a long spacing can be compared, and the effect of mud-cake and wall rugosity may be evaluated and compensated for.

If the response of a short spacing (6–8 in.) is compared to the response of a long spacing system (12–18 in.) as the thickness of mud cake is increased from 0 in., the amount of mud cake of a given thickness examined by each system is proportional to the spacing length. On the other hand, the total volume

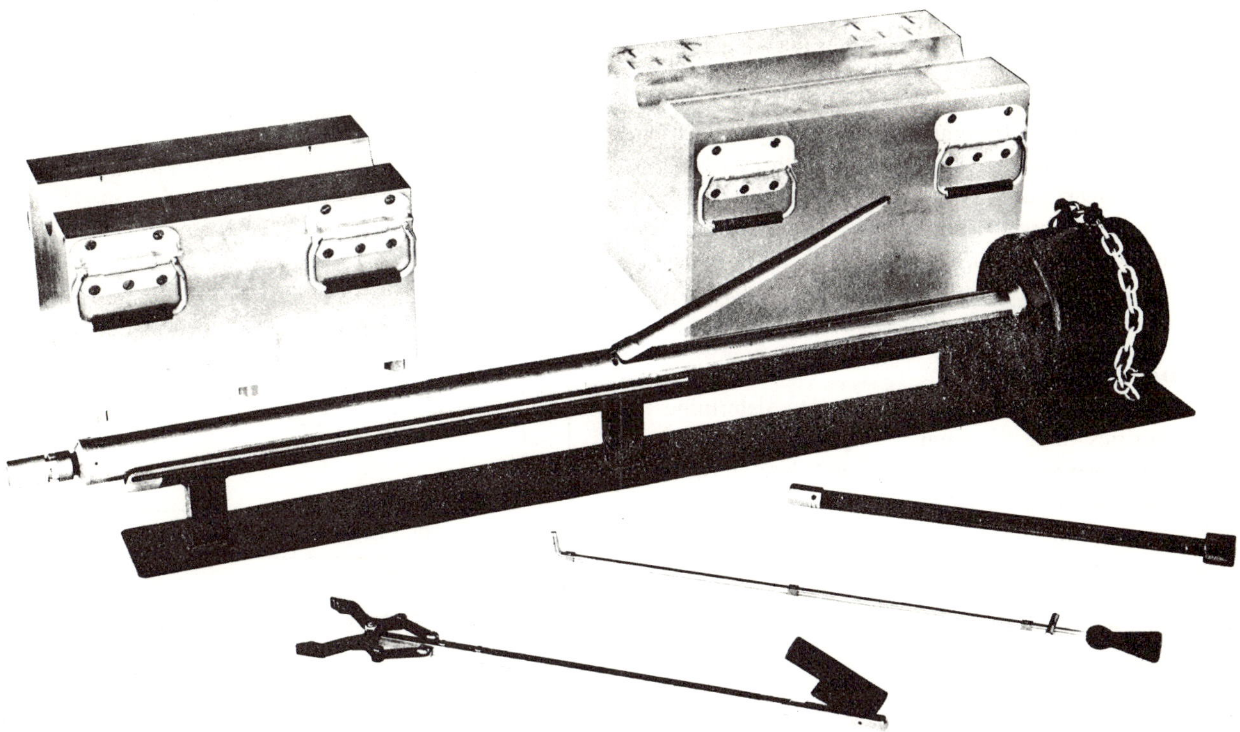

FIG. 5–47

Mineral-type density tool (courtesy Mt. Sopris Inc.)

examined by each system is a function of the square of the spacing. Thus, the mud cake is a larger proportion of the total volume of the short spacing than of the long spacing. Therefore, as the mud-cake thickness increases, since the mud cake is usually of a lower density than the formation, the counting rate of the short spacing increases more rapidly than that of the long spacing. If this response is plotted, the results are as shown in Fig. 5–48. Note that the effect of weighted mud and mud cake is opposite to that of light mud cake.

This dual spacing system is usually one source and two detectors (Fig. 5–49). This is the standard density system in petroleum logging but is seldom used in nonpetroleum logging. It probably will be used soon because a compensation for borehole rugosity is very badly needed. In coal logging it was initially assumed that the hole wall is smooth and to gauge in the coal. Therefore, a compensated system was not needed. Experience has shown this to be a faulty premise. Also, if the short spacing detector response is made identical to, or a function, of the long spacing response, the effect of the natural gamma radiation could be compensated for. This would make this type of scattered gamma-ray logging feasible in a high-radiation environment, such as in a uranium-bearing formation.

The response of the dual spacing density system in terms of ρ_b, formation bulk density, is:

$$\rho_b = a - b \log C_L + c \log C_s \qquad (5.51b)$$

Where:

C_L and C_s = Counting rates of the long spacing and short spacing systems, respectively

a, b, and c = Best-fit coefficients

A standoff from the wall of the hole, such as caused by wall rugosity or mud cake, causes an error, $\Delta\rho_b$, in the calculated bulk density. A function of the amount of material between the tool and the wall of the hole is V:

$$V = f + \log C_L - g \log C_s \qquad (5.51c)$$

Where:

f and g = Best-fit coefficients

$$\Delta\rho_b = \frac{c}{g} V \qquad (5.51d)$$

$$V = k\,T\,(\mu_t - \mu_f) \qquad (5.51e)$$

Where:

T = Effective thickness of the rugose portion or mud cake

μ_t and μ_f = Effective linear absorption coefficients of the rugose volume and the formation volume, respectively

A second type of compensated density system uses dual detectors and dual sources, both of which are gated on and off. With suitable combining the effects of all near measurements can be cancelled and a discrete volume within the formation can be measured. This type has not been used extensively because of the difficulty of gating an isotopic source.

Fluid Density

Gamma-ray scattering is also used to determine fluid densities of drilling mud. Mud logging procedures make use of them routinely, and they are also used downhole. In general, fluid systems use a source and detector and measure the transmitter gamma radiation through the free-flowing fluid between them. Occasionally, they use backscattering but not often. The principle of the operation is similar to the backscattered systems.

■ Litho-Density*

The Litho-density* system is a system from which improved density and photoelectric cross-section information may be obtained. It is used as a lithology indicator and is especially effective when used in combination with other lithology systems (Fig. 5–50).

Gamma rays traversing a material, such as a formation material, undergo both Compton scattering and photoelectric absorption. The Compton scattering is nearly linear with respect to the atomic numbers of the elements within the material. Thus, Compton scattering is excellent for determining bulk density.

Gamma rays may undergo photoelectric absorption if the gamma energy is low or if the element atomic number is high. The probability that such a reaction may occur is indicated by the photoelectric cross section, τ:

$$\tau = KZ^{4.6} \quad (5.52a)$$

Where:
K = Energy dependent coefficient
Z = Atomic number

If a window of energies is examined, an average K, $\overline{K}$, may be used. If both sides are divided by Z, then:

$$\tau/Z = \overline{K}Z^{3.6} \quad (5.52b)$$

The Schlumberger system arbitrarily makes:

$$\overline{K} = 10^{-3.6} \quad (5.52c)$$

*A mark of Schlumberger

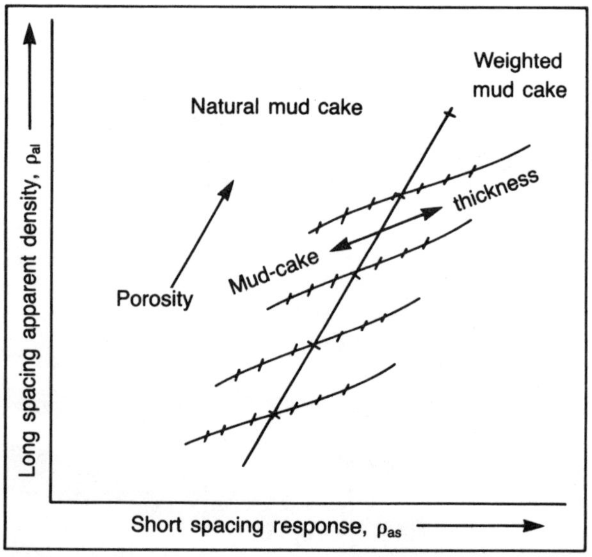

FIG. 5–48
Dual-spacing density response

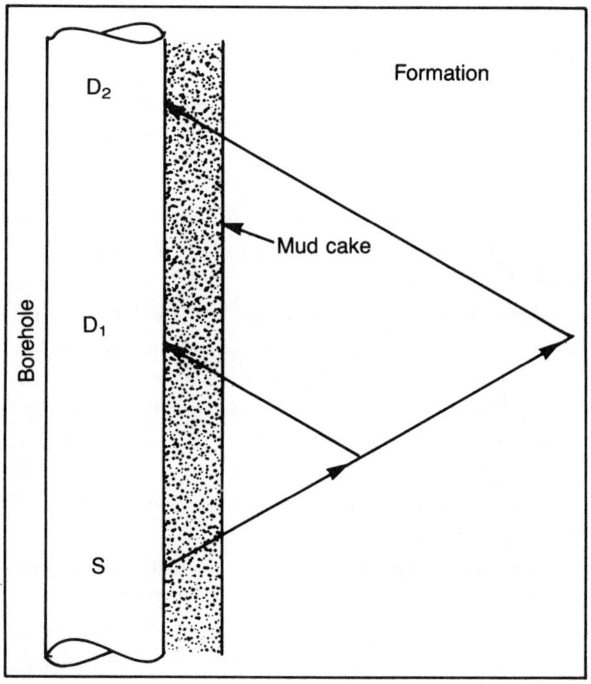

FIG. 5–49
Dual-detector density

Therefore:

$$\tau/Z = P_e = (Z/10)^{3.6} \quad (5.52d)$$

Where:
P_e = Index of the effective photoelectric absorption cross section, barns/electron

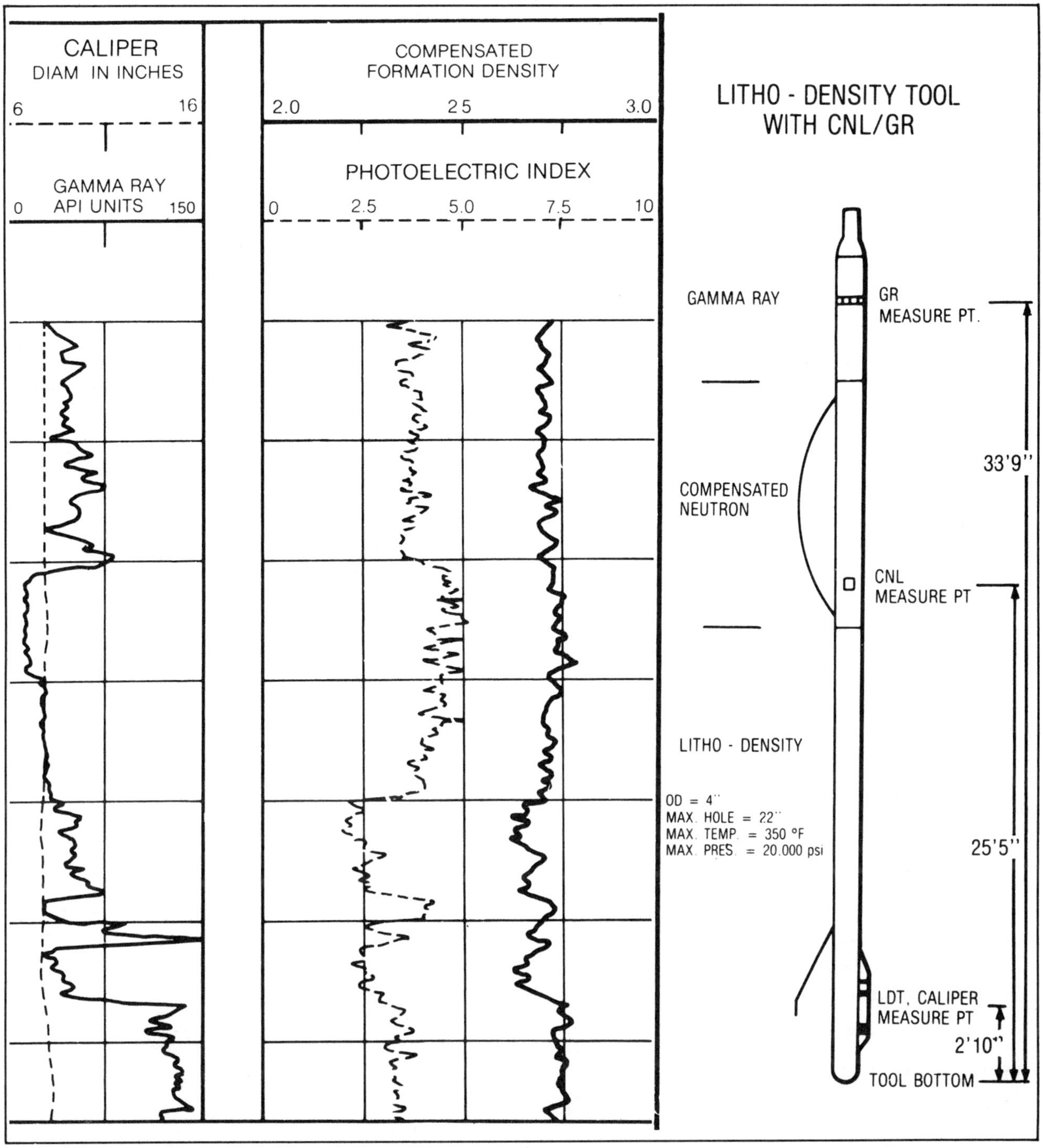

FIG. 5–50

Litho-Density* CNL tool and log (*mark of Schlumberger)

The bulk density, ρ_b, is calibrated to:

$$\rho_b = 1.0704\rho_e - 0.1883$$
$$\rho_b = (\rho_b + 0.1883)/(1.0704) \text{ electrons/unit volume}$$
$$U = P_e\rho_e \text{ barns/unit volume}$$
$$\rho_b = \phi\rho_f + (1 - \phi)\rho_m$$
$$U = \phi U_f + (1 - \phi)U_m$$

Where:
f = Fluid
m = Matrix

The value of P_e can be used in the following manner:

- Alone, as a lithology indicator not greatly affected by porosity
- In combination with ρ_b to determine a two-mineral porosity system
- In combination with ρ_b and ϕ_N for more complex lithology determinations

P_e is effective because it is little affected by porosity.

In hard-rock environments, the density systems are invaluable. One must be careful if sedimentary-type tools are used. These tools are seldom calibrated beyond 2.7 g/cc. In addition, scales often cut off at 3.00 g/cc. Although these two things frequently make the commercial systems useless above 3.00 g/cc, a density log calibrated and scaled high enough and run at rather sensitive scales makes a fine log to use in a hard-rock system.

The most valuable use of the density tools is in combination with other systems. One excellent way to combine measurements of two or more tools is to crossplot them. Crossplotting is plotting points from the same level on two axes, one or a combination on the x axis and the other or a combination on the y axis. Groupings and trends yield information that is difficult to obtain other ways.

The porosity-density relationship is:

$$\rho_b = \phi\rho_f + (1 - \phi)\rho_m \quad (5.52e)$$

This may be rewritten as:

$$\rho_b = \rho_m + (\rho_f - \rho_m)\phi \quad (5.52f)$$

The last equation is linear, so ρ_b versus ϕ may be plotted on a linear grid with ρ_b on the y axis and ϕ on the x axis. Points plotted from logs of a porous, nonhydrocarbon-bearing zone form a locus that is a straight line. The slope of the line is the value of $\rho_f - \rho_m$. Since ρ_m is usually larger than ρ_f, the slope is negative. That is, the porosity is inversely proportional to the bulk density. If ϕ is made equal to zero—that is, if the line is extrapolated to ϕ equals zero—the intercept of the y axis is the value of ρ_m, the rock density. If the line is extrapolated to a value of ϕ equals 100%, the values of ρ_m cancel and ρ_b is equal to ρ_f, the fluid density. Thus, both of these constants may be evaluated, and the rock and fluid can be identified. If the formation fluid is gas in places instead of water, those places are indicated by a skewing of those points downward toward a lower density.

■ Uses of Density Measurements

The conversion of the density-tool counting rate to bulk density has been covered. The density of the formation material is of importance directly. The low-density range of coal, for example, is definitive of the presence of coal. Likewise, high density readings can be indicators of massive, nonporous formation material or extra-dense contaminants. Table 5–5 lists densities of various geological materials. It can be seen that the ranges of densities are narrow enough to allow an identification of many lithologies, and the density measurement is especially valuable in hard-rock environments.

The density systems were originally adopted in petroleum logging as a measure of porosity. The presence of a mud cake opposite the interesting sands was a source of inaccuracy with the single spacing tools, however. In spite of this, the density tool was an excellent one for porosity because it allowed an accurate porosity determination even in the presence of shale. The reason is that the shale density in a deep hole is very near that of the sand-grain density. Table 5–6 shows Gulf Coast shale densities as a function of depth.

The problem of the inaccuracies caused by mud cake led to the development of the compensated formation density systems. An accurate determination is quite possible with the compensated systems even in poor mud conditions.

The relationships for fractional porosity, ϕ, are:

$$\rho_b = \phi\rho_f + (1 - \phi)\rho_g \quad (5.52e)$$

Table 5–6

Typical Gulf Coast Shale Densities

Depth, ft	Depth, m	Bulk Density, g/cc
0	0	1.8
2,000	600	2.20
4,000	1,200	2.34
6,000	1,800	2.44
8,000	2,400	2.52
10,000	3,000	2.57
12,000	3,600	2.60
16,000	4,800	2.63

$$\rho_b = \phi\rho_f + V_s\rho_s + V_{sh}\rho_{sh} \qquad (5.53)$$

Where:

ρ_b = Bulk density
ρ_f = Fluid density
ρ_g = Solid or grain material density
ρ_s = Sand density
ρ_{sh} = Shale density

If the bulk or corrected density is used, the others must also be corrected. If the apparent density is used, the others must be apparent or uncorrected densities. While density readings can be used effectively by themselves, their value is greatly increased when they are used with other curves, as in crossplotting.

■ Calibrations

Calibrations of the various scattered gamma-ray methods are as important and critical as with the natural gamma-ray systems and somewhat more involved. The absolute accuracy possible with these systems is probably an order of magnitude better than with natural gamma-ray gross-count systems. Therefore, the calibrating methods must be at least as good. The response of any tool is affected by many factors, many of which are best determined empirically. Literally, no quantitative use of a density system is possible without calibrations. With them the results are precise, and the possibilities are endless.

Density calibrations are done initially in blocks of known density. These blocks must be simple, predictable materials. The block primary material and the trace elements must be determined by chemical analysis and must be taken into account to determine a real density. Then the Z/A ratio of each must be accounted for, and an apparent density is determined. Several blocks of densities covering the range expected should be chosen and analyzed, then the apparent density determined. The materials commonly chosen are aluminum, magnesium, and water. Sometimes various plastics or sulfur are used. Plastics give a problem in that the real hydrogen content is difficult to determine with sufficient accuracy.

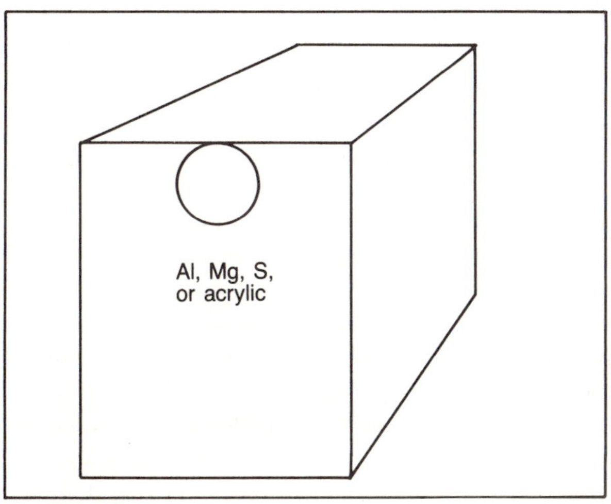

FIG. 5–51
Calibrating block for density

Aluminum has a density of 2.70 g/cc. Its Z/A ratio is 0.4818, so its apparent density is 2.60 g/cc. Aluminum, like many commercial metals, is seldom pure. It has various impurities and alloyed metals. A typical chemical assay for an aluminum block may be as shown in Table 5–7.

The real bulk density of this block of aluminum and alloyed metals is:

$$\rho_b = \Sigma\left[\left(\frac{\text{wt\%}}{100}\right)\left(\begin{matrix}\text{bulk density of the component}\\ \text{element}\end{matrix}\right)\right] \qquad (5.54)$$

ρ_b = 2.82 g/cc

The apparent density is:

$$\rho_a = \Sigma\left[\frac{\text{wt\%}}{100}\left(\frac{Z/A}{0.5}\right)\left(\begin{matrix}\text{bulk density of the}\\ \text{component element}\end{matrix}\right)\right] \qquad (5.55)$$

ρ_a = 2.71 g/cc

The total Z/A ratio is:

$$Z/A = \Sigma\left[\frac{\text{wt\%}}{100}\ (Z/A \text{ of the component element})\right] \qquad (5.56)$$

Z/A = 0.4810

Magnesium typically has large amounts of aluminum, copper, and silicon in it, but the procedure is the same as for aluminum.

Water is an excellent material to use for the low-density end because it can usually be obtained with negligible impurities (<500 ppm). It does have a Z/A ratio of 0.55 and an apparent density of 1.111

Table 5–7
Aluminum Block Assay

	Wt%	Bulk Density	Apparent Density	Z/A Ratio
Al	94.00	2.70	2.60	0.4818
Mg	3.02	1.74	1.73	0.4975
Cu	2.08	8.92	8.14	0.4564
Fe	0.40	7.86	7.37	0.4687
Si	0.30	2.40	2.39	0.4985
Ca	0.15	1.50	1.50	0.4990
All others	0.05	7.40	7.40	0.5000 assumed

g/cc, however. It can have salts dissolved in it to raise the density. Sodium chloride, for example, can be used to raise the apparent density of the water solution to more than 1.4 g/cc.

If the front-to-back ratio of the density tool is greater than 200, no correction need be made for borehole effects. However, if it is less, calibrations must be made with both air- and water-filled boreholes. This procedure does not consider photoelectric absorption. It can be a significant factor for elements of higher atomic number than 20 (calcium). The PE absorption with iron, nickel, copper, molybdenum, and such appear heavier than they really are.

For example, iron has a counting rate about 11% too low when a cesium-137 source is used because of photoelectric absorption. Nickel has a low counting rate of 14%; copper, 17%; and molybdenum, about 37% too low. Fortunately, these elements usually appear only as trace alloys. If they are present in more than trace amounts, a correction must be made.

A further effect that must be taken into account is entrapped gases in aluminum and magnesium. These gases can be in significant amounts (2–5 vol%). Also, they often are dispersed so finely that they are virtually undetectable. If possible, the block should be forged after casting and before machining. Finally, a machined sample may be taken from the borehole space for density determinations and chemical assaying. If possible, the entire block should be weighed accurately, and the density should be calculated.

The calibration block, or tank in the case of water, must be large enough that the effect of external materials is negligible. If we assume that the source is cesium-137 at 660 KeV energy and we wish less than 0.1% effect, the radius of the block must be greater than 16.4 cm for example (Fig. 5–51). We assumed the gamma photon would travel, maximum, from the source to the outside edge and back to the detector. Therefore, the relationship used was:

$$x = \frac{\ell n\,(I/I_o)}{-2\mu\rho_a} \qquad (5.57)$$

Where:

- x = Minimum distance through the block from the borehole to the outside wall (radius)
- I/I_o = Relative intensity at the detector, 0.1% or 0.001 in this case
- μ = Mass absorption coefficient for 660 KeV gamma rays, 0.078
- ρ_a = Apparent density of aluminum, 2.71 g/cc

For water the minimum radius must be 15 in. (38 cm).

Once three reliable points have been determined for the tool response, a curve of counting rate or logarithm of counting rate can be plotted against the apparent density. The curve on semilog paper should be nearly a straight line with a slight droop at the high- and low-density ends. An example is shown in Fig. 5–52a. A best-fit equation can be determined if a general calculator or computer solution is desired.

■ Activation

There are several gamma-ray activation systems, some of which may be very effective. Most of these systems are surface systems. The main problem discouraging the use of activation by gamma rays has been the fact that most reactions involving gamma-ray activation require high energy gammas. Most easily available sources have maximum energies of 2 MeV or less. Electronic sources have been suggested, however, and appear promising. Also, an iron reaction with 6 MeV gammas appears quite promising. No activation methods are in commercial use at present.

■ X-Ray Fluorescence

A valuable reaction, which is now being offered commercially, is X-ray fluorescence. With this reaction a gamma energy is chosen that has sufficient energy to excite the K- or L-level orbital electrons. These excited electrons then decay back to their former energy, emitting X-ray photons of a characteristic energy.

One typical laboratory system uses an electronic source of X-rays to excite the electrons and a diffraction grating to separate the emitted X-rays by energy so their intensities may be read. This type of analysis can be extremely effective. It is as accurate, in many cases, as a wet chemical analysis, and it is much less trouble and is faster.

A portable field unit uses an isotopic source and a pair of filters chosen for a specific element. The sample is excited, and the emission intensity is read through a filter whose absorption edge is just above the emission energy of the chosen element. A second reading is taken through a filter whose absorption edge is just below the characteristic energy. The difference in readings is the fluorescent emission and has an intensity that is a function of the amount of the subject element present in the sample. For several reasons these instruments are better suited to qualitative measurements in raw field conditions. They are, however, very effective in controlled conditions such as a probe tower.

X-ray fluorescent systems are not often used downhole because the characteristic X-ray usually is of a

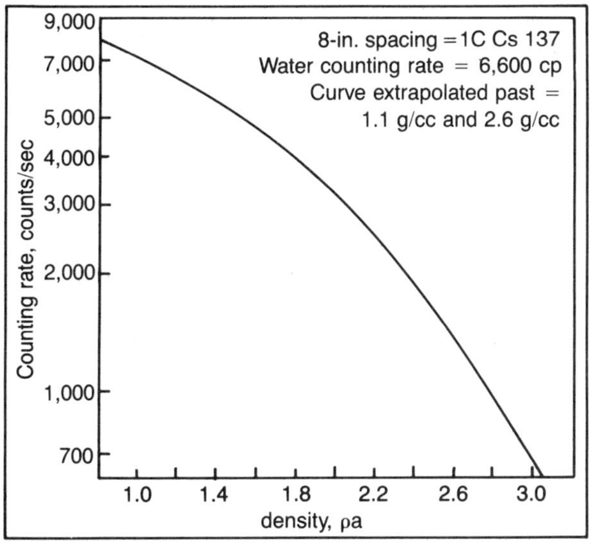

FIG. 5–52a

Typical density calibration curve
(courtesy Century Geophysical Corp.)

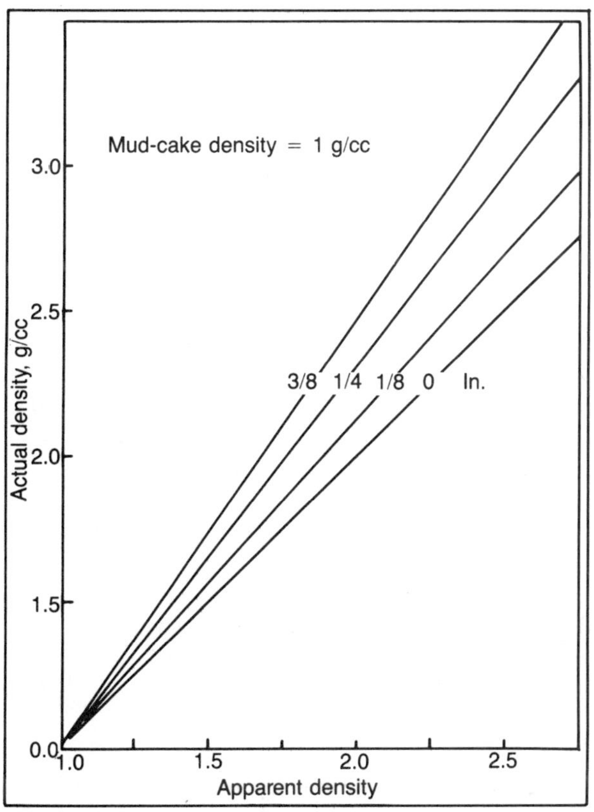

FIG. 5–52b

Standoff error of the density tool

low energy. It is difficult to have a low-mass housing required to pass the low energy X-rays. Steel is usually needed to protect the instrument from the borehole environment.

X-ray fluorescence methods, in general, are excellent for both identification and quantity. Where they are suited, they can rival chemical analysis in quality. They are quite sensitive to sample preparation, particularly grain size and moisture content.

■ Gamma Assaying

Gamma assaying is used to try to determine, quantitatively, the amount of radioactive material (usually uranium) in a sample from a core or cuttings. It suffers from the same drawbacks and has the same advantages as an in situ natural gamma determination. It has the advantage of a tightly controlled environment but the disadvantage of a very small sample—it is not representative statistically. These two factors must be taken into account.

A spectrographic gamma assay is appreciably better than a gross-count assay. Usually, a sample can be examined for a long enough time and with a thin enough sample that the statistical problems and Compton scattering are lessened. With a suitable detector, such as a germanium detector, it is possible to read the gamma emission from daughters that are so close to the uranium in time that they give a quantitative assay, independent of disequilibrium.

A *closed-can assay* is a means of determining disequilibrium. In this method a sample of uranium mineral is sealed in a thin-walled metal can. Randon and the subsequent daughter products are allowed to buildup over a period of about 30 days. Three or more gross-count gamma-ray readings are taken during the 30-day period to define the buildup curve. This curve is extrapolated to determine a level to which it is asymptotic. That level is assumed to be the in situ gamma-ray level. The uranium determination from this level is compared with a chemical or X-ray analysis of the same sample to determine the actual uranium content and the disequilibrium. These readings can also be compared to the natural gamma-ray log. However, the difference in sample sizes must be considered in this latter case.

■ Beta-Gamma Method

Another laboratory technique is the *beta-gamma method.* This method also has been attempted downhole but has the same problems, perhaps more severe, that the downhole X-ray fluorescence has, with the tool housing shielding a beta particles from the detector. In addition, sample preparation is more critical; water content, grain size, and sample surface character are important. Disequilibrium problems are present.

The beta-gamma method assumes that the gamma emission from uranium compounds is from the lower daughter products, i.e., bismuth-214 and lead-214. It also assumes that the daughter products whose atomic number are greater than 87 are emitters of a large proportion of the beta particles. If these assumptions are correct, the ratio of beta-plus-gamma emission to gamma only, is a function of the uranium content of the sample. Fig. 7–28 is a diagram of the uranium decay series, showing the alpha and beta emissions.

The beta-gamma system must use samples that are carefully dried and of uniform texture because of the low penetrating power of the beta particles. Also, a particular instrument and detector must be calibrated for a given type of ore and checked with representative samples frequently. The beta-gamma method can be very effective under the right circumstances. However, there are many pitfalls.

NEUTRON LOGGING

Neutron logging in its most common use is designed to measure formation porosity. It is, however, sensitive to chlorine and trace element content of the borehole and formation. It has gained additional favor because it is an effective cased-hole porosity device. Most casing is quite transparent to the neutrons. Noncentered casing and PVC casing can cause trouble in interpretation.

Neutron porosity logging makes use of the efficient moderation of neutrons by hydrogen. The epithermal neutron population, the thermal neutron population, or the capture gamma-ray intensity is interpreted in terms of the hydrogen content of the formation. Since most of the formation hydrogen is in the fluids, the neutron porosity systems measure the relative amount of fluid in the formation. Thermal neutrons are neutrons that have an energy equal to the kinetic or thermal energy of the surrounding material. This is about 0.033 eV. Epithermal neutrons have energies up to about 100 eV. Fast neutrons have energies above this.

There are, of course, aberrations to these processes, as with the water content of clays. These can interfere and give a false picture of porosity. The aberrations also can be used to emphasize and quantify these departures, especially when the neutron log information is combined with that of some other device. Neutron systems are effectively used with other types of porosity tools and are covered under multiple-curve methods.

Neutron systems are effectively used as normalizing systems for some other methods, such as the DFN tool. Neutron capture cross sections and the resulting capture gamma emission can be effectively used for elemental and lithological determination. These are promising new fields. The neutron population or reaction can be interpreted in terms of the mechanical or chemical composition of the formation and the probabilities of various reactions.

Neutron logging circuitry is similar to other types of radioactivity circuitry, except for the detector. The most popular detector at this time for neutron logging is the $^{3}_{2}He$ detector.

Neutron methods have been gaining popularity in recent years, both in petroleum and nonpetroleum usage. The various methods include the porosity (hydrogen determination) tools, chlorine (cross-section) detection, lithology (cross-section) determination, decay-time and live-time (activation) methods, elemental (activation) analysis, and temperature (spectrographic) measurements.

Principles

There are virtually no free neutrons in the earth's formations. Therefore, we can assume that any neutrons present during logging are the result of the neutron logging process. This is convenient because it means the elimination of a background component as a variable to be evaluated.

All neutron logging tools contain a source of neutrons. There are three general types. The first source is isotopic and undergoes spontaneous fission with the emission of neutrons. The major representative of this type of source in logging is californium-252 ($_{252}Cf^{98}$). It has a 2.63-year half-life and an output of about 4.4×10^9 neutrons/sec/curie at 2.3 MeV average energy (6 MeV maximum). It is a moderate gamma emitter. It emits about 2×10^6 neutrons/sec/microgram, so it is a very efficient emitter of neutrons and almost a point source. Cf-252 is a reactor product and has been dropping in price rapidly. It is competitive with many other source types. It does, however, have a half-life that is short for analog applications. It has a somewhat broader, lower, average energy neutron output than other common sources.

The second type of source is the alpha-beryllium source. It is an isotopic alpha-particle source. The alpha particles are reacted with beryllium or sometimes lithium or boron:

$$^{4}_{2}\text{He} + ^{9}_{4}\text{Be} \longrightarrow ^{12}_{6}\text{C} + ^{1}_{0}\text{n} + Q \qquad (5.58)$$

Some of the isotopes used in this reaction are listed in Table 5–8. The most common one at this time is americium-241. It has a reasonably long half-life (433 years) and a low-energy gamma output. Thus, decay corrections and shielding problems are not difficult.

The third type of source is the linear accelerator or neutron generator. This source usually uses the deuterium-tritium reaction:

$$^{2}_{1}\text{H} + ^{3}_{1}\text{H} \longrightarrow ^{4}_{2}\text{He} + ^{1}_{0}\text{n} \qquad (5.59)$$

The resulting neutrons have an energy of 14 MeV. Occasionally, a deuterium-deuterium reaction is used, yielding 8 MeV neutrons, but it has a lower yield.

The generator uses deuterium gas, which is boiled from a zirconium filament. A high-intensity field ionizes the gas, which is then accelerated by an intense field gradient and focused magnetically into a target containing tritium. The reaction (Eq. 5.59) takes place at high particle velocities. Alternate processes may use tritium accelerated onto a deuterium target or a mixture of deuterium and tritium at both ends.

The accelerator sources may be operated in a pulsed or a continuous mode. Outputs range from 5×10^{7} to 10^{13} neutrons/sec, depending on the tube design and power. The big advantages to the generator-type source are that it may be turned off when not in use, it may be gated on and off in use, and its output can be varied electrically. Fig. 5–53 is a diagram of one type of neutron generator.

Regardless of which process or reaction is used to generate the neutrons, they enter the formation at high velocity or energy. Within the formation the neutrons undergo collisions with the formation atoms and lose part of their energy to them. The loss of energy, ΔE, for head-on collisions only is given by:

$$\Delta E = E\left[1 - \left(\frac{A-1}{A+1}\right)^{2}\right] \qquad (5.60)$$

Thus, for the common geologic materials up to an atomic mass of 40, the maximum fractional loss per head-on collision and the mean number of collisions to thermalize high energy neutrons is as follows:

	Z	A	$\frac{\Delta E}{E}$	Mean No. Collisions to Thermalize
H	1	1.0079	1.00	18
He	2	4.0026	0.64	40
Li	3	6.941	0.44	
Be	4	9.0122	0.36	50
B	5	10.81	0.31	
C	6	12.011	0.28	110
N	7	14.0067	0.25	
O	8	15.9994	0.22	
F	9	18.998	0.19	
Ne	10	20.179	0.18	
Na	11	22.9898	0.16	
Mg	12	24.305	0.15	
Al	13	26.9815	0.14	
Si	14	28.086	0.13	
P	15	30.9738	0.12	
S	16	32.06	0.12	
Cl	17	35.053	0.11	
A	18	39.948	0.10	
K	19	39.098	0.10	
Ca	20	40.08	0.10	

Obviously, every neutron does not have a head-on collision. However, this does serve to show that the energy transfer with hydrogen is extremely efficient.

Table 5–8

Common Neutron Source Materials

Sources Alpha-Berylium	Yield/Curie	Energy Average, MeV	Half-Life
$^{254}_{84}$Po-Be	2.5×10^{6} n/s	4.2	138d
$^{241}_{95}$Am-Be	2.2×10^{6} n/s	4.0	458y
$^{226}_{88}$Ra-Be	1.5×10^{7} n/s	3.9	1,600y
$^{239}_{94}$Pu-Be	2.2×10^{6} n/s	4.5	24,400y
$^{238}_{94}$Pu-Be	2.5×10^{6} n/s	4	86y
Spontaneous emitters			
$^{252}_{98}$Cf	4.4×10^{9} n/s	2.3	2.65y
Fission products			
$^{235}_{92}\text{U} + \text{n} \longrightarrow$ fission fragments + prompt neutrons + delayed neutrons			
$^{17}_{8}\text{O} + \text{n} \longrightarrow ^{17}_{7}\text{N}$ + Neutrons + γ			
$^{2}_{1}\text{H} + ^{3}_{1}\text{H} \longrightarrow ^{4}_{2}\text{He}$ + Neutron			
$^{2}_{1}\text{H} + ^{2}_{1}\text{H} \longrightarrow ^{3}_{2}\text{He}$ + Neutron			

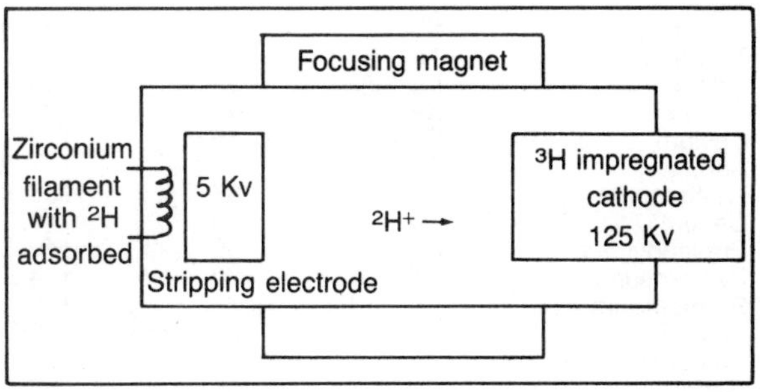

FIG. 5–53

Neutron accelerator

Table 5–9

Elements, Atomic Numbers and Weights, Valence, and Thermal Neutron Capture Cross Sections

Element	Symbol	Atomic Number, Z	Atomic Weight, A	Valence	Thermal Neutron Capture Cross Section, barns
Hydrogen	H	1	1.008	+1, −1	0.33
Helium	He	2	4.003	0	0.0000
Lithium	Li	3	6.939	+1	0.0376
Beryllium	Be	4	9.013	+2	0.009
Boron	B	5	10.811	+3	758.86
Carbon	C	6	12.011	+2, +4, −4	0.0034
Nitrogen	N	7	14.007	+1, +2, +3, +4, +5, −1, −2, −3	0.075
Oxygen	O	8	15.999	−2	0.0002
Fluorine	F	9	18.998	−1	0.0098
Neon	Ne	10	20.183	0	0.0038
Sodium	Na	11	22.898	+1	0.400
Magnesium	Mg	12	24.120	+2	0.0625
Aluminum	Al	13	26.982	+3	0.232
Silicon	Si	14	28.086	+2, +4, −4	0.1638
Phosphorous	P	15	30.974	+3, +4, −3	0.190
Sulfur	S	16	32.065	+4, +6, −2	0.49
Chlorine	Cl	17	35.457	−1, +1, +5, +7	33.338
Argon	Ar	18	39.946	0	0.668
Potassium	K	19	39.102	+1	2.152
Calcium	Ca	20	40.080	+2	0.455
Scandium	Sc	21	44.96	+3	25.000
Titanium	Ti	22	47.90	+2, +3, +4	6.401
Vanadium	V	23	50.95	+2, +3, +4, +5	4.800
Chromium	Cr	24	52.01	+2, +3, +6	3.074
Manganese	Mn	25	54.94	+2, +3, +4, +7	13.300
Iron	Fe	26	55.47	+2, +3	2.514
Cobalt	Co	27	58.94	+2, +3	37.5
Nickel	Ni	28	58.71	+2, +3	4.264
Copper	Cu	29	63.54	+1, +2	3.812
Zinc	Zn	30	65.38	+2	1.06
Gallium	Ga	31	69.72	+3	1.147
Germanium	Ge	32	72.60	+2, +4	1.470
Arsenic	As	33	74.92	+3, +5, −3	4.300
Selenium	Se	34	78.96	+4, +6, −2	12.0
Bromine	Br	35	79.916	−1, +1, +5	2.798
Krypton	Kr	36	83.80	0	31.0
Rubidium	Rb	37	85.48	+1	0.7
Strontium	Sr	38	87.63	+2	1.2
Yittrium	Y	39	88.92	+3	1.3
Zirconium	Zr	40	91.22	+4	0.18
Niobium	Nb	41	92.91	+3, +5	1.2
Molybdenum	Mo	42	95.95	+6	2.7
Technetium	Tc	43	97.00	+4, +6, +7	22.0
Ruthenium	Ru	44	101.1	+3	2.6
Rhodium	Rh	45	102.91	+3	150.0
Paladium	Pd	46	106.4	+2, +4	8.0
Silver	Ag	47	107.70	+1	96.00
Cadmium	Cd	48	112.41	+2	7,000.00
Indium	In	49	114.82	+3	190.0
Tin	Sn	50	118.70	+2, +4	0.62
Antimony	Sb	51	121.76	+3, +5, −3	5.7
Tellurium	Te	52	127.61	+4, +5, −2	4.7
Iodine	I	53	126.91	−1, +1, +5, +7	6.20
Xenon	Xe	54	131.30	0	35.0
Cesium	Cs	55	132.91	41	28.0
Barium	Ba	56	137.36	+2	1.20
Lanthanum	La	57	138.92	+3	8.9
Cerium	Ce	58	140.13	+3, +4	0.73
Praseodymium	Pr	59	140.92	+3	3.90
Neodymium	Nd	60	144.27	+3	46.0
Prometheum	Pm	61	145.0	+3	
Samarium	Sm	62	150.35	+2, +3	5,600.0
Europium	Eu	63	152.0	+2, +3	1,400.0
Gadolinium	Gd	64	157.26	+3	46,000.0
Terbium	Tb	65	158.93	+3	46.0
Dysprosium	Dy	66	162.50	+3	930.0
Holmium	Ho	67	164.94	+3	65.0
Erbium	Er	68	167.20	+3	173.0

Source: Gearhart Industries Inc.

Table 5–9 *continued*

Element	Symbol	Atomic Number, Z	Atomic Weight, A	Valence	Thermal Neutron Capture Cross Section, barns
Thulium	Tm	69	168.94	+3	127.0
Ytterbium	Yb	70	173.04	+2, +3	37.0
Lutetium	Lu	71	174.99	+3	112.0
Hafnium	Hf	72	178.49	+4	105.0
Tantalum	Ta	73	180.95	+5	21.0
Tungsten	W	74	183.86	+6	19.2
Rhenium	Re	75	186.2	+4, +6, +7	86.0
Osmium	Os	76	190.2	+3, +4	15.3
Iridium	Ir	77	192.2	+3, +4	440.0
Platinum	Pt	78	195.09	+2, +4	8.8
Gold	Au	79	197.0	+1, +3	98.8
Mercury	Hg	80	200.61	+1, +2	380.0
Thallium	Tl	81	204.39	+1, +3	7.0
Lead	Pb	82	207.19	+2, +4	0.17
Bismuth	Bi	83	208.99	+3, +5	0.034
Polonium	Po	84	209.0	+2, +4	0.003
Astatine	At	85	210.0	−1	
Radon	Rn	86	222.0	0	0.72
Francium	Fr	87	223.0	+1	
Radium	Ra	88	226.0	+2	20.0
Actinium	Ac	89	227.0	+3	510.0
Thorium	Th	90	232.04	+4	7.4
Protractinium	Pa	91	231.0	+4, +5	260.0
Uranium	U	92	238.03	+3, +4, +5, +6	2.7

Therefore, the primary moderator, or slowing-down agent, in the formation is hydrogen. It is six times more likely to thermalize neutrons than the next closest common geological element, carbon. A list of thermal neutron cross sections is shown in Table 5–9.

Moderation also depends on the likelihood of a collision or cross section. Cross section is literally the target size of an atom. It is a probability function and is in barns. One barn is 10^{-24} sq cm. Cross section, σ, is defined as:

$$\sigma = (N/n_o n_t) \tag{5.61}$$

Where:

N = Number of reactions per square centimeter of target area
n_o = Number of incident particles per square centimeter of target area
n_t = Number of nuclei per square centimeter of target area

As the neutron slows down, there are several reactions with various isotopes it may undergo. However, when it has reached thermal energy, there is a high probability it will react with a number of things. Primarily though, the neutron population at thermal and epithermal energies is a function of the amount of hydrogen in sedimentary formations and the original or source neutron flux rate. Fig. 5–54a is a typical neutron population log that reflects the porosity of the formation material.

The thermal neutron population also depends on the thermal neutron reactions. Since there are many of these, the thermal population may be depleted by other materials in the formation (i.e., chlorine, boron, gadolinium, and other high cross-section isotopes).

When a neutron reacts with a nucleus, part of the mass appears as energy, usually as gamma energy.

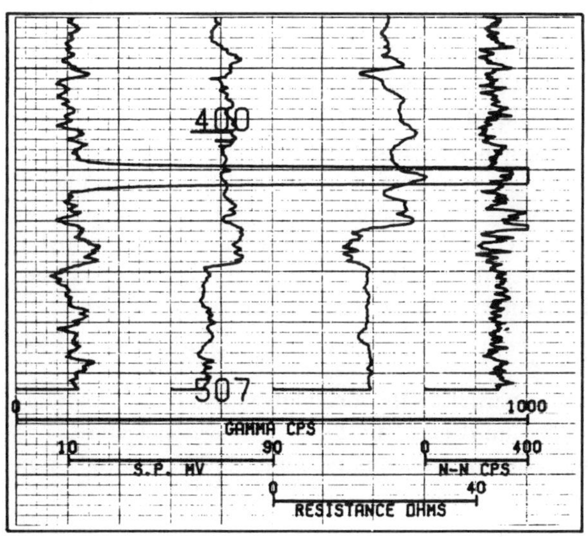

FIG. 5–54a

Typical neutron log
(courtesy Century Geophysical Corp.)

The likelihood of absorption and reaction is a function of the reciprocal of the neutron velocity. Thus, thermal neutrons are most likely to be absorbed. A thermal neutron has a mean free path of about 1 cm in most liquids and solids.

The capture gammas may be considered to be a function of the thermal neutron population. In a high hydrogen environment, as in a porous sand, the gamma intensity is therefore a function of the hydrogen present in the formation. However, gamma detection for porosity determination must consider the chlorine and trace element concentration of the formation and borehole.

The gamma energies resulting from neutron absorption are characteristic of the element or isotope reacting. Therefore, spectrographic detection of capture gammas provides a means of elemental identification. This is a very important new technique in mineral logging.

Also, at thermal energies neutrons easily cause fission in uranium-235 and thorium-232. This is used in the prompt fission neutron (PFN) and delayed fission neutron (DFN) techniques.

Detectors

As mentioned, the neutron systems resemble other radioactivity circuits quite closely. Since a neutron has no charge and no electromagnetic phenomena associated with it, the detector must be somewhat different from gamma and charged particle detectors. These neutron detectors employ materials that have large cross sections to neutrons.

In logging, there are three major types of detectors in use. The early neutron porosity equipment (some still in use today; Schlumberger GNT series, for example) and all of the spectrographic equipment (elemental analysis) make use of gamma-ray detectors. The response of these must take the formation chemistry into account.

A second detector is of the scintillation type, using a material that has a high neutron cross section. Lithium scintillators use the reaction:

$$^{6}_{3}Li + ^{1}_{0}n \longrightarrow ^{3}_{1}H + \alpha + 4.78 \text{ MeV} \quad (5.62)$$

The α particle is the ionizing agent in this scintillator.

The third class are the tube or gas detectors. The most common of this type and the most common in present logging equipment is the helium-3 detector:

$$^{3}_{2}He + ^{1}_{0}n \longrightarrow ^{3}_{1}H + P + 765 \text{ keV} \quad (5.63)$$

Boron hexafluoride gas is also used in a proportional counter:

$$^{10}_{5}B + ^{1}_{0}n \longrightarrow ^{7}_{3}Li + \alpha + 2.78 \text{ MeV} \quad (5.64)$$

Sometimes the counters are lined with uranium to increase their efficiency. The reaction is:

$$^{235}_{92}U + ^{1}_{0}n \longrightarrow \text{fission fragments} + \sim 160 \text{ MeV} \quad (5.65)$$

Fig. 5–54b shows a typical log made with tube or gas detectors.

All of these reactions are sensitive to epithermal as well as thermal neutrons. The relative sensitivity to epithermal neutrons can be increased by surrounding the detector with cadmium. Cadmium has an extremely high cross section to thermal neutrons but a much lower one to epithermal neutrons. Therefore, it absorbs thermal neutrons and passes most of the epithermal neutrons.

All of the presently used neutron detectors have pulse outputs, with one pulse representing the passage of one neutron through the system. Efficiencies are usually high, and backgrounds are negligible. Therefore, in spite of often low counting rates, reliable measurements may be made with neutron detectors. With the gamma detectors, the natural gamma background must be taken into account.

Neutron Porosity Systems

As a neutron diffuses through a sedimentary formation, the hydrogen in the formation is the most efficient agent for slowing the neutron. There are few chances of absorption of the neutron before it reaches epithermal energy. Thus, the population of epithermal neutrons is a function of the efficiency of moderation of the neutrons. An epithermal neutron detector with a source of neutrons can be used as a measuring device to determine the amount of hydrogenous material in the formation and borehole. This type of detection, however, leaves a rather low counting rate. Therefore, it is not used as much as a mixture of thermal and epithermal detection.

When the neutrons have been slowed further to thermal energies, the population also depends on the cross section of other materials in the formation. These materials absorb thermal neutrons and deplete their population. A moderate correction for formation composition must be made with thermal neutrons. Fig. 5–55 shows this type of correction for a mixture of thermal and epithermal neutron detection. Fig. 5–56 shows a typical family of correction curves for a thermal-epithermal system.

The detection of the capture gamma rays has the greatest amount of variance when it is used for porosity measurements. The capture gamma ray intensity depends on the thermal neutron population, the average capture cross section of the formation, the

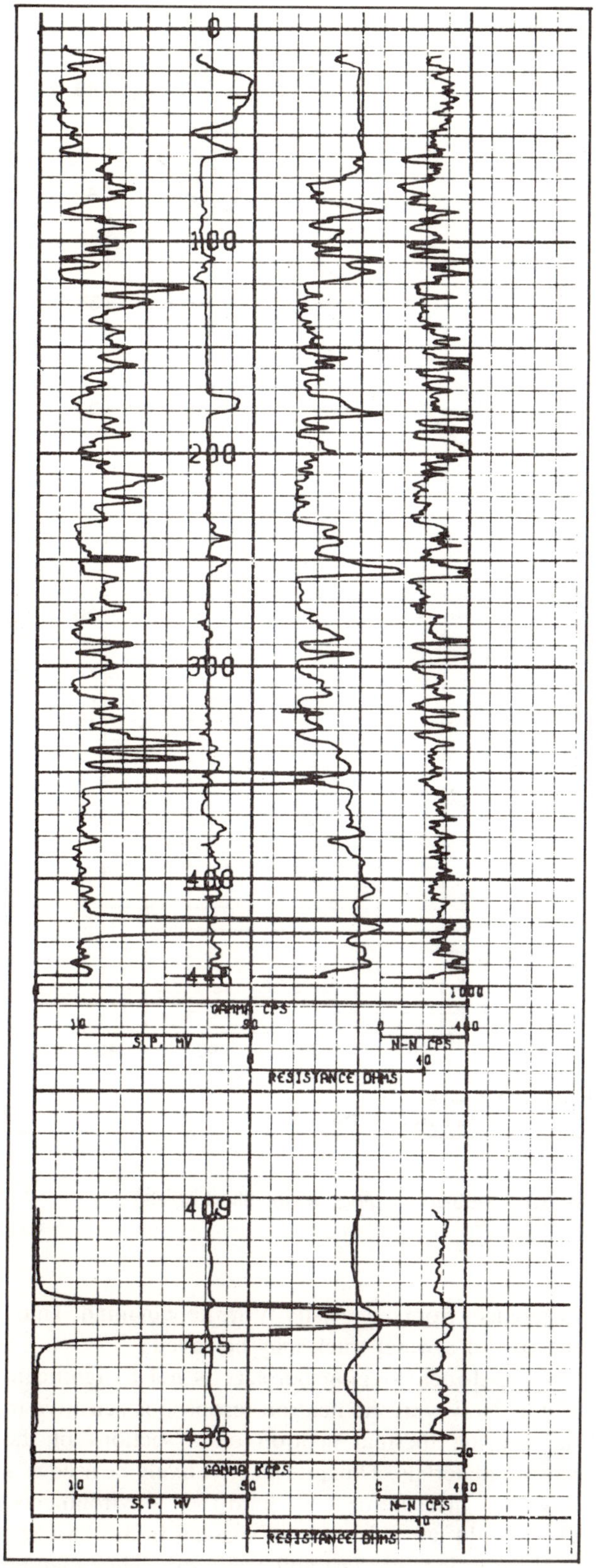

FIG. 5–54b

Typical shallow neutron porosity log
(courtesy Century Geophysical Corp.)

borehole fluid amount and type, and the sensitivity of the detector to natural gamma rays. Typical gamma detection curves for neutron porosity equipment are shown in Fig. 5–57.

As the neutrons from the source diffuse away from the source, they are slowed down at a rate that depends inversely on the atomic weights of the formation atoms. Thus, there is a Gaussian distribution of thermal neutrons around the detector, forming a spheroidal shell whose radius depends inversely on the porosity of the formation. In a 35% porosity sandstone, about 90% of the signal for thermal neutron detection comes from a depth of less than 10 in. Therefore, the peak of the thermal neutron distribution lies about 4–5 in. from the source. As the porosity is decreased to zero for a carbonate formation, this distance of the peak increases approximately 2.5 times that or to 10–12 in. Since the detectors on most modern logging tools are on the order of 15–20 in. from the source, they read the variation on the far slope (Fig. 5–58). Thus, the counting rate is an inverse function of the porosity.

Most porosity logging tools are omnidirectional devices that ride the wall of the hole by gravity. One must be careful with such tools that the probe does not float away from the hole wall because of the Bernoulli effect. This effect is a danger with any hole whose deviation is less than about 2°.

Some neutron porosity tools are sidewall tools. The borehole effect is minimized by applying the source detector assembly to the wall of the hole and by shielding the source and detector from the borehole. This type of tool is seldom used in nonpetroleum work because it usually is large (≃4-in. diameter).

Neutron porosity systems are usually interpreted for porosity by means of empirically determined charts. These are shown in Fig. 5–56 and Fig. 5–57. The general form for the neutron deflection, ND, can be closely approximated by:

$$ND = A + B \log \phi \qquad (5.66)$$

Where:

A = Intercept at 100% porosity
B = Slope of the curve and a function of hole size
θ = Water filled porosity

This is illustrated in Fig. 5–59.

In general, the neutron porosity tool is a good system to measure porosity. It must be remembered, however, that it will see the bound water in clay or shale as apparent porosity. An independent shale assessment should be made to obtain the effective porosity.

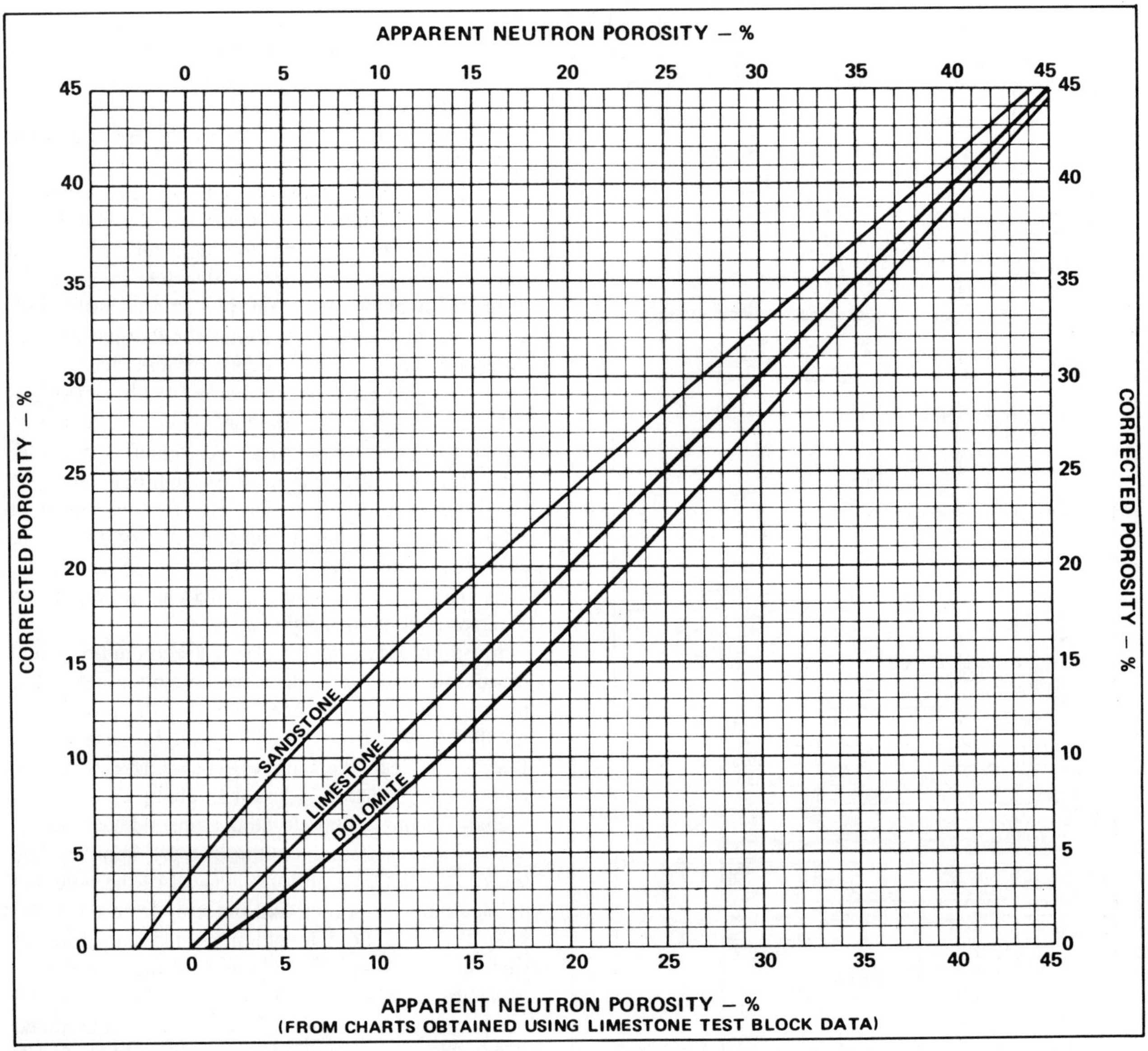

FIG. 5–55

Estimated neutron correction for formation chemistry effects (neutron-neutron logging, water-filled holes)
(courtesy Gearhart Industries Inc.)

The neutron porosity system is a relatively shallow investigation device. Because of the sensitivity to the formation rock composition, the neutron porosity tool is an excellent one to use in crossplotting for lithology determinations.

In mineral exploration a borehole is often drilled with air and is water filled only to record a resistivity curve. Since both the resistivity and neutron curves are porosity curves, the neutron porosity can often be used in place of the resistivity. The neutron tool is effective in cased holes and air-filled holes if the differences from the resistivity are kept in mind. Thus, the cost of a water truck may be saved.

In hard-rock logging porosity is not a significant factor. However, the neutron response is sensitive to the varying thermal neutron cross sections of the various materials. Thus, it can effectively be used as a lithology measurement in hard-rock environments. One should use a fairly long spacing in this case.

Neutron logs are fast and easily run. They may be run in any borehole, gas or liquid filled, open or cased hole. The neutron log resembles the resistivity logs

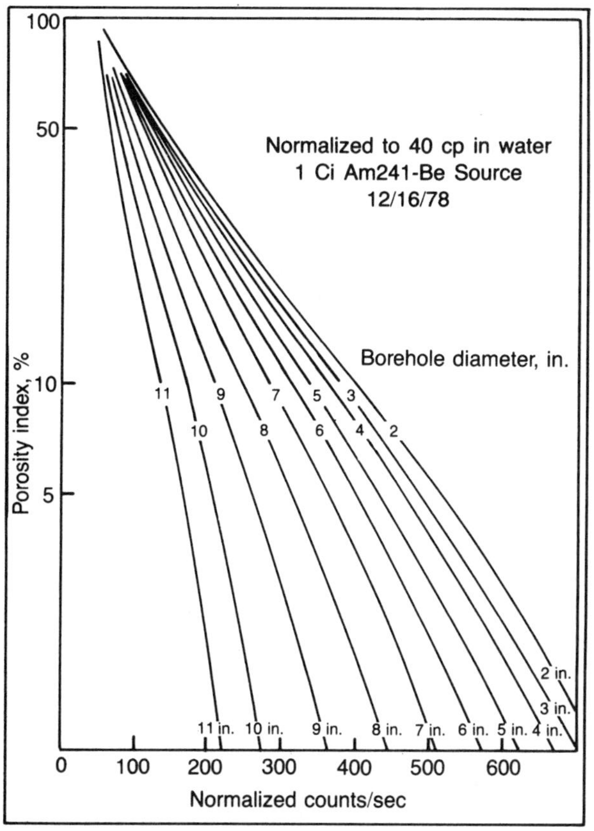

FIG. 5–56

Neutron porosity index in limestone as a function of normalized counts/second
(courtesy Century Geophysical Corp.)

because both are functions of the porosity. The neutron tool responds to the hydrogen content of the formation material. Since in a sediment the hydrogen is mostly contained in the interstitial water and hydrocarbon, the neutron system is a porosity tool.

Eq. 5.66 shows the response of the neutron porosity system. The assumption that the response is a linear one is only an approximation, but it is a good one. The term A is an artificial one. This is the intercept that a plot would have if the response were linear. It would be the system response in a solid-rock environment. At the other extreme, if the assumed linear response is extrapolated to 100% porosity with the slope B, the response of the system would be somewhat in error for a 100% pure-water medium. The value of B depends on the borehole size and material. Between the porosity limits of 5–45%, the error is small enough that the simplification is a valuable one.

In any given borehole situation the plot of ND versus ϕ gives a locus of points for each rock type. The relationship from Eq. 5.66 is:

$$ND = A + B \log \phi \tag{5.67}$$

If values for a horizon are plotted on a semilog grid with porosity on the logarithmic scale and the deflection on the linear scale, an extrapolation to 100% porosity gives the value of A. The slope gives the value of B. Do not try to use these as the rock and water values because they will be in error.

Because of its sensitivity to hydrogen, the neutron systems show an apparent high porosity in clay minerals. This is because of the usual large amount of bound water in the clay. In deep shales (below about 2,000 ft or 600 m), this will be 40–45% of the shale. In shallow shales, especially clays, a value well above 45% is not at all unusual. This uncertainty in the water content makes the 2-point calibrating system difficult to use in shallow holes. The uncertainty is compounded by the sensitivity of most systems to the chlorine content of the formation.

The neutron log must be corrected for the clay content of a shaly sand or carbonate. To do this, determine the probable clay content, V_{sh}, with the gamma-ray log as described. Then determine the apparent neutron porosity in the shale or clay, ϕ_{Nsh}. Multiply ϕ_{Nsh} by V_{sh} and subtract the product from the apparent neutron porosity in the shaly sand:

$$\phi_N = \phi_{Na} - V_{sh}\phi_{Nsh} \tag{5.68}$$

Neutron porosity logs run by petroleum logging contractors often present the log scaled in limestone porosity units. These porosity values are only correct when the horizon is porous limestone. Any other rock type results in erroneous porosity values. Nevertheless, the device is a valuable one. First, it gives a quick picture of the approximate porosity. Second, the errors tell much about the formation material. For example, an erroneously low porosity reading can indicate the presence of gas in the pore space. The hydrogen density of gas is quite low (Table 5–10). A porosity reading that is too high can indicate some degree of shaliness. Porosity readings in a carbonate that go below zero indicate some degree of dolomitization.

■ Neutron Activation Systems

When a neutron interacts with an atom, there is an exchange and radiation of energy. Under some circumstances the quantitative detection of that radiated energy can be used to identify the involved atoms. These reactions usually involve thermal neutrons

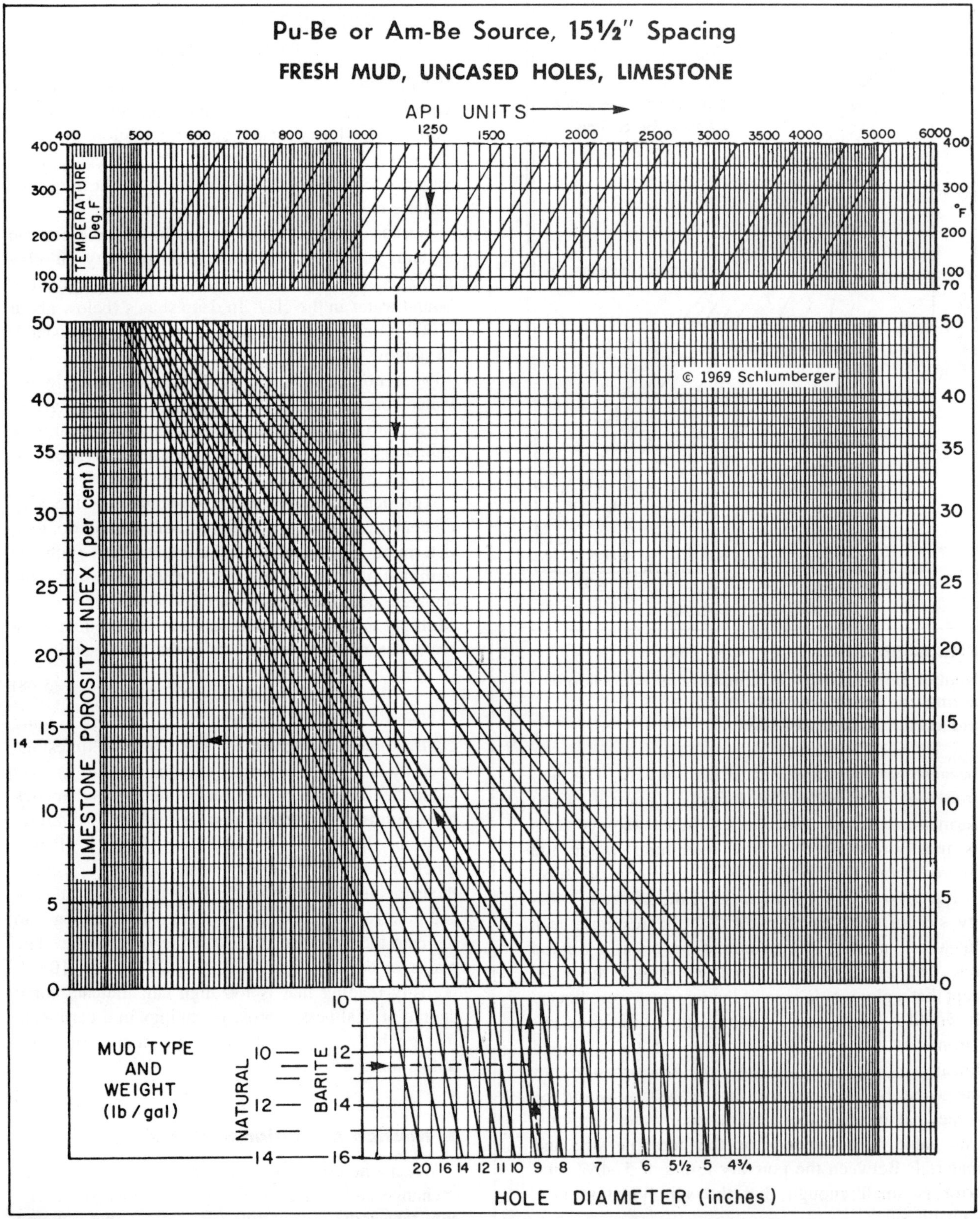

FIG. 5–57

Neutron departure curves—GNT, F, G, or H (courtesy Schlumberger Well Services Inc.)

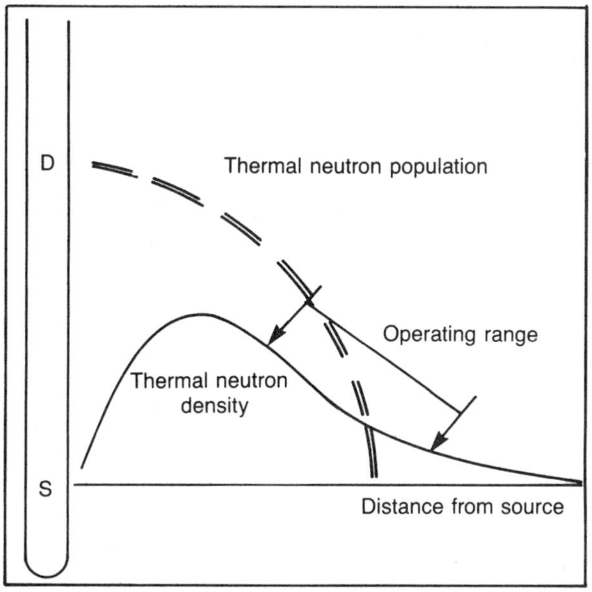

FIG. 5–58

Thermal neutron population radially away from the source

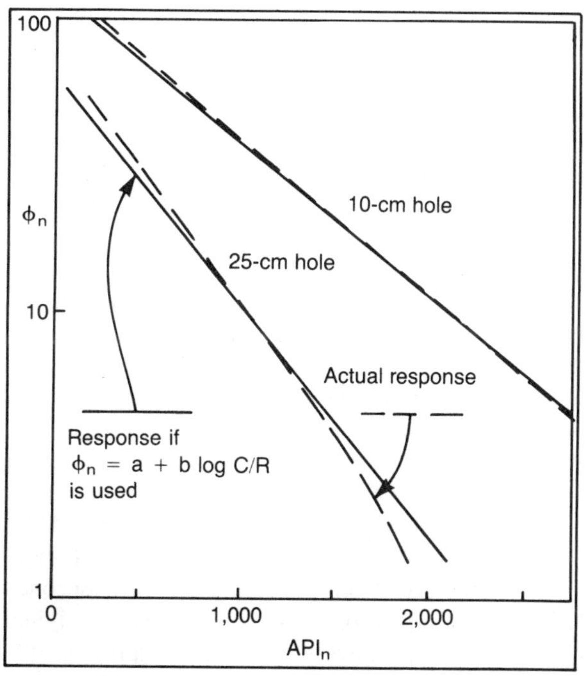

FIG. 5–59

Approximation of neutron porosity response

because of the high probability of thermal neutron capture.

There are three major categories of this class of system: direct uranium measurements, elemental analysis, and lifetime measurements.

Table 5–10

Hydrogen Content of Various Substances

Substance	Hydrogen Atoms $\times 10^{23}$/cc	Hydrogen Index*
Pure water		
60°F, 14.7 psi	0.669	1
200°F, 7,000 psi	0.667	1
Salt water, 200,000 ppm NaCl		
60°F, 14.7 psi	0.614	0.92
200°F, 7,000 psi	0.602	0.90
Methane, CH_4		
60°F, 14.7 psi	0.0010	0.0015
200°F, 7,000 psi	0.329	0.49
Ethane, C_2H_6		
60°F, 14.7 psi	0.0015	0.0023
200°F, 7,000 psi	0.493	0.74
Average natural gas		
60°F, 14.7 psi	0.0011	0.0017
200°F, 7,000 psi	0.363	0.54
N-pentane, C_5H_{12}		
68°F, 14.7 psi	0.627	0.94
200°F, 7,000 psi	0.604	0.90
N-hexane, C_6H_{14}		
68°F, 14.7 psi	0.645	0.96
200°F, 7,000 psi	0.615	0.92
N-heptane, C_7H_{16}		
68°F, 14.7 psi	0.658	0.99
200°F, 7,000 psi	0.632	0.95
N-octane, C_8H_{18}		
68°F, 14.7 psi	0.667	1.00
200°F, 7,000 psi	0.639	0.96
N-nonane, C_9H_{20}		
68°F, 14.7 psi	0.675	1.01
200°F, 7,000 psi	0.645	0.97
N-decane, $C_{10}H_{22}$		
68°F, 14.7 psi	0.680	1.02
200°F, 7,000 psi	0.653	0.98
N-undecane, $C_{11}H_{24}$		
68°F, 14.7 psi	0.684	1.02
200°F, 7,000 psi	0.662	0.99
Bituminous coal, 0.8424 (C) 0.0555 (H)	0.442	0.66
Carnallite	0.419	0.63
Limonite	0.369	0.55
Cement	about 0.334	about 0.50
Kernite	0.337	0.50
Gypsum	0.325	0.49
Kainite	0.309	0.46
Trona	0.284	0.42
Potash	0.282	0.42
Anthracite coal	0.268	0.40
Kaolinite	0.250	0.37
Chlorite	0.213	0.32
Kieserite	0.210	0.31

Source: Gearhart Industries Inc.

*Comparison to fresh water (under equal pressure and temperature in the case of fluids)

The number of hydrogen atoms per cubic centimeter of a substance may be determined as follows:

- Determine the molecular weight of the substance
- Divide the molecular weight by the density
- Divide Avogadro's Number (6.025×10^{23}) by the above quotient
- Multiply the above quotient by the number of hydrogen atoms per molecule of the substance

The density variation for crude oils with temperature and pressure used above was estimated from charts published by D.L. Katz and M. Muskat.

Table 5–10 *continued*

Substance	Hydrogen Atoms $\times 10^{2}$53/cc	Hydrogen Index*
Serpentine	0.192	0.29
Nahcolite	0.158	0.24
Glauconite	0.127	0.19
Montmorillonite	0.115	0.17
Polyhalite	0.111	0.17
Muscovite	0.089	0.13
Illite	0.059	0.09
Biotite	0.041	0.06

Direct Uranium Measurements

The thermal neutrons that result from moderation in the formation can react easily with $^{235}_{92}U$. When the neutron is absorbed, the atom becomes unstable and decays into several smaller fragments. It undergoes fission—that is, uranium-235 is a fissile material. When the fission of uranium-235 takes place, two or more prompt neutrons are emitted. These neutrons moderate to epithermal energy where they may be detected [the prompt fission neutron technique (PFN)]. The fission fragments are also unstable and decay, emitting more delayed neutrons. The average half life of these fragments is less than 22 sec. The neutrons from both of these reactions can be detected when they have reached thermal energies. This is done with the delayed fission neutron technique (DFN).

There are two DFN systems in use. One was designed by Mobil and uses a pulsed generator as a source of neutrons. The other was invented by Eberline and Shreve, run by Kerr-McGee, and improved by IRT Corporation. It uses a californium-252 isotopic source of neutrons, of about 5×10^9 neutrons/sec. The IRT version carries the source on a moving belt to irradiate the formation for a longer time, while moving the probe continuously. The PFN and the Mobil DFN both use electronic generator sources of 14 MeV neutrons. At about 8 MeV the neutrons can react with the $^{17}_{8}O$ in the formation and borehole, causing a fission and a production of neutrons. These must be taken into account on both of these systems as a background reading. Uranium-238 is also fissile at about 8 MeV. However, since the relative composition of uranium is fixed in most natural deposits, the neutrons from uranium-238 aid detection rather than hinder it. The IRT system does not have either of these reactions, since it uses a ^{252}Cf source of 2.3 MeV neutrons, with a maximum energy of 6 MeV.

All three systems appear to be workable. Fig. 5–60 diagrams the three systems. However, all three use complex processes and, therefore, have complex corrections to make. All three are sensitive to the formation porosity and borehole fluid. The IRT system is at a slight advantage in that it does not have the oxygen-17 background to contend with. However, it must use a larger source to compensate for the loss of the uranium-238 contribution.

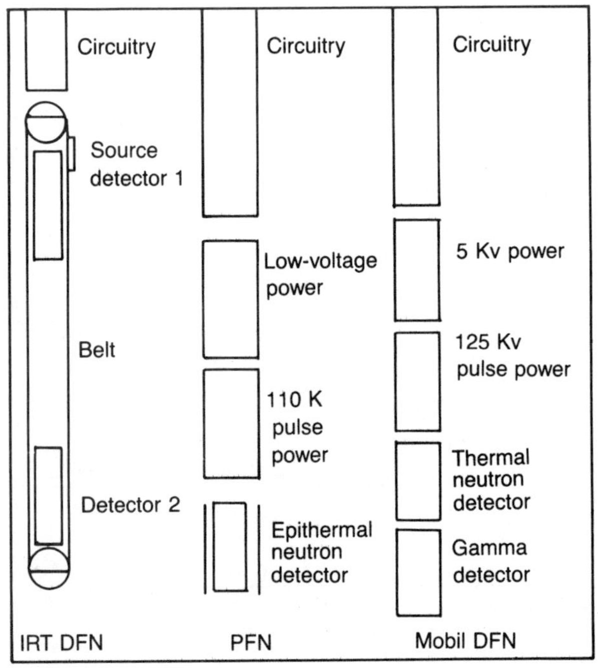

FIG. 5–60
Diagrams of three direct uranium measurement systems

The generator-type sources have the dual advantages that they can be pulsed and can be turned off; the californium-252 source cannot be. On the other hand, the neutron generator has a life of 10^6–10^7 pulses or approximately 5×10^{14} neutrons. The californium-252 source has a half life of about 4.2×10^{17} neutrons.

The PFN system pulses the neutron generator 100–200 times a second for a total neutron output of about 10^9 neutrons/sec. Each pulse is a few microseconds long. After the neutron pulse is over and the original or source neutrons have reacted and disappeared, the epithermal neutron detector is gated on and the population of epithermal neutrons caused by the fission reaction is measured. The original generator neutrons die away in 50–100 μsec. Any neutrons remaining are a function of the amount of fissile material in the formation.

With the Mobil DFN system a similar generator is used, but it is only pulsed two to ten times a second for a total output of about 10^8 neutrons/sec. Detection is delayed until all primary and prompt fission neutrons are gone. Detection times are 100–500 ms. Substantially higher counting rates are obtained than with the PFN system.

The IRT DFN system exposes a portion of the formation with a californium-252 source of about 10^9 neutrons/sec for a period of time. The source is then moved mechanically to another exposure location, and a detector reads the delayed neutrons at the first location.

Handling of all three systems is done by computers in the truck. The general procedure is to run a conventional gamma-ray log. The PFN or DFN is then run through any anomaly on the log, regardless of size. A neutron log is also run to assess the porosity of the formation. With the PFN and Mobil DFN a correction is made for oxygen-17 background count. This is subtracted from the total (corrected) count. The remainder is a function of the uranium and thorium content, borehole size, borehole fluid type, and formation porosity. Fig. 5–61 and Fig. 5–62 show the interpretation curves for the Mobil DFN system. Some trace elements, such as boron and gadolinium, cause errors. However, at present no correction is made for these. Their presence is uncertain, but evidence appears to indicate amounts present which are large enough to cause errors.

Sensitivities of 100 ppm uranium (0.01 wt%) seem to be practical for the DFN systems. The lower sensitivity for the PFN system is unknown.

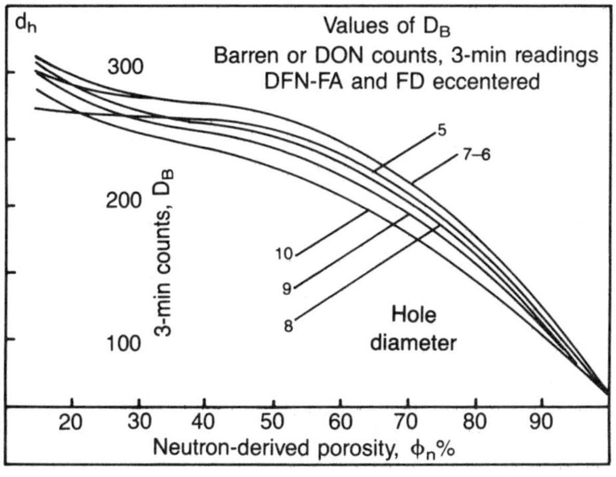

FIG. 5–61

Values of D_B

(courtesy Century Geophysical Corp.)

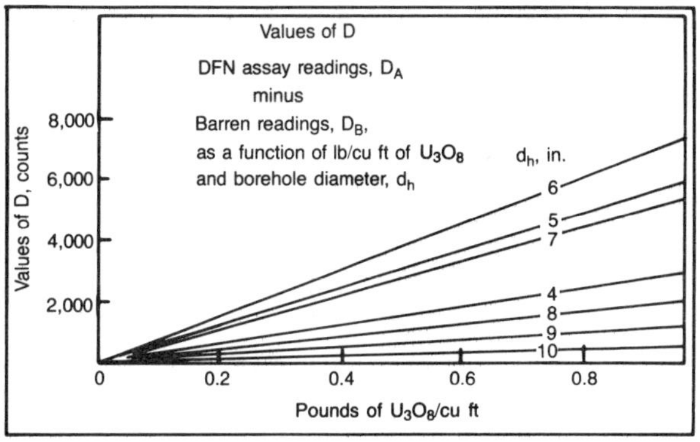

FIG. 5–62

Values of D

(courtesy Century Geophysical Corp.)

■ Elemental Analysis

If the neutron source is combined with the gamma-ray spectrograph, the capture gamma photons can be quantitatively measured at each energy. Capture gamma photons result from the absorption of neutrons by atoms. The absorption takes place easily at thermal energies of the neutrons. The resulting combination of the neutron and atom has less mass than the sum of the two. The difference plus the energy of the incident neutron appears as the energy of the emitted gamma ray. Since the energy of a thermal neutron is low (less than 1 eV) the emitted gamma-ray energy is essentially characteristic of the isotope involved in the absorption. Therefore, identification of the capture gamma energy and intensity provides a means of identifying the type and abundance of the isotope present.

The neutron activation elemental analysis system can use to advantage the high resolution of the germanium detectors. However, a fair amount of work, especially in the laboratory, is done with the lower resolution sodium iodide detectors. This technique can use any source of neutrons. There are several advantages to the pulsed neutron source. Besides the advantage of being able to turn it off to store a pulsed neutron source of 14 MeV, neutrons can cause reactions with iron and oxygen isotopes at about 8 MeV. Since these are both abundant geological constituents, a knowledge of their abundance is valuable. The feature of the possibility of gating off allows the examination of prompt reactions. The isotopic sources must be moved mechanically. Therefore, usually only the long half-life reactions ($T_{1/2} > 1/2$ sec) can be examined.

At present, this technique is slow and requires strong sources. It is, however, one of the most promising of the new mineral techniques. It has, for instance, given trace element analysis in coal, overburdens, and fluvial muds that are quite good.

Some forms of elemental analysis are being offered commercially. One form uses a PGT intrinsic germanium detector assembly and a californium 252 source. An accuracy of ±20% is reported for in situ measurements of 40 ppm silver and 5 ppm gold (gold has a greater dètection sensitivity). An activation gamma-ray spectrum for a gold analysis is shown in Fig. 5–63a, b, and c.

Scintrex reports similar accuracies in copper at a grade of 0.5% in nickel at a grade of 1%. Neutron activation logs of the ratios of Ca/Si and S/O are run routinely and commercially. These give valuable information about the rock type in the presence of hydrocarbons.

■ Neutron Lifetime Logs*

A valuable commercial petroleum logging system is the Neutron Lifetime* or Thermal Decay Time log.** This system puts a short (about 10 μsec) burst of high energy neutrons into the environment around the neutron generator source. Because of the proximity of the hydrogenous borehole, the neutrons interact with the borehole hydrogen first. As they travel outward, they begin to react with the hydrogen of the formation pore space. If the detector is left off for a short period after the initial pulse of neutrons, the borehole reaction is allowed to die away. The reaction that then is occurring is in the formation. At this time the detector may be gated on, and the borehole effect is substantially ignored. Within the pore space the fluid is mostly hydrogenous. The water, however,

*Trademark of Dresser-Atlas.
**Mark of Schlumberger.

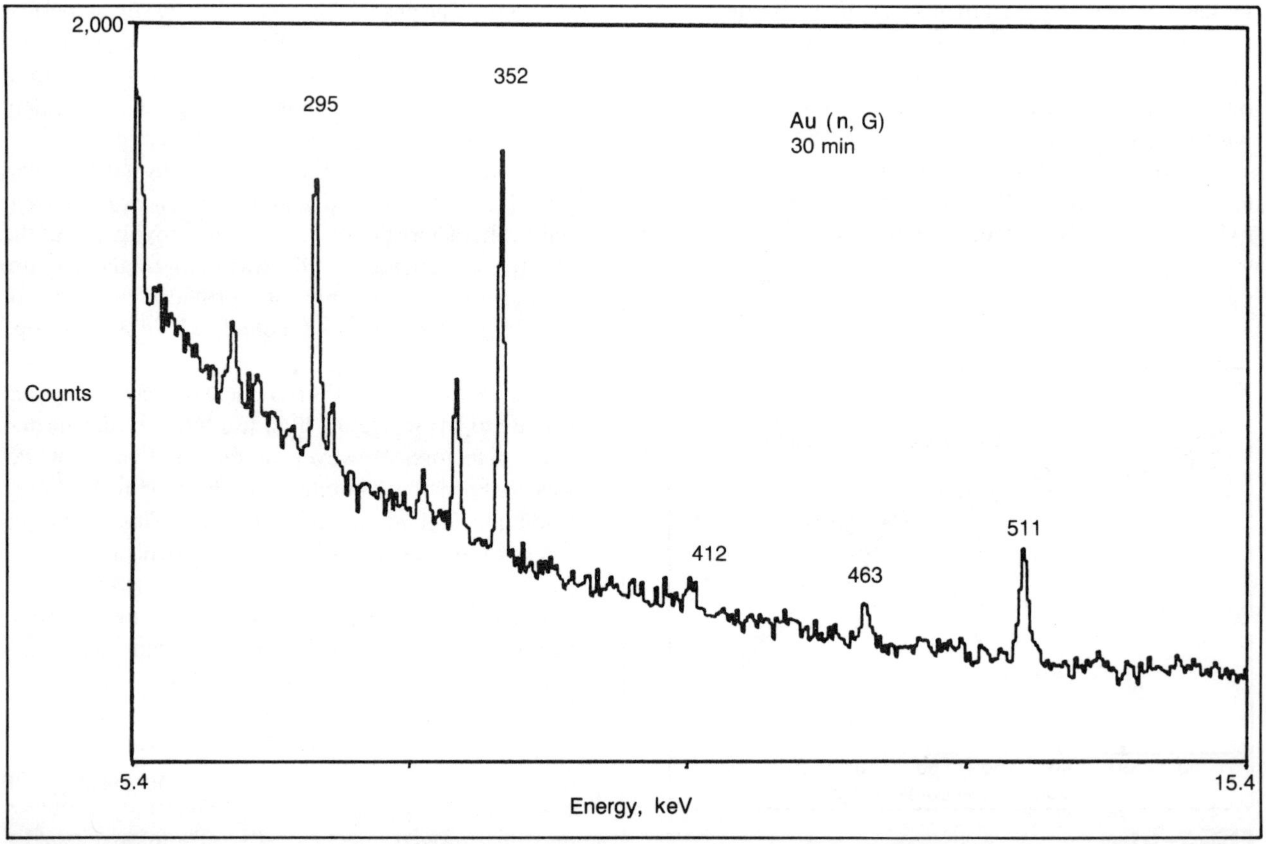

FIG. 5–63a
PGT gold study (courtesy Princeton Gamma Tech.)

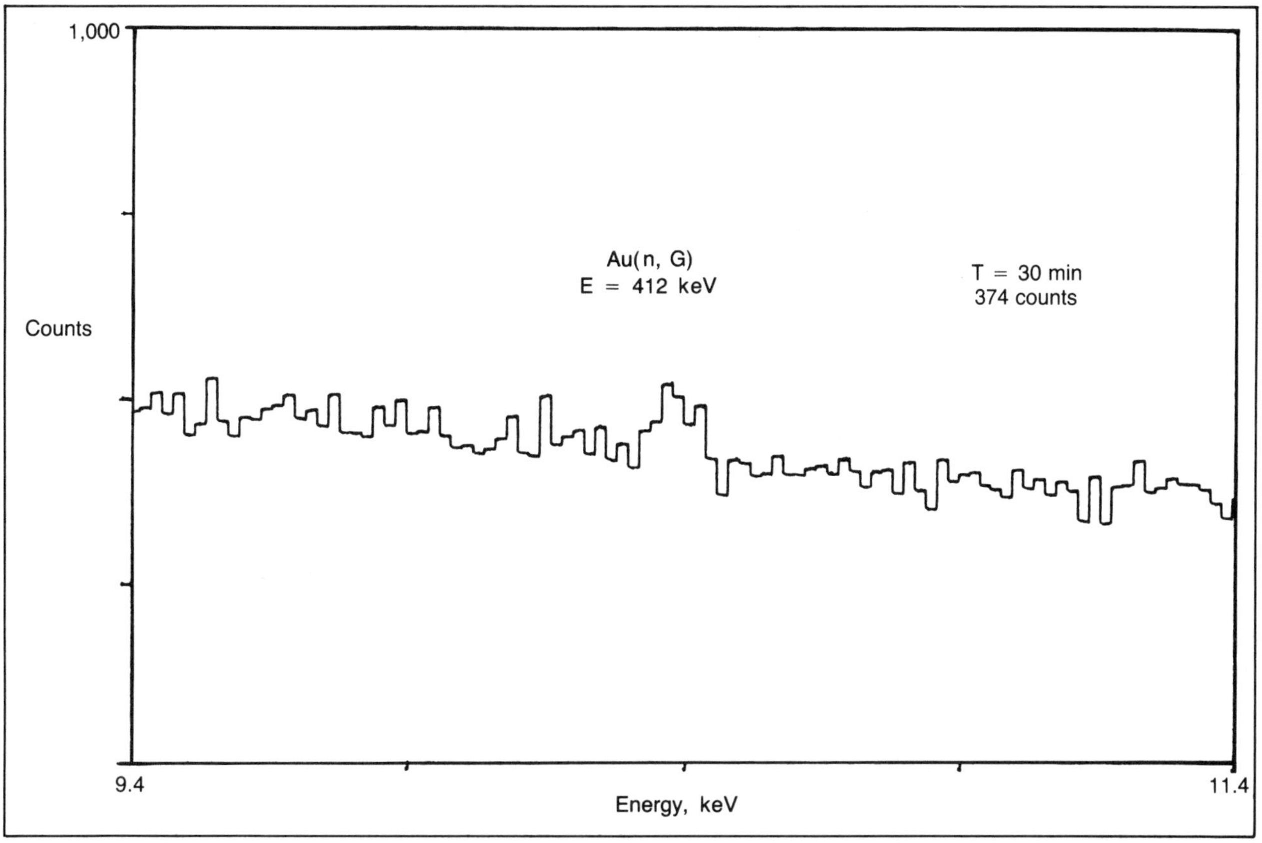

FIG. 5–63b

PGT gold study (courtesy Princeton Gamma Tech.)

contains some chlorine, which increases its thermal neutron capture cross section. Oil on the other hand, even though it has nearly the same thermal neutron capture cross section as pure water, contains some gas. The gas reduces the cross section, Σ. Thus, in any practical situation the water and oil have different cross sections. The relationship is:

$$\Sigma = \Sigma_w \phi S_w + \Sigma_h \phi (1 - S_w) + \Sigma_m (1 - \phi)$$

Where:
- ϕ = Porosity
- Σ = Macro cross section
- w = Water
- h = Hydrocarbon
- m = Rock matrix
- S = Relative pore-space saturation

The main value of the lifetime systems is that the casing of the hole is quite transparent to neutrons. Thus, the system may be used with equal effectiveness in open or cased holes. They are strongly affected by invasion.

■ Thermal Neutron Temperature Logging

The IRT Corporation operates a system that detects the energy of thermal neutrons. Since the thermal neutrons are in thermal equilibrium with their surroundings, the average energy of them is a function of the average temperature of the surrounding material. If the borehole contribution is eliminated by suitable timing, the average energy of the thermal neutrons reflects the temperature of the near-formation material. The measurement is in contrast to the conventional temperature tools, which measure borehole temperature only. These are offered by Dresser-Atlas and have a potentially useful place in nonpetroleum work.

The general spectrometry tool (GST*) is another neutron activation system that can also spectrometrically detect capture gamma rays and inelastic gamma rays. This tool is specifically set up to detect hydrogen, calcium, silicon, chlorine, sulfur, iron, and

*Mark of Schlumberger.

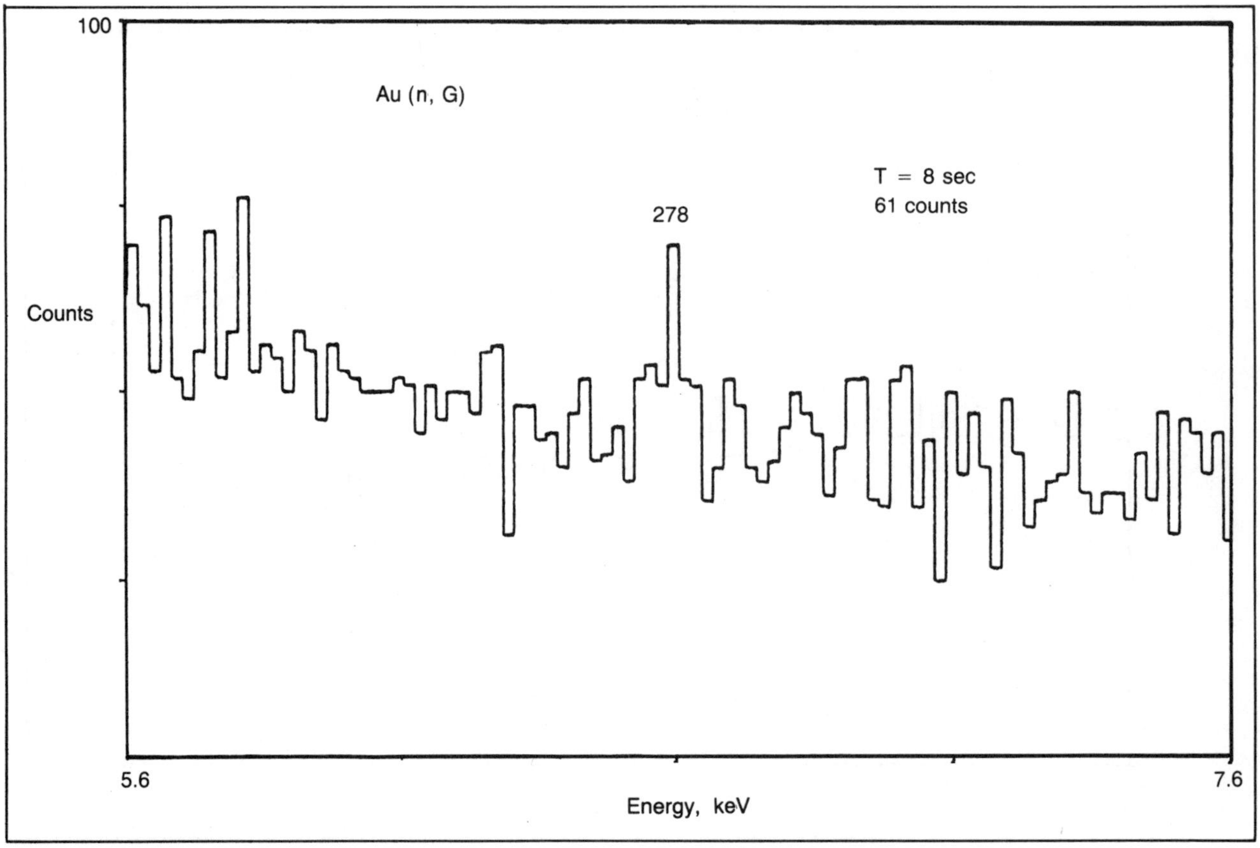

FIG. 5–63c

PGT gold study (courtesy Princeton Gamma Tech.)

carbon. By a suitable combination of measurements of specific elements, many common compounds and situations may be identified.

The technique involved in using these tools is best illustrated by the technique employed with the Dresser-Atlas Carbon/Oxygen Log. This system records the ratio of the nominal amounts of carbon to oxygen. It also records the ratio of the nominal amounts of calcium to silicon or silicon to calcium. If C/O is high and Ca/Si is low, a hydrocarbon-bearing sand is indicated. If both ratios are high, a calcareous zone is indicated. This analysis, which is simplified in this text, is extended in practice to many more situations for both the Carbon/Oxygen system and the GST* system.

■ Cased-Hole Logs

The neutron systems measure a relatively small, finite volume around the source detector assembly. The sensitivity of the system falls off in a nonlinear manner as the distance from the assembly increases. Within this volume is the borehole, which is filled with mud or air and which may be cased. The borehole and casing are near the source-detector assembly where the sensitivity of the system is high. While this situation is no different in principle than with other types of tools, there are consequences that affect the measurement and must be considered in an interpretation.

The presence of hole casing is not a severe problem for the neutron systems. Steel is relatively transparent to high-velocity, thermal neutrons. Therefore, steel casing would be thought to cause slightly less interference than an uncased hole. However, there are three possibilities for interference.

Steel is often made from scrap steel. If an appreciable part of this is boron from borosilicate steel, the boron presents a large cross section to thermal neutrons. Boron is also present in borosilicate glass used in fiberglass casing. The result is an error toward the lower counting rate or higher porosity direction. This is usually not a serious problem.

The casing may be made of polyvinylchloride (PVC), a plastic. PVC has a large quantity of chlorine, which has a large cross section to thermal neutrons. This effect can be corrected if it can be evaluated. There are different thicknesses of PVC pipe, which present different amounts of chlorine. It is necessary to measure the pipe thickness. The schedule of the pipe is not adequate. Also, it does appear that there may be different formulations of PVC with different amounts of chlorine. This may be evaluated if the driller is asked to save a length of pipe to use in a water-tank calibration measurement.

The third effect is more difficult to evaluate. Non-centered casing may be anywhere in the borehole. If the borehole is not large, this is not a serious problem. However, if the borehole is large, errors up to 100% may be made because of the uncertainty of the tool position within the borehole. The problem is identical, except for a greater magnitude, to that of gamma-ray logging in cased holes. This type is

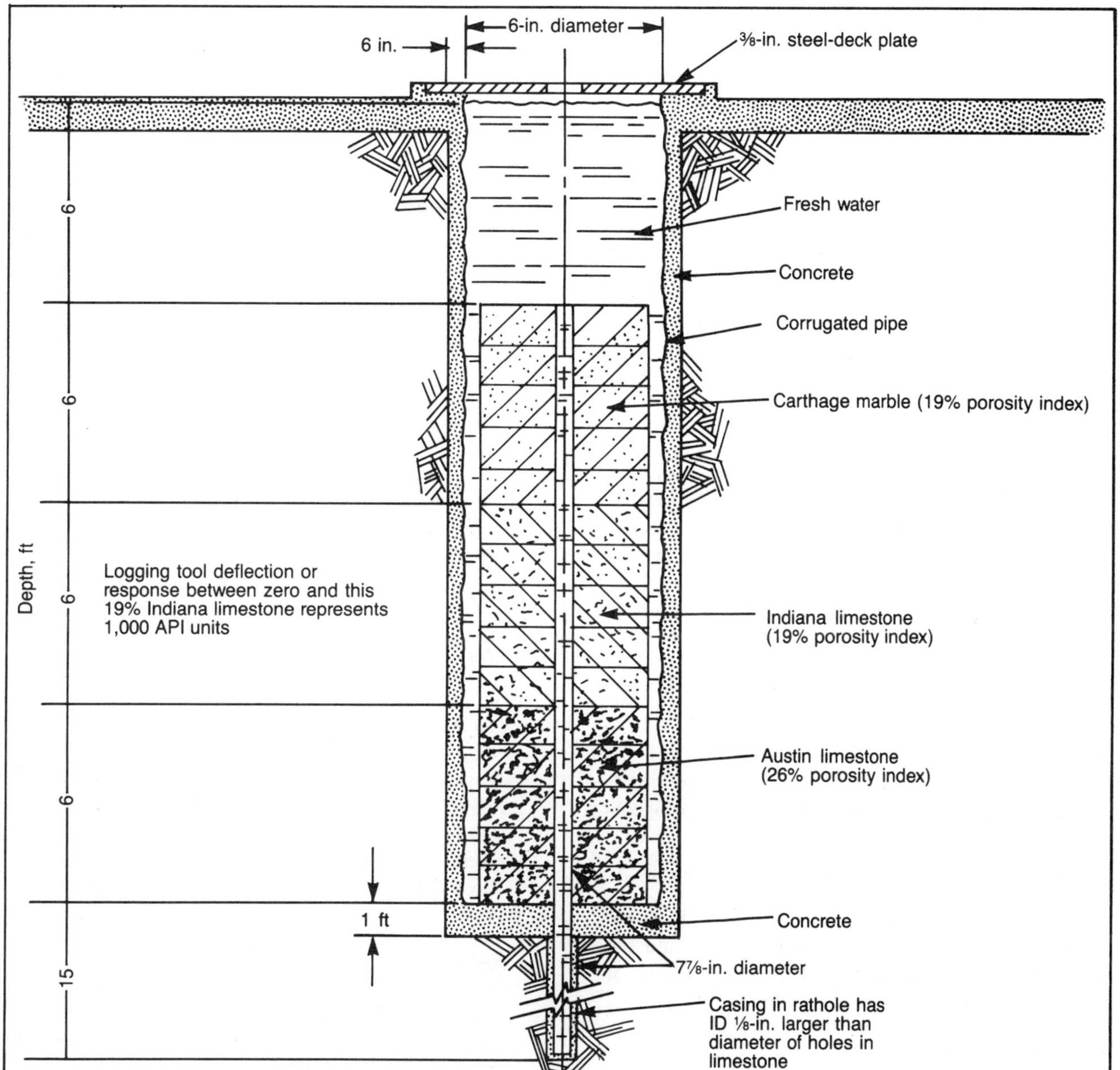

FIG. 5–64

API neutron calibration model

found in the temporary casing of mineral boreholes. Permanent casing is usually centered.

The presence of casing requires additional data in the log heading. The following information is very necessary: casing diameters and wall thickness (not schedule), casing material, manufacturer (if possible), centering, cementing, top of cement, type of joints, lengths of individual pipe pieces, type of fluid inside pipe, and resistivity of mud. These data can literally make the difference between a usable and a useless log. In addition, if a length of casing is available, a measurement should be made in a tank of water or a creek with and without the casing. This information on the log heading identifies the casing correction.

Cement in the casing borehole annulus may be treated as if it were water. The casing factor and the hole-size factor for the open hole must both be applied. The effect of the casing factor is to convert the reading to an open-hole situation. The hole-size factor converts the reading to an air-filled hole the same size as the probe. As an alternative, the corrections may be made directly from curves. In either event the amount of correction depends on the tool design, the casing type, the borehole fluid type, and the tool position within the hole.

Laboratory Methods

Laboratory neutron methods are generally restricted to neutron activation techniques at this time. Irradiation may be done with an isotopic source or a generator. Readings are usually taken with a gamma-ray spectrograph.

Laboratory activation measurements have several advantages over borehole measurements:

- The sample geometry may be prepared for optimum size and shape
- Irradiation times may be long and precise
- Equipment size is no problem, so large, high-output generators may be used with cryogenically cooled detectors
- There are no borehole effects to contend with

The major disadvantages to laboratory measurements are the uncertainty of sample origin and the turnaround time. However, a judicious mixture of measurements probably should include this type.

Calibrations

Neutron system calibrations are at least as important as with other systems. Since the transport phenomena of the neutron from the source to the detector are so complex, the problem of calibration is a difficult one.

The only public facility for neutron calibration at this time is the API model at the University of Houston. The U.S. Department of Energy has been considering calibration models for neutron porosity systems and has models for the direct uranium and uranium-thorium systems. Several logging contractors have their own models.

At best, calibration information for neutron systems is very sketchy and hard to obtain. One generally must take the few measurements available and combine vertical and horizontal smoothing to obtain a reasonable set of curves. Normalization of the data to the calibration situation at the time of use is imperative. In all probability if one is planning to use neutron systems regularly, it would pay to consider one's own calibration model, especially a water tank.

REFERENCES

Alder and Poessler. *Introduction to Probability and Statistics*. 5th edition, San Francisco: W.H. Freeman and Company, 1972.

American Petroleum Institute. *Recommended Practice for Standard Calibration and Format for Nuclear Logs*. API RP33, 3rd edition, April 1974.

Belknap, et al. "API Calibration Facility for Nuclear Logs." unidentified API Document circa mid-1960.

Birks, J.B. *Scintillation Counters*. New York: McGraw-Hill.

Bivens, et al. "Pulsed Neutron Uranium Borehole Logging with Epithermal Dieaway." Albuquerque: Sandia Laboratories, 1976.

Burke, J.A.; Schmidt, A.W.; and Campbell, R.L. Jr. "The Litho-Porosity Crossplot," *Log Analyst,* volume 6, 1969.

Cork, James M. *Radioactivity and Nuclear Physics*. New York: D. Van Nostrand Company, 1957.

Crew, Merle E. "Multipit, A Method for Calibration of Logging Systems." Grand Junction: U.S. DOE, 1978.

Crew, M.E., and Berkoff, E.W. "Twopit, A Different Approach to Calibration of Gamma Ray Logging Equipment." Grand Junction: U.S. DOE, April 1969.

Czubek, Jan A. "Natural Selective Gamma Logging." *Nukleonika,* volume XIII, No. 1, 1968.

Czubek, J.A. *Quantitative Interpretation of Gamma Ray Logs*. Krakow, Poland: Institute of Nuclear Research.

Czubek, J.A. "Recent Development of the Nuclear Geophysics Methods to the Mineral Prospection and Exploitation." Houston: ACS Paper 15, February 22, 1970.

Daniels, Scott, and Blackmon, Starkey. "Borehole Geophysical Investigations in the South Texas Uranium District." *Journal of Research of the USGS,* volume 5, No. 3, May–June 1977, 343–357.

Darnley, A.G. "Airborne Gamma Ray Survey Techniques." Geological Survey of Canada, 1971.

Dodd, P.H. "Quantitative Logging and Interpretation

Systems to Evaluate Uranium Deposits.'' *SPWLA Proceedings,* 1966, Paper P.

Dodd, P.H., and Droullard, R.F. ''A Logging System and Computer Program to Determine Rock Density of Uranium Deposits,'' U.S. AEC Paper Sm-68/33.

Dodd, P.H., and Eschliman, Dennis H., NATO. ''Borehole Logging Techniques for Uranium Exploration and Evaluation.'' *Methods of Prospecting for Uranium Minerals,* London: October 21, 1971.

Edmundson, H.N., and Raymer, L.L. ''Radioactive Logging Parameters for Common Minerals.'' *Log Analyst,* volume XX, No. 5, September–October 1979.

Edwards, J.M.; Ottinger, N.H.; and Haskell, R.E. ''Nuclear Log Evaluation Potash Deposits.'' *Proceedings SPWLA,* 1967, Paper L.

Felder, B., and Boyeldieu. ''The Lithodensity Log.'' *6th European SPWLA Symposium Transactions,* 1979, paper O.

Fertl, Walter H. ''Gamma Ray Spectral Data Assists in Complex Formation Evaluation.'' *Log Analyst,* volume XX, No. 5, September–October 1979.

Fishel, K.W., and Mayer, R. Jr. ''Extremely High Resolution Density Coal Logging Techniques,'' *2nd International Coal Exploration Symposium Transactions,* 1978.

''Formation Evaluation Data Handbook.'' Gearhart-Owen Industries, 1974.

Gardner, J.S., and Dumanoir, J.L. ''Litho-Density Log Interpretation.'' *21st SPWLA Symposium Proceedings,* 1980.

George, et al. ''A Borehole Gamma Ray Spectrometer for Uranium Exploration.'' BFEC, DOE GJBX-82(78), May 1978.

Givens, W.W., et al. ''Uranium Assay Logging using a Pulsed 14-Mev Neutron Source and Detection of Delayed Fission Neutron Source and Detection of Delayed Fission Neutrons.'' *Geophysics,* volume 41, No. 3, June 1976, 468–490.

GOI Bulletin 116-U, Gearhart-Owen Industries. ''API Calibration of Gamma Ray Logs.'' Document, January 5, 1967.

Granberry, R.J.; Jenkins, R.E.; and Bush, D.C. ''Grain Density Values from Gulf Coast Formations and Their Importance in Formation Evaluation,'' *SPWLA Symposium Transactions,* 1968.

Grasty, R.L., and Holman, P.B. ''Optimum Detector Sizes for Airborne Gamma Ray Survey.'' Resource Geophysics and Geochemistry Division, Geological Survey of Canada, Paper 74-1, part B.

Hallenburg, J.K. ''Uranium Ore Body Analysis Using the DFN Technique.'' *SPWLA Logging Symposium Transactions,* Paper T, 1978.

Hallenburg, James K. ''Interpretation of Gamma Ray Logs.'' *WGA Earth Science Bulletin,* September 1973.

Hartman, Ronald R. ''Airborne Gamma Ray Spectrometry.'' Philadelphia: Aero Service, August 1967.

Heslop, A. ''Porosity in Shaly Sands.'' *SPWLA Transactions,* 1975.

Jonas, J. Thomas. ''Digital Data Processing Techniques Applied to the Natural Gamma Ray Log.'' thesis, Golden: Colorado School of Mines, 1975.

Kellogg, William C. ''Aerial Radioactivity Surveying. Techniques of Successful Application.'' Pasadena: Lockwood, Kessler, and Bartlett, February 1968.

Killeen, P.G. ''A Gamma Ray Spectral Logging System.'' Geological Survey of Canada, 1977.

Kowalski, E. *Nuclear Electronics.* New York: Springer-Verlag, 1970.

Lapp, R.E., and Andrews, H.L. *Nuclear Radiation Physics.* Kensington: Prentice-Hall, 1949.

Moore, Donald C. ''Interpretation of Radiometric Logs in Thin and Dipping Beds.'' BFEC Publications, 1978.

''National Resource Evaluation, Preliminary Report.'' Energy Resource and Development Authority, June 1976.

Nicholson, P.W. *Nuclear Electronics,* New York: John Wiley and Sons, 1974.

Owen J.D. ''A Review of Fundamental Nuclear Physics Applied to Gamma Ray Spectral Logs.'' *Log Analyst,* September 1966. SPWLA Reprint Volume: *Gamma Ray, Density, and Neutron Logging,* 1978.

Princeton Gamma Tech. ''The PGT Probe.'' 1978.

Rosholt, John N. Jr. ''Natural Radioactive Disequilibrium of the Uranium Series.'' *USGS Bulletin,* 1084-A.

Scott, J.H. ''Computer Analysis of Gamma Ray Logs.'' *Geophysics,* Society of Exploration Geophysicists. volume XXVIII, No. 1–6, 1963.

Scott, J.H. ''The Grade Computer Program for Calculating Uranium Ore Reserves.'' AEC RME 145.

Scott, J.H. ''The Gamlog Computer Program.'' AEC RME 143.

Senftle, F.B., et al. ''In-situ Analysis of Coal Using a ^{252}Cf-Ge(Li) Borehole Sonde.'' *Mining Engineering (AIME),* volume 30, No. 6, 666–674.

Serra, O.; Baldwin, J.; and Quirein, J. ''Theory, Interpretation, and Applications of Natural Gamma Ray Spectroscopy.'' *21st SPWLA Symposium Transactions,* 1980.

Sherman, H., and Locke, S. ''Depth of Investigation of Neutron and Density Sondes for 35 Percent Porosity Sand.'' *SPWLA Symposium Transactions,* 1975. SPWLA Reprint Volume: *Gamma Ray, Density, and Neutron Logging,* 1978.

Smith, Orsino C. *Identification and Qualitative Analysis of Minerals.* New York: D. Van Nostrand Company, 1953.

Steinman, D.K., and John, Joseph. ''^{252}Cf-Based Logging System for In Situ Assay of Uranium Ore.'' IRT Corporation report Intel-RT 7019-005, circa 1977.

Stromswold, D.C. ''Comparison of Scintillation Detectors for Borehole Gamma Ray Logging.'' *21st SPWLA Symposium Transactions,* 1980.

Tittman, J. ''Radiation Logging.'' Petroleum Engineering Conference, University of Kansas, April 2–3, 1966, reprinted in SPWLA Reprint Volume: *Gamma Ray, Density, and Neutron Logging,* 1978.

Tittman, J., and Wahl, J.S. ''Formation Density Logging (Gamma-Gamma) Principles and Practice,'' *Proceedings of the Nuclear Geophysics Conference,* Krakow, Poland: 1962. SPWLA Reprint Volume, *Gamma Ray, Neutron and Density Logging,* March 1978.

Wahl, J.S., et al. *Journal of Petroleum Technology,* December 1964; *Transactions AIME,* volume 249, SPWLA Reprint Volume, *Gamma Ray, Neutron and Density Logging,* March 1978.

Woolson, W.A., and Gritzner, M.L. ''Borehole Model Calculations for Direct Uranium Measurements with Neutrons.'' Grand Junction: Science Applications, U.S. Department of Energy, January 1977.

Woolson, W.A., and Gritzner, M.L. ''Evaluation Models of Active Neutron Logging Tools for Direct Uranium Measurement.'' Grand Junction: Science Applications, U.S. Department of Energy, September 1978.

CHAPTER

6 Magnetic Methods

There has not been much activity in downhole magnetic methods. This is in contrast to the surface and airborne systems where there is great activity and many good methods are available. It is probable that downhole methods may be used in the future. Work has been done in the past by USGS and the U.S. DOE. However, instrumentation problems such as temperature drift have been unsolved. The contractors who build their own equipment have discussed the problems but have not offered the service. Only Scintrex offers a borehole modification of the Turam system.

Surface and airborne magnetic methods should certainly be considered for exploration in combination with other geophysical methods. They offer wide area coverage and are valuable when used in combination with borehole logging. Two magnetic downhole systems have been covered elsewhere in this book. They are the borehole deviation devices and the induction log.

MAGNETIC SUSCEPTIBILITY

Magnetic susceptibility is a fundamental physical property. The measurement of it is a means of identifying many materials, especially certain states of some materials. The techniques are useful in uranium development and in hard-rock exploration.

Physical Basis

Magnetic susceptibility is the index to the bulk magnetization of matter in an external magnetic field. It is defined as:

$$M = \chi H \tag{6.1}$$

Where:
M = Magnetization
χ = Susceptibility
H = Magnetic intensity

In practice, the magnetic susceptibility is usually measured with a Maxwell-Wien bridge. The relationship is:

$$B = \mu H \tag{6.2}$$

Where:
B = Magnetic flux density
μ = Magnetic permeability

The permeability is:

$$\mu = (\mu_0 + \chi) \tag{6.3}$$

Where:
μ_0 = Permeability of free space

The relative permeability, K, is often expressed as:

$$K = \mu/\mu_0 \tag{6.4}$$

■ Uses

Magnetic susceptibility logging is useful because of a negative correlation with the presence of uraniferous compounds and because of the frequent occurrence of iron compounds with other redox-sensitive metals. Use is made of the large contrast in susceptibilities of iron in the oxidized and the reduced states. The systems are usable in gas- or water-filled boreholes and in cased or open holes, as long as the casing material is not iron or other metal. A typical magnetic susceptibility log is shown in Fig. 6–1. Table 6–1 lists magnetic susceptibilities of minerals.

Downhole probes are frequently patterned after the Schlumberger induction log. The signal phase, which is not used in induction logging, is detected. These systems, however, have a high mutual coupling, which results in a large constant signal. Large temperature effects plague many systems. Some systems use surface current coils to overcome some of the difficulties.

■ MAGNETIC FIELD STRENGTH

Magnetic field strength measurements are closely related to susceptibility characteristics. From the preceding section we see that the magnetic flux density, B, is related to the susceptibility, χ:

$$B = (\mu_0 + \chi)H \tag{6.5}$$

Flux density, susceptibility, and intensity may be derived from one another. Therefore, the same com-

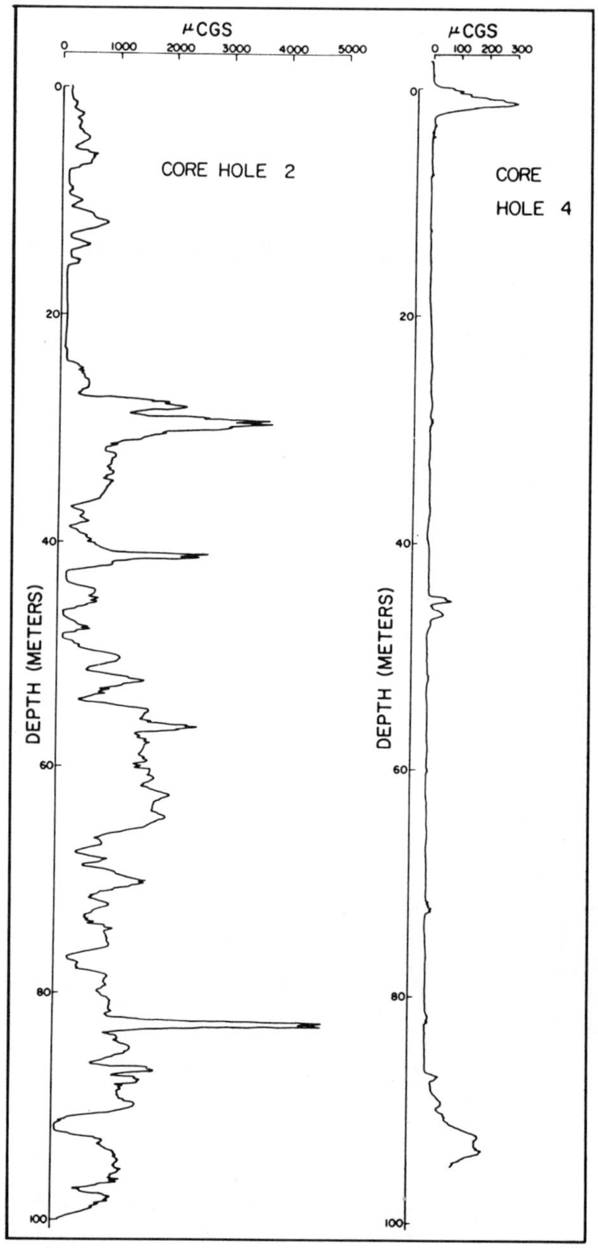

FIG. 6–1
Typical magnetic susceptibility log
(courtesy DOE, Grand Junction, and Bendix Field Engineering Corp.)

ments may be used for these measurements as for susceptibility.

Surface field strength measurements of various kinds are important in mineral exploration. These may be induced or natural. In general, field strength measurements are made on a grid or line pattern. Isomagnetic lines are then plotted. Increases of magnetic susceptibility, caused by concentrations of high

Table 6–1

Magnetic Susceptibilities

Material	Temperature, °C	Susceptibility, 10^{-6} ergs
Aluminum, Al	18 230	0.65 0.64
Aluminum oxide, Al_2O_3		−0.098
Calcium carbonate, $CaCO_3$		−0.382
Carbon, diamond, C	20 200	−0.49 −0.50
Carbon, graphite, C	20 300	−3.5 −2.7
Chromium, Cr	18	3.6
Chromic oxide, Cr_2O_3	18	25.5
Copper, Cu	18	−0.086
Cupric sulfate, $CuSO_4 \cdot 5H_2O$		5.9
Cupric sulfide, CuS	17	−0.20
Cuprous sulfide, Cu_2S	18	−0.18
Ferric chloride, $FeCl_3$	20	86.2
Ferrous chloride, $FeCl_2 \cdot 4H_2O$	19	60.1
Magnesium carbonate, $MgCO_3 \cdot 3H_2O$		−0.525
Nickel hydroxide, $Ni(OH)_2$		48.3
Nickel sulfate, $NiSO_4 \cdot 7H_2O$	19	16.0
Silicon dioxide, SiO_2		−0.493
Uranium, U	18	2.6
Uranium dioxide, UO_2	17	7.5
Uranium oxide, U_3O_8	15	0.95
Uranium trioxide, UO_3	16	1.08
Vanadium oxide, VO_2	13	3.73
Vanadium petoxide, V_2O_5	15	0.85
Vanadium trioxide, V_2O_3	15	13.9

Source: Geometrics

susceptibility minerals or ferromagnetic minerals results in a sharp magnetic field gradient, changes from the average field, and a clustering of the iso-magnetic lines, or a sharp change in direction of the lines (See Table 6–2). Sometimes, for greater clarity the first derivative of the magnetic field is plotted. Fig. 6–2 shows a map of the total intensity of the earth's magnetic field. Surface and airborne field strength measurements detect the sharp deviations from this. (Refer to the references following this chapter.)

Table 6–2

Typical Maximum Anomalies of Common Objects*

Object	Near Distance	Far Distance
Automobile, 1 ton	30 ft 40 gammas	100 ft 1 gamma
Ship, 1,000 tons	100 ft 300–700 gammas	1,000 ft 0.3–0.7 gammas
Light aircraft	20 ft 10–30 gammas	50 ft 0.5–2 gammas
File, 10 in.	5 ft 50–100 gammas	10 ft 5–10 gammas
Screwdriver, 5 in.	5 ft 5–10 gammas	10 ft 0.5–1 gamma
Revolver, 38 special or 45 automatic, induced approximately equal to permanent	5 ft 10–20 gammas	10 ft 1–2 gammas
Rifle	5 ft 10–50 gammas	10 ft 2–10 gammas
Ball bearing, 2 mm	3 in. 4 gammas	6 in. (0.5 ft) 0.5 gamma
Fenceline	10 ft 15 gammas	25 ft 1–2 gammas
Pipeline, 12 in. diameter	25 ft 50–200 gammas	50 ft 12–50 gammas
DC train	500 ft 5–200 gammas	1,000 ft 1–50 gammas
Cow magnet (½ in. W, 3 in. L)	10 ft 20 gammas	20 ft 2 gammas
Well casing and wellhead	50 ft 200–500 gammas	500 ft 2–5 gammas

*Anomalies are only representative and may vary by a factor of 5 or even 10 depending on the many factors described.

SURFACE AND AIRBORNE MAGNETIC METHODS

Some of the surface and airborne methods available are the total field strength method, low-frequency electromagnetics, and magnetotellurics. Total field strength measurements use flux-gate coils, proton precession detectors, and rotating coils.

Flux-gate systems consist of three coils (occasionally two coils) oriented at mutual right angles to each other. Each of the three components of the earth's magnetic field is balanced out or nulled by a coil current. When the field is nulled, the nulling current strength is a function of the strength of that field component. Thus, both intensity and direction are measured.

The proton precession systems align hydrogen nuclei (or helium) with a strong magnetic field. The time required to realign with the earth's field is a measure of the field strength. These devices do not measure the direction of the earth's field but do have a high sensitivity and a great range.

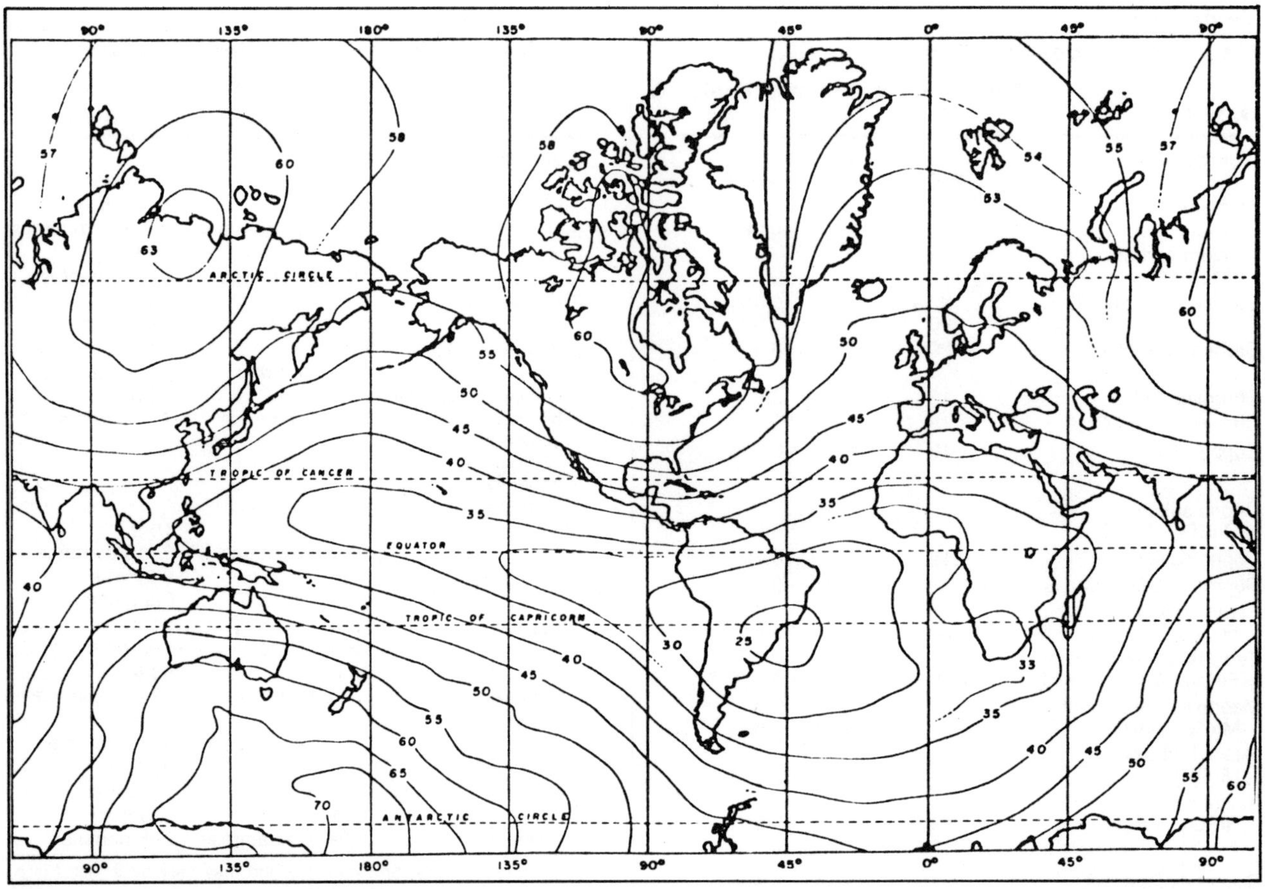

FIG. 6–2

The total intensity of the earth's magnetic field (courtesy Geometrics)

The electromagnetic or EM systems use an alternating source field and measure the in-phase and quadrature resultant fields. These are functions of the electrical and magnetic properties, respectively, of the sample. There are many variations: horizontal loop, vertical loop, Turam, Turair, VLF, and multiple frequency systems. They are used both on the surface and airborne. These are excellent systems to be used in combination with borehole measurements.

The magnetotelluric systems use moving magnetic fields, natural or artificial, to generate currents within the formations. The voltage fields resulting from the induced currents are detected with electrode systems and voltmeters. The signals are functions of the conductivities of the formations.

All of these methods are potentially valuable because they respond to the simple and complex resistive and magnetic properties of the earth. They are potentially applicable to borehole use and, certainly, in combination with other borehole methods.

REFERENCES

Anderson, Wallace L. "Theory of Magnetic Susceptibility Measurements with Coil Pairs." *Geophysics*, volume 33, No. 6, December 1968, 962–971.

Breiner, S. "Applications Manual for Portable Magnetometers." Geometrics publication, 1973.

Collett, L.S., and Bell, C.K. "AFMAG Use in Geological Interpretation," *72nd Annual CIM Transactions*, volume LXXIV, 1971.

Ellis, John R.; Austin, S. Ralph; and Droullard, Robert F. "Magnetic Susceptibility and Geochemical Relationships as Uranium Prospecting Guides." U.S.A.E.C. Publication AEC-RID-4, Grand Junction: U.S. DOE.

Hallenburg, J.K., et al. "Borehole Logging for Uranium Exploration, A Manual," Technical Report Series No. 212, Vienna: International Atomic Energy Agency, 1982.

Keller, G.V., and Frischkneckt, Frank C. "Electrical Methods in Geophysical Prospecting." Pergamon Press, 1966.

Scintrex Catalogue of Geophysical and Geochemical In-

strumentation and Services.
Sears, F.W., "Electricity and Magnetism." Cambridge: Addison-Wesley Publishing, 1946.
Telford, W.M., et al. "Applied Geophysics." Cambridge: Cambridge University Press, 1966.
Zablocki, C.J. "Magnetic Assays from Magnetic Susceptibility Measurements in Taconite Production Blast Holes." *Geophysics*, volume 39, No. 2, April 1974, 174–189.

CHAPTER

7 Mechanical Methods and Sampling Methods

Foremost among the mechanical methods are the acoustic measurements. These methods put a mechanical pulse into the formation to determine mechanical information. This class also includes borehole calipers, deviation measurements, and temperature sampling. These are usually methods that we consider peripheral to the main measurements. However, these give vital information that often is indispensable to the interpretation. These methods are as important as a gamma ray, resistivity, or other measurement.

METHODS/ACOUSTIC

Acoustic measurements use mechanical energy in the form of pulses or continuous waves to investigate the surrounding material. They include sonic absorption, acoustic velocity, and acoustic amplitude systems. They were originally developed as porosity devices. Later usage has demonstrated their usefulness for determining other parameters. Some of these are cement bonding of casing, determination of elastic parameters, lithology determinations, fracture location, and casing and borehole wall evaluation. They are useful in open and cased holes and in sedimentary and hard-rock environments.

Principles

The acoustic systems all put mechanical energy into the formation material, usually via the borehole and usually in the form of pulses of acoustical frequencies. The absorption tools, however, use continuous mechanical waves. All require liquid-filled boreholes. The transmission velocities, wave types, attenuations, and the required driving energies all depend on the mechanical properties of the media through which the energy travels.

A mechanical wave can only exist as a compressional or pressure wave in a liquid. A compressional or p-wave is one where the molecules of the transmitting medium move parallel to the direction of wave propagation. In a rigid solid the energy can propagate as a p-wave or as a shear wave (s-wave). An s-wave is one where the molecules move normal to the direction of propagation. P-waves and s-waves travel at different velocities, both of which depend on the mechanical properties of the medium.

The velocity of a p-wave is V_p:

$$V_p = [(B + 4/3\ G)/\rho]^{1/2} \tag{7.1}$$

Where:
ρ = Material bulk density
G = Shear modulus
B = Bulk modulus

The shear wave velocity, V_s, is also related to the density and elastic properties of the medium bearing the wave:

$$V_s = (G/\rho)^{1/2} \tag{7.2}$$

The interval travel time, Δt, is frequently used in place of the velocity:

$$\Delta t = (10^6)/V \tag{7.3}$$

System Types

The simplest form of the acoustic velocity tool is the single-transmitter, single-receiver tool. When this tool is centered in the hole, the acoustic pulse travels from the transmitter to the formation through the mud as a p-wave. At the formation it enters and travels as a p-wave and an s-wave. The pulse also travels in the mud column as a direct or tube wave and along the wall of the hole as a boundary wave. At this time no use is made of these two modes. This discussion will be confined to the pressure (compressional) wave and the shear wave (p-wave and s-wave).

Each material has its own characteristic narrow range of velocities. These are from about 5,000 ft/sec (200 μsec/ft) in water to about 26,000 ft/sec (38 μsec/ft) in dunite. The p-wave and s-wave travel through the formation at the average velocity for the mixture of materials. The waves are propagating in all directions at velocities faster than the direct wave in the hole, and they radiate back into the hole at a point nearest the receiver. They travel through the mud from the wall of the hole to the receiver as two p-waves and are detected by the receiver. Therefore, the total time, Δt_T, is:

$$t_T = t_{f1} + t_m + t_{f2} \tag{7.4}$$
$$t_T = t_m + 2t_f$$

Where:
t_{f1} and t_{f2} = Travel times through the borehole fluid at the transmitter and receiver, respectively
t_m = Average travel time through the formation

The tool is assumed to be centered in and parallel to the axis of the borehole. Fig. 7–1 shows the principle of the acoustic velocity tools.

With the single receiver system, the travel time through the mud remains an unknown quantity because of the uncertainty of the tool position. In order to overcome this problem, two receivers are commonly used. Then, electronically the travel time at the first receiver is subtracted from that at the second:

$$t_T = t_{m2} + 2t_{f1} - t_{m1} - 2t_f \tag{7.5}$$
$$t_T = t_m - t_m$$

$$\Delta t = t_T/d \tag{7.6}$$

Where:
d = Distance between the two receivers

This way, if the tool is positioned properly, the travel time through the mud is cancelled out. If the distance between the receivers is divided into the travel time, the interval travel time, Δt, is obtained. Fig. 7–2 illustrates several types of acoustic velocity tools.

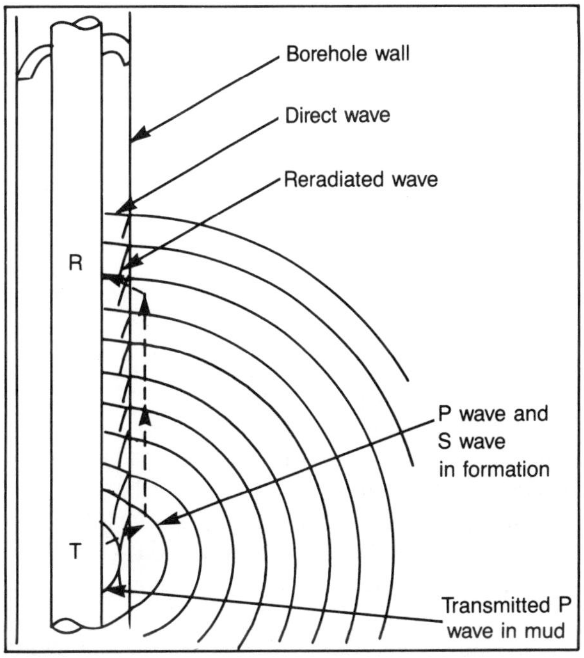

FIG. 7–1
Principle of the acoustic logging tool

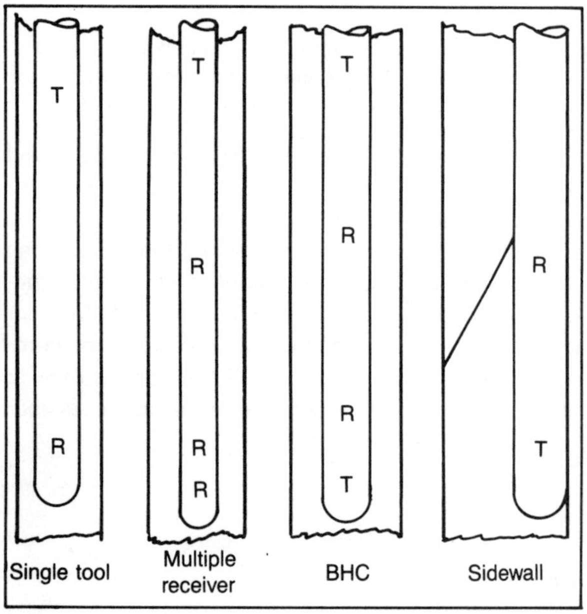

FIG. 7–2
Acoustic tool forms

Acoustic Velocity Tools

When the two-receiver system is used, the assumption is made that the transmitter and both receivers are the same distance from the borehole wall. This is not always true. The probe can be canted in the borehole, or the borehole diameter may be irregular. The Borehole Compensated Sonic system (BHC) was developed to overcome this difficulty.* This system uses two transmitters and two receivers. One transmitter is above the receivers, and one is below. The pulse from each transmitter is handled as with the two receiver instrument. Then the responses caused by the two transmitters are averaged:

$$\Delta t_1 = (t_{R_1T_2} - t_{R_1T_1})/d \tag{7.7}$$

$$\Delta t_2 = (t_{R_2T_2} - t_{R_2T_1})/d \tag{7.8}$$

$$\Delta t_T = (\Delta t_1 + \Delta t_2)/2 \tag{7.9}$$

It is also feasible to use a single transmitter and two receivers to obtain compensation. This is done by using an on-site computer to combine readings based on precise positioning of the array in the borehole. This process is used commercially at Dresser-Atlas.

The acoustic velocity tools described—single receiver, two receiver, and BHC systems—use two main types of recording systems. The original and still the most widely used system in petroleum logging records only the travel time of the first arrival of the p-wave. This arrival is usually detected by means of an amplitude discriminator whose level may or may not be adjustable. This type of system is usually analog. The second system records the full wave train arriving at the receiver. Detection and sorting into component waves may be done optically (photographic recordings), or it may be done digitally (sensing relative time frames, apparent frequency, and phase relationships). The first type is usually recorded in log or strip-chart form. The second method recording depends on the system and the use intended.

*A mark of Schlumberger

If the full pulse is sent to the surface in analog equipment, the pulse is usually used to modulate the amplitude or intensity of a triggered oscilloscope. Fig. 7–3 shows examples of this type of recording. The initial pulse triggers the oscilloscope. The sweep rate is picked to have a definite time scale (i.e., 50 μsec/in. Sometimes, the initial portion of the sweep, from the triggering pulse until just before the first arrival, is suppressed. This allows the interesting detail to be expanded without losing data. The recording is made on a strip film, which is drawn past the oscilloscope screen as a function of depth and logging speed.

If the system is a digital one, the wave may be sampled downhole and an encoded signal sent up, an analog signal may be sent to the surface and reduced at the surface, or the p-wave and s-wave may be separated downhole. In any of the three digital options, the first arrival of the p-wave is determined by discrimination. The s-wave, however, arrives later because it usually has a lower velocity and is added vectorially to the p-wave. Therefore, the s-wave can be detected by a sharp change in amplitude and an apparent frequency and phase change and shift.

The full wave or the p-wave/s-wave recording is done from a single receiver. Therefore, the travel time through the mud to and from the borehole wall must be determined. This can be done when one has two or more receivers by comparing the actual distance to one receiver with the calculated travel time distance:

$$\text{Travel time in mud} = \text{total travel time} - \text{span} \times \Delta T$$

There are several important modifications of this system for special uses. The amplitude of the p-wave can be recorded. This is usually done in analog log form but may be one of the digital variations. The resulting log is a function of the attenuation of the signal caused by scattering, reflection, refraction, or compression losses.

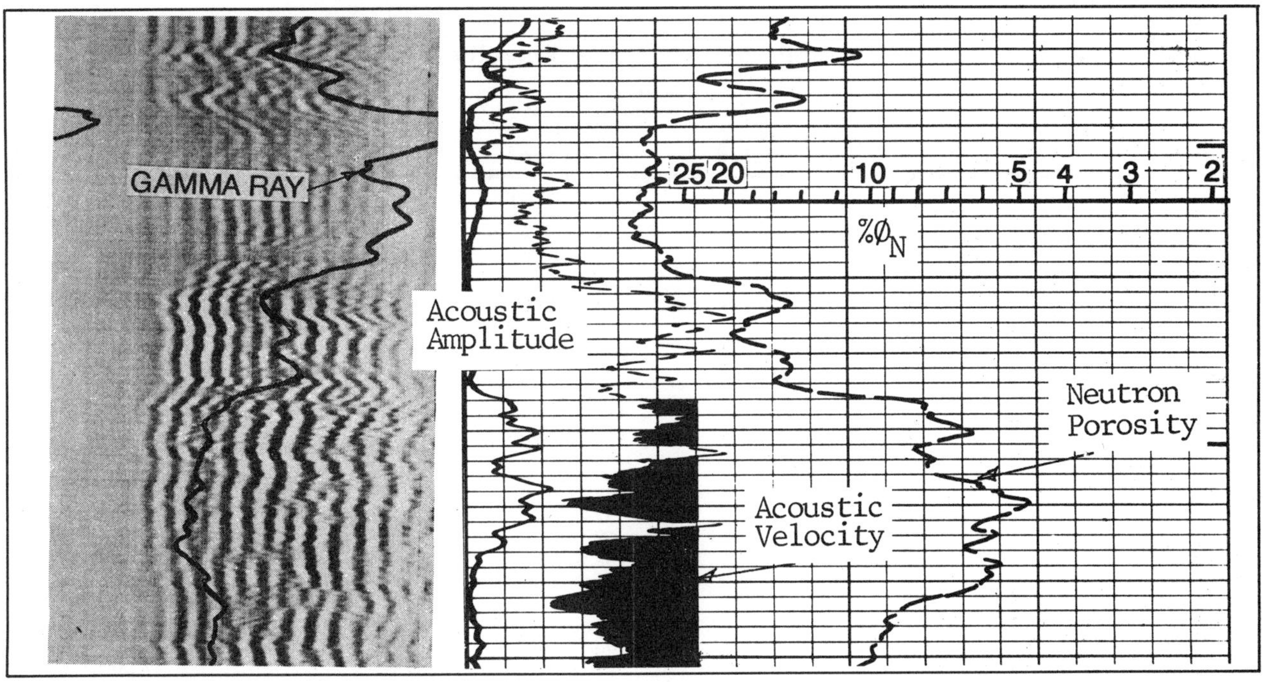

FIG. 7–3

A composite acoustic velocity log (courtesy Welex, a Halliburton Co.)

Laboratory equipment is used routinely to determine the acoustic velocities in cores. These methods have been standard for reference for many years. The principle is the same as downhole logging, but the geometry is different, of course. Helander and Myung feel, with good reason, that the in situ (or log) measurements probably are more reliable than the laboratory measurements.

Surface seismic techniques are not discussed, except to state that they are valuable in combination with the acoustic velocity logs. Refer to a standard text on surface seismic techniques for additional information.

■ Uses

The initial use of the acoustic velocity tool in petroleum logging was as a porosity-measuring device. Each of the components of a formation has a characteristic travel time (see Table 7–1). From Wyllie the travel time through the formation, Δt, is the sum of the fractional contributions of the components:

$$\Delta t = \phi \Delta t_f + (1 - \phi) \Delta t_m \qquad (7.10)$$

Where:

ϕ = Fractional porosity

Δt_f and Δt_m = Travel times in the fluid of the pore space and the rock matrix, respectively

Eq. 7.10 is Wyllie's time average equation. A more general case involves mixtures:

$$\Delta t = V_f \Delta t_f + V_s \Delta t_s + V_{sh} \Delta t_{sh} + V_a \Delta t_a \qquad (7.11)$$

Where:

s = Sand
sh = Shale
a = Any other component
V = Fractional volume
$V_f = \phi$

Note that these two equations resemble the corresponding equations for the density closely.

The Wyllie relationship has been used successfully for many years. It does have some problems of nonlinearity or poor correspondence under some conditions. Unfortunately, many of these conditions are important to nonpetroleum logging. The Wyllie equation tends to give porosity results that are too low at very low porosities, very high porosities, and in unconsolidated formations.

Raymer et al., have proposed a combination of three algorithms to represent the sandstone response curve of the pressure-wave transit time-porosity

Table 7–1

Formation and Minerals Sonic (Compressional Wave) Velocity/Interval Time Catalog (stp)

Material	Matrix Travel Time, μsec/ft Average Value	Matrix Travel Time, μsec/ft Range	Matrix Velocity, ft/sec Average Value	Matrix Velocity, ft/sec Range
Dunite	38.2	(34.7–41.1)	26,174	(24,305–28,807)
Gabbro	42.4	(42.2–47.6)	23,586	(20,998–23,586)
Hematite		42.9		23,295
Dolomite	44.0	(40.0–45.0)	22,727	(22,222–25,000)
Norite	44.1	(43.5–49.0)	21,683	(20,400–22,967)
Diabase	44.6	(44.0–46.0)	22,435	(21,746–22,730)
Anorthosite		45.4		22,016
Calcite	46.5	(45.5–47.5)	21,505	(21,053–22,000)
Aluminum		48.7		20,539
Anhydrite		50.0		20,000
Albitite	50.2	(49.5–50.6)	19,916	(19,752–20,212)
Granite	50.8	(46.8–53.5)	19,685	(18,691–21,367)
Granite		50.8		19,700
Steel		50.8		19,686
Limestone	52.0	(47.7–53.0)	19,231	(18,750–21,000)
Langbeinite		52.0		19,231
Iron		52.1		19,199
Gypsum	53.0	(52.5–53.0)	19,047	(18,868–19,047)
Serpentine		53.9		18,702
Quartzite	55.0	(52.5–57.5)	18,182	(17,390–19,030)
Quartz	55.1	(54.7–55.5)	18,149	(18,000–18,275)
Sandstone	57.0	(53.8–100.0)	17,544	(10,000–19,500)
Casing* (steel)		57.1		17,500
Basalt		57.5		17,391
Polyhalite		57.5		17,391
Shale		60.0–170.0		5,882–16,667
Aluminum tube*		60.9		16,400
Trona		65.0		15,400
Halite		66.7		15,000
Sylvite		74.0		13,500
Copper		78.7		12,700
Carnallite		83.3		12,000
Cement (wide variation)	95.0	(83.3–95.1)	10,526	(10,526–12,000)
Anthracite coal	105.0	(90.0–120.0)	9,524	(8,333–11,111)
Concrete (wide variation)	95.2	(83.3–125.0)	10,500	(8,000–12,000)
Bituminous coal	120.0	(100.0–140.0)	8,333	(5,906–10,000)
Sulfur		122.0		8,200
Lignite	160.0	(140.0–180.0)	6,250	(3,281–7,143)
Lead		141.1		7,087
Water, 200,000 ppm NaCl, 15 psi		180.5		5,540
150,000 ppm NaCl, 15 psi		186.0		5,375
100,000 ppm NaCl, 15 psi		192.3		5,200
pure		207.0		4,830
Glacial ice		87.1		11,480
Rubber (neoprene)		190.5		5,248
Kerosene, 15 psi		214.5		4,659
Oil		238.0		4,200
Methane, 15 psi		626.0		1,600
Air, 15 psi		910.0		1,100

Source: Gearhart Industries Inc.
To convert meters to feet, multiply by 3.2808.
To convert feet to meters, multiply by 0.3048.
*Value of extensional wave (first arrival wave) in thin rods.

relation empirically. The proposed algorithms are, 0–37% porosity range:

$$V_1 = (\rho_{ma}/\rho)^{1/2} (1 - \phi)^{1.9} V_{ma} \quad (7.11a)$$

$$V_1 = (1 - \phi)^2 V_{ma} + V_f \quad (7.11b)$$

$$\Delta t = \frac{10^6}{V}$$

The 37–47% porosity range:

$$\Delta t = \frac{0.47 - \phi}{0.1} \Delta t_1 + \frac{\phi - 0.37}{0.1} \Delta t_2 \quad (7.11c)$$

$$\Delta t = \frac{0.47 - \phi}{0.1} \Delta t_1 + \frac{\phi - 0.37}{0.1} \Delta t_f \quad (7.11d)$$

The 47–100% porosity range:

$$\Delta t_2 = \left[\frac{\rho\phi\Delta t_f^2}{\rho_f} + \frac{\rho(1-\phi)\Delta t_m^2}{\rho_m}\right]^{1/2} \quad (7.11e)$$

Where:
V = Pressure wave velocity
ρ = Density
Δt = Travel time
f and ma = Fluid and matrix, respectively

These last relationships seem to reflect the rock character. In the range above 50% porosity, the fluids tend to be suspensions or slurries.

As pointed out by Raymer et al., the Wyllie relationship has some problems. The empirical solution to these problems for years has been to use the relationship:

$$\Delta t = C + D\phi,$$

Where:
C = Travel time intercept, equal to the rock matrix travel time
D = Slope of the water-filled travel time versus porosity locus of points

In sandstones, D is equal to $(\Delta t_f - \Delta t_m)$ and results in Wyllie's equation. The values of C and D are:

Rock Type	C, μsec/ft	D, μsec/ft/decimal porosity unit
Sandstone	50–56	130–300
Limestone	45–50	60–120
Dolomite	42–48	60–120

The acoustic measurements are shallow. The frequencies used are high (25–100 kHz). The components detected are those that enter the formation at the critical angle (see Snell's Law) and travel parallel to the wall of the hole.

Acoustic waves are quite sensitive to the presence of clays. The clay velocities are low. This may be because of the plastic rather than rigid nature of clays. Any gas or other compressible material severely absorbs energy and attenuate the acoustic wave.

The velocity of a mechanical wave within a material depends on the density and the elastic moduli of that material. The determination of the elastic moduli is an excellent use for this system, which has the potential of large savings in coring costs. The bulk density, p-wave, and s-wave are all required.

Young's modulus, E, is the modulus of compression:

$$F/A = (Ee)\ell_0, \quad (7.12)$$

Where:
F = Force applied
A = Cross-sectional area
e = Elongation (positive or negative)
ℓ_0 = Original length

$$E = 2\mu(1 + P) \quad (7.13)$$

Where:
P = Poisson's ratio (Eq. 7.14)

$$P = \frac{\tfrac{1}{2}(V_p/V_s)^2 - 1}{(V_p/V_s)^2 - 1} \quad (7.14)$$

Poisson's ratio is the ratio of transverse contraction to elongation.

The shear modulus, μ, is the modulus of the change of shape with transverse force:

$$(F/A) = \mu\theta \quad (7.15)$$

Where:
θ = Shear angle

From the logs μ is:

$$\mu = \rho V_s^2 \quad (7.16)$$

Where:
ρ = Bulk density

The bulk modulus, B, is the modulus of the change of volume with external force:

$$\Delta P = B\frac{\Delta V}{V_o} \quad (7.17)$$

Where:
ΔP = Change in external force or pressure
V = Change of volume
V_o = Original volume

From the logs:

$$B = \rho V_p^2 - 4/3\, V_s^2 \quad (7.18)$$

The shear wave does not exist in a liquid, so in a liquid:

$$B = \rho V_p^2 \quad (7.19)$$

A very important use of acoustic velocity logs is in correcting the values used in seismic surveys. This is called a seismic reference service when it is combined with other services, such as a geophone measurement of surface shots. This allows a more accurate, detailed assessment than either one alone. Synthetic seismograms can also be made from acoustic logs (Fig. 7–4).

Another acoustical method that is important is the cement bond log. The impedance match between steel casing and drilling mud is poor. Therefore, this interface is a highly reflecting surface to high-frequency mechanical pulses. However, if cement is bonded to the pipe in the annulus, the interface impedance match with the steel is much closer. As a result, when the cement is bonded properly, energy is lost from the pipe to the cement and formation, and less is available at the receiver. Therefore, a drop in amplitude indicates good bonding.

Borehole Televiewer

The borehole televiewer is a device with high-frequency pulses that are rapidly repetitive and tightly focused in a beam to the borehole wall and that scan rotationally through 360°. The reflected signals are examined for both amplitude and transit time.

The return signals are handled by an on-site computer in one excellent version of the device (Amoco's Volumetric Scanning Device). They are presented on a cathode-ray tube screen whose x axis is the azimuthal position of the pulse transmitter, scaled as North, East, South, West, and North. The y axis is the borehole depth. Fig. 7–5 shows some of the results of this technique. In the top example well 2 is being drilled to intersect well 1. The lower photograph shows a plan view of the two holes.

Other methods that are only referenced are acoustic holography and standard acoustic methods for mine-face logging, high-resolution seismics for location of coal, and sonar methods for underwater use. One final technique is fracture location. Because of the sharp discontinuity affected by a fluid-filled fracture, reflection from a fracture is high. This usually results in a great loss of amplitude and cycle skipping.

Cycle skipping is caused by a pulse amplitude that is low enough so the discriminator rejects the first arrival and detects the second. The result on the log is an anomalously long travel time.

Calibrations

The calibration of the acoustic velocity tools is fairly straightforward. The circuitry can be checked for timing with suitable instruments. The tool can be run in porosity calibration models, but this is not always satisfactory because of the presence of cracks and unconsolidated material. A water-filled model can be built and used with ease.

In the laboratory calibration is again fairly simple. Machined blocks of material, such as aluminum can be put into the system in place of the usual core samples. But, again, the principle determining factor is the precise timing of the system.

Interpretation

If the value of the interval travel time is plotted against the porosity for multiple levels, the relation is:

$$\Delta t = \Delta t_m + D\phi$$

The value of ϕ may be from the cores or from another of the porosity determining devices. If the porosity, ϕ, is made zero, $D\phi$ becomes zero and:

$$\Delta t = \Delta t_m$$

This way, travel time for any horizon may be determined, and perhaps the rock type can be identified from the catalog.

In most media:

$$D = \Delta t_f - \Delta t_m$$

If the value of ϕ is made 1.0 (100% porosity), the measured travel time is:

$$\Delta t = \Delta t_f.$$

Thus, the fluid of the formation horizon may be identified.

The presence of gas in the pore space skews the plotted points greatly in a longer travel-time direction. The presence of clay skews the points in a longer travel-time and a lower porosity direction. Fig. 7–6 shows this type of crossplot.

Cement Bond Logs

Improved methods of evaluating the quality of the cementing of production casing have been introduced recently (Havira and also Gollwitzer and Mason). These methods may be valuable for in situ leach and similar projects. The Cement Evaluation Tool (CET*) detects casing thickness, bond quality, and the presence of channelling in the cement. The Cement Bond Tool (CBT) is an improved version of the older cement bond system. The CBT incorporated improved calibration, better centering, greater immunity to poor centering effects, and self-calibration.

CALIPERS

Calipers or borehole calipers are used to determine the condition of the borehole, to provide correction

*Mark of Schlumberger.

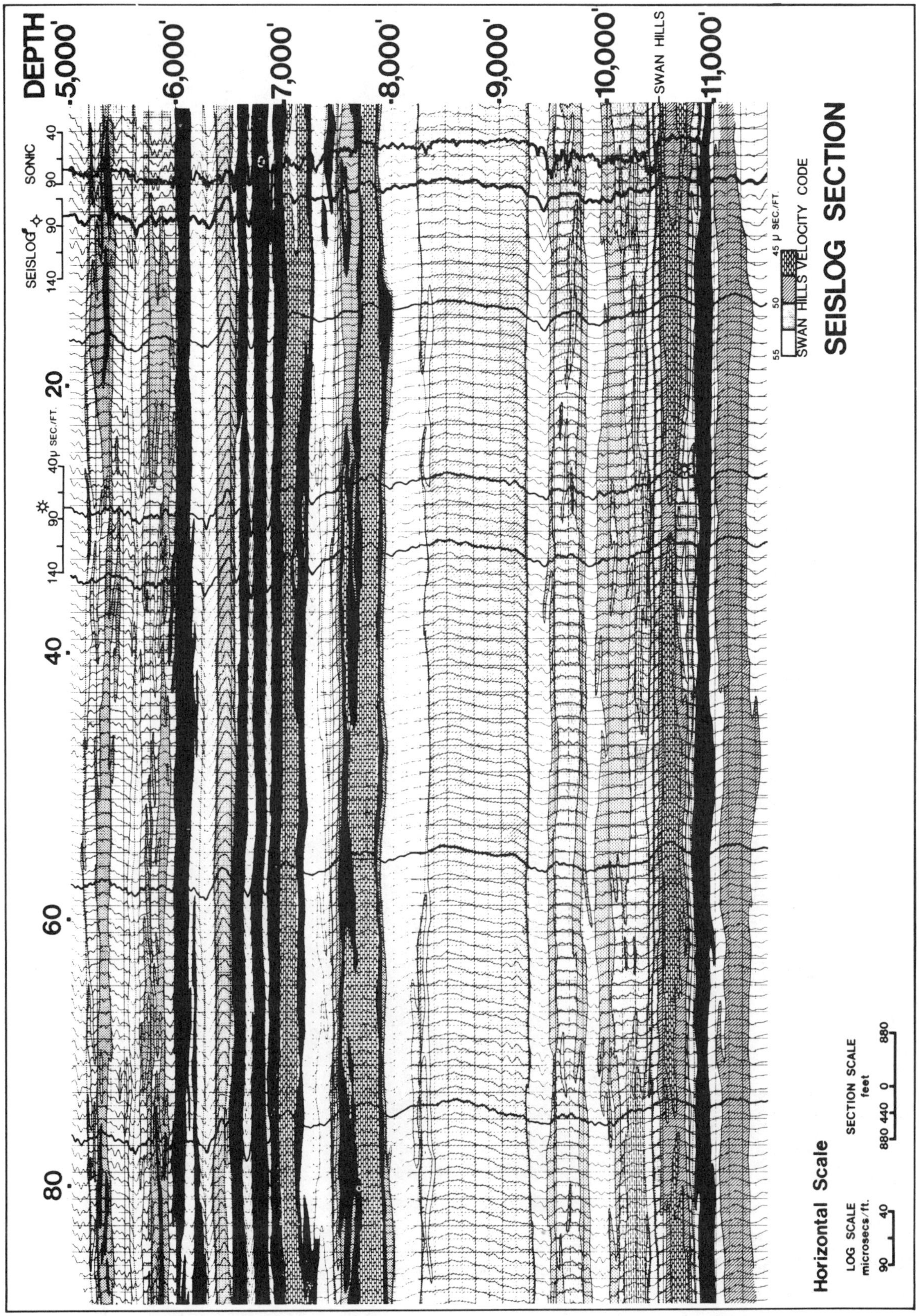

FIG. 7–4
Synthetic log section

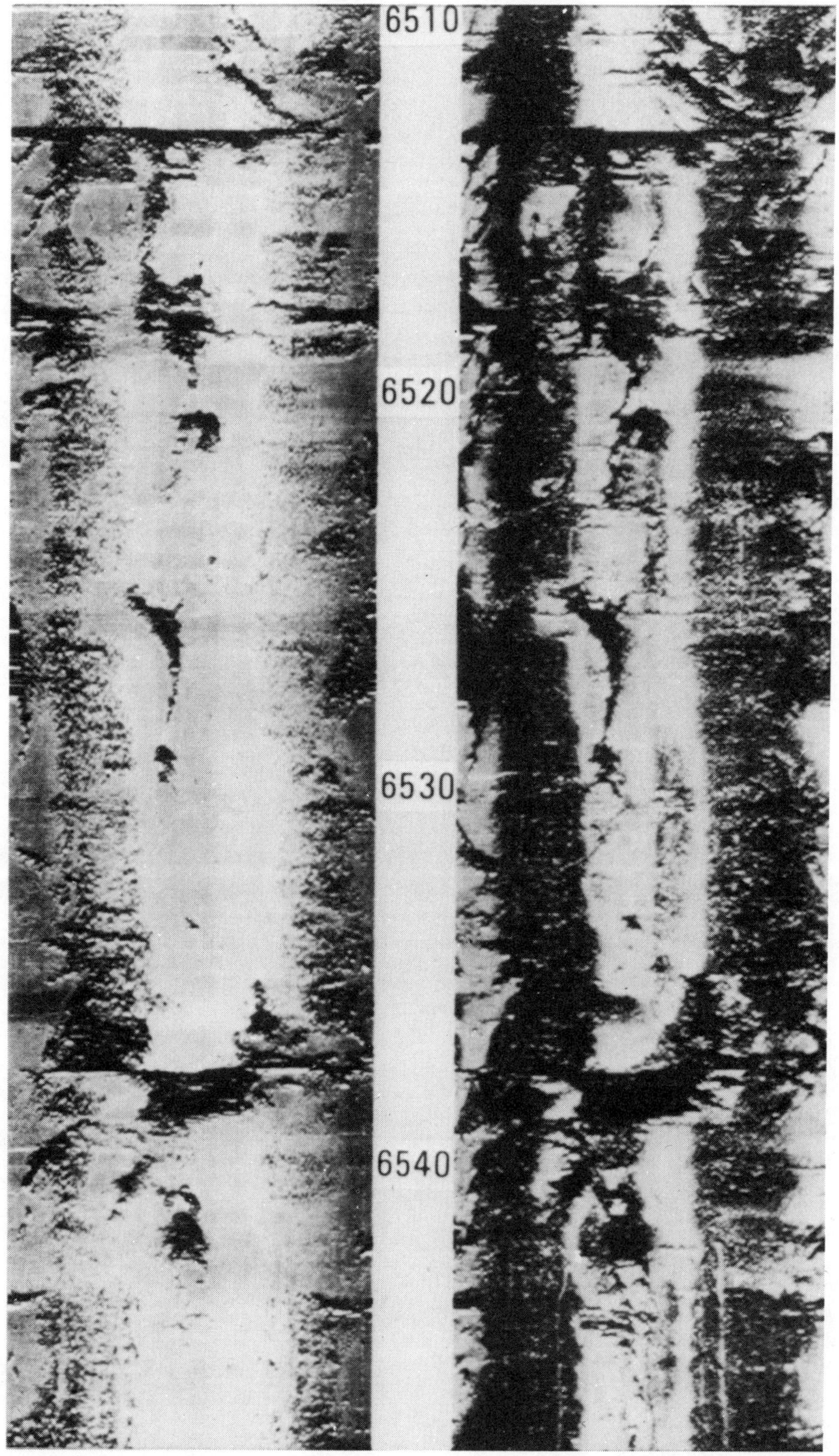

FIG. 7–5a

Volumetric scan display (after Broding, "Volumetric Scan Well Logging," *Log Analyst,* volume XXIII, No. 1, 1982)

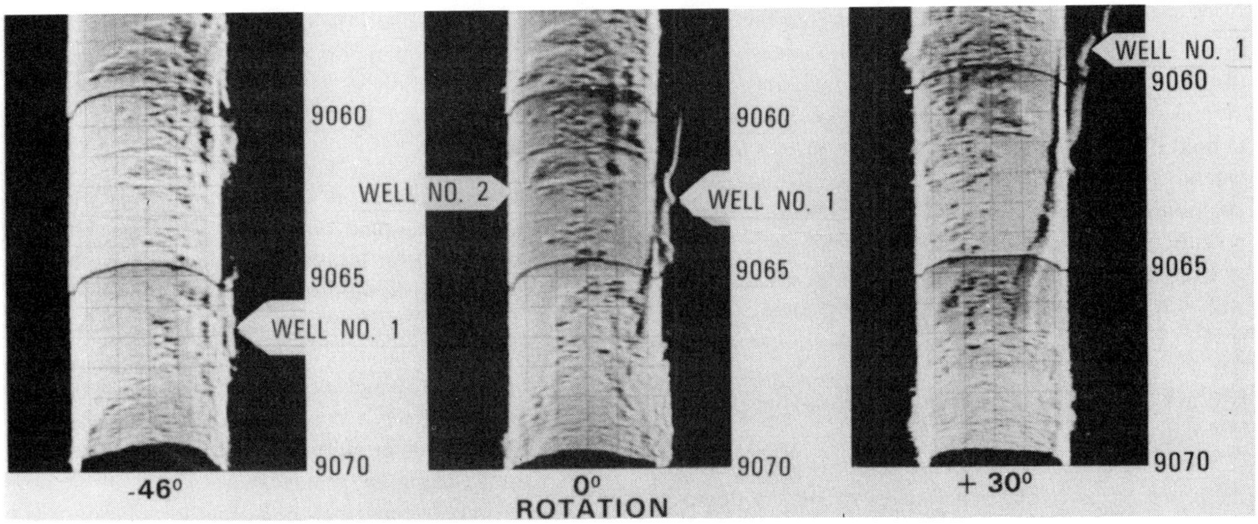

FIG. 7–5b

Volumetric scanning images on intersecting well (after Broding, "Volumetric Scan Well Logging," *Log Analyst,* volume XXIII, No. 1, 1982)

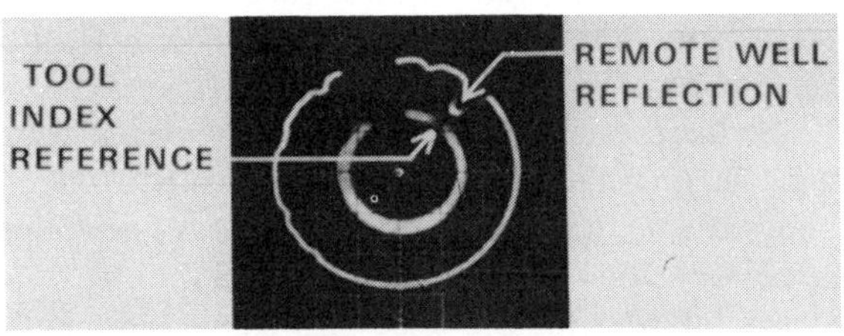

FIG. 7–5c

Volumetric scanning images of an intersecting well (after Broding, "Volumetric Scan Well Logging," *Log Analyst,* volume XXIII, No. 1, 1982)

for other measurements, and to determine borehole volume for cementing purposes. Several types of calipers are available. Single-arm calipers are often run with other services. Multiple, ganged-arm devices are most common. Multiple, independent-arm devices are gaining favor because of the better-quality information. Acoustic devices, similar to a sonar, are occasionally used and may soon be found incorporated in other tools.

■ Explanation, System Types, and Circuits

Borehole calipers are devices to measure the inside diameter of the borehole. This can be done in several ways and for several reasons. In general, the measurement of borehole diameter is made with one or more arms against the wall of the borehole. If one arm is used, the tool body is assumed to be against the wall opposite the arm. The angle between the arm and tool body is measured. This angle, θ, is a function of the borehole diameter:

$$d_h = d_s + L \sin \theta \quad (7.20)$$

Where:
d_s = Tool diameter
L = Length of the caliper arm

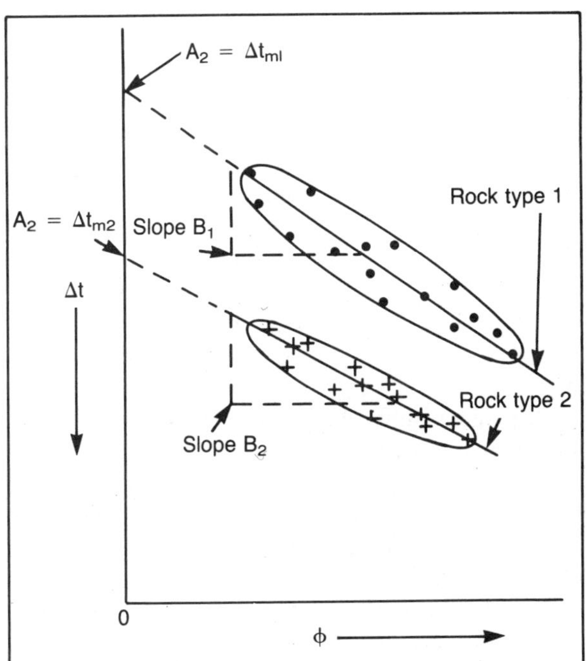

FIG. 7–6

Typical transit time-porosity crossplot

There are several methods of measuring this function. The most common method is to measure the displacement of the arm through a rack-and-pinion system that also transmits a spring force to the arm to hold it out. This same mechanism is further used to extend and retract the arm for use or storage. The position of the rack is sensed with a linear or rotary potentiometer which gives an electrical signal, which is a function of sin θ. The signal can be made linear with circuit parameters, computer programs, or log scales. An analog system, though, usually has a nonlinear scale that is a sine function. This type of system is often found on sidewall tools, such as the formation density tool.

Two-arm caliper systems are seldom used because there are no particular advantages to such a device and several disadvantages. A two arm system is somewhat unstable in a borehole. It tends to measure the major diameter of a noncircular borehole.

Three-arm caliper systems are very common. A three-point measurement, with arms at 120°, is quite stable in a borehole. A three-arm caliper tends to measure the minor diameter of a noncircular borehole.

A borehole often does not have a circular cross section; it is ovaloid. With a one- or two-arm system, it is difficult to tell whether the system is measuring the major or minor axis or something in between. Systems with three or more independent or semi-independent arms have been designed and used in petroleum logging. These give much better values for hole diameter and shape. Four-arm calipers are common with opposite arms ganged. Birdwell runs a 12-arm caliper.

A caliper dipmeter was suggested at one time, but this system is not in commercial use. It had three independent arms and outputs. The detail of the outputs reflected the small washings and cracks in the borehole wall. These could be correlated with formation dip. The cutting marks from the drill bit, which were independent of dip, showed up as background noise.

One microcaliper device is used on the face of a formation density system. It is assumed that the pressure of the face of the tool is low so it rides on top of the mud cake. The force on the microcaliper is high, and its area of contact is small. Thus, it is presumed to cut through the mud cake and measure the mud-cake thickness. This device is operated by Birdwell.

Acoustic calipers are being used occasionally. These devices put a directional, high-frequency, mechanical pulse into the mud column, normal to the tool axis. The travel time is a function of the mud unit travel time and the distance to and from the hole wall. High frequency is used to minimize penetration into the formation.

■ Uses

Single-arm borehole calipers are of restricted use, so they are designed for special purposes. One usually sees them on tools designed primarily for other purposes, such as the formation density tool. Here, it indicates the quality of the density curve by giving an idea of the character of the wall of the hole: smooth or rough, to gauge or caved, and the mud-cake thickness. For these purposes it is quite good. There is always an uncertainty whether the single-arm caliper is measuring the major or minor axis of the borehole or in between, or if the hole is circular or oval or ellipsoid in cross section. Also, there is always an uncertainty about whether the measured feature is present on the opposite wall where the other measurement is being made. The same thing is true of two-armed calipers. Therefore, these should never be used to calculate borehole volume. The problem is illustrated in Fig. 7–7.

The three-ganged-arm caliper is somewhat better than the preceding ones because it tends to give a more realistic measurement. There is less uncertainty as to the position of the tool and the shape of the hole. It also indicates the presence of a feature around the periphery. If the arms are made independent or are ganged in pairs, the shape of the hole can be realistically determined. If the caliper has four arms, the assumption can be made that they are measuring the major and minor axis. This is not always true, as shown in Fig. 7–7, but the error in this assumption is small. If the caliper has more arms, the actual shape of the cross section can be determined. The maximum reading and minimum reading can be taken as the major and minor axis, respectively. In this case borehole volume is V_h:

$$V_h = \pi abz \tag{7.21}$$

Where:

a = Maximum reading
b = Minimum
z = Depth interval

If the borehole is caved asymmetrically, the volume, V_h, can be approximated by assigning a regular geometric shape to the cave, in addition to the regular shape of the borehole. Thus:

$$V_h = V_{cave} + V_{borehole} \tag{7.22}$$

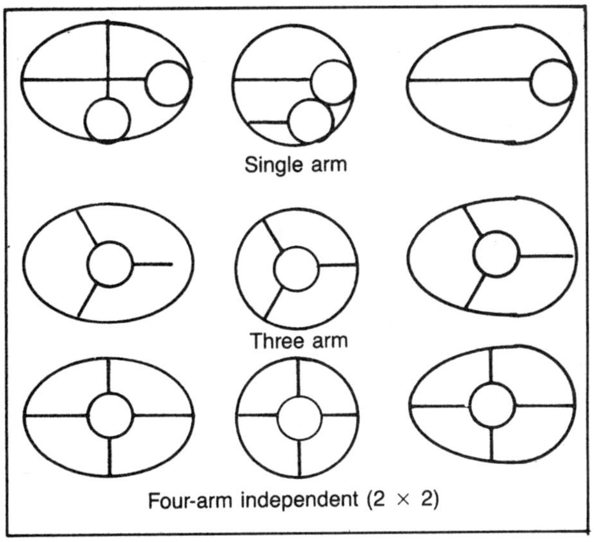

FIG. 7–7
Possible caliper measurement situations

If the borehole is to be cased and the casing cemented, there are several ways the amount of cement can be determined. If a computer is not available, the best method is probably to solve the problem completely on a short-length basis. If V_c is the volume of cement:

$$V_c = \Sigma(A_h - A_p)i \qquad (7.23)$$

Where:
i = Convenient depth interval
A_h and A_p = Cross-sectional areas of the borehole and the outside of the casing, respectively

Computer programs have more flexibility but may be handled in a similar manner. If only an analog trace is available, a planimeter may be used to determine the average borehole diameter. Then the volume of the hole is:

$$V_h = \pi \bar{r}_h^2 \Delta d \qquad (7.24)$$

The cement volume:

$$V_c = \pi(\bar{r}_h^2 - \bar{r}_p^2)\Delta d \qquad (7.25)$$

Where:
$\bar{r}_h$ = Average radius of the borehole
$\bar{r}_p$ = Average radius outside the casing

Calibrations

Caliper calibrations are rather simple but quite necessary, especially with analog equipment, because of the drifts of power supplies caused by temperature changes and the wear and bending of caliper arms. If these variations were reliably eliminated, frequent calibration would not be needed. However, this is seldom possible. A calibration should be made any time arms are repaired or changed. Any ring made of a piece of casing, pipe, or other rigid circular material can be used as a calibrating device. Two or more diameters should always be used to detect nonlinearities. Fig. 7–8 is a photograph of a typical 3-arm caliper with a calibration jig in place.

FIG. 7–8
Three-arm caliper with calibration jig
(courtesy Bendix Field Engineering Corp. and DOE, Grand Junction)

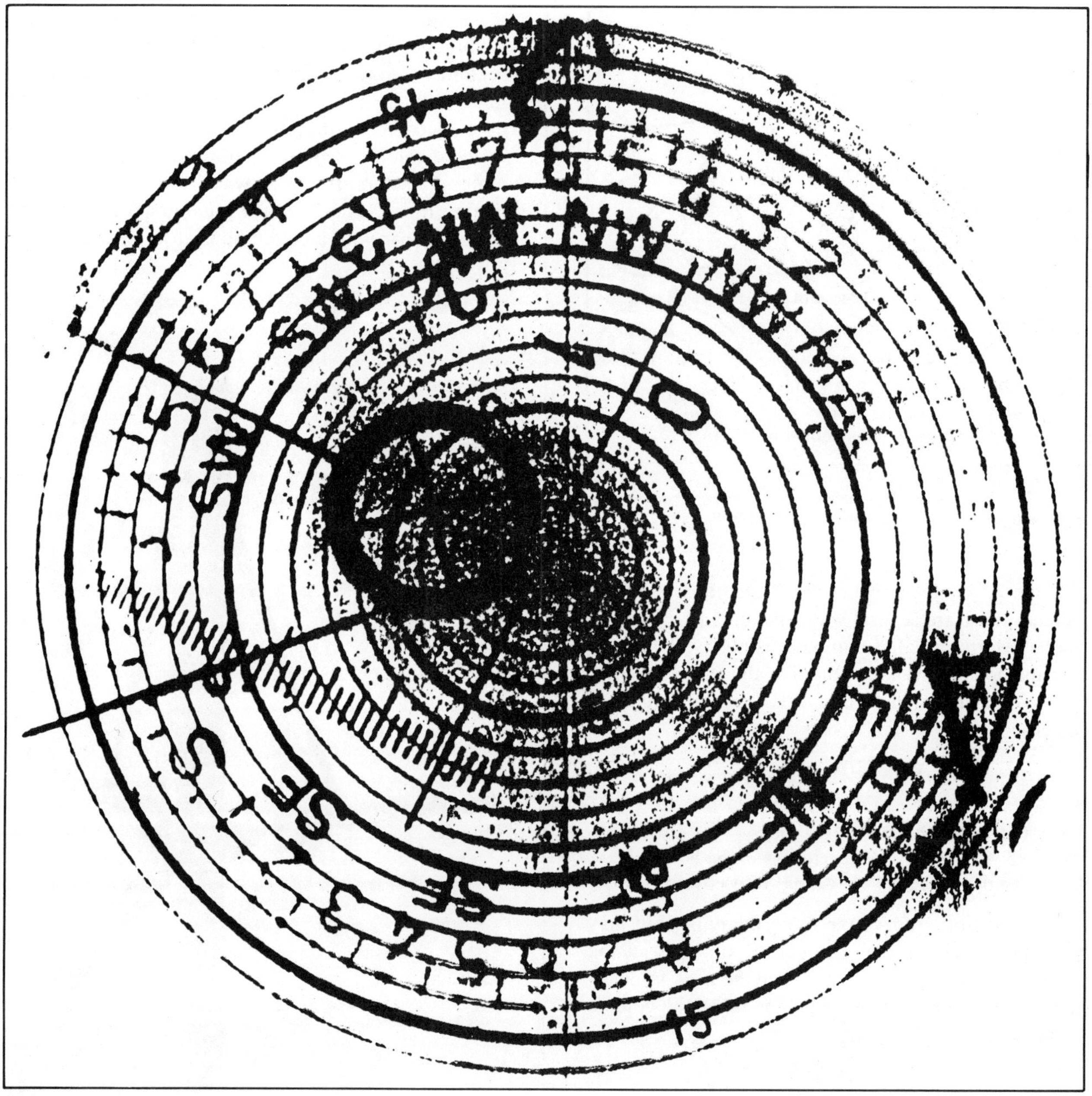

FIG. 7–9

Multishot image

BOREHOLE DEVIATION METHODS

Deviation measurements are used to correct other measurements, to define deposit locations, to satisfy legal and contractual requirements, and to spot problems. Borehole deviation measurements are made independently or in combination with other systems. Several types of deviation tools are available. The surface-readout, continuous systems appear to be superior in every way to single-shot and photographic methods. Calculations are tedious but are not difficult. There is a distinct advantage to the service of an on-site computer.

Borehole deviation measurements fall generally into two catagories: those that use the earth's magnetic field and those that use an inertial reference. The earth's magnetic field is stable enough and well-enough known that it can be measured to reference borehole direction to a geographical direction. It can be measured in a horizontal plane or two or three vectors downhole.

■ Station Type

The simplest directional tools are the so-called one-shot tools. The use of these is mainly restricted to drillpipe to satisfy contractual obligations. This type of system has a circular card mounted on a magnetic compass. The card is kept oriented to magnetic north by the compass. The compass responds approximately to the horizontal components of the earth's magnetic field. On command or by timing, a pen marker on a plumb bob puts a mark on the card indicating the tilt and direction of tilt of the tool.

The second class of tools was used for many years in the petroleum and mineral business and is still used regularly. This is the multishot photographic device. There are several minor modifications of the multishot system. However, they all photograph a magnetic compass through a tilt indicator (a plumb or a ball in a watch-glass or a bubble-type spirit level) so the two images are on one diagram showing degree and direction of tilt. Fig. 7–9 shows a typical image.

The multishot devices photograph at depth intervals or stations in the borehole, which are chosen and recorded manually by the operator. The tool operation at the station is controlled either by timing (with the downhole instrument having self-contained batteries) or by electrical control and power from the surface. Fig. 7–10 is a diagram of this type of tool. The operator inserts a pattern of deliberate bad or inconclusive shots to identify the position in his notes. After the survey has been run, the film is developed and the patterns are matched to depths and read.

Modern multishot tools are small and compact and incorporate a high degree of damping. Thus, they may be run in combination with other devices by merely pausing for two or three seconds during a logging of some other curve to take a station reading.

Because of the high damping, the instrument settles down quickly. This is not true of all instruments and must be checked for each type. These devices have two serious drawbacks. First, the film must be developed and the calculations made long after the tool comes out of the hole. Second, the limited time in the borehole and the limited film capacity require long intervals between stations that can cause major inaccuracies.

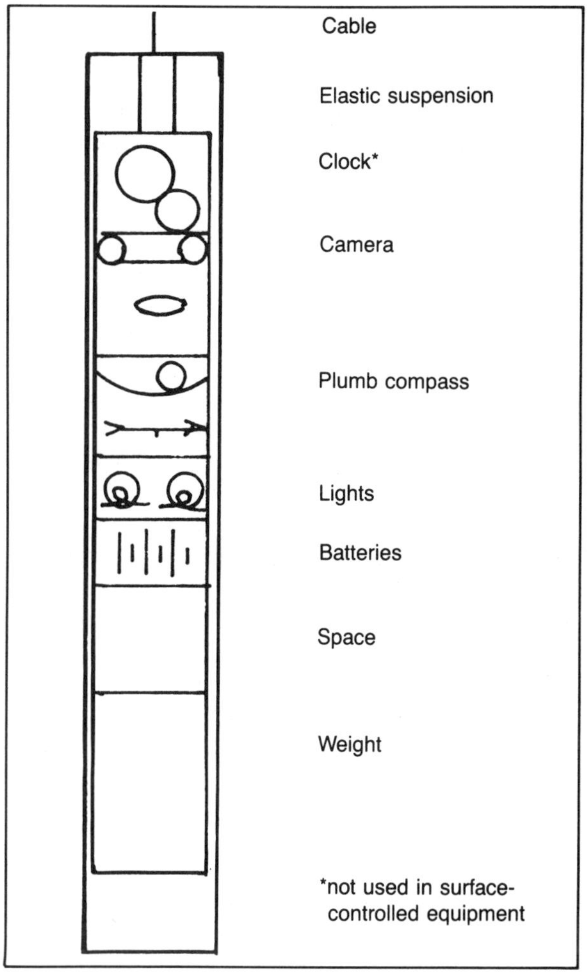

FIG. 7–10
Photographic deviation tool

■ Continuous Type

Surface readout tools were developed to circumvent the drawbacks of the photographic tools. There are two major types of these tools in use. One type, the Humphrey system, uses a two-plane fluxgate magnetometer to read the horizontal direction of the earth's magnetic field as two vectors. A resultant signal from this operates a servo system to orient a cage N-S continuously. The cage contains two plumbs that indicate the N-S component of tool tilt and the E-W component. These two values define the position of the tool in space unambiguously. The position of the tool is assumed to coincide with the axis of the borehole. This type is a tremendous improvement over the photographic systems.

Since the signals are sent to the surface continuously, any station length may be used. The readings may be taken while the tool is moving. Therefore, the interval between measurements may be made so small as to introduce negligible error. Also, the signals can be handled by a computer, so a finished plot of borehole deviation is available shortly after the survey is finished. Fig. 7–11 is a diagram of this type of device.

Another system in common use uses a three-plane fluxgate to read the direction of the earth's magnetic field to determine the tool position within that field. Also, two fixed-axis accelerometers read the earth's gravitational field to define the tilt of the tool. All five of these signals are digitally encoded, sampled periodically, and transmitted to the surface. There, the computer solves the orientation problem and plots hole deviation on site, within a few minutes after the finish of the measurement. One instrument of this type is made by Owl Technical Associates Inc. Samples of the readout of it are shown in Table 7–3, and the Century Geophysical system is shown in Table 7–2 and Fig. 7–12.

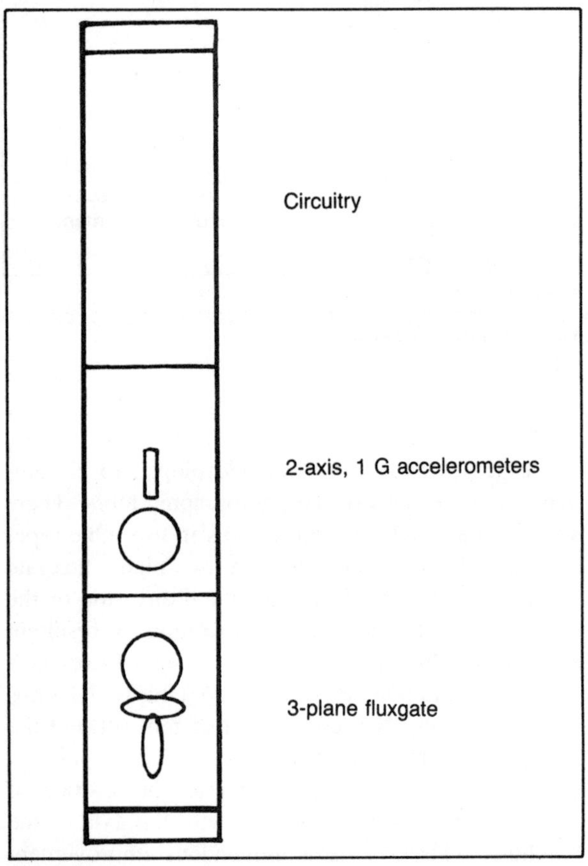

FIG. 7–11

Typical continuous deviation tool—fluxgate and accelerometers fixed with respect to tool

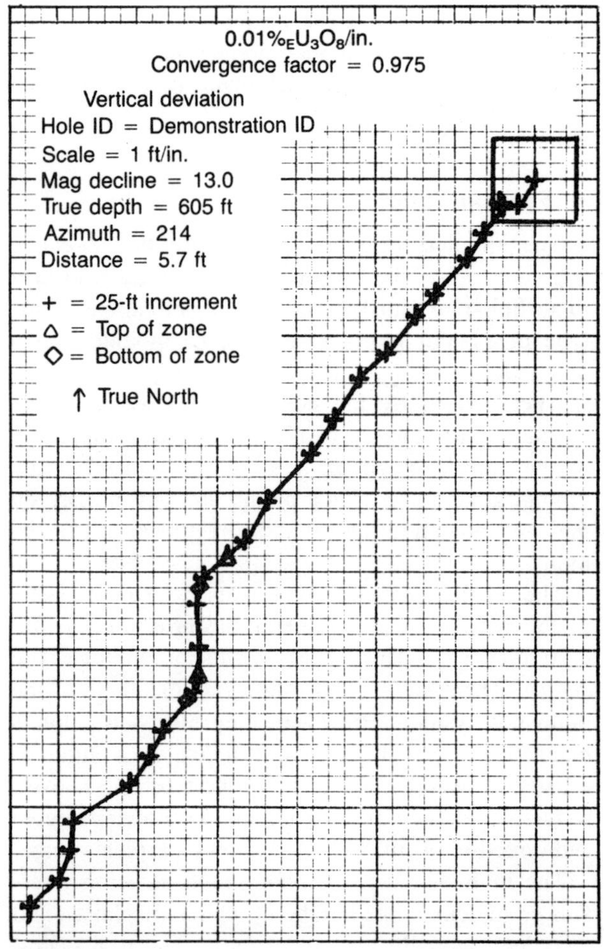

FIG. 7–12

Continuous deviation log
(courtesy Century Geophysical Corp.)

■ Inertial Type

None of the magnetic field tools can be used within a steel pipe, such as casing. Steel or magnetic mineral bodies near a borehole can distort the earth's field seriously. This problem can be overcome by using an inertial device, such as a gyroscope, to furnish a continuous, spatial position reference, in place of the three-plane fluxgate (and gravity sensors, usually). Fig. 7–13 diagrams this type of tool.

Inertial tools can be handled in exactly the same way as magnetic tools. However, they have several drawbacks that limit their use severely. They are expensive and delicate. They are subject to precession because they maintain a fixed position in space, regardless of the motion of the earth. They are subject to drift caused by bearing and skin friction, and the

Table 7–2

Hole Deviation Data

CABLE DEPTH	PROBE INCLN	PROBE DIRECTION	TRU VERT DEPTH	FEET EAST	FEET NORTH	CLOSURE DIST	NET BEARING
0	0.1	154	0.0	0.0	0.0	0.0	0.0
10	0.1	154	10.0	0.0	-0.0	0.0	153.5
20	1.1	76	20.0	0.1	0.0	0.1	81.8
30	1.1	34	30.0	0.2	0.1	0.3	64.1
40	2.0	60	40.0	0.4	0.3	0.5	57.3
50	2.1	68	50.0	0.8	0.4	0.9	59.9
60	2.2	71	60.0	1.1	0.6	1.3	62.7
70	2.4	75	70.0	1.5	0.7	1.6	65.1
80	2.5	80	80.0	1.9	0.8	2.1	67.6
90	2.6	83	90.0	2.3	0.9	2.5	70.0
100	2.7	92	99.9	2.8	0.9	2.9	72.7
110	2.8	95	109.9	3.3	0.8	3.4	75.5
120	3.2	101	119.9	3.8	0.8	3.9	78.5
130	3.5	106	129.9	4.4	0.6	4.4	81.7
140	4.1	110	139.9	5.0	0.4	5.0	85.1
150	4.4	114	149.9	5.7	0.2	5.7	88.5
160	4.6	114	159.8	6.4	-0.2	6.4	91.5
170	4.8	116	169.8	7.2	-0.5	7.2	94.1
180	5.3	118	179.7	7.9	-0.9	8.0	96.5
184.3	5.2	118	184.0	8.3	-1.1	8.4	97.5

Source: Century Geophysical Corporation

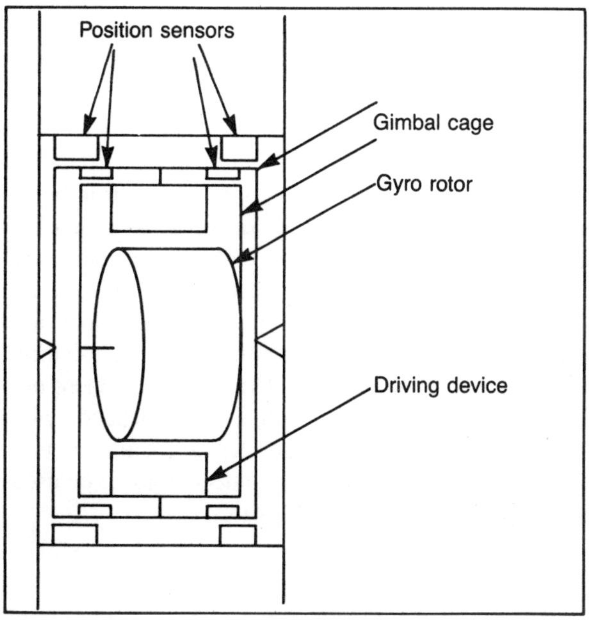

FIG. 7–13

Inertial deviation tool

small size required by logging tools is difficult to handle. These problems can be compensated for, but one must be aware that they exist.

■ Calculations

Calculation of the drift or deviation of a borehole from deviation surveys can be done in one of two manners. In the mineral industry it is customary to use a tangential method—a linear extrapolation from each station or calculation point. This method is excellent if the distance between calculations or stations is short or if the borehole deviation is small and variable. Fig. 7–14 shows the effect of station length on a hole-deviation determination. With the linear extrapolation the angle of deviation and direction of deviation are assumed to be constant from one station or calculation to the next. The vector components of one station are added to the next, with the assumption that they are unchanged.

A better method is employed in the petroleum industry when the slant angle of the hole is changing. The radius of curvature method assumes the hole is

Table 7–3
Sample Readout

```
     BEARING      97.7
  HORIZ DIST       8.4
  VERT DEPTH     183.9
```

Footage	Inclination	Direction
184.3	5.2	118
180	5.3	118
170	4.8	116
160	4.6	114
150	4.4	114
140	4.1	110
130	3.5	106
120	3.2	101
110	2.8	95
100	2.7	92
90	2.6	83
80	2.5	80
70	2.4	75
60	2.2	71
50	2.1	68
40	2.0	60
30	1.[illegible]	34
20	1.1	76
10	0.1	154
0.[illegible]	0.1	154

Source: Owl Technical Associates

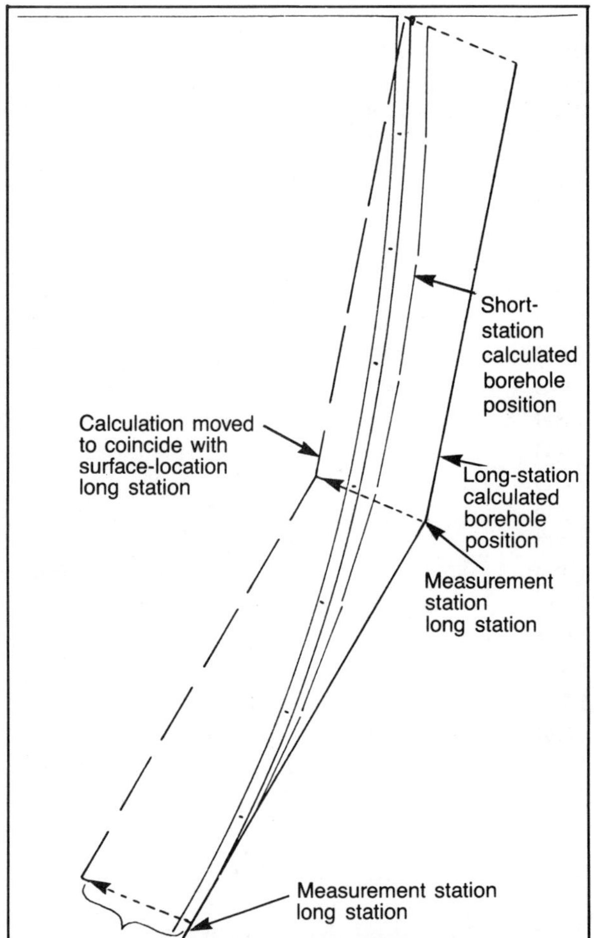

FIG. 7–14
Error of deviation calculation resulting from the use of long stations

curved if the upper and lower station slant angles are not the same. Thus, the error is minimized.

Hole-deviation methods allow one to locate where the bottom of the hole is. This is important in non-petroleum logging where a pulldown method instead of drill collars is often used to put weight on the drill-pipe. A pulldown puts force on the top of the flexible drillpipe, causing it to deviate because the drillpipe will bend slightly. When drillcollars are used, the weight is directly on the bit and the drillpipe is under tension. Fig. 7–15 illustrates this problem.

Hole-deviation methods also allow calculation of the true vertical thickness of a formation. This is necessary in order to calculate reserves properly. Neglect of a 17° deviation causes an overestimation of reserves by 5%.

Hole-deviation measurements are difficult in boreholes that are nearly vertical. In this case the movement of the fluid past the tool causes it to float away from the hole wall because of the Bernoulli effect. When this happens, the deviation device does not measure the hole drift, but only the probe position within the borehole (Fig. 7–16). This movement occurs at hole deviations of less than 1–3°, depending on the logging speed, the mud density, and the tool weight. One should insist that a decentralizer be used on the tool in near vertical holes.

TEMPERATURE MEASUREMENTS

Temperature measurements are potentially quite valuable. However, their use is limited by the ex-

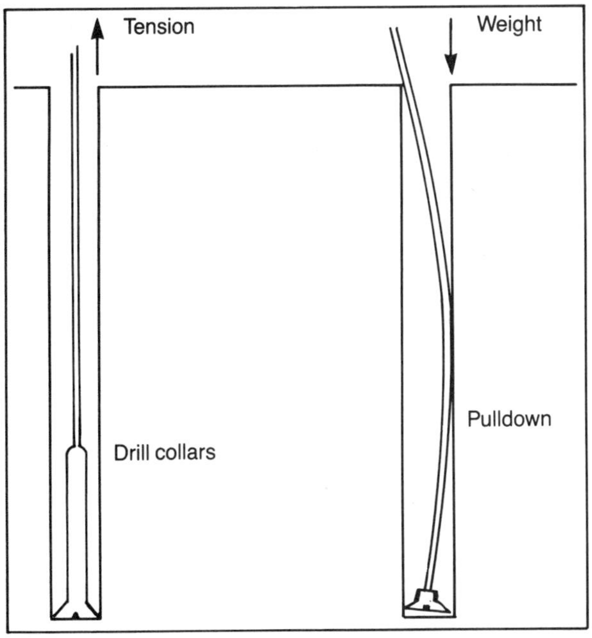

FIG. 7–15

Effect of pulldown on drillpipe

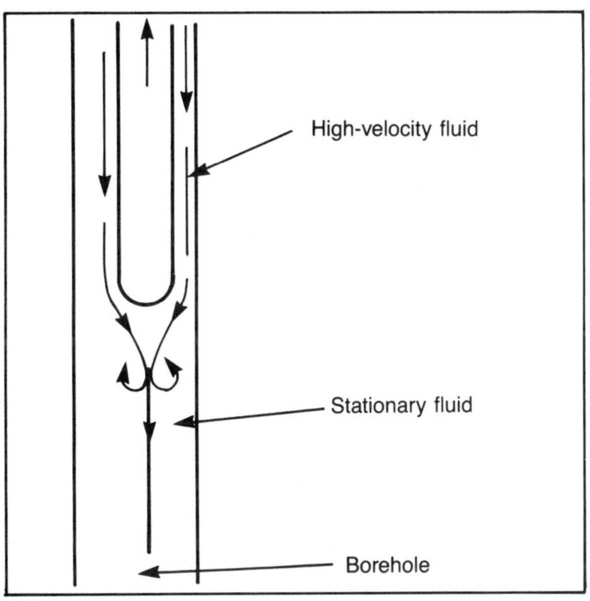

FIG. 7–16

Flow of borehole fluid around moving sonde

treme thermal disturbance of the formations during drilling. Primarily, temperature logs can be used in gas- and fluid-entry evaluation, corrections for resistivities, location of trouble, cement location, and as a substitute for resistivity curves under some circumstances (Fig. 7–17a, b).

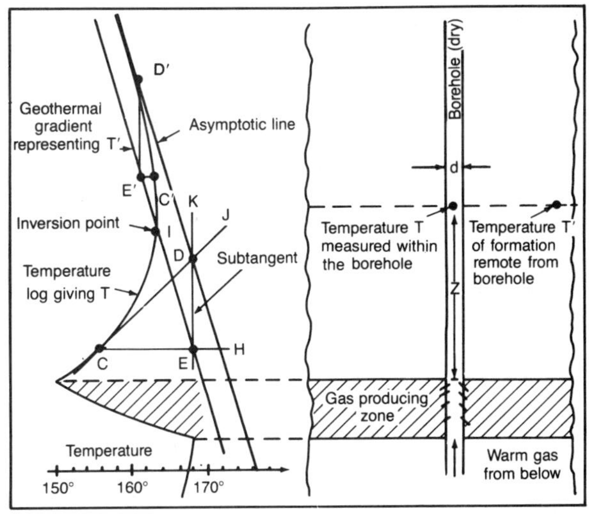

FIG. 7–17a

Temperature log above a gas producing zone

(after Kunz and Tixier, "Temperature Surveys in Gas Producing Wells," *Journal of Petroleum Technology,* volume 205, 1955)

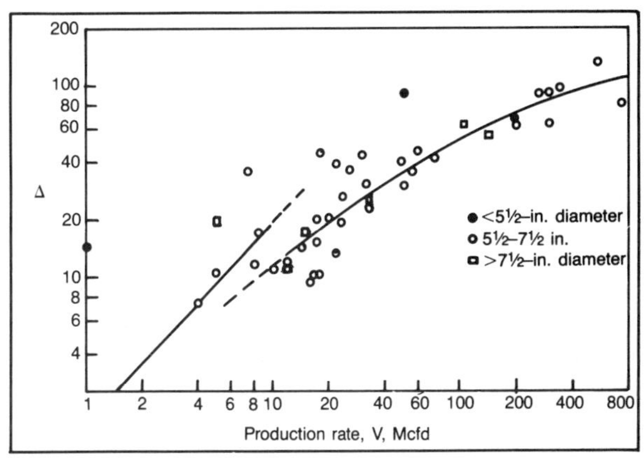

FIG. 7–17b

Value of Δ (or subtangent) vs. production rate of gas (after Schonblom, courtesy Birdwell)

■ Principles

With a hot interior and a cold exterior, the earth has a continuous flow of heat outward to space. Local surface variations as a result of solar heating are minor. A map of the average annual surface temperature in the United States is shown as Fig. 7–18. Therefore, there is a temperature gradient through the crust of the earth where we do virtually all of our logging. For any particular region this gradient, except for some local anomalies, can be predicted quite accurately.

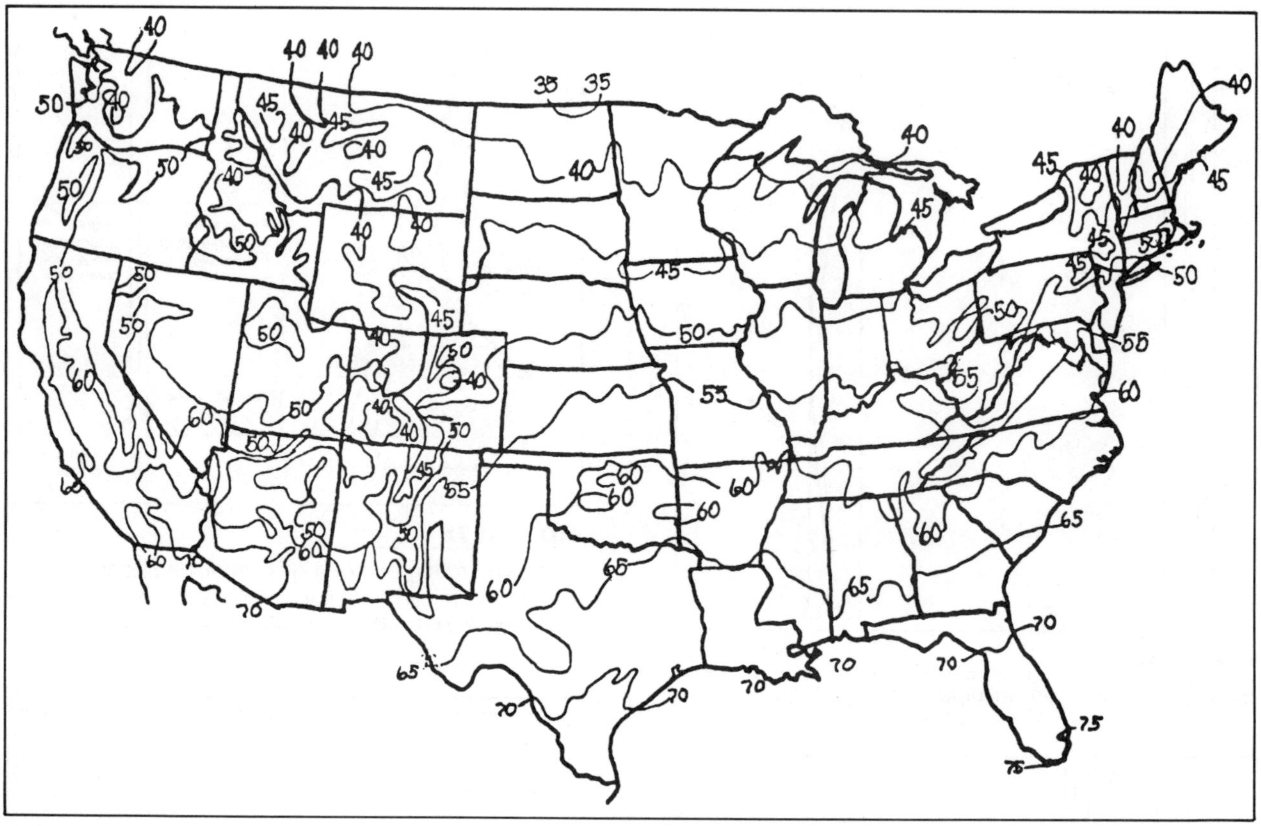

FIG. 7–18

Mean surface temperature in the United States (courtesy Barry Hallenburg)

The temperature at any point, and the gradient through any distance depends on the temperature difference, the thermal conductivity of the material, and the rate of heat flow. These are related by Fourier's Law:

$$\frac{dQ}{d\theta} = -KA\frac{dT}{dx} \tag{7.26}$$

Where:

$dQ/d\theta$ = Quantity of heat per unit time, the instantaneous rate of heat flow
A = Cross-sectional area
dT/dx = Temperature gradient along the path of flow
K = Conversion constant or thermal conductivity

A chart of thermal conductivities is shown in Table 7–4. Notice the close relationship to the expression for electrical current flow through a conductive medium:

$$\frac{dq}{d\theta} = -\sigma A\frac{dV}{dx} \tag{7.27}$$

Where:

$dq/d\theta$ = Instantaneous electrical current flow
A = Cross-sectional area
dV/dx = Voltage gradient along the direction of flow
σ = Conversion constant or electrical conductivity, closely related to the thermal conductivity, K.

In a borehole where the system has returned to equilibrium, we can measure the thermal gradient,

Table 7–4

Thermal Conductivities

The mechanism of heat flow is similar to that of fluid flow.

Darcy's Law of fluid flow:	$\frac{dQ}{dt} = A\frac{KdP}{\mu dL}$
Fourier's Law of heat flow:	$\frac{dq}{dt} = Ak\frac{dT}{dL}$
Radial heat flow loss from well to formation:	$q = 2\pi K\frac{T_t - T_w}{\ell n(r_e r_w)}dh$
If the fluid flow rate is V_f and its density D_f:	$q = V_f D_f C_f dT_f$

Where:
C_f = Heat capacity of the fluid

Equating and solving for V_f:

$$V_f = \frac{2\pi K}{D_f C_f \ell n(r_e/r_w)}\ \frac{T_f - T_w}{dT_f/dh}$$

determine the heat flow, and calculate the thermal conductivity. This can be used as effectively as the electrical conductivity.

The usual situation during logging, however, is that the thermal environment has been greatly disturbed by drilling. The radial heat flow from a well to the formation (or the other way) is a serious thermal disturbance. However, in a permeable formation the invading fluid also disturbs the formation temperature. As a result, the uses to which temperature measurements can be put quantitatively are usually limited. This is illustrated in Fig. 7–19.

Since we are usually working in a thermally disturbed environment, the thermal situation is changing rapidly. We can make effective use of differential temperature measurements, as well as absolute measurements. The simplest way to make differential temperature measurements is to place two or more thermal sensors along the probe and measure the absolute temperature of each and the difference between two of them. Then the temperature gradient or differential temperature, dt/dx, becomes:

$$\frac{dt}{dx} = \frac{t_2 - t_1}{x} \tag{7.28}$$

Where:

t = Absolute temperature

x = Distance between the sensors

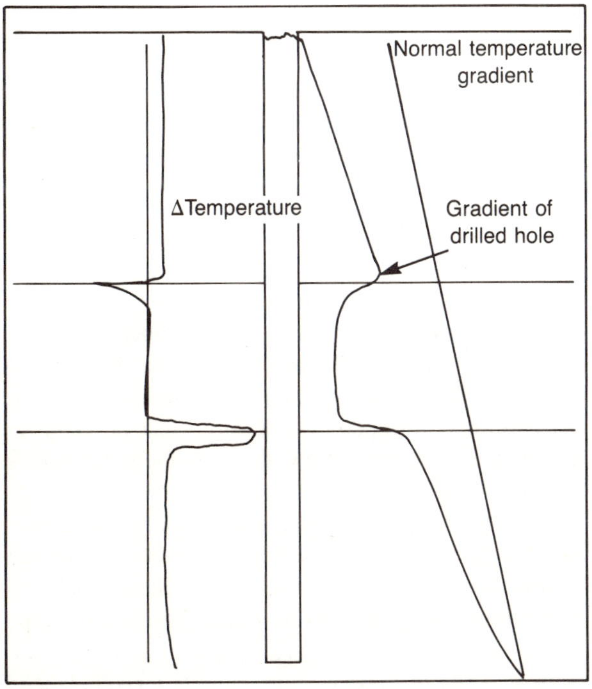

FIG. 7–19

Normal temperature gradients

The temperature and differential temperature can also be measured by recording the temperature with a single sensor on tape, or memorizing it, and comparing it with a later measurement at a fixed distance, x, along the borehole. Time differential measurements can also be used. In this case the temperature is stored for a particular length of time and compared with a later time. This can be done with a tape or memory system, or it can be done with a capacitive-resistive time-constant system. Several types of differential temperature systems are shown in Fig. 7–20. The main purpose of the differential temperature systems, whether distance or time differentials, is to allow enough sensitivity to be able to detect small temperature changes, without undue interference from the relatively constant normal gradient and without a severe reduction of sensitivity to keep a relatively high-temperature measurement curve on scale.

Thermal measurements have time constants that can affect the measurement, much as the gamma-ray system is affected. The longest time constant is the thermal time constant of the sensor, however, not in the following circuitry. The greatest offender in lengthening the thermal time constant is usually the housing over the sensor. The sensor usually must be protected from the electrically conductive, corrosive mud by a metal protective cover placed over the sensor. This results in a much longer thermal time constant than the bare sensor.

The other offender, which is more difficult to overcome, is the improper flow of borehole fluid across the sensor. The sensor must be protected from mechanical damage in the borehole, so a heavy steel cage is usually put around the sensor. It is easy to block or distort the flow of fluids or to allow a collection of mud cake along the sensor. These, of course, alter the thermal characteristics drastically. The mass of steel near the sensor can also alter its thermal time constant. A typical temperature sensor assembly is shown in Fig. 7–21.

Temperature sensors can be of several forms. Each type has its advantages and disadvantages. The earliest type used for logging (by Kelvin) was a platinum resistor in a Wheatstone bridge circuit. A platinum resistor has a very linear temperature coefficient of electrical resistivity. This type is still used for precise measurements. Unfortunately, the resistivity of platinum is not high nor is the coefficient. Therefore, a fairly large amount of fine platinum wire must be used, and the change or signal is not large. Several types of circuit diagrams are shown in Fig. 7–22.

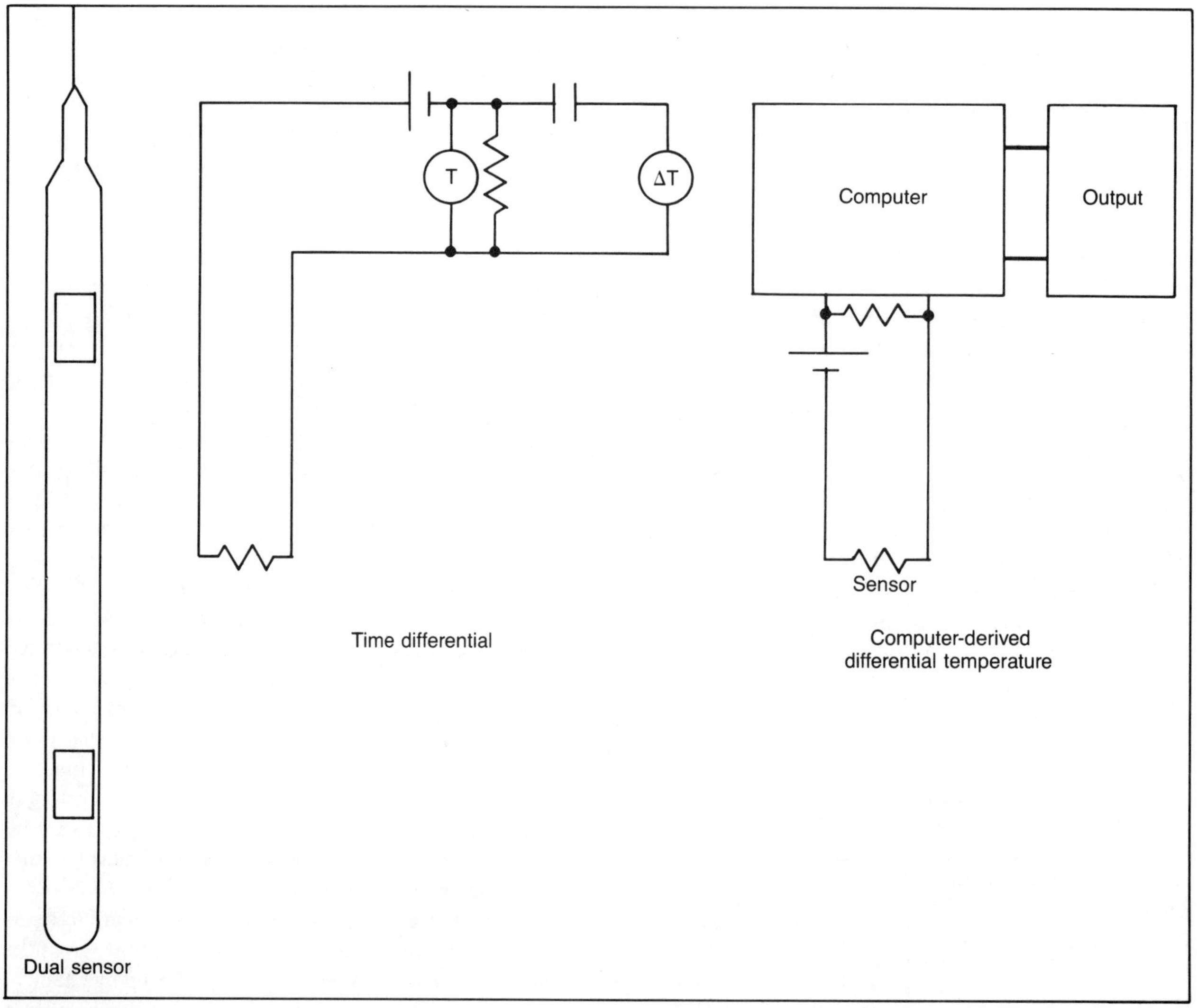

FIG. 7–20
Types of differential temperature tools

Thermistors have largely replaced platinum resistors for logging use. These, again, are temperature variable resistors. They have much higher temperature coefficients of resistivity than platinum. They also are much cheaper. However, the thermistors are nonlinear in their response. Their response is made nearly linear by appropriate resistance in a network. Also, they are usually materials with rather low thermal conductivities. Therefore, they are more difficult to work with than platinum.

Thermocouples are sometimes used in logging. A thermocouple is the interface between two dissimilar metals, such as iron and copper. A potential difference (electrical) appears across this interface as a result of a temperature, which is a function of the absolute temperature of the interface. In practice, this potential is often compared to the potential of an identical couple in a known temperature—the reference junction. Thermocouples can be slow—they have long time constants—if the assembly is not carefully designed. Great care must be taken to minimize the current flow through them to reduce resistive heating.

Usually, the resistive-type sensor (platinum resistor and the thermistor) is used in a Wheatstone bridge circuit. In this circuit the sensor resistance is compared to a stable, known resistance. The voltage-type sensor (thermocouple) is often used in a comparison circuit. Either of these circuits may be servo balanced to operate with a minimum of power drain. Either type may be used as the controlling element of an oscillator circuit. In this case the frequency is a

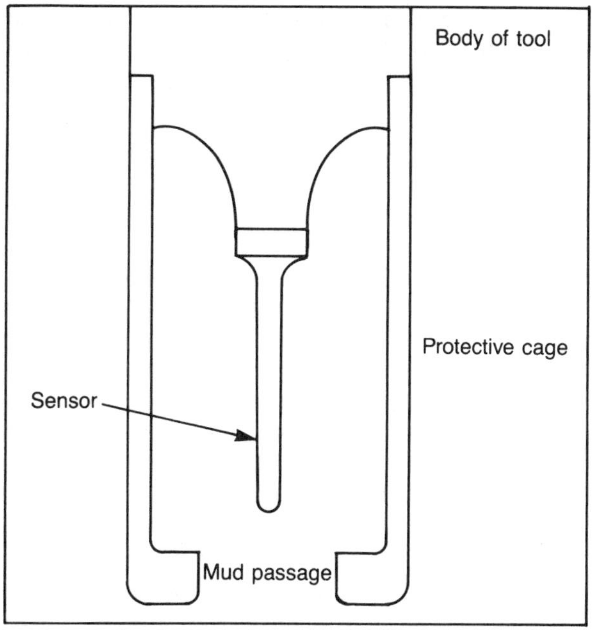

FIG. 7–21
Typical temperature tool assembly

function of the temperature, and the surface instrument can be a ratemeter or digital circuitry.

Table 7–5 catalogs thermal conductivities of geological materials. Temperature measurements used downhole allow thermal conductivities calculation and lithology identification.

■ Measurement Uses

The most common use of downhole temperature measurements is to correct the values of resistivity for temperature. The conductivity of an ionized material in water solution is a function of the absolute temperature. Thus, the resistivity (conductivity) of a mud sample measured on the surface at surface temperature must be corrected to the formation temperature before it can be used. Usually, this is done by means of charts that plot resistivity or conductivity of a sodium chloride solution for any given concentration as a function of solution temperature. Fig. 7–23 shows such a chart. For exacting work or special cases, the NaCl solution charts are inaccurate for

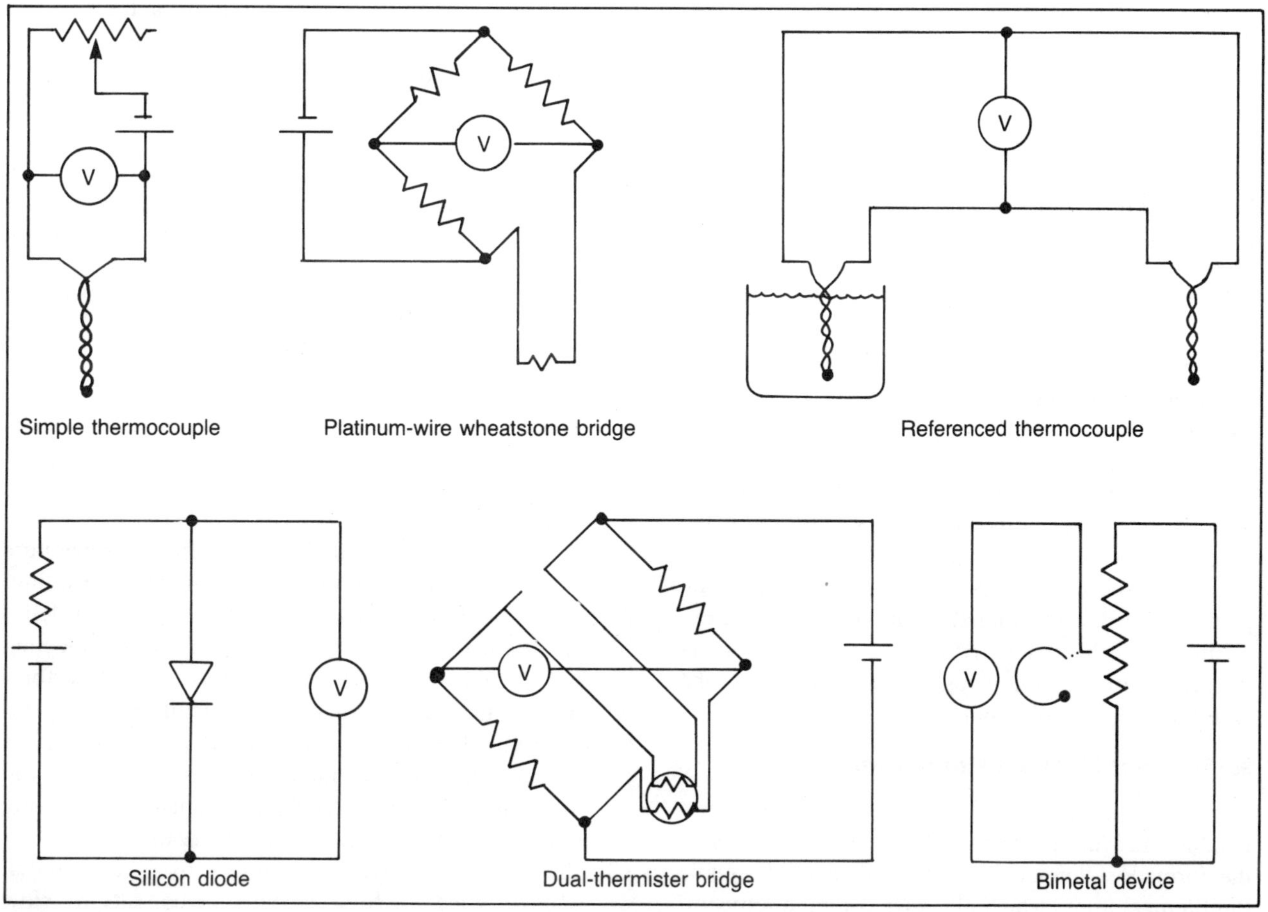

FIG. 7–22
Temperature circuits

Table 7–5

Thermal Conductivities

Material	Thermal Conductivity, 10^{-3} (cal/cm²-sec)/(°C/cm)
Coal-lignite	0.33–1 at 68
Cement (Portland)	0.71 68
Water	1.39 32
	1.43 68
	1.60 167°F
Clay	2–3
Chalk	2–3
Gypsum	2–4
Shale	2–4
Sandstone	3–12.2
Limestone	2.4–8
Polyhalite	3.7
Basalt	4–7
Granite	5–8.4
Calcite	5–13
Feldspar	5.8
Slate	6
Quartz	6–30
Salt	14.3 (8–15)
Dolomite	9.3–11.9

non-NaCl ion species. Corrections for other ion types must be made.

Downhole temperatures for resistivity correction purposes are measured in a variety of ways. A maximum reading mercury thermometer is often encased in a steel housing and is attached to the cable head. The maximum reading is assumed to be the bottom-hole temperature (BHT), and the gradient to the surface is assumed uniform. Either of these assumptions may be faulty in actual practice. Care must be taken with a bottom-hole thermometer to ensure that its thermal time constant is not too great to get an accurate reading.

A dry, clean thermometer in a steel case has a time constant of many minutes to as much as an hour. The case should be partly filled (*not* completely filled) with water, graphite, metal powder, or some other thermally conductive material to give good contact with the borehole fluids. If water is used to fill the case, two precautions must be observed. The case must never be completely filled. If it is, the outside pressure can be transmitted to the mercury column, raising it mechanically. Also, if the temperature at the bottom of the hole is high, vapor pressure in the case may do the same thing.

■ Thermal Neutron Measurements

The IRT Corporation has been operating a temperature-measuring device that tends to measure the formation temperature, rather than the borehole fluid temperature, which all other common temperature devices do. The thermal neutron tool detects the energy of the thermal neutrons in the formation near the borehole. High energy neutrons are put into the formation from a source on the logging tool. Presumably, the source neutrons are emitted in a short pulse. The detector is gated on after the borehole effects have died away. The mean energy of the thermal neutrons then reflects the temperature of the formation near the borehole.

The temperature log and the differential temperature log are used to detect fluid entry into the borehole. Fig. 7–24 is an example of the temperature and differential temperature log. The formation fluid is seldom at the same temperature as the borehole fluid. Thus, fluid entry can be detected by a sharp, anomalous temperature change. Gas entering the hole flows into a zone of lower pressure and undergoes adiabatic (expansion) cooling. This is a common method of evaluating gas production zones. The adiabatic cooling of the gas entering the hole results in a negative anomaly on the temperature curve, away from the normal gradient (Fig. 7–24). The rate at which the gas returns to the normal gradient and formation temperature depend on the rate of flow of the gas. To calculate this, first calculate a value of ΔT (usually called simply delta):

$$\Delta T = T_o - T_g(dT_g/dD)^{-1} \qquad (7.29)$$

Where:

T_o = Undisturbed or normal temperature at the zone of interest, determined by drawing the normal gradient through the anomaly or the temperature that would be read if no gas were entering

T_g = Actual temperature of the gas at that zone

dT_g/dD = Change of gas temperature with the change in depth or the slope of the curve

The value of the gas production rate, GPR, in millions of cubic feet per day is given:

$$GPR = 7.58\ e^{0.05\Delta T} \qquad (7.30)$$

Borehole temperature and differential temperature measurements are also used in temperature profiling, a cased-hole technique. Temperature profiling deliberately causes temperature anomalies by injecting hot or cold water into the formation through perforations in the casing. Successive logs of both temperature and differential temperature are made, first before injection and then at intervals after. The initial run after injection shows temperature changes indicating where the water went. Runs at intervals for 24–48 hr show zones of high permeability by a large initial change, zones of high capacity or porosity by slow change back to normal, and communication between

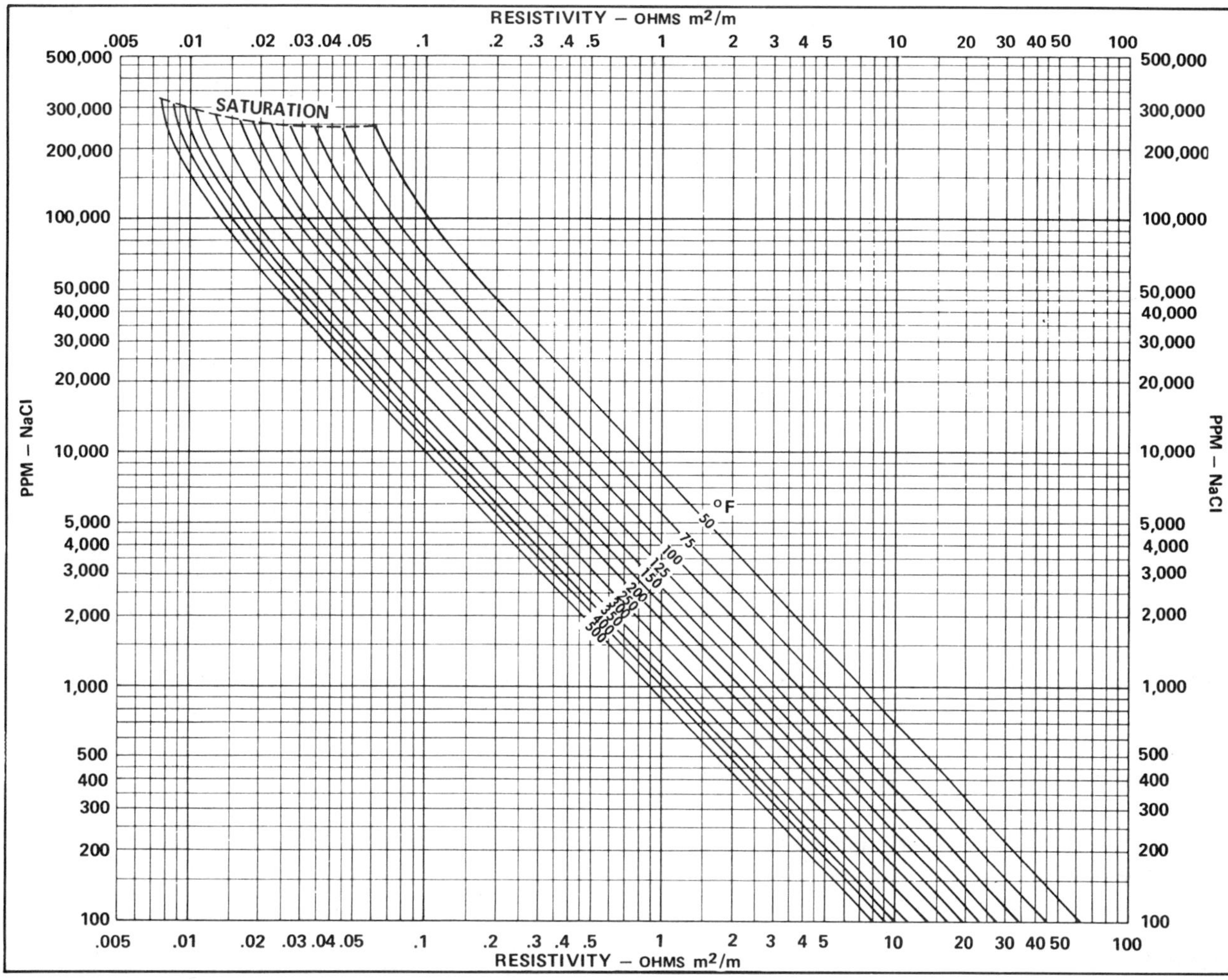

FIG. 7–23

Resistivity vs. temperature—NaCl solutions (courtesy Gearhart Industries Inc.)

zones by changes in zones not injected. The maximum temperature in the hole can give an idea of tool performance, while anomalous readings or gradients can indicate geological phenomena.

Laboratory Methods and Thermoluminescence

Laboratory methods of temperature measurement are straightforward. Geologic materials have characteristic heat capacities (specific heats) and thermal conductivities, which are related to electrical conductivities. Melting points and boiling points are often specific for many materials and mixtures. These data can be used to identify core and cuttings samples.

Thermoluminescence is caused by damage to the structure of a solid by radiation. Upon heating, the damaged portion returns to normal in many materials, with the release of heat (or light). This property is evident in silicas and can be used to identify samples which have been irradiated. The measurement procedure is to place the sample in an insulated chamber and raise the temperature evenly. Radiation damage becomes evident by anomalous temperature excursions superimposed on the record of the smooth temperature increase.

Calibration

Calibration of temperature tools is simple and uncomplicated. With a temperature device it is usually sufficient to obtain a reading in a mixture of ice and water and another in boiling water (corrected for elevation). Differential temperature devices must be calibrated with a change of temperature. The same

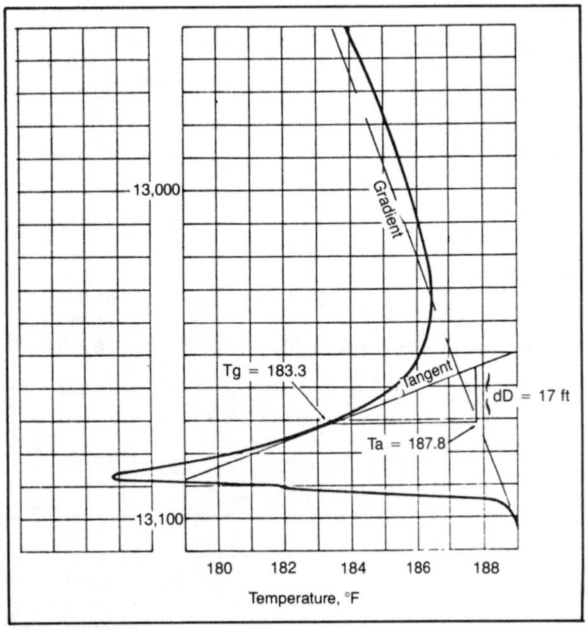

FIG. 7–24

Temperature and differential temperature log

mixtures as for the absolute measurement can be used. That is, no temperature change on successive readings should result in a zero differential or gradient. The probe should then be quickly moved to another temperature.

SAMPLING

Sampling methods are any of the numerous methods that actually take a sample of the formation material through which the borehole passes. There are several methods available. They include coring, cuttings sampling, sidewall sampling, percussive sidewall sampling, fluid sampling, swabbing, flowing, bailing, and sniffing. Most of these methods are not used except for special purposes. At that time discuss the problem with the driller or the logging contractor. Some, however, should be used routinely, such as coring, which especially must be a part of the evaluation effort.

Cores

Coring drill holes is one of the older methods of obtaining quantitative data from the drill hole. With the introduction of rotary drilling, it became a valuable and precise data-gathering method. Coring is done with a drill bit that cuts a ring, leaving a solid cylinder inside the drillpipe. Usually, a core barrel is put directly above the drill bit to receive, hold, and protect the core.

Coring bits may use an imbedded diamond surface, carbide cutters, or steel roller cones. The type of bit used depends on the formation type. Core barrels may be tubular or split. The tubular barrels are often used in sand/shale sequences because they are cheaper and easier to use. The problem of lost or out-of-sequence cores is much greater than with a split barrel. The core is removed from a tubular barrel by pushing or blowing it out. A split barrel opens lengthwise to allow access to the core. Then the core can be evaluated before it is disturbed.

Cores may be taken in short or long lengths. Long lengths increase the danger of lost or ambiguous cores, but short lengths take time. Therefore, a compromise must be made. Core diameters vary, also. The larger diameter a core is, the more representative it is likely to be. However, coring costs, especially time, increase with the cross-sectional area of the bit.

Cores are direct samples of the formation. As such, they have become the standard of comparison with other data-gathering methods. For this purpose (and others) they are invaluable. However, it must always be remembered that:

- A core has been physically cut from the formation by force. It probably has sustained mechanical and thermal damage in the process.
- A core has been washed and invaded with drilling mud of a known or unknown composition. It almost certainly has been contaminated chemically.
- A core has been lifted out of a high-temperature, high-pressure, liquid-saturated environment up to the surface: The surface environment is very different from the downhole one.
- A core has an entirely different geometry from the undisturbed formation and the logged volume. It also has a smaller size. Therefore, its statistical variations are greater than the logged volume. Its radiation characteristics also are different from the formation and the log.
- Finally, a core is handled many times by persons who may or may not be familiar with coring problems and practices, and unfortunately, sometimes do not care.

The depths, lost sections, and order, may be correct, or they may be entirely wrong. There has been little control over the way many cores have been handled from the formation through to the final data sheet.

Cores can be used for some types of analysis that are used with logs. With proper care and careful attention to differences in geometry and handling, the electrical, mechanical, radiological, and chemical characteristics can be compared with the logs. Cores that are to be correlated with logs should be logged on a core log or lithology log sheet as soon as they come out of the core barrel and, preferably, before going into the core box. They should then be run through a logging device, such as a core-gamma recorder. The log obtained should then be matched in depth with the downhole log. Missing intervals should be noted on the core gamma log and on the core log.

Cores and logs should be compared in similar intervals after depth adjustment of the core. Comparison with the gamma-ray curve, for example, should be done in running 3-ft core intervals. Fig. 7–25 shows one way this may be done. The weighting and averaging are done over seven sections of core, each 0.5 ft long. The center section has a weight of 16. One-half foot above and below this section have a weight of four; one foot above and below the second sections have a weight of two; one and one-half foot above and below that have a weight of 1.4. Thus:

$$\text{Average core grade} = \frac{\Sigma\,(G_d W)}{\Sigma W} \qquad (7.31)$$

Where:

d = Depth interval in half-foot increments
G_d = Grade at depth d
W = Weighting factor

This weighting approximates an exponential drop-off of contribution. Besides a weighted average the core is a smaller sample than the gamma ray. Therefore, we would expect the statistical variation of the core values to be greater than that of the log values by the square root of the volume differences. Thus, if the gamma ray examines a 3.5-ft diameter sphere and the core is 3.0 in. diameter by 3.5 ft long, the volumes are 48,792 cu in. for the log and 254 cu in. for the core. This is a difference in volume of a factor of 152. Thus, we would expect to find 12 times the statistical variation in core values as in the logged values.

If the log is run by a computer and a deconvolved calculation of the mineral grade is made, the core and log can be matched on a 1:1 basis, making allowances for statistical variations of each and after careful depth corelation is done with both.

Notice, though, that unless a deconvolution program is used, such as the AEC GAMLOG program, the comparison cannot be made on a 1:1 basis. The

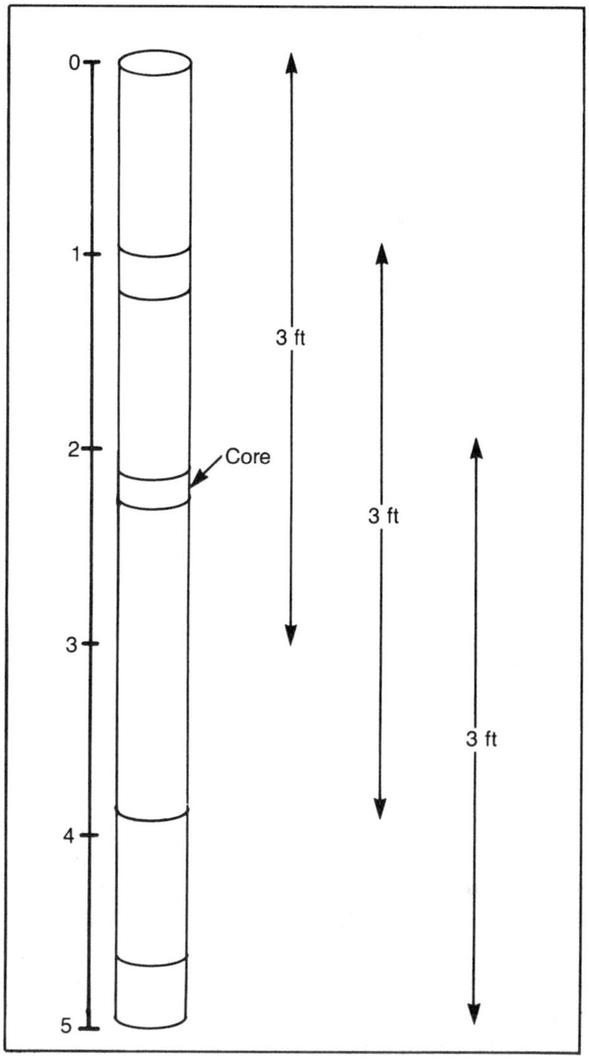

FIG. 7–25

Averaging cores for comparison with gamma ray

cores must be averaged over the same interval as the log. At this time only one contractor is running a deconvolved eU_3O_8 analysis in the field, despite claims to the contrary. The programing of even this simple deconvolution is complex and is not done without a computer.

Finally, it should be remembered that mineral grades determined from core assays are often given in weight percent of dry weight. Most logs are volumetric measurements. The gamma-ray log determination of eU_3O_8 is an exception. It is given in dry weight percent. A measurement such as the DFN, however, gives the uranium content in pounds per cubic foot. In order to convert to weight percent, the pounds per cubic foot must be converted. The relationship to convert pounds per cubic foot, ρ, to weight percent is:

$$\% = \frac{1.602P}{\rho}$$

Where:
ρ = Dry bulk density of the formation.

The dry bulk density, ρ, is:

$$\rho = \frac{\rho_m(\rho_{wet} - \rho_f)}{\rho_m - \rho_f}$$

Where:
m = Matrix
wet = Fluid saturated
f = Fluid

Also:

$$\rho = (1 - \phi)\rho_m$$

Cuttings

Cuttings or drill cuttings are the pieces of solid material cut from the formation by the drill bit. One must use them with caution, but they can be very valuable.

Cuttings suffer most from the lack of awareness of their importance by the drilling crew. As a result samples are often collected carelessly or deliberately incorrectly. The lag in cuttings, the time from cutting to appearance on the surface, is difficult to assess. Therefore, there is always some depth uncertainty in the cuttings log. Mixing can occur during the trip to the surface. Heavy particles may travel more slowly than fine ones. Clay tends to disappear, as does any soluble materials. Nevertheless, if these things are taken into account, much valuable information can be gathered from cuttings.

Sidewall Samples

Sidewall samples may be taken with a drillpipe sampler or a percussion-type sample gun. Either way, the sample is small and damaged, and often, it is only better than nothing.

Fluid Samples

Fluid samples in petroleum logging are usually taken with a wire-line fluid sampler until the well is cased. These tools are too big for the usual mineral hole. Therefore, in mineral holes the fluid sample is usually taken with the drillpipe. To ensure a good sample, a packer should be used to block off the mud. However, a high rate of pumping through the drillpipe can often result in a respectable sample. The driller can help immensely on this. He usually has the skill to get good samples.

GAS MEASUREMENTS

Gas measurements are primarily surface techniques. They are used to detect the mercury vapor associated with mineral deposits, the methane and carbon dioxide associated with petroleum, and the radon and helium associated with radioactive minerals. The presence of the gas is usually correlatable with one or more of these mineral deposits. For example, the detection of radon for health protection is important.

The methods of sampling the gas are the main barrier to using the methods downhole. They usually must be used in air-filled holes. The samples of gas are small, and statistical problems are an impediment. Some of the better methods deliberately use a long time constant to achieve a reliable statistical sample.

Mercury, Radon, and Helium Sniffers

The sniffer-type gas-measuring systems take samples of gas, either in a discrete sample or continuously during the measurement. The discrete sample methods are seldom used. They consist of taking a sample of gas from a borehole, a mine, or the general atmosphere and bottling it for transport to a laboratory. Often, the sample is under pressure. This method is difficult to use effectively because it usually involves a statistically small sample and the normal statistical variations of any measurement often make the results ambiguous. Any of the normal measurement methods can be used.

The continuous methods employ a pump to carry a continuous gas sample over a detector or a concentrating mechanism. For borehole work the hose to the pump is usually lowered downhole. The circulating and collecting system is seldom lowered.

Mercury is usually collected on gold foil. A large surface can be exposed with foil and mercury amalgamates easily with gold. Once the mercury has been collected, there are several sensitive, quantitative chemical and optical tests for mercury. If the volume of gas exposed to the gold foil is recorded, the sniffer methods can be semiquantitative. Usually, however, relative amounts are examined to pinpoint anomalies.

Radon is detected through its alpha emission. Gas is drawn through or across a zinc sulfide screen, and the scintillations are counted with conventional photomultipliers and circuitry. This method may be used downhole, on the surface with a probe, or in the atmosphere. Occasionally, especially for health monitoring purposes, the gas is drawn across a cotton or paper filter. This traps and concentrates the radon by adsorption. Radon detection is a valuable technique

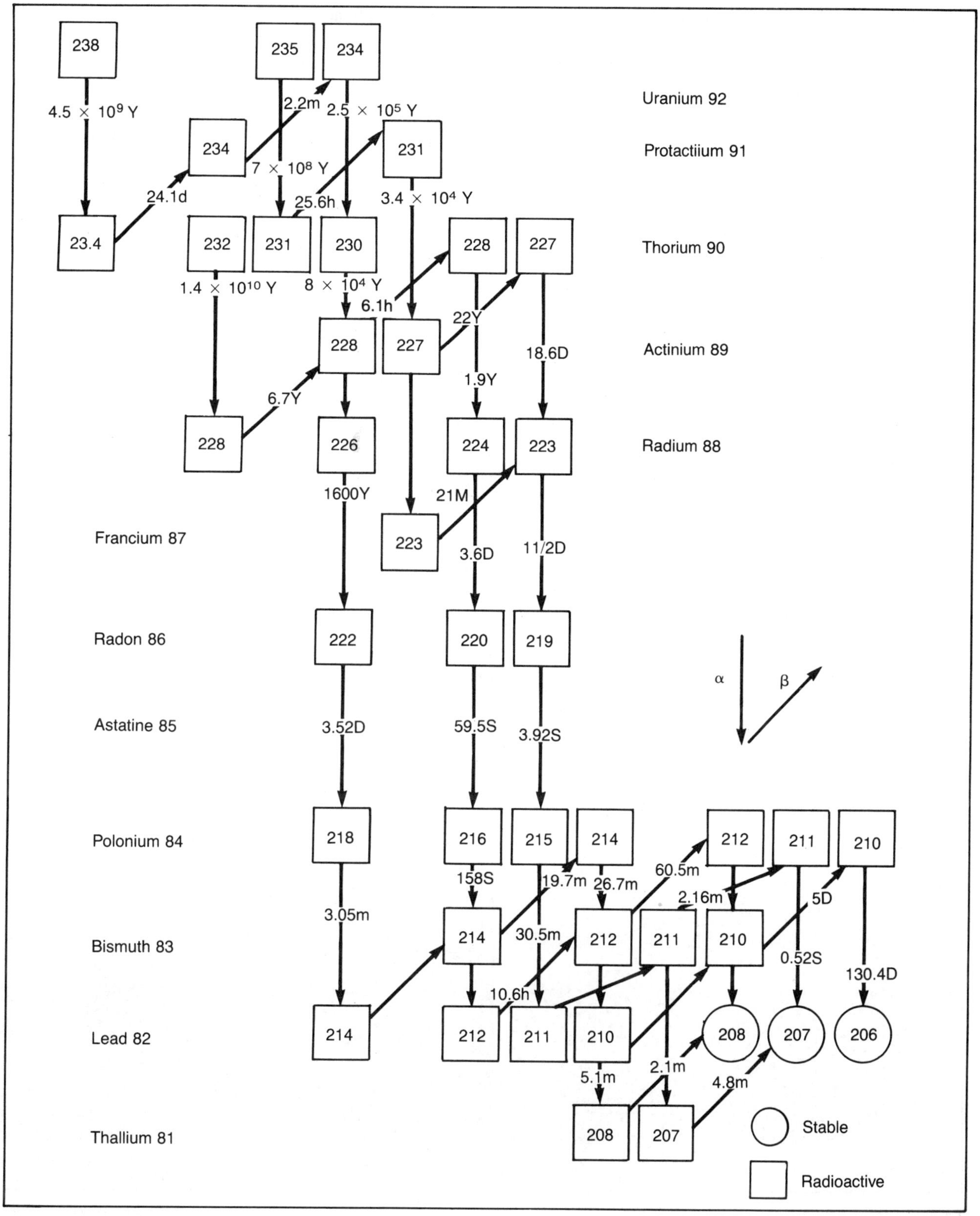

FIG. 7–26

Uranium decay series

because it is a daughter product of uranium and thorium. Since the longest half-life of any of the radon isotopes is 3.8 days, its presence indicates uranium or thorium nearby. Radon-222 is the 3.8-day half-life daughter of both uranium-238 and thorium-234 (also of ^{234}U, ^{232}Th, ^{230}Th). Two other radon isotopes may be present, ^{220}Rn (T/2 = 54 sec) and ^{219}Rn (3.9 sec). These also are daughters of uranium and thorium. See the decay series in Fig. 7–26.

Radon gas can be transported by migration (including bubbling), diffusion, solution in water, methane, or air. It is likely to migrate along faults, up boreholes, and in underground or surface water. Radon migration and diffusion may be responsible for the radioactive halos near uranium or thorium deposits. These halos are often the gamma emitters detected by ground and airborne surveys.

The atmospheric gamma-emitting daughter products of uranium and thorium are detected during airborne gamma surveys by shielding the gamma detector from the ground radiation. The reading is assumed to be indicative of the atmospheric radon content.

Helium sniffers use a modification of a mass spectrograph for detection. The gas sample is ionized, accelerated by an electric field through a magnetic field for separation by mass. A charge detector reads the amount of helium present in the sample. Helium is present near radioactive deposits because the alpha

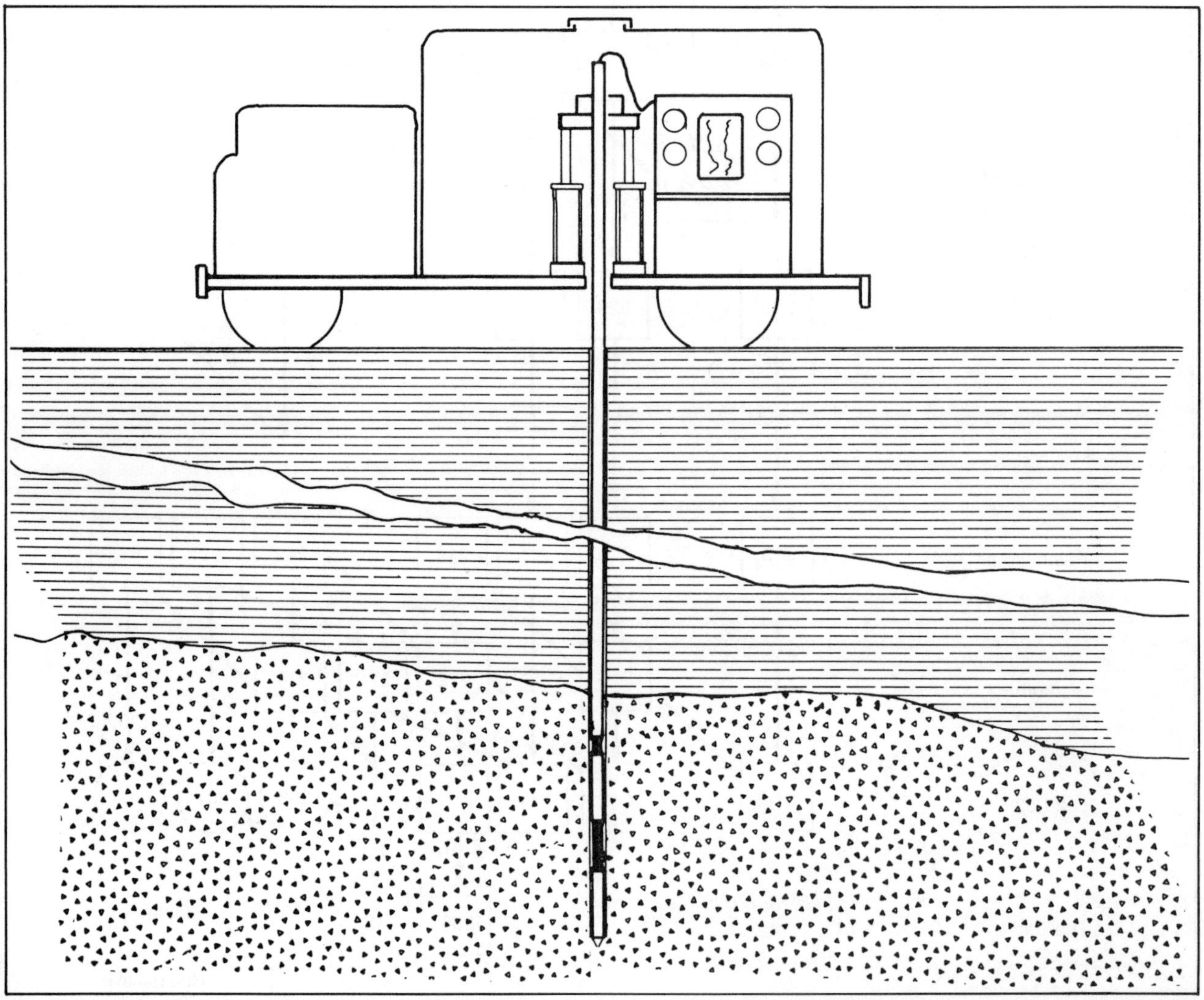

FIG. 7–27

Typical penetrometer (courtesy Ertec Western Inc.)

particles from them are helium nuclei. Again, the presence of helium concentrations may indicate radioactive or hydrocarbon deposits nearby.

■ Radon Cups

Radon cups are alpha detectors that are inexpensive and can be buried for days to months at a time to obtain a statistically valid reading (as a result of a long time constant or collecting time) of migrating or diffusing radon gas. There are two types of radon cups. One uses a plastic that is damaged by the passage of an alpha particle. The other uses a foil. Both types can be used effectively for mapping radon and radioactive trace-element anomalies.

The plastic type is in most general use at this time, as the track-etch method. A cup containing the sensitive plastic is buried, usually in a hole 2–3 ft deep, covered, and left for 20–40 days. Any radon that migrates to the surface emits alpha particles, which are detected by the unit. The detector indicates the presence of Rn-220, Rn-222 and Rn-219. Only Rn-222 has a long enough half-life (3.8 days) to migrate appreciable distances. Therefore, unless the cup is close to the mineral deposit, it is relatively insensitive to Rn-220 (55 sec), a daughter of thorium, and Rn-219 (4 sec). Rn-220 also can be filtered out.

These devices have the advantage of a large sensitivity and a long integration time. Thus, probable errors and statistical variations are minimized.

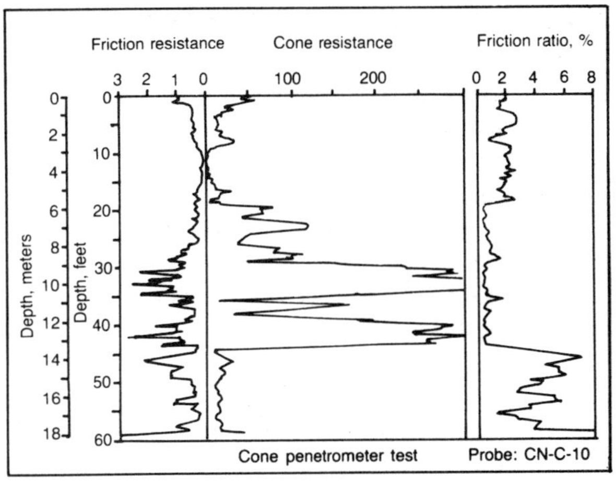

FIG. 7–28

Typical CPT sounding log

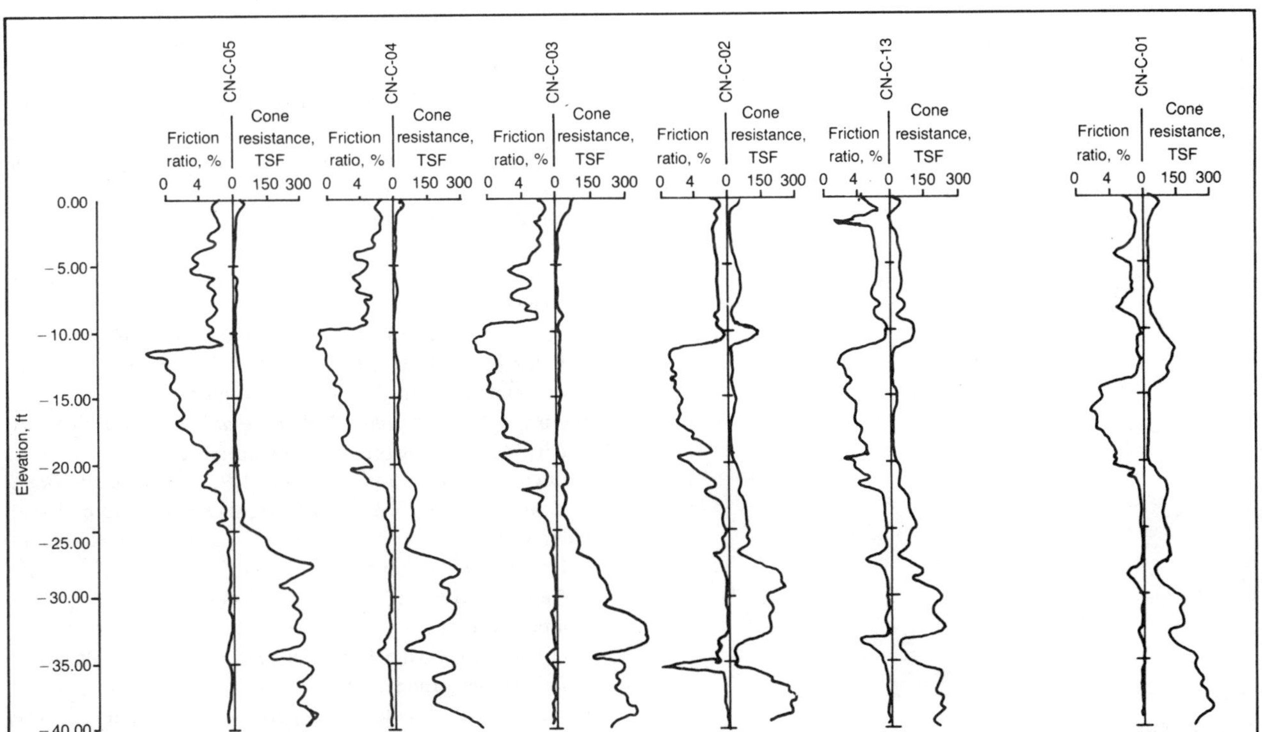

FIG. 7–29

Typical cross section utilizing CPT (courtesy Ertec Western Inc.)

ELECTRIC CONE PENETROMETER

The electric cone penetrometer (ECP) is a device used primarily by engineering projects. It is not a wire-line method. However, the results and the practice are both so similar to wire-line methods that it should be mentioned here.

The ECP is a cone that is pushed into the ground, and the force required is measured. Some of the mechanical properties of the sediments can be determined from this method.

Principles

The electronic cone penetrometer is a steel cone, usually about 35 mm diameter, which is pushed into the soil or filled by a hydraulic system on the logging truck. The force required is measured as a function of depth of penetration. The friction of the soil on a collar above the cone is also measured. The resulting log is an indication of the mechanical properties of the zones penetrated.

Fig. 7–27 diagrams a typical penetrometer. Fig. 7–28 shows a log made with the device. Fig. 7–29 is a cross section of a fill, and Fig. 7–30 shows the relation of the force of penetration and the friction to the soil types and consolidation.

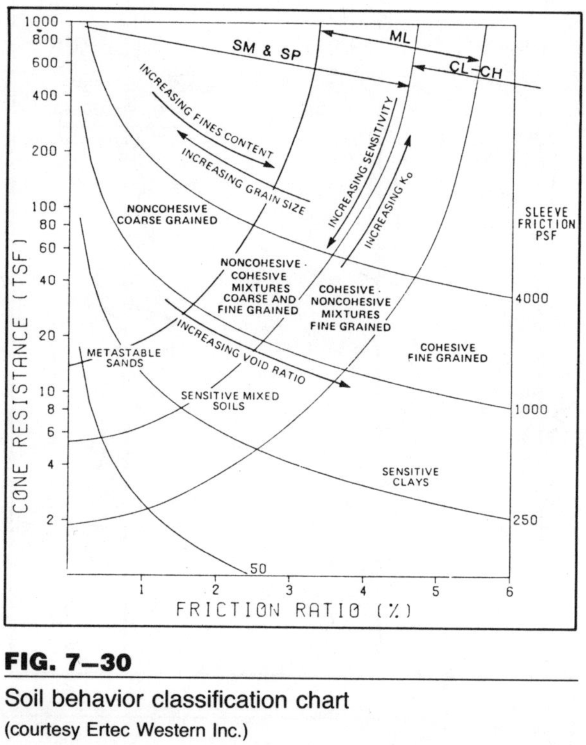

FIG. 7–30

Soil behavior classification chart

(courtesy Ertec Western Inc.)

REFERENCES

"A New Approach to Sonic Logging and Other Acoustic Measurements." Schlumberger Publication.

"An Introduction to Single Element Differential Temperature Logging." GO International document.

Blythe, E.J. Jr. "Computing Accurate Directional Surveys." *World Oil,* August 1, 1975.

Brenden, Byron B. "A Comparison of Acoustical Holography Methods." *Acoustical Holography,* volume 1, 1969.

Broding, R.A. "Volumetric Scanning Well Logging," *22nd SPWLA Symposium Proceedings,* 1981.

Catalog, San Diego: Humphrey Inc.

Catalog 789, Houston: Sperry-Sun.

Chaney, Preston E.; Zimmerman, Carl W.; and Anderson, Warren L. "Some Effects of Frequency upon the Character of Acoustic Logs." Society of Petroleum Engineers of AIME, Paper SPE 1326, November 1965.

Christensen, D.M. "A Theoretical Analysis of Wave Propagation in Fluid Filled Drill Holes for the Interpretation of the 3 Dimensional Velocity Log." *5th Annual Logging Symposium Transactions,* SPWLA, May 1964.

Christensen, D.M. "The 3-D Velocity Log." Birdwell, Division of Scientific Software Corporation, October 1963.

Cullity, B.D. "Fundamentals of Magnetostriction." *Journal of Metals,* January 1971.

Douglas, B.J., and Olsen, M. "Soil Classification Using the Electric Cone Penetrometer." Internal document, Ertec Western Inc.

Esring, M.I.; Kirby, R.C.; and Murphy, B.S. "Advances in General Effective Stress Method for Prediction of Axial Capacity for Driven Piles in Clay." *Journal of Petroleum Technology,* September 1981.

"Formation Evaluation Data Handbook." Gearhart-Owens Industries, 1974.

"Formation Evaluation Data Handbook." Gearhart-Owens Industries, 1978.

Gingrich, J.E., and Fisher, J.C. "Exploration for Uranium Utilizing the Track Etch Technique." Sydney: 25th International Geological Congress, August 1976.

Havira, R.M. *Ultrasonic Cement Bond Evaluation.* SPWLA 23rd Annual Logging Symposium, 1982.

Gollwitzer, L.H., and Masson, J.P. *The Cement Bond Tool.* SPWLA 23rd Annual Logging Symposium, 1982.

Hilpert, L.S., and Bunker, C.M. "Effects of Radon in Drill Holes on Gamma Ray Logs." *Journal of Economic Geology,* volume 52, 1957, 438–455.

Holmes, Charles S., and Smith, Samuel C. "Calculation of Circulating Mud Temperatures." *Journal of Petroleum Technology,* June 1970.

Itten, Howard. "Velocity Logging, a New Family." *The Petroleum Engineer,* September 1959.

Jenkins, R.E. "Accuracy of Porosity Determinations." *Log Analyst,* volume VII, No. 2, 1967.

McAdams, Williams H. *Heat Transmission*. New York: McGraw-Hill, 1954.

Mitchell, J.K., and Lunne, T.A. ''Cone Resistance as a Measure of Sand Strength.'' *Journal of the Geotechnical Division,* Proceedings of the American Society of Civil Engineers, July 1978.

Myung, J.I., and Sturdevant, W.M. ''Introduction to the Three Dimensional Velocity Log.'' Birdwell Division of Scientific Software Corporation, July 1970.

Myung, John I., and Helander, Donald P. ''Correlation of Elastic Moduli Dynamically Measured by In-Situ and Laboratory Methods.'' *13th Annual Logging Symposium Transactions,* SPWLA, May 1972.

Owl Technical Associates Inc., Longmont, Colorado.

Pedersen, C.P.; Dunbier, J.; and Gingrich, J.E. ''Experience with the Track Etch Method for Uranium Exploration in Northern Australia.'' Terradex Corp., 1976.

Pirson, S.J. *Handbook of Well Log Analysis*. Englewood Cliffs: Prentice-Hall, 1963.

Raymer, L.L.; Hunt, E.R.; and Gardner, J.S. ''An Improved Sonic Transit Time to Porosity Transform.'' *21st SPWLA Symposium Proceedings,* 1980.

''Recommended Instrumentation for Uranium and Thorium Exploration.'' Technical Report Series Number 158, Vienna: International Atomic Energy Agency, 1974.

Sears, Francis W. *Electricity and Magnetism*. Reading: Addison-Wesley, 1955.

Silby, Samual M. ''Standard Mathematics Tables.'' 17th edition. The Chemical Rubber Company, 1969.

Stripting, A.A. ''Velocity Log Characteristics.'' *AIME Petroleum Transactions,* September 1958.

Tixier, M.P.; Alger, R.P.; and Doh, C.A. ''Sonic Logging.'' *Journal of Petroleum Technology,* May 1959.

Tixier, M.P.; Alger, R.P.; and Tanguy, D.R. ''New Developments in Induction and Sonic Logging.'' *Journal of Petroleum Technology,* May 1960.

U.S. Department of Transportation. ''Guidelines for Cone Penetration Test Performance and Design.'' FHWA-TS-78-209.

Wichmann, Paul A. ''Acoustic Log.'' Dresser Atlas Technical Bulletin.

Wilson, G.J. ''Radius of Curvature Method for Computing Directional Surveys.'' *SPWLA Symposium Proceedings,* 1968.

Zemausky, Mark W. *Heat and Thermodynamics*. New York: McGraw-Hill, 1943.

Zimmerman, D.W. ''Dating of Ancient Pottery by Thermoluminescence.'' Oxford: Research Laboratory for Archeology, University of Oxford.

CHAPTER 8 Analog and Digital Systems

Throughout this text the modes of data gathering and data processing have largely been ignored. The processing of data by computer has been discussed somewhat, but by and large, the descriptions have been kept away from the type of data system used. This was done purposely to emphasize the importance of the mechanism in the formation and that in the detector system. Analog, hybrid, and digital systems, however, are alike. Each has its advantages and disadvantages. Each has its characteristics. One should pick the data system as well as the log type for log suites.

The original logging systems were all analog systems. The term, or description, *analog,* comes from the fact that these systems produce electrical signals for transmission and for processing that correspond in some obvious way (or are analogous to) the parameter being measured. The SP signal is a direct-current analog whose amplitude varies directly as do the potentials at the electrode. A gamma-ray system sends up one electrical analog pulse for each photon detected. The recorded neutron signal is from a direct current out of the ratemeter that is directly proportional to the neutron flux rate at the neutron detector. The signal at any place in the system is said to be the analog of the thing being measured.

An analog system has many advantages. It is simple in concept. A person who knows little about information theory can design and use most analog logging systems. Maintenance personnel can examine the signal anywhere in the system and can relate it to the causal phenomenon. This system usually requires few components, especially active components. Analog-type electrical-measuring equipment has been advanced tremendously from Galvo's frogleg to modern servo recorders and servo control systems, but it is still similar in principle. Analog systems have the big advantage that they are usually straightforward in concept.

Analog systems have some disadvantages, too. The care and precision with which they are designed, built, maintained, and used must be great. Any small error, aberration, or nonlinearity anywhere in the system appears in the output. Often, this error is not noticed because it may be of the same order of magnitude as the signal, itself. Worse, components can change gradually with time, temperature, pressure, or moisture and can cause corresponding changes in the output. Many elaborate feedback schemes have been devised to help overcome these gradual drifts. These schemes may be circuit devices, mechanical devices, techniques or measurements, and monitoring. Even so, we have millions of feet of bad logs because the systems have been out of calibration, and the observers were not aware of it.

Analog systems usually handle signals continually. This makes many of the sampling problems much simpler. Sample size usually only means the sensitive volume of the detector or the length of time (or distance) the signal is averaged. On the other hand, since analog signals are usually continuous, a wire, a circuit, or a channel must be dedicated to one or a few signals. Each analog processing system in general must be dedicated to one signal flow. The range over which the circuits can vary is usually quite limited. The difference between extremes—the highest and lowest voltage, the greatest and least counting rate, and so forth—is usually 1,000 to 1. This is called the dynamic range. Occasionally, in an exceptional system it may be 10,000 to 1, but, the price for this extension of range is usually great.

The drawbacks of analog systems are especially apparent with radioactivity measuring circuits and other random-signal systems. Usually, processing must be done in real time. Averaging can usually be done only on the basis of past time. Distance averaging is rare. Sampling adjustments (time constant, range, etc.) are usually left to the discretion of the operator, who may be poorly trained or simply may lack enough data to make a viable choice.

Many design variations have evolved to overcome—often very successfully—many of these problems. Multiplexing of signals on a time-shared basis have been quite successful. Schlumberger's Pulsator is an excellent example of this. The normal resistivity and the lateral resistivity systems time-shared wires, current circuits, and electrodes. Multiple-signal types share wires. A dc-supply power often is phantomed on a pair of wires carrying ac or pulse-type signals. Positive and negative pulses are routinely combined on one wire and are separated by surface discriminator circuits. Filter networks also allow the same wire to carry power. A typical gamma-ray, neutron, collar-locator system operates (without downhole batteries) on one wire. However, these systems represent complex extremes. Frequency modulation systems have been used to send up multiple signals, but they are always limited by the bandwidth of the circuitry, mostly the cable.

Bandwidth is the width of range of the frequencies a system can handle. It is a direct measure of the amount of data that can be transmitted. Bandwidth affects every analog signal directly. A device, like a cable, does not simply transmit up to a certain frequency and then quit. Usually, the higher the frequency is, the more the signal is attenuated. For example, if a cable had a bandwidth of 10 hertz, one could send 10 pulses or 10 bits up every second. If he tried to send 20 pulses, he would have to increase the power greatly in the pulses to make up for the attenuation. On the other hand, if the cable had a bandwidth of 1,000 hertz, he could send 1,000 bits of information up without undue attenuation. This is 1,000 times as much information. Also, the cable with 1,000 hertz bandwidth can handle four channels of information of 250 hertz apiece.

In addition to these frequency losses, the dc signals, such as the SP and the supply power, suffer attenuation because of the electrical resistance of the cable. All of these losses, ac and dc, appear as heat in the circuit. A typical analog circuit is shown in Fig. 8–1.

Recently, commercial digital systems have gained in usage because they can be designed to overcome many of the analog limitations. A digital system con-

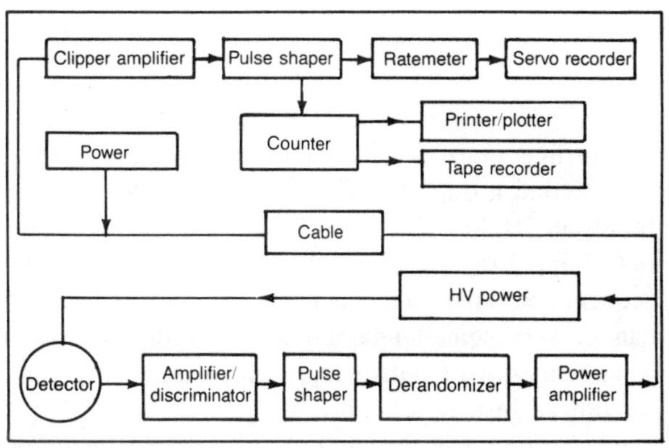

FIG. 8–1

Derandomizer analog with digital recording

verts the information of a signal into a coded signal. A simple, digital code represents a portion of the more complex system. For example, suppose we have a system that transmits 50,000 pulses/sec with a reasonable amount of design effort. With a straightforward periodic pulse system (one where the pulses are uniformly separated), 50,000 pulses/sec could be transmitted. Thus, one channel of 50 kHz could be transmitted, or two of 25 kHz, ten of 5 kHz, etc. If the signal were a random one, as it would be with an unprocessed gamma-ray system, the limit would be 16,667 pulses/sec, if a 50% dead-time correction is used as the upper limit of favorable probability. The full 50-kHz limit could be regained by using a derandomizing system downhole.

If a digital encoding is used, for example, a BCD 1248 code, and sampling is done each tenth of a second, 2,778 channels of 999,999 pulses/sec, each, are theoretically possible.

A code, like the BCD 1248, represents a number by pulses separated in time, space, or both. BCD stands for *binary-coded-decimal*. BCD 1248 indicates that there are four channels of binary information combined to represent a decimal number. A pulse on channel 1 has a value of one; a pulse on 2 has a value of two; a pulse on channel 3 has a value of four; and a pulse on channel 4 has a value of eight. The sum of one plus two plus four plus eight is fifteen. If we consider no pulses on any channel as zero, then there are sixteen combinations of pulses and channels, more than enough to represent any decimal number. Thus, if any number is picked at random, it can be represented by a BCD 1248 code. The number 176,295, for example, in BCD 1248 would be:

		1	7	6	,	2	9	5
BCD	1	1	1				1	1
	2		2	2		2		
	4		4	4				4
	8						8	
		1	7	6		2	9	5

The first pulse on channel 1 represents 100,000. The second group of pulses on channels 1, 2, and 3 represent 70,000. The third group of pulses on channels 2 and 3 represent 6,000 and so on. By sending a sequence of pulses on 4 channels the number 176,295 can be represented unambiguously with 11 pulses. That sequence of pulses represents that number and no other. Instead of 176,295 pulses, only 11 are needed. The tradeoff is in time, of course. The signal on an analog system is continuously representing 176,295. On the digital system of our example, that number is read only every 1/10th of a second. On the other hand, if the analog signal accidently drops in amplitude 10%, it erroneously reads 158,665. If the digital signal drops by 10%, it still represents 176,295. As long as the digital signal can be detected, it is interpreted as the correct number.

Digital systems have many advantages. Several of them have been discussed. They also have some disadvantages. The circuitry is complex. Even a simple digital system contains many active components. The old vacuum-tube gamma-ray probes had seven vacuum tubes. In a digital probe one chip (of many) has 10,000 transistors in it. The signal at any place in the circuit must be decoded to determine if it is of the correct form. It is almost impossible for a technician to repair a digital system in the field, which may really be a big advantage. One advantage is that when a digital system malfunctions, it usually produces nonsense. Therefore, a failure is evident.

Digital systems can use other averaging systems than time. Also, the data are recorded before processing in any good system. Thus, data points are known and are available symmetrically about the measure point. The averaging can be symmetrical, rather than the asymmetrical averaging usually used in analog systems. Since distance averaging can be used, higher logging speeds do not distort the signatures of the logs, as they do in analog systems. If one wants to increase the logging-speed capability of a digital system, he merely increases the detector sensitivity to increase the sample size. A big advantage to a digital system is that the information is in a form easily used by a computer. Digital systems are usually designed to relieve the operator of routine duties where the probability of making an error is high. They keep track of scales, zeros, and changes. They demand calibrations, they remind operators of options, and they force correct procedures. This, alone, makes a digital system worth extra cost.

On-site computers present data in meaningful parameters, instead of arbitrary ones that must be converted. They also preserve data in unprocessed form to allow further or later processing. A typical digital system is shown in Fig. 8–2.

There are a number of mixtures or hybrids. The most common is the digital readout. One or more digitizing networks are inserted in the data processing portion of an analog system. Basically, the system is still an analog system, although the data processing is in some degree digital. The digital data processing may merely be a digital printout of analog data, a digital calculation, or a full computer capability.

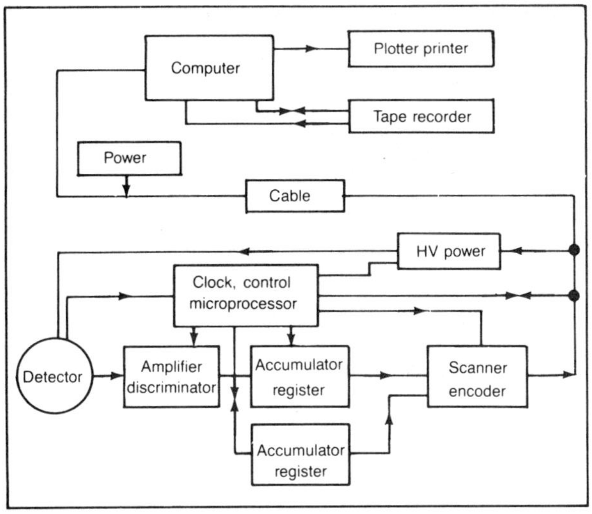

FIG. 8–2

A typical digital system

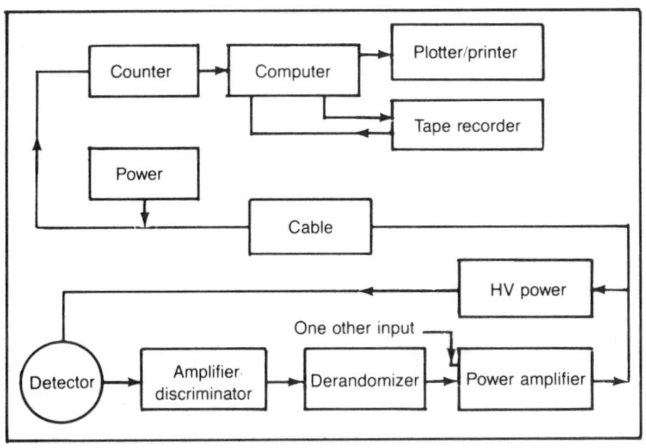

FIG. 8–3

Hybrid analog/digital system

There is an infinite number of variations. Take care to know what is being used. Get an independent opinion because, sometimes, even the contractor does not know the possibilities and capabilities. A number of the better hybrid schemes are shown in Fig. 8–3.

In general, analog systems are good for portable use, experimental use, and for single measurements. Digital systems are ideal for production and other high-volume work. Hybrid systems and digital systems are effective for permanent records. Digital systems with on-site computers are probably the best commercial logging systems designed at this time.

CHAPTER

9 Interpretation Methods

We have seen that we can use each of the various curves—the gamma ray, resistivity, acoustic velocity—quantitatively. Each curve can be used by itself to calculate one or more parameters that can be used to aid the exploration or development of a project.

Excellent single-curve interpretation methods also are available that should be considered and used. The big advantage, however, comes when several methods are combined. This is primarily because most of the measurements we use are not direct measurements. We measure density or resistivity. We do not measure porosity. Therefore, there are aberrations in each determination we make because we are deriving the desired parameter from the precise measurement. The measurement of porosity by two or three different, independent methods allows us to make corrections not possible with only one curve. The response of the density curve to a shale is different from the response of a sonic curve to a shale. The measurement of density or acoustic velocity are correct, but the responses are different from each other. The differences allow corrections to be made, and these corrections can also be used analytically. Since different lithologies affect each type of log differently, the differences identify the type of lithology. These methods can be used in the field or in the office. The advent of on-site computers has aided the field use of these data tremendously.

The effective use of logs—single or multiple curves—usually requires that the curve or curves be carefully calibrated. Qualitative methods have been included, and they are valuable and quite valid. They are, however, in the minority. Most of the methods assume that the curve has been calibrated as well as present technology allows.

Some of the interpretive methods covered in earlier chapters have been repeated here. This is especially true of the single-curve methods. However, there probably is an advantage in having all of the interpretive methods together for comparison. Some petroleum methods are included as well as nonpetroleum. Sometimes, they are very similar. If we are familiar with interpretation methods, we can adapt existing methods to a particular use. Thus, there is an advantage to having some knowledge of both types of methods.

With computers these single curves should certainly be presented in usable parameters, such as density or porosity, not counts per second or millivolts. Individual curves should be presented, especially at the site, in parameters such as density, porosity, and percentage of sand, not in millivolts or counts per second. However, it is imperative that the raw data always be recorded and preserved.

Interpretation methods can be subdivided into several groups as follows:

- *Conventional methods*. These include qualitative and some quantitative methods. A single curve can be used to show sand/shale sequences; two curves show invasion depth or calculate porosity; three curves determine an invasion profile or roll-front orientation.
- *Pattern recognition*. These are qualitative to semiquantitative methods. They usually use several curves to show depositional history or trends. They include the use of the dipmeter to identify fault features and structure type. These methods can be handled by computer.
- *Crossplotting*. These are quantitative, graphical methods ideally suited to computer handling. They can, however, be used manually. They are used to determine in detail lithology, porosity, fluid content, fluid type, permeability, hydrocarbon saturation, and many other specific factors.
- *Equation methods*. These are usually computer methods and are closely related to the crossplotting methods. In fact, there is an advantage to determining the equation set for a crossplot. These methods are all quantitative and may be rigorous or empirical. They include the percentage component plots on logs and the tabular reserve listings.

INITIAL DATA TREATMENT

Before the logs are used, they must be evaluated for quality. If they prove to be faulty and there is a chance that they will be, the methods of correction or salvage must be determined. There is little point in wasting time processing faulty data. It is probable, however, that the data can be salvaged.

The first step is to see that all of the heading information is in place. If it is not, the logger should be contacted to see if he can furnish it. This is valuable, necessary information, and it is the responsibility of the logging operator to see that it is accurately listed.

The validity of the heading and logged information should be assessed. The values should be reasonable and should probably correlate from curve to curve. Keep in mind at all times that there is a high density of information on the log. Even a careful logging operator is going to make an occasional mistake.

Once the quality of the log has been determined, there are some things that can be done to improve faulty data. One of the best is to normalize the data. Normalizing, in this case, is to smooth the data statistically on the assumption that there is only a small probability that there will be abrupt changes. To do this on a single log or on the data from a whole file, use the following procedure:

- Select an appropriate interval for examination. Pick an interval that is not too large but that allows an overall view. Make the assumption, unless one knows otherwise, that there will be a smooth transition from one point to another. This is especially true horizontally in a sedimentary basin. It is much less true vertically or in a hard-rock environment. Eliminate anomalous zones or points. Even if the purpose of the log is to locate mineral deposits such as uranium, the elimination of the anomalous zones tends to indicate the background situation.
- List the data. Digitize them if possible. Use the raw data when they are appropriate. When the corrected data are presented by the contractor, use them. Look for anomalous values and sudden changes in the listed data. A standard, cored hole is valuable.
- Histograms and frequency plots of the data for a hole and for nearby holes help to pinpoint errors. In a given horizon, the signatures of the curves should agree from hole to nearby hole. It may be necessary to shift the diagrams in depth or elevation to get them to correlate. Usually, a log-normal distribution can be assumed.
- Crossplotting techniques are valuable. Use the mineral lines and areas on the crossplots. Shift

the values to agree with the minerology or lithology. Overlay the object log with a standard one or a nearby offset log. Mark the known values.

- Compute the values. Make appropriate corrections to the logged values. Compare the values with core values by graphing.
- Map the corrected values. Determine a best fit surface by regression analysis. Variances are probably log errors.
- Preserve the original data. List the statistical data for the record and experience.

Such a procedure may not be necessary. However, it may well result in a better picture of the project. Table 9–1 indicates the applications for each type of logging system.

SINGLE-CURVE METHODS

Single-curve interpretation methods can be used under some special conditions. They are simpler and easier to facilitate than other methods. They may be used, however, only when the environment is simple and known.

Table 9–1
Applications for Logging Systems

Logging System	Lithology	Stratigraphy Correlation	Porosity Permeability	Reserves Mineralization	Engineering	Alteration Diagenesis	Fluid Movement Hole Conditions	Structure Deposition	Fluid Character
Gross-count gamma ray	* × # ×	* × # ×	* × # ×	* × # ×	* × # ×	* × # ×	* + #	* + # ×	
Gamma-ray spectrograph	* × # ×	* #	* + # +	* × # ×		* + # +	* + # +	# +	
Resistance	* 0	* ×		* 0	* 0	* 0	* 0 # 0	* × #	* 0
Unfocused resistivity	* × #	* ×	*	*	* ×	* ×	* ×	* × # ×	* ×
Spontaneous potential	* ×	* ×	*	*		* ×	* ×	* ×	* ×
Density	* × # ×	* # +	* × # +	* × # +	* × # ×	* × # ×	* #	* # ×	* × #
Neutron porosity	* × # 0	* + # +	* × # 0	* × #	* × #	* × #	* #		* × # ×
Neutron cross section and activation	* × # ×			* × # ×		* × #			* × #
Acoustic methods	* × # ×	* × # ×	* #		* × # ×	* × #		* × # ×	* × #
Magnetic susceptibility	* × # ×	* #	#	* × # 0	*	* × # ×	*	* #	
Induced polarization	* #			* × # ×		* × # ×			
High-resolution dipmeter		* ×			* ×			* ×	
Temperature			*				* × # ×		
Logging tool usage:	* = Sedimentary use # = Hard-rock use × = Proven and excellent use + = Has good possibilities 0 = Approximation only								

Gross-Count Gamma Ray

The gross-count gamma-ray systems (GCGR) are quite useful for determining the amount of radioactive material in the formation surrounding the detector. This includes clays, shales, ash, potash, uranium, and thorium. The GCGR curve is important in all phases of logging. It is run by itself but usually in combination with almost any other curves.

When using these methods one should always keep in mind that the GCGR shows apparent concentrations of radioactive material, not actual concentrations. This is for several reasons. In the case of clays and shales, the GCGR is detecting the combination of potassium-40, uranium, and thorium daughter products. Clays or shales do not always have the same composition from one sample or location to another. In fact, many clays are not even potassium compounds. We make the assumption that the compositions of average clays or shales are uniform. This is not always true. In the case of ash in coal, we make the further assumption that the ash is shale. Again, this is not entirely true. Surprisingly, though, these methods work quite well.

With potash, the problem is contamination with other radioactive materials. We assume that the potash is pure potassium carbonate. In reality, there may be other potassium compounds or other radioactive materials present. Even the same chemical compound can exist in different forms. Potassium carbonate can exist in two crystalline forms, $K_2CO_3 \cdot 2H_2O$ and $2K_2CO_3 \cdot 3K_2O$ that have slightly different responses.

Uranium and thorium account for most of the quantitative use of the GCGR curve. The problem of whether the parent radioactive mineral is uranium or thorium is always present. Also, disequilibrium of uranium with its daughter elements is severe. No attempt is usually made to separate these problems until cores or samples can be chemically analyzed. Then an average disequilibrium and composition correction is made.

Clay, Shale, and Ash Content

The usual approach on clay or shale content is to compare the shaly or clayey sand with a pure shale or clay and a clean sand. Pick a good, uniform shale and a thick, clean sand. Then the assumption is made that the sand and shale in the beds being examined have the same compositions and responses as the pure reference formations. The readings may be in API units or in counts per second. If in counts per second, a correction for borehole size and fluid must be made. The curve need not be calibrated since only relative readings are used. The calibration is actually done by reference to the pure sand and pure shale. The expression for relative shale content, V_{sh}, of a shaly sand is:

$$V_{sh} = f(I_{sh}) \tag{9.1}$$

$$I_{sh} = \frac{\gamma - \gamma_s}{\gamma_{sh} - \gamma_s}$$

Where:

V_{sh} = Fraction of shale by volume in the sand
γ_{sh} = Gamma-ray reading in a thick shale
γ_s = Gamma-ray reading in a clean sand
γ = Gamma-ray reading of the shaly sand in question

Fig. 9–1 shows a log illustrating this procedure where V_{sh} is 10% in a sand zone. Also see Fig. 5–28.

If the formation in question is a carbonate, γ_s should be a clean carbonate reading. The accuracy of this system can be improved by appropriate cross plotting. Note that Eq. 9.1 is of the form:

$$y = a + bx \tag{9.2}$$

Where:

$y = \gamma$
$a = -\gamma_s$
$b = \gamma_s - \gamma_{sh}$

Thus, a number of points of sand, shale, and sandy shale may be plotted, and a best fit line can be determined by linear-regression analysis. Projection or extrapolation of the best fit curve to V_{sh} = 0% and V_{sh} = 100% indicates the accuracy or validity of the sand and shale readings, respectively.

Coal Ash

Much the same procedure as used in shaly sands is used to determine the ash content in a coal. The assumption is made that the ash in the coal closely resembles the nearby shales. Then the nearby shale can be assumed to be 100% ash. This is plotted as a point on a linear graph of ash versus the gamma-ray reading. Since it is difficult to find a coal that is sufficiently ash free to give the intercept value, γ_s, the line may be drawn through zero counts for zero ash as a first approximation and corrected later from core results. As an alternative, an arbitrary value of the y intercept may be chosen as the residual radioactive content of the coal on the basis of experience. This, again, may be corrected later from core assays. This is a graphical solution of the shaly sand problem.

Both of these methods are fairly reliable, especially for field use. Experience aids in picking values of sand, shale, and residual coal activity, and cross

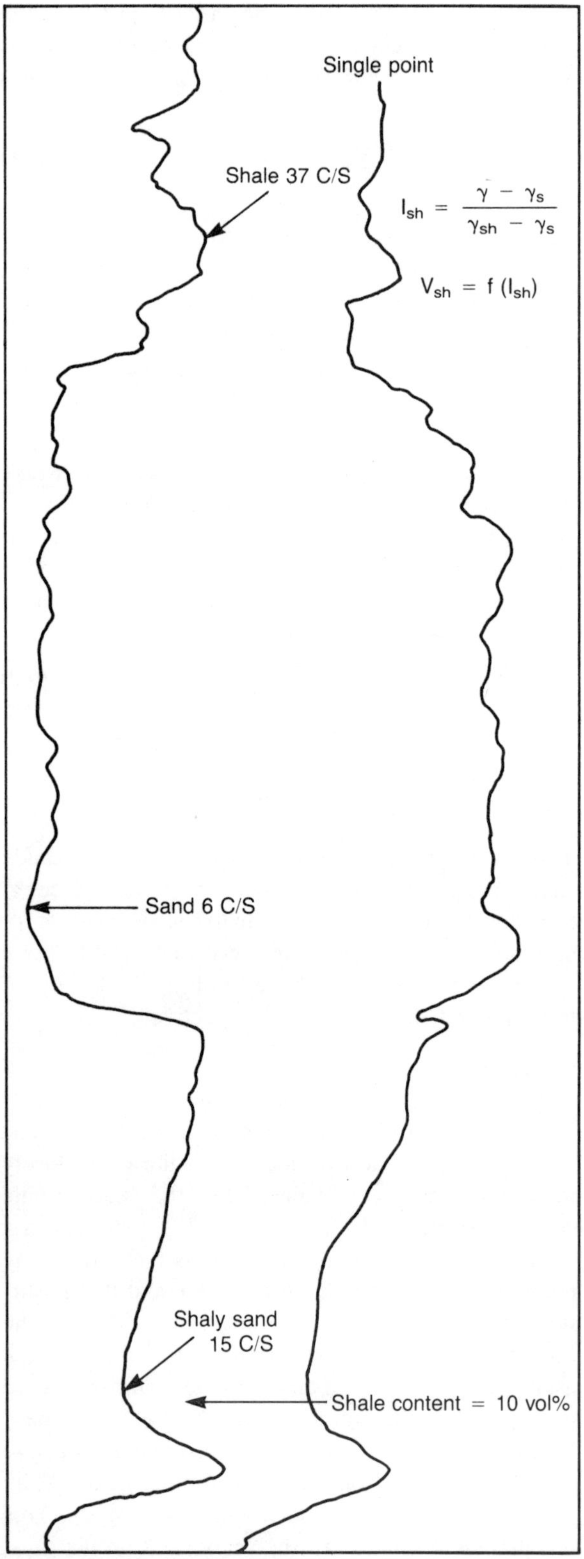

FIG. 9–1
Calculation of shale content from gamma ray

plotting greatly increases accuracy. In any event these methods are much better than guessing.

One must be careful in either coal-ash or shaly-sand determinations, to stay away from the bed boundary. In the bed-boundary region it is difficult to determine a correct value for the gamma-ray reading because of the response characteristics of a GCGR tool going across the boundary. The chart in Fig. 5–31 shows the variation of sample volume with density of the medium. One can see that in a coal the sample volume of a gamma-ray system may be as much as 2.5–3 ft in radius. In a shale or sand the radius is about 1.5 ft. In a dolomite the radius may only be 1 ft. This then is the distance the GCGR detector must travel to detect completely the transition across a bed boundary. If we have a shale adjacent to a coal, about 2.0–2.5 ft of GCGR curve in the coal must be discarded and about 1.5 ft in the shale. This is illustrated in Fig. 9–2. The full transition at 261.2 ft requires a total of 3.5 ft (1.3 ft in the lower, coaly shale and 2.2 ft in the upper coal).

Once the coal zone has been determined, the ash content of it may be estimated rather closely by the procedure outlined for Fig. 5–28 and 5–29.

Potash, Uranium, and Thorium

When logging in potash, uranium, or thorium, there are several methods used to determine the radioactive mineral content of the formation. Since there is no reference zone of 100% potash, uranium, or thorium in a formation as there is for shale, the conversion factor for the tool, the logging system, and the interpretation system must be known. Most of these methods use the relationship between the area under the gamma-ray anomaly and the radioactive content of the formation. These areal methods could be applied to shaly-sand and coal-ash problems as well but usually are not. Actually, the methods used in sand and coal can be considered special cases of the areal methods.

To use the areal methods, several assumptions must be made. First, it must be assumed that the area under the curve on the log is a function of the amount of radioactive material in the formation. This is a safe assumption and can be proven for simple cases. Second, it must be assumed that the amount of gamma radiation from the formation is proportional to the amount of radioactive mineral in the formation. This is not a good assumption. In fact in the case of uranium particularly, we often have little idea of the relationship of the counting rate to the real amount of radioactive material present or disequilibrium during the

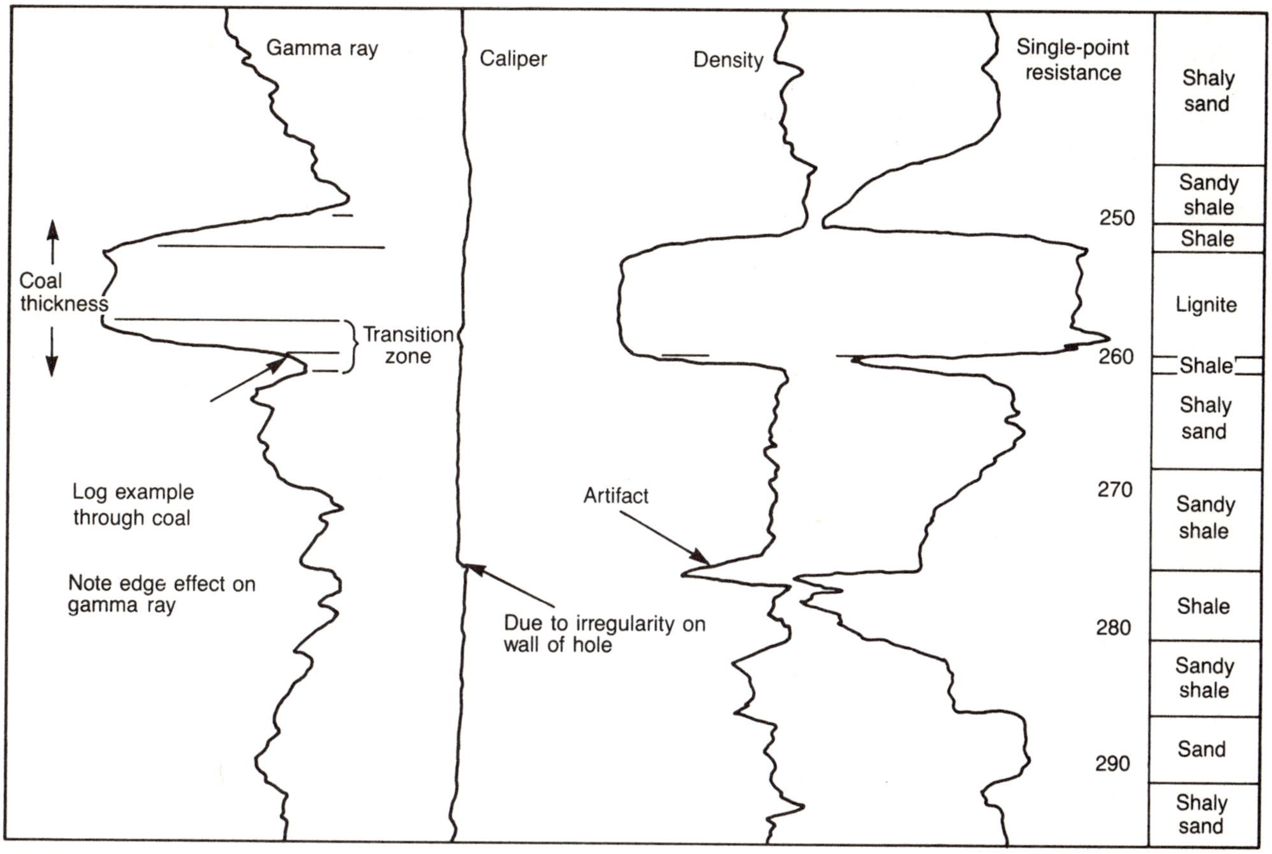

FIG. 9–2

Log example through coal

logging, even during the initial calculations. Later, when individual or average disequilibria are known, they may be applied to correct the assumed values. Third, the assumption is made that the amount of radiation seen by the detector is proportional to that from the formation. The validity of this assumption depends on the care with which hole size and other corrections are made.

The borehole represents a noncontributing volume and an attenuating volume for gamma rays if it is fluid filled. Corrections for this must be made. Fig. 9–3 gives corrections for 1.75-in. diameter scintillation detector tools. If the log is scaled in API gamma units, the calibration has been made in a borehole 4.9 in. diameter, water filled. For any other size and any other fluid such as air, a correction must be made. In all probability, this correction has not been made on the API format log. If the log is scaled in counts per second, a borehole correction must also be made. A uranium tool is calibrated in a 4.75-in. diameter air-filled hole. For any other hole diameter or fluid

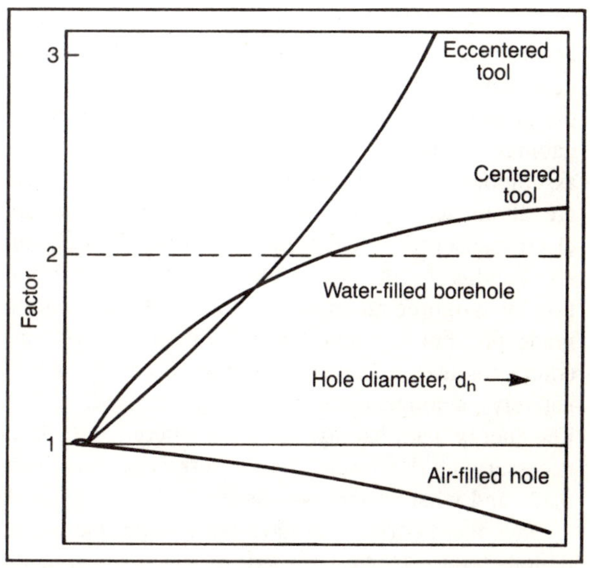

FIG. 9–3

Typical borehole correction factors for gamma ray

(water), a correction must be made. If the log is scaled in eU_3O_8 units, the chances are that hole size corrections have been made. However, this must be checked with the operator or contractor because it is by no means certain. These corrections can be substantial, as shown Fig. 9–3. It is probable that a correction should be made for mud weight. However, information in the form of response curves and of actual mud weights in the borehole are seldom available. Therefore, usual practice is to assume that the mud characteristics are the same as water.

If the counting rate in the anomaly goes very high as it might in a good, high-grade, thick uranium or thorium deposit, a correction must be made for dead-time effects. This correction is seldom needed for potash deposits when modern equipment is used, and it is never needed when logging shales, sands, or coals. The corrected counting rate, N, is:

$$N = \frac{n}{1 - \mu n} \tag{9.3}$$

Where:
n = Recorded counting rate
μ = Measured dead time of the system, sec

In a computer plotted log any of these corrections may have been applied. The operator should be questioned, and the information should be on the log heading for future reference.

A correction must also be made for the Z effect, which may have been applied by the contractor during calibration. This effect also must be checked at the time the log is run. The correction is required because the amount of uranium in a high-grade deposit is sufficient to cause excessive scattering of the gamma rays within the formation. The scattered gamma rays have lower energy than predicted by the assumptions of this method. Therefore, more of them are attenuated by the tool housing and by photoelectric absorption within the formation. This is usually handled by the contractor at the time of calibration by assigning an artificially low grade to the high-grade pit. For example, in a 2.73% U_3O_8 pit one contractor uses an apparent grade of 1.91%. Alternatively, a longer dead time may be used, or the low-energy gamma rays may be filtered out. It is necessary to determine if this correction has been made and what system was used.

Once these corrections have been made, the log or digital printout can be used to calculate mineral grade. If the log is in API units, the conversion is simply:

$$1 \text{ API} = 2 \times 10^{-5} \text{ wt\% } eU_3O_8 \tag{9.4}$$

Average the readings each half foot or each foot through the anomaly, starting 1.5 ft before one bed boundary and ending 1.5 ft after the other. This obviously limits the method to beds thicker than 3.0 ft.

In potash and sylvite deposits the same method can be used. In this case:

$$11.2 \text{ API} = 1 \text{ wt\% potash} \tag{9.5}$$

$$13.5 \text{ API} = 1 \text{ wt\% sylvite} \tag{9.6}$$

In these evaporites the radius of investigation is about 20 in. Therefore, the zone must be more than 40 in. thick before a valid calculation can be made. Again, average readings should be taken at regular intervals throughout the valid portion of the anomaly.

If the log is scaled in counts per second, the same method may be slightly modified. The expression for grade of eU_3O_8 by weight percent, G, is:

$$G = K\,N\,F \tag{9.7}$$

Where:
K = Conversion factor determined in the U.S. DOE models*
N = Dead-time-corrected, average counting rate within the valid zone
F = Correction factor for water-filled hole diameter

All of these evaluation methods require thick, uniform beds. The bed must be thicker than the diameter of the measurement volume. Readings from within the distance from the bed boundary of the radius of the measurement volume must be discarded. In attempting to overcome some of the difficulties of the preceding methods, several more rigorous methods have been evolved for calculating uranium grade, as eU_3O_8 weight percent. The first of these is the AEC method. The AEC method results in an average grade thickness product, GT, for the bed. It can be carried from bed boundary to bed boundary safely.

If a computer or a programmable calculator is available, one is much better off using one of the deconvolution methods of calculating mineral grades. These methods are the GAMLOG method and the Jonas proposal.

Spontaneous Potential and Resistivity

The spontaneous potential and resistivity measurements constitute the electric log. They are often run in combination with the gross-count gamma-ray curve

*The K factor used here is the universal K factor, K. It does not have incremental width factor, w, included in it. It is related to the half-foot K factor, K_m, usually used in the U.S., by:

$$K = \frac{K_m}{0.5} = \frac{K_m}{w}$$

in nonpetroleum work. They are the standard type of log suite in petroleum logging in most areas.

Spontaneous Potentials

The spontaneous potential curve (SP) is used to determine the amount of shale and siltiness of a sedimentary formation. In doing this, it can be used to differentiate between sands and shales in sedimcntary sequences. As the grain size of the sediment decreases, the membrane potential increases. This causes the spontaneous potential to approach the shale potential more nearly. Thus, the spontaneous potential curve can be used as a qualitative indicator of grain size.

Depositional history can be deduced by the relative SP deflection. For example, a clean, coarse sand gives a large sand deflection (Fig. 9–4a). This sand is deposited under a high flow rate condition. As the water flow rate decreases, the particle sizes becomes smaller and the SP deflection approaches the shale value. Fig. 9–4b shows the response of the SP curve in various types of formations and beds.

The amount of spontaneous potential deflection from the shale value can also be equated to the relative salinity of the formation water. The expression for the SP deflection, E_{DA}, in millivolts is:

$$E_{DA} = -K \log(R_{mf}/R_{we})$$
$$K = (0.133°F + 60°) \qquad (9.8)$$

Since E_{DA} can be determined from the log and R_{mf}, the mud filtrate resistivity, can be measured, the value

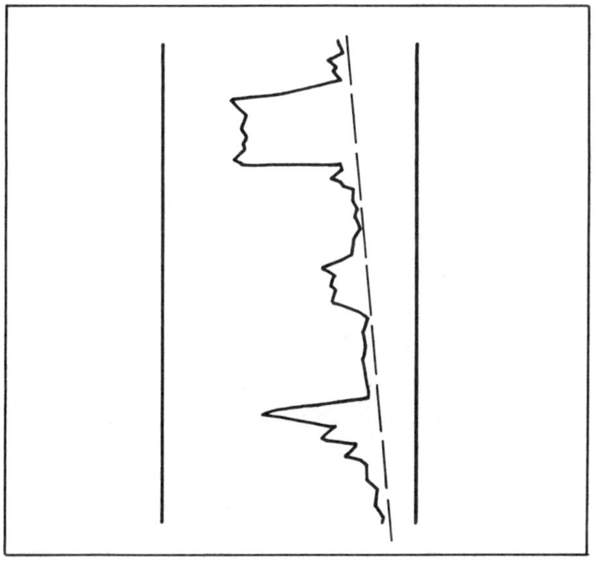

FIG. 9–4a
SP curve base-line drift

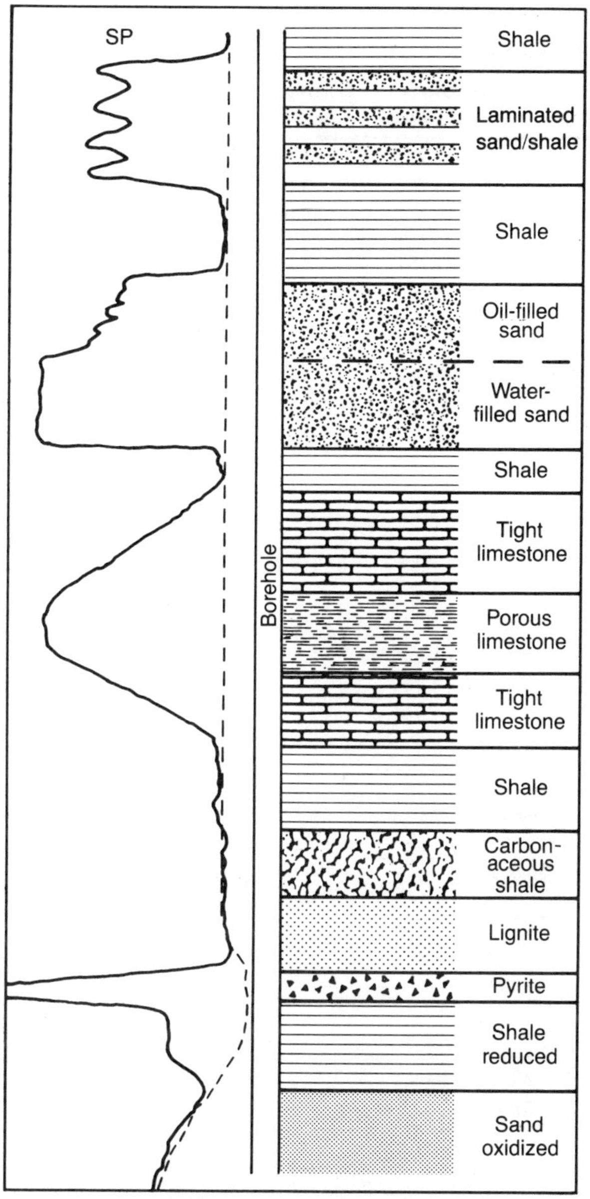

FIG. 9–4b
The SP for a sedimentary sequence

of R_{we}, the resistivity of the formation water if the dissolved solids are all sodium chloride (NaCl), can be calculated.

The value of K is −70.7 millivolts if the formation is a clean sand in contact with a solid shale. This is the potential difference from the shale or membrane potential (−59.1 millivolts) to the clean sand diffusion absorption potential (−11.6 millivolts) or:

$$K = (K_D + A_{DA}) = (-11.6 - 59.1) = -70.7 \qquad (9.9)$$

Other values of K_D for various types of ions are shown in Table 9–2. Values of A_{DA} are shown in

Table 9–2

Values of K_D

Ion	Mobility, cm^2/ohm	Dilute, mvs	Salt	0.1 Mole, mvs
Na^+	43.5	−11.6	NaCl	−12.7
K^+	64.6	−0.4	KCl	−0.1
Ca^{++}	51.6	−19.6	$CaCl_2$	−26.4
Mg^{++}	45.0	−22.5	$MgCl_2$	−29.1
H^+	315.0	44.6	H_2CO_3	—
Cl^-	65.5	37.9	HCl	38.3
SO_4^{--}	67.9	−11.7	$MgSO_4$	−10.7
HCO_3	40.7	2.2	$NaHCO_3$	—
CO_3^{--}	60.0	7.5	$NaCO_3$	13.5
HS^-	42.5	44.0	H_2S	—

Table 9–3

Values of A_{DA}

Rocks	A_{DA}, mvs
Clean sands and poorly cemented sand	−5 to +5
Sands with trace amounts clay	2–10
Cemented sands with trace of clay	7–10
Carbonate cemented sand	5–15
Sandy red beds	0–10
Shale red beds	5–20
Shales	20–70
Clean limestones	−7 to +5
Limestone with 2–10% shale	−5 to +15
Shaly limestone	10 to 40
Marls	20 to 70
Shaly dolomites	20 to 55

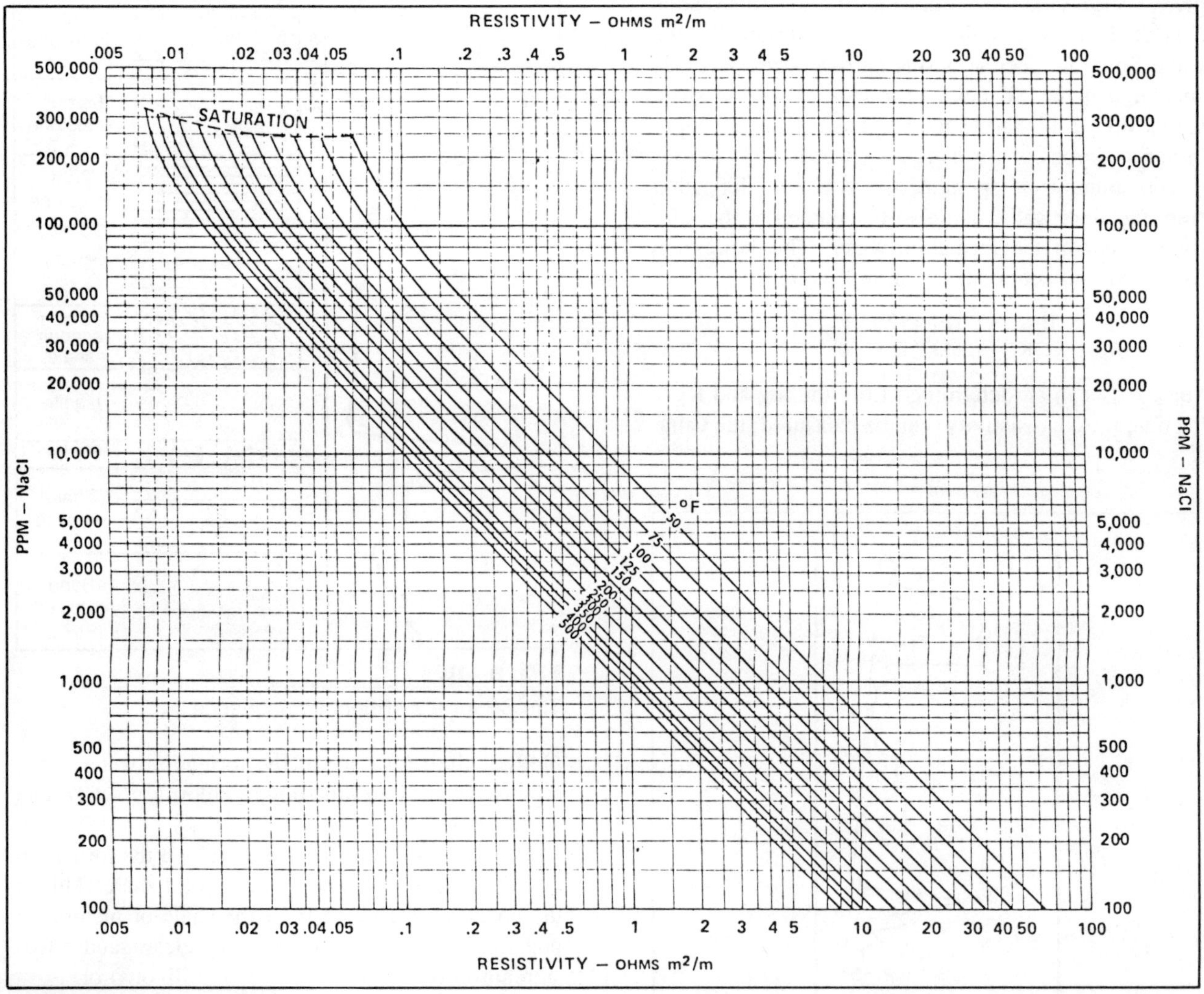

FIG. 9–5

Solution resistivity-salinity (courtesy Gearhart Industries Inc.)

Table 9–3. Note that a correction must be made for the formation temperature (Fig. 9–5). The value of R_{we} is only valid if the formation water is a solution of NaCl. A correction must be made for any other ion type. Fig. 9–6 may be used if the ion types are unknown. If the ion types are known, Fig. 9–7 may be used to make a correction.

The log in Fig. 9–8 shows SP deflections opposite sands. Note that the sand at 1,523–1,544 ft has the same SP deflection as that from 1,572–1,644 ft. They have, however, different resistivity deflections. These two horizons have the same formation water salinity. Since this deflection is −125 millivolts, the value of R_{we} is 0.013 ohmm if the mud resistivity is 0.5 ohmm

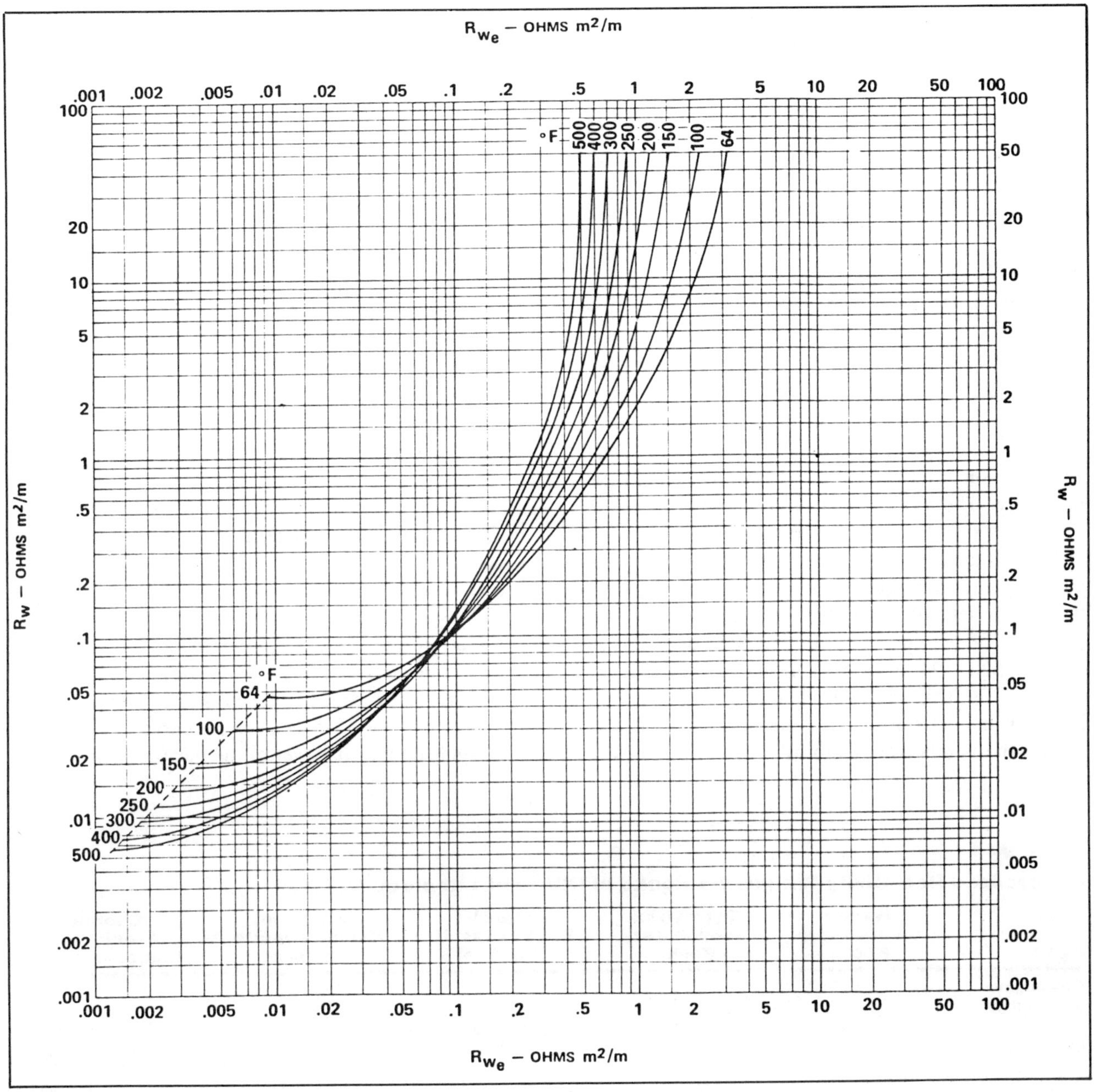

FIG. 9–6

R_w vs. R_{we} at various temperatures (courtesy Gearhart Industries Inc.)

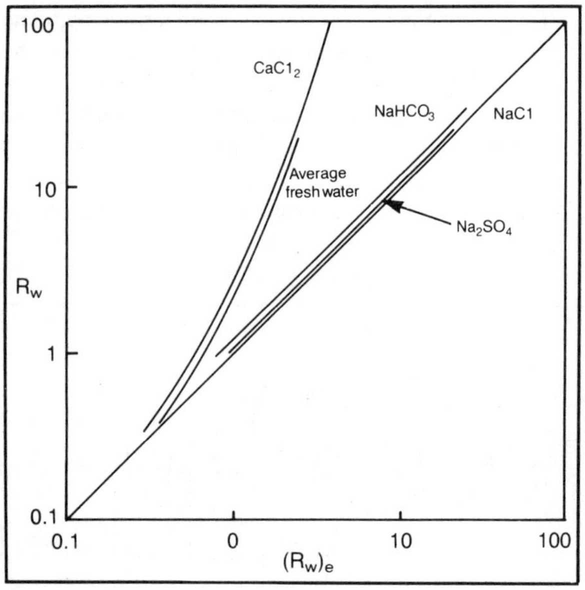

FIG. 9–7

R_w vs. $(R_w)_e$ for solutions of several salts

and the formation temperature is 100°F. This value of R_{we} can be corrected to an extremely high value if we use the Tixier correction chart, Fig. 9–7. In this case if we are interested in the total dissolved solid content and ion types, a chemical analysis would be necessary. The TDS content is quite low. Fig. 9–9 may be used instead of calculating R_{we}.

The spontaneous potential curve can also be used in a mapping technique that appears to be related to the redox state of the formation (Fig. 9–10). The base line of the spontaneous potential curve in the sand at 580–594 ft is more negative than that at 532–580 ft. This indicates that the lower sand is probably more oxidized than the upper one. If the upper sand is relatively uniformly less oxidized over a wide area, it can be used as a comparison standard for the lower sand, and the relative redox states can be mapped as an adjunct to the map of hematite staining from the lithological log.

Resistivity

Resistivity curves can be used in much the same fashion qualitatively as SP curves. This is because the resistivity of the sedimentary formation is affected by the grain size through sorting (porosity) and grain-size effects. In fine grain sands the resistivities are always lower than would be predicted by Archie's Law. Thus, as with the SP, the resistivity curve can be used to some extent as an indicator of relative grain size and clay content. Depositional history can be deduced from it. Also, the resistivity of formations is related to the geological age through porosity and salinity (Fig. 9–11 and Table 9–4).

Probably the greatest use of resistivity curves in the mineral industry is in correlation from one hole to another. Cross sections and profiles are the main use of the mineral industry resistivity curves. This results partly because of the commonly uncalibrated presentation of the resistivity curve (single point). The single-point curve is an excellent one for correlation purposes. It is symmetrical and has short, regular boundary effects. It does not have to be calibrated nor quantified to be used for these purposes. The induction log, focused resistivities, and microcurves can be used for correlation. The normal curves and lateral curves have some bed-boundary effects and asymmetries (lateral) that interfere with their use in profiling and cross sectioning. Be sure to take these into account. Fig. 9–12 shows a typical correlation sequence using the SP and resistivity curves. The actual use of the resistivity curve quantitatively depends on the character of the measurement system.

Table 9–4

Generalized Resistivity Ranges for Rocks of Different Lithology and Age

Age	Marine Sedimentary Rocks	Terrestrial Sedimentary Rocks	Extrusive Rocks (Basalt, Rhyolite)	Intrusive Rocks (Granite, Gabbro)	Chemical Precipitates (Limestone, Salt)
Quaternary and Tertiary age	1–10	15–50	10–200	500–2,000	50–5,000
Mesozoic	5–20	25–100	20–500	500–2,000	100–10,000
Carboniferous Paleozoic	10–40	50–300	50–1,000	1,000–5,000	200–100,000
Early Paleozoic	40–200	100–500	100–2,000	1,000–5,000	10,000–100,000
Precambrian	100–2,000	300–5,000	200–5,000	5,000–20,000	10,000–100,000

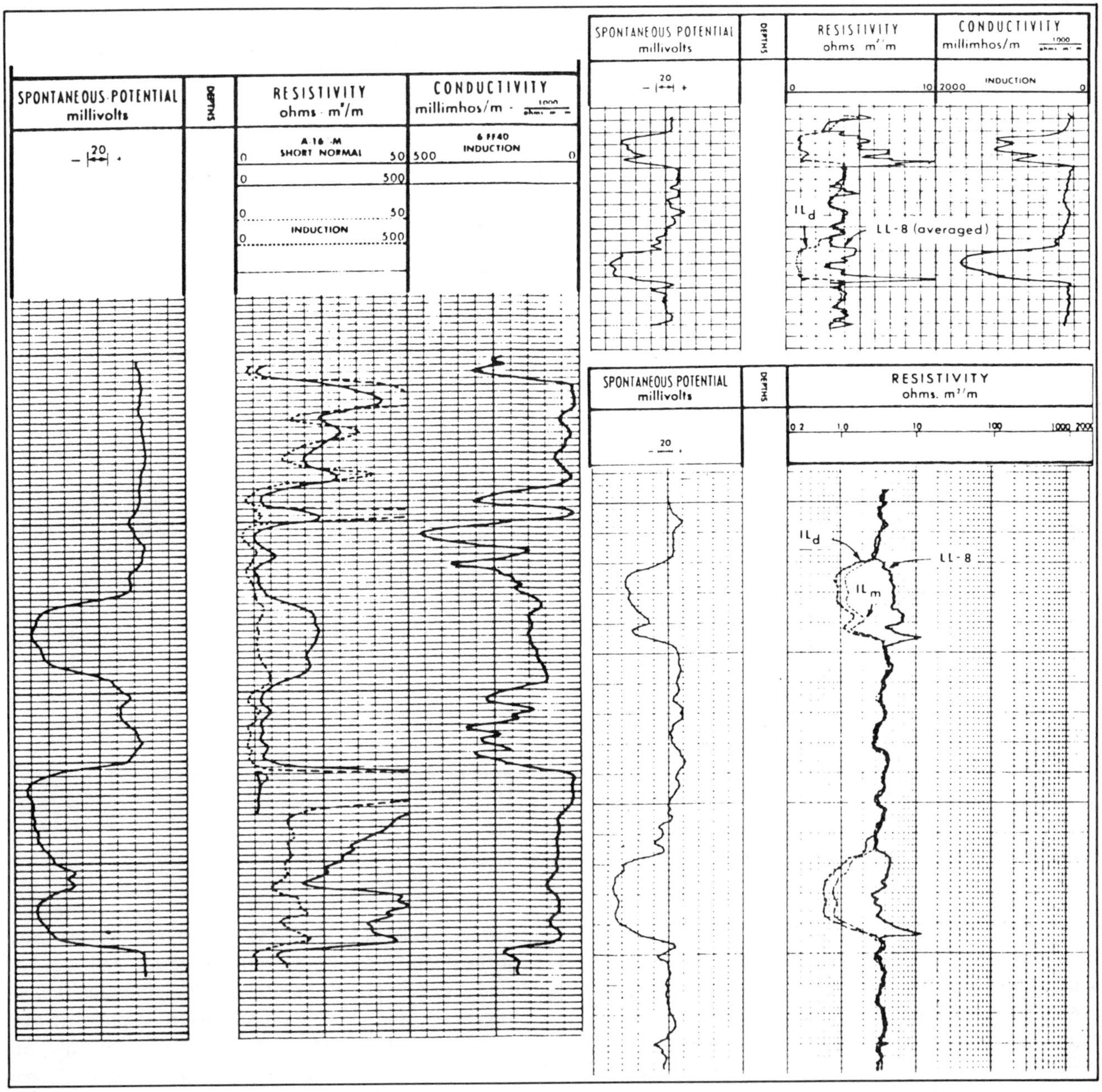

FIG. 9–8

Schlumberger electric logs (courtesy Schlumberger Well Services Inc.)

A single-point curve can be scaled in resistivity units if the electrode dimensions are known:

$$r = \frac{R}{2\pi L} \ell n \frac{2L}{d} \quad (9.10)$$

Where:

r = Grounding resistance or reading of the single-point system if the return is a great distance away and has a negligible resistance

R = Bulk or average resistivity of the medium surrounding the electrode

L = Length of the electrode, m

d = Diameter of the electrode, m

This expression is valid only when the formation is thick and homogeneous and the mud has the same resistivity as the formation. This is the type of response we would get with the single-point system in a large tank of water.

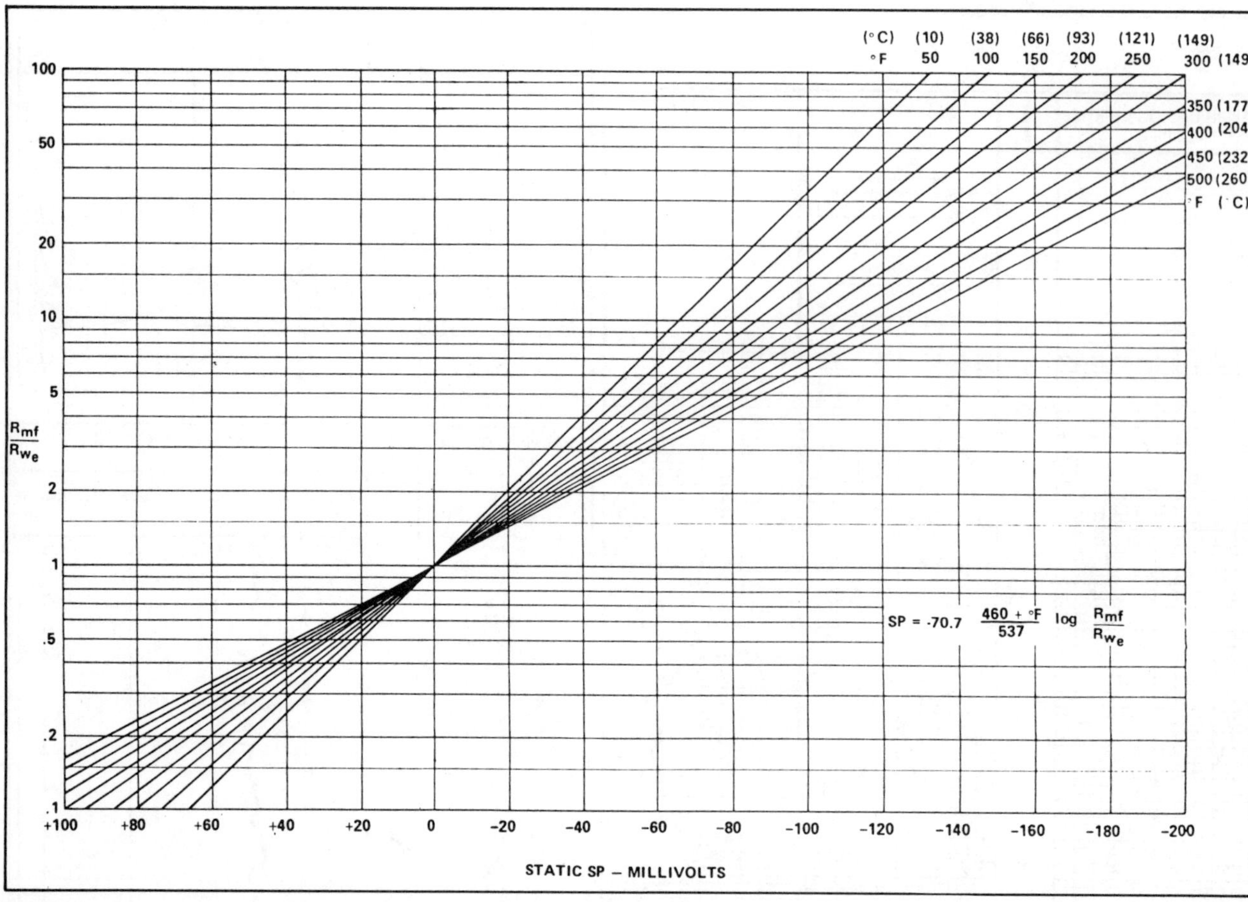

FIG. 9–9

Equivalent formation water resistivity from SSP (courtesy Gearhart Industries Inc.)

When the formation resistivity and the mud resistivity differ or the formation is thin and bounded by adjacent beds of different resistivities, the situation becomes more complex. The departure from the above expression can be described by departure curves. However, these must be determined as they have not been published. Response and departure curves have been published for normal and lateral curves and for the focused systems. Notice that the readings and other resistivity values are usually normalized to the mud resistivity, R_m. The value of R_m must be noted on the log heading, even for the single-point resistivity system.

The shallow investigation resistivity curves can be chosen to read the resistivity of an invaded zone of a porous formation in preference to other zones. The choice of the system depends on the character of the formation and the drilling system. A mineral exploration or development project using natural muds in porous sands has deep invasion. Thus, almost any of the shallow focused or unfocused systems can be used: single point, short normal, microlaterolog, MicroLog, and even the three-electrode guard if the invasion is deep. The value derived from any of these curves after borehole and departure corrections have been made is R_{xo}, the flushed-zone resistivity, or R_z, the invaded-zone resistivity.

To use the value of any resistivity curve quantitatively, the resistivity of the filling fluid, R_{fluid}, must be known. In the case of R_{xo}, the fluid is the mud filtrate. The mud filtrate resistivity, R_{mf}, can be measured with a mud resistivity meter and a mud press, or it can be derived with some sacrifice in accuracy or confidence from Fig. 9–13. The value of the ratio of R to R_{fluid} is the formation factor:

$$R_{xo}/R_{mf} = F_s \qquad (9.11)$$

The deeper investigation curves, such as the three-

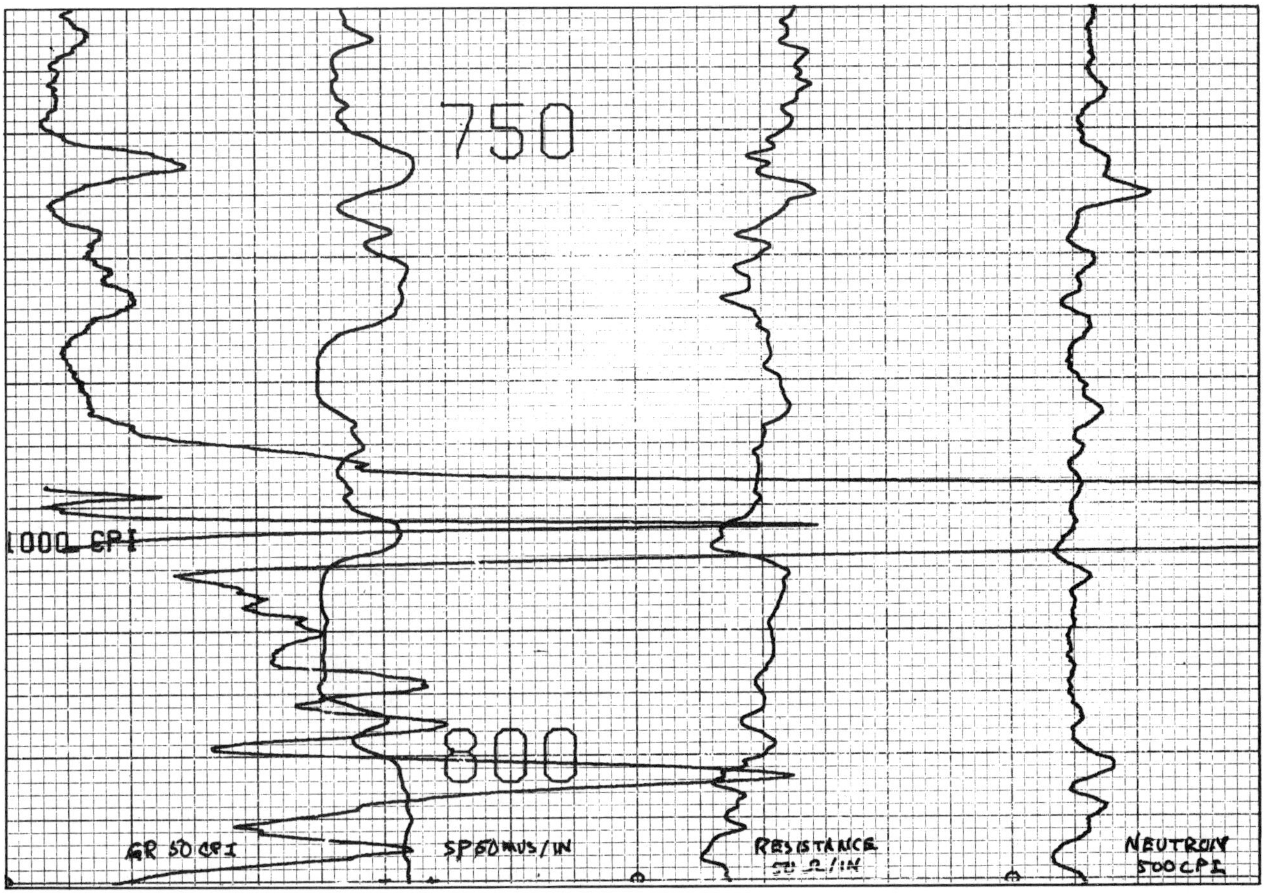

FIG. 9–10

Log showing the redox shift of the SP curve

electrode guard or the short normal if low water-loss mud is used in a very permeable formation, begin to see past the flushed zone into the mixed zone. In other words they may read the entire invaded zone resistivity, R_z. This zone is more difficult to use, since it is not easy to determine the distribution of values. Therefore, the techniques used in this zone are not covered.

Induction logs, laterals, and the long focused resistivity curves, such as the seven-electrode guard logs, eight-electrode guard logs, and even three-electrode guard logs in high-porosity, high-permeability formations, read values from deep into the formation past the invaded zone. After correction is made for the invaded zone and hole-fluid resistivity (none is needed for the induction log) the value of the undisturbed formation resistivity, R_t, may be determined. The ratio of R_t to the resistivity of the fluid filling the formation, R_w, also gives the apparent formation factor:

$$R_t/R_w = F_a \tag{9.12}$$

If there are no hydrocarbons in the formation, the two values of F—this deep one, F_a, and the shallow-curve value, F_s—should be equal. If $F_a > F_s$, there probably are hydrocarbons in the formation, assuming all corrections are proper.

Archie's Law relates the value of F to the porosity of the formation:

$$F = a\phi^{-m} \tag{9.13}$$

Where:

a = Formation-factor constant
m = Cementation factor
ϕ = Fractional porosity

Values of a may be determined from m with Fig. 9–14, if desired.

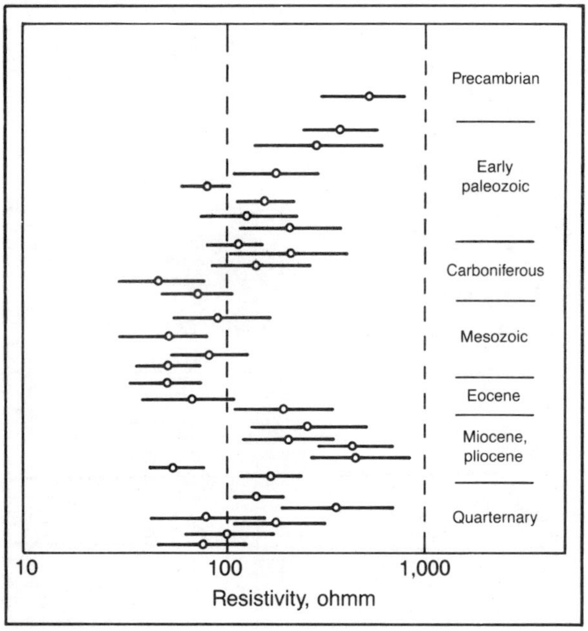

FIG. 9–11

Average resistivity for groups of sedimentary rocks with similar characteristics. Each circle indicates an average resistivity determined from 35–250 measurements about radio stations. The bar through each circle indicates the range within which 95% of the values for that group of data fall

(after Keller and Frichknecht, *Electrical Methods in Geophysical Prospecting,* courtesy Pergamon Press, 1966)

If the value of F_s does not equal F_a, the resulting value of ϕ will be different. If the value of the shallow porosity, ϕ_s, is less than that of the deeper curve, ϕ_a, there probably has been an incorrect choice of constants or an incorrectly measured value. If the value of ϕ_s is greater than ϕ_a, the difference is the amount of hydrocarbon in the undisturbed formation. This is usually given in terms of the amount of water in the pore space, S_w, the water saturation, which is the fraction of pore volume occupied by water. The oil saturation would be:

$$S_{oil} = (1 - S_w) \tag{9.14}$$

The water saturation is a measure of the amount of water and, thus, oil or gas in place in the formation. The water saturation can be calculated by Eq. 9.15:

$$S_w = [F(R_w/R_t)]^{1/n} \tag{9.15}$$

In Eq. 9.15 F has been obtained from shallow devices: resistivity, density, neutron porosity, or acoustic velocity.

There are many variations of the water-saturation determination method. All of these methods depend on the modification of the deep resistivity, R_t, by the low conductivity hydrocarbon.

The value of F can be obtained from any of the nonresistivity porosity devices, such as the density, neutron porosity, and acoustic velocity systems. The density-derived porosity, ϕ_d is close to the real porosity in a shaly sand. The neutron-derived porosity, ϕ_n is closer to the clean sand value. The acoustic-velocity-derived porosity, ϕ_s is most effective in a clean sand or limestone. The porosity value is best if two or more of the above curves are used to obtain a cross-plotted porosity value. All three of these devices are shallow devices. The relationship for the value of F is Eq. 9.13.

The movable oil plot (MOP) measures the amount of oil that can be removed from the formation compared to that adhering to the grain structure, which cannot be removed. This system calculates the total amount of oil in the formation using R_t and R_w, the flushed-zone residual oil using R_{xo} and R_{mf}, and the total porosity from the density or acoustic velocity system.

Neutron, Density, and Acoustic Velocity

The neutron-porosity, density, and acoustic-velocity tools were all developed originally as shallow-investigation porosity-measuring systems. All three measure 12 in. or less into the formation. The neutron and density systems must use specific charts or relations to convert from counts per second to porosity index for the neutron and density for the density system. The acoustic velocity measures directly in velocity or travel-time parameters. The density and acoustic velocity values can be used to calculate apparent porosity:

$$\Delta t = \phi \Delta t_f + (1 - \phi)\Delta t_m \tag{9.16}$$

$$\rho_b = \phi \rho_f + (1 - \phi)\rho_m \tag{9.17}$$

Where:
Δt = Travel time
ρ = Density
ϕ = Fractional porosity
f = Fluid
m = Matrix material

Some peculiarities of measurement must be kept in mind when using each curve by itself. The neutron systems are sensitive to the amount of hydrogen in the formation. Therefore, shales and clays, with their large amount of bound water, show large apparent porosities. Neutron systems are also sensitive to some

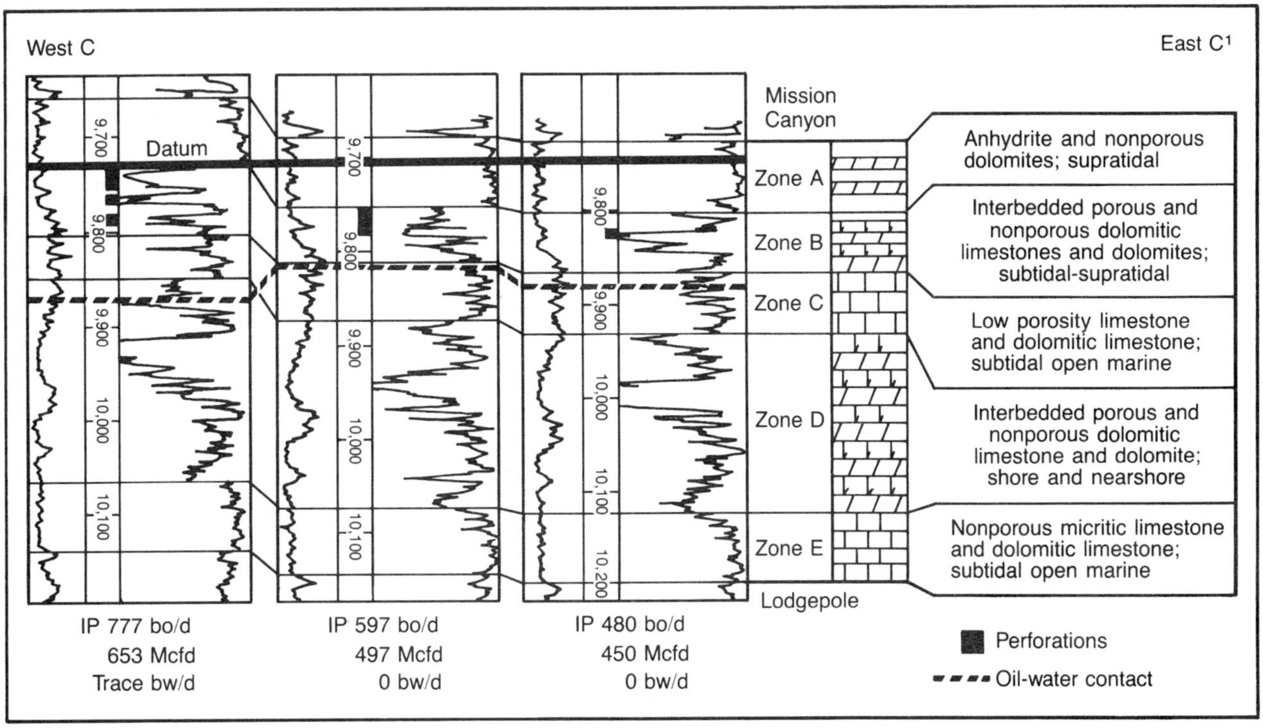

FIG. 9–12

Structural cross section c-c′ Little Knife Field, North Dakota (after Wittstrom and Hagemeier, "A Review of the Little Knife Field Development," WGA, *Bulletin,* volume 11, No. 4, 1978)

formation trace materials. Thermal neutron and capture gamma-ray systems are sensitive in varying degrees to the chlorine, boron, and gadolinium contents. Epithermal systems are not as sensitive. Acoustic velocity systems show slow travel times in shales. They are also adversely affected by gas in the formation, as is the neutron system. The density systems show dense materials as too heavy and hydrogenous materials as too light because of photoelectric absorption and Z/A distortions, respectively.

All three systems can be used for lithology identification, especially in nonporous media. Each material has, in general, a specific, narrow range of thermal neutron capture cross section (Σ), bulk density (ρ_b), and acoustic velocity (V). These can be used to identify formation minerals. This is particularly valuable in hard-rock environments. Fig. 9–15 shows typical logs and the appearance of the anomalies.

It is easy to be misled, however, when using a single curve. For example, the ash in coal is substantially more dense than the coal itself by about 2:1. Therefore, the density reading in a coal should be a function of the ash content. At first glance it appears that a plot of density or density-counting rate against ash content should result in a usable relationship. Fig. 9–16 shows such a plot from an actual situation. Note that there is an excellent straight-line relationship between ash content and density. However, when the spread of points is examined, it is found to be quite great. The density of a sample of coal depends on more components than just the ash, and an oversimplification as shown in Fig. 9–16 results in an intolerable uncertainty.

■ Dipmeter, Caliper, and Deviation (Drift)

The dipmeter systems are valuable for determining formation dip and for presenting patterns that give information about depositional and geological history. The recent introduction of small-diameter, high-resolution dipmeters increases the use of these tools in nonpetroleum logging.

The caliper systems fall into three general classes. The single-arm systems are used to tell wall and borehole quality; the three-arm systems are used for general corrections of hole-size sensitive tools, such as the gamma ray; and the multi-independent-arm

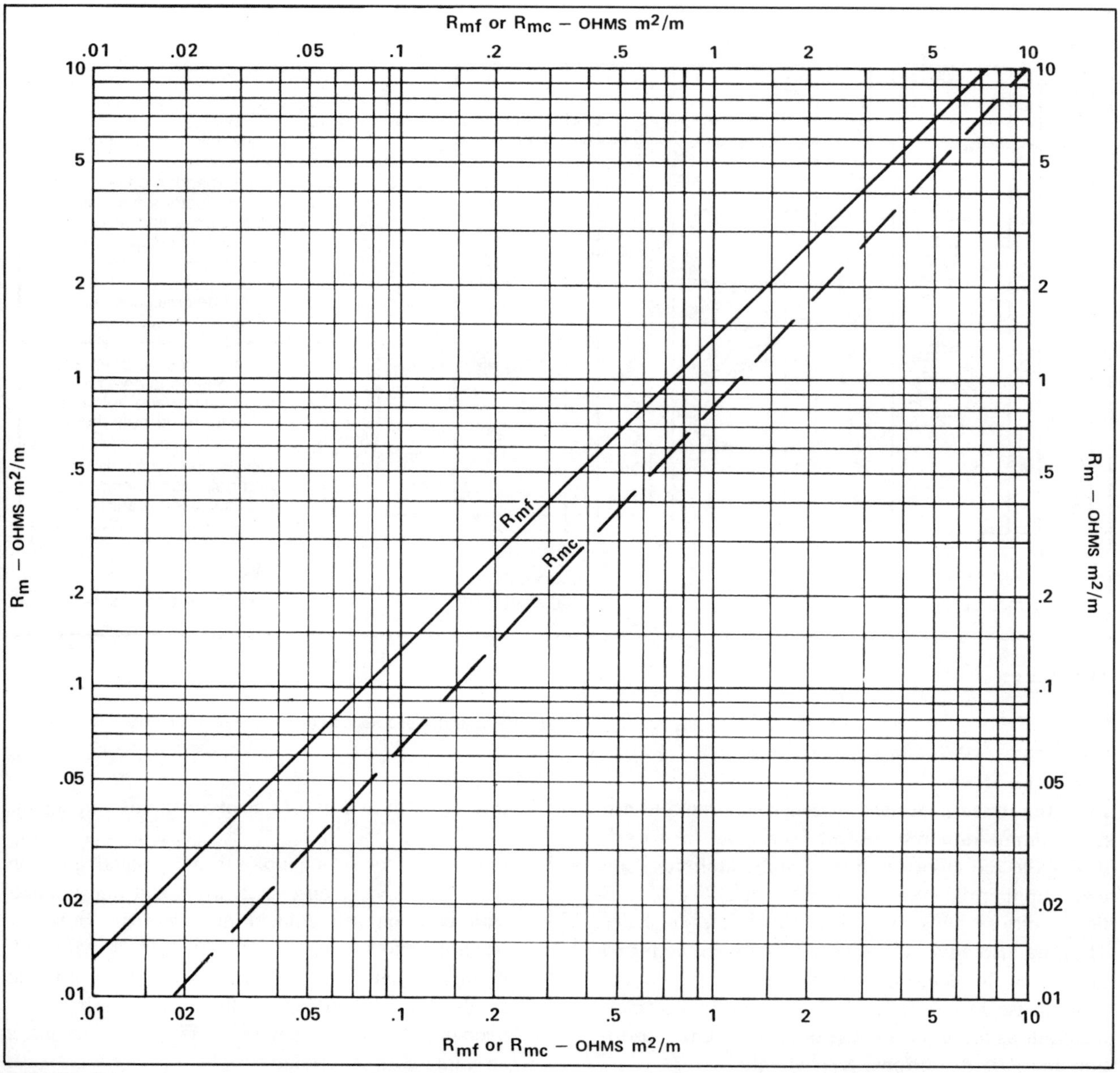

FIG. 9–13

R_m-R_{mf}-R_{mc} relationships (courtesy Gearhart Industries Inc.)

tools are used for calculating hole volume. Hole-deviation tools must be used with the dipmeter tools, but deviations tools also are used by themselves. Mostly, the solo use of the deviation tools is for legal purposes.

■ MULTIPLE CURVES AND LOG SUITES

In this section we point out some of the multitudinous combinations and uses of measurements so we can begin to specify suites and interpretation techniques and variations. For example, the uranium suite may be best, but, it helps to know that.

■ Choosing Combinations

By far, the most effective way of using logs and other geophysical information is by combining the information from two or more measurements. This has been done regularly in the past, but the use of calculators, programmable calculators, and comput-

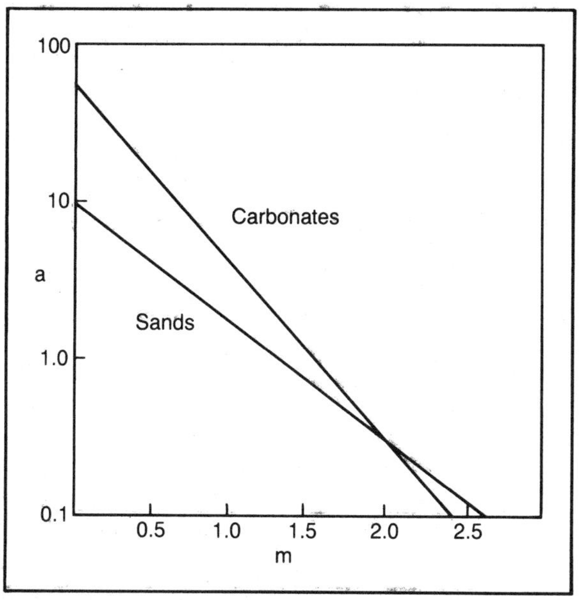

FIG. 9–14

Chart for determining *a* or m (after Gomez-Rivero)

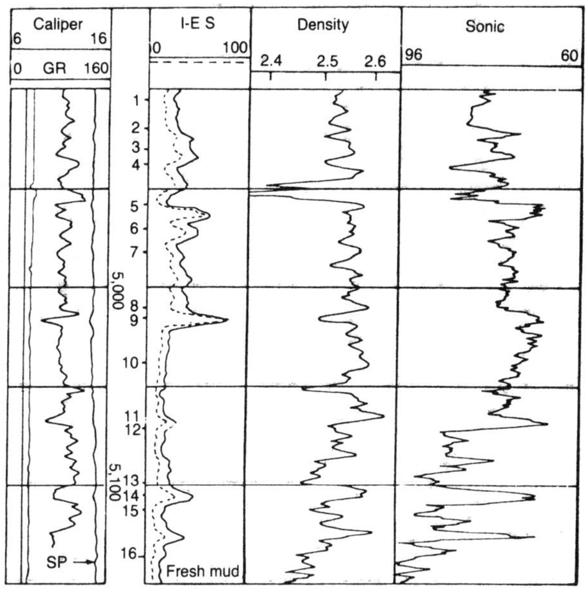

FIG. 9–15

Logs of several of the common systems

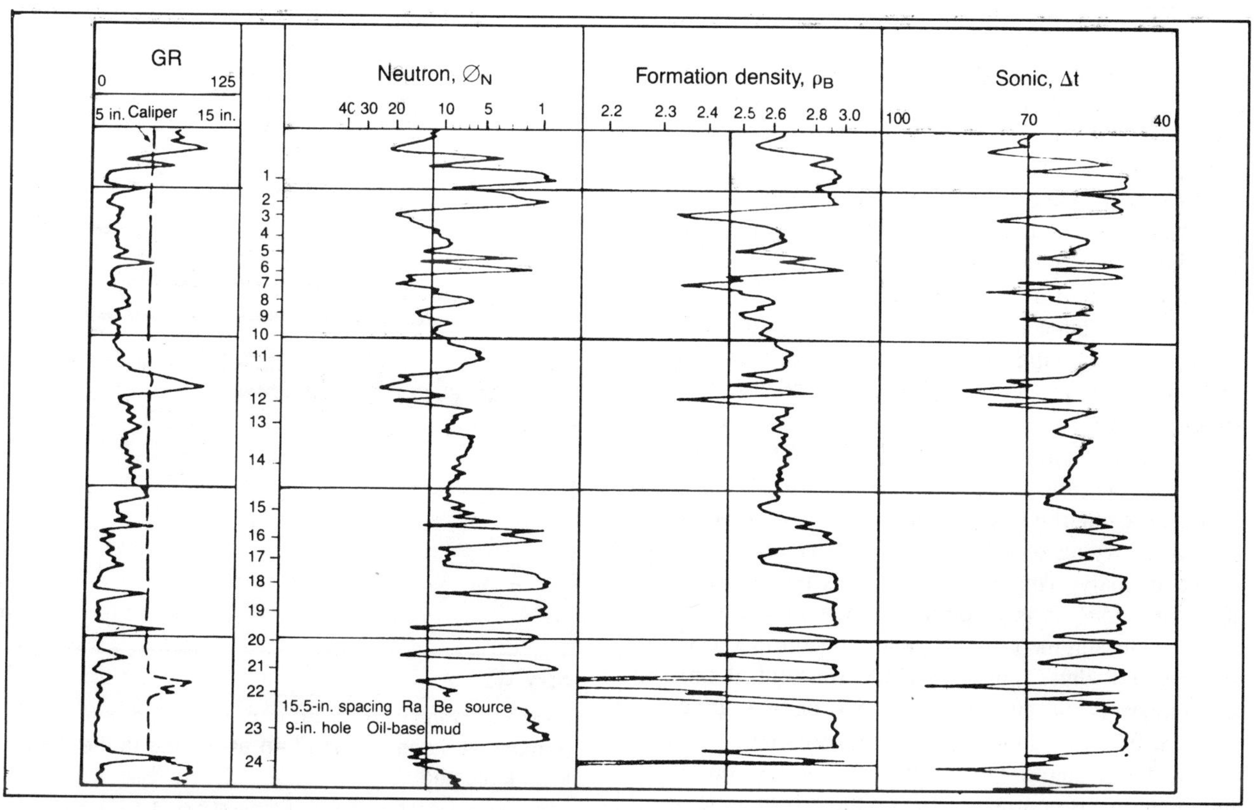

FIG. 9–15 *continued*

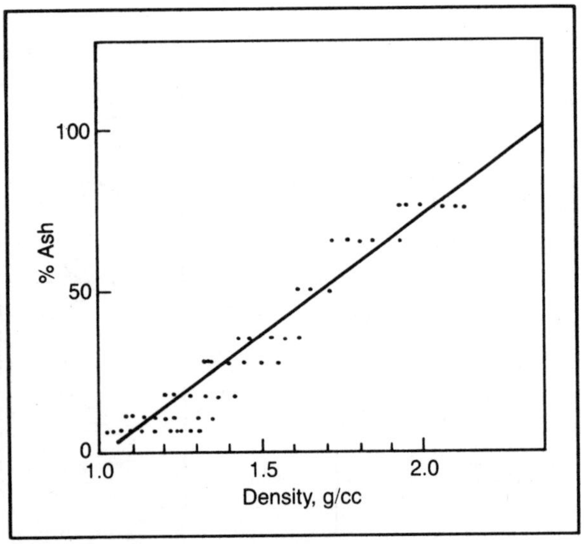

FIG. 9–16

Determination of the ash content in coal using the density values

ers has made the presentation of finished or processed information and multicurve analysis much easier. As with the single curves, the raw data must be recorded in permanent form. However, after recording raw data, the curves can be combined and plotted or printed out in an almost infinite variety. Instead of counts per second, we can have a curve of Young's Modulus or of fractional parts of sand, shale, and water, or of gas content.

The task for the user is to specify the proper combinations of measurements to best do the calculations. We have gotten into the habit, in nonpetroleum logging, of using a standard suite of logs regardless of the appropriateness of that suite. This is the direct result of not knowing enough about the measurements to specify the proper logs for a specific use. We should be familiar with the geology and lithology of an area in a general sense before we go into it. This eliminates the all-too-common problem of trying to make sense out of a single-point curve in a bituminous coal or of trying to measure the density in a radioactive anomaly. This is not in any way poking fun at these attempts. It is to point out that unfamiliarity with the measurement characteristics results in a waste of time and, thus, money. Equally wasteful of time and money is the attempt to use one or two curves where a full suite should have been recorded.

It is dangerous to believe that one can rely on the judgment of his contractor to pick the proper suite of logs. This may be the best way, and certainly, his advice should be considered. However, more than likely, the prober and sometimes his manager know less about the logs and their uses than we do.

■ Patterns

The first multiple curve application to be covered is a qualitative use. It is the recognition of curve patterns and the reasons for them. This can be a tremendously valuable application. The SP and resistivity curves are discussed mostly because they are convenient. However, others can be substituted effectively.

The patterns of signatures of the logged curves can be used to determine the extent of the various beds or strata. This can be done because the resistivity, SP, gamma-ray, and other curves show the similarities of physical characteristics of a given bed from one hole to another. Also, they are all sensitive in one way or another to the sand/shale content of the formation. For example, Fig. 9–17 shows the correlation of one type of curve, probably a gamma ray, along a

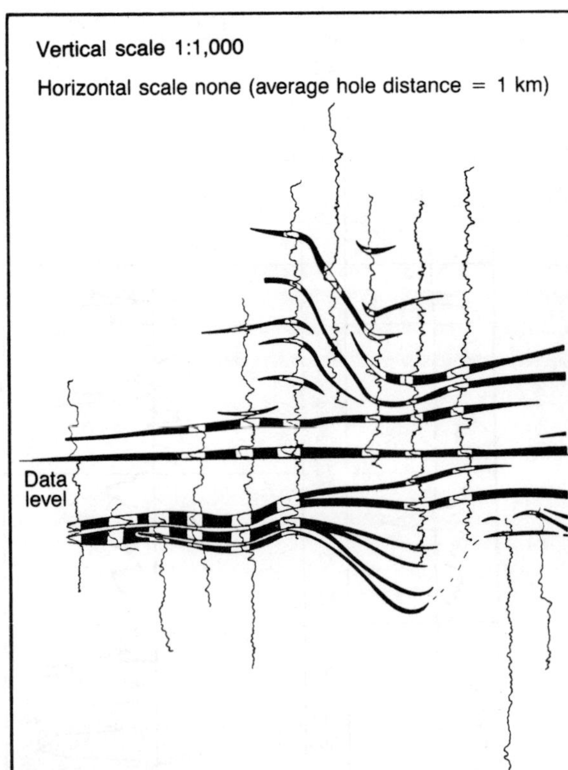

FIG. 9–17

Example of log correlation in complicated coal formations. Logs are lined up according to the data level indicated

(after Lavers and Smits, "Recent Developments in Coal Petrophysics," *Coal Exploration,* 1976)

cross section of the earth. Notice that the similar anomalies are connected. Also, notice that anomalies are not identical in signature but are similar from one hole to another. The second example, Fig. 9–18, shows a similar situation using two curves—the SP and resistivity. Notice that recognition and correlation are much easier with two curves than with one. Both of these examples were adjusted in depth to a single formation (the data level). If the surface elevation is known, a more informative cross section can be made by comparing all of the logs at the correct elevations; that is, hanging them on a constant elevation. Once the cross sections are made, ore-body and isopach maps can be made, as shown in Fig. 9–19.

More data can generate more information. Fig. 9–20 shows a profile across a uranium geochemical cell that uses three curves and the cuttings log. The description of that cell profile is quite complete and detailed.

Logs that are combinations of different curves can be chosen to enhance the measurements being sought. The next series of diagrams, Fig. 9–21 to Fig. 9–23 shows the SP curve, which is sensitive to ion content and grain size, and the resistivity curve, which is sensitive to grain size through electrical conductivity. These two curves have been combined to emphasize the differences between sands and shales. The first illustration, Fig. 9–21 shows what occurs if the borehole is assumed vertical when it really is slanted, which it usually is with shallow drilling equipment. The two curves, spontaneous potential and resistivity, both read the same sand under identical conditions except for hole deviation. Notice that the slant hole makes the sand and ore reserve appear thicker. A log always reads the hole distance and not the true or vertical depth. The result in a uranium project is to increase the GT product falsely. If the bed is horizontal, that is, if its dip is zero, the real thickness of the bed is t:

$$t = h \cos \theta \qquad (9.18)$$

Where:
h = Logged thickness of the bed
θ = Angle of hole deviation

If the sand has an appreciable dip, the real thickness, t_o, depends on the angle of the hole with respect to the component of dip in the direction of deviation.

$$t_o = h[\cos \theta + \cos \phi \cos(\alpha - \beta)] \qquad (9.19)$$

Where:
ϕ = Dip angle
α = Azimuth of dip
β = Azimuth of deviation

The correction, however, does not really become significant (1%) until the angle, θ, or the resultant angle (ϕ and θ) is greater than 8°. For reserve calculations hole deviation measurements are very necessary.

The addition of the high-resolution dipmeter to the above suite of logs is uncommon. It is, however, a tremendous aid to this type of problem. It also opens the path for an increasing determination of exploration geology.

Many stratigraphic features can be detected with logging methods even though they may not appear directly on the logs.

Fig. 9–22 and Fig. 9–23 show boreholes intersecting two types of faults. If the area is known to be faulted and a sand appears in the first and third of three inline boreholes but not in the second, one should consider the presence of a normal fault. If the signature of the sand in the second hole is repeated, one should consider the presence of a reverse fault, as in Fig. 9–22. It is often difficult to tell the exact thickness of a sand in a faulted area. The presence of anomalous radioactivity, especially with no indication on the spontaneous potential or resistivity curves, may help indicate the fault intersection.

Fig. 9–24 shows a folded sand. If only the gamma ray, SP, and resistivity logs in hole number 2 were available, it would be impossible to tell the thickness of the sand. The three logs together indicate either a folded or lenticular sand. A dipmeter would help distinguish between a folded or lenticular sand. Fig. 9–25 shows a folded sand and the role of the dipmeter.

The gamma-ray, SP, and R measurements are all sensitive to the presence of clays and the SP and R are sensitive to grain size and sorting. Therefore, they indicate something of depositional history. Fig. 9–26 diagrams a distributary channel. The change of grain size, sorting, and the increase of clay as the channel is filled and the flow rate is slowed affects the curve shapes of the GCGR, SP and R. As radiation increases, the SP trends toward the shale line, and the R decreases in the direction of decreasing flow rate. This can be applied to beach sands (Fig. 9–27), a valley fill (Fig. 9–28), turbidity sands (Fig. 9–29), and periodic flooding (Fig. 9–30). In each case the deflections of the curves can be interpreted in terms of flow rate. Rouff has utilized this technique in a computer-mapping system (Fig. 9–31a–e). A comparison of his depositional map and the computer map are shown in Figs. 9–31c and 9–31e.

The general procedure in any of these signature

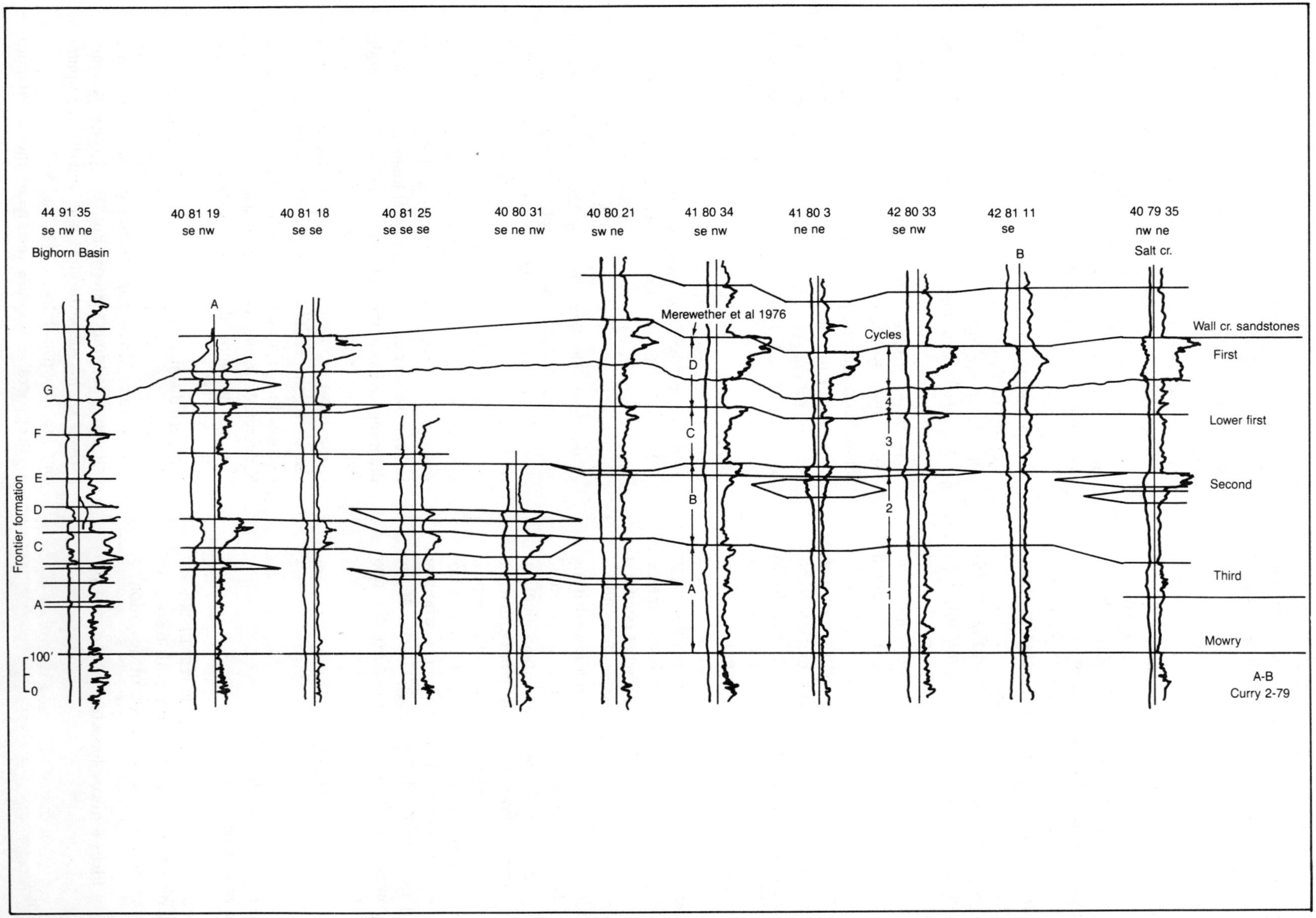

FIG. 9–18

Cross section A-B. Well log correlations from southeast flank of Tisdale anticline (A) to the northeast flank of the anticline (B). A western well log in the Big Horn Basin and eastern well log at Salt Creek are shown for reference

(after Curry, "Type Section of Wall Creek Sandstone Member of the Frontier Formation," WGA, *Earth Science Bulletin,* volume 12, No. 2, 1979)

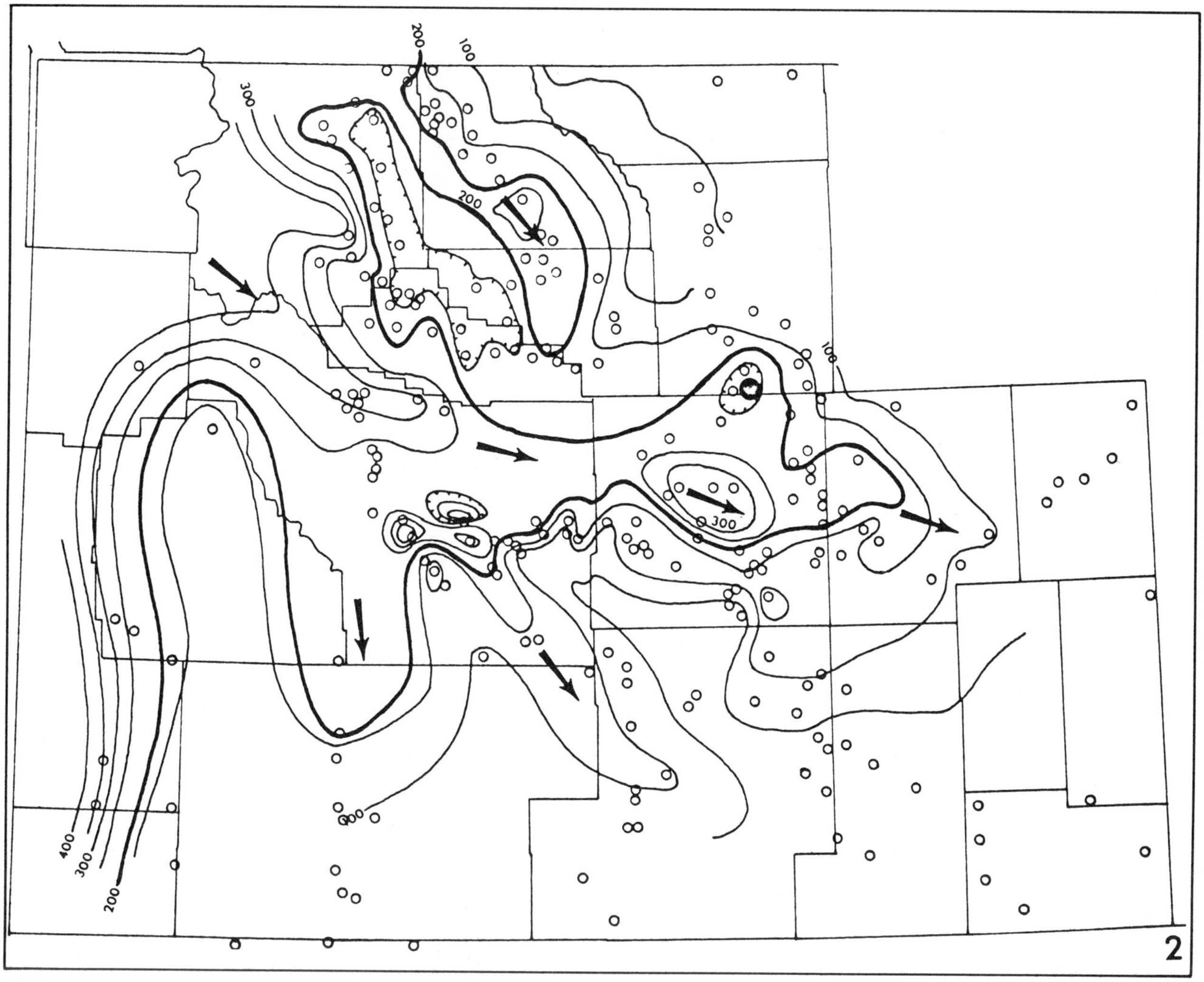

FIG. 9–19

Isopach map of Frontier sandstones above the Mowry formation and below the first frontier. (after Goodell) Arrows show interpretation of delta distributaries to four main delta lobes. Area shown is most of the state of Wyoming; county boundaries are also shown. (after Curry, "Type Section of Wall Creek Sandstone Member of the Frontier Formation," WGA, *Earth Science Bulletin*, volume 12, No. 2, 1979)

identifications is to try to picture what would make the curves act as they do. Once this has been pictured, many possibilities become available for this technique.

■ Crossplotting

The use of two or more curves to correct the faults or aberrations of each has been alluded to many times. The neutron-porosity system, for example, gives a high apparent porosity to shales. An acoustic-velocity tool gives a low velocity to shales. A density system, on the other hand, detects a shale density that is not far from that of the sand matrix of silica and feldspar. Thus, the density reading is much less severely affected, but each one is somewhat in error when used to calculate porosity.

Fig. 9–32 shows a hypothetical example of plotting a shaly-sand value from the density and neutron curves. The ordinate and abscissa are the apparent porosity from the density curve and the apparent porosity from the neutron curve. Three independent points are identifiable for a given sand-shale mixture: the zero porosity point for 100% sand and for 100% shale points, and the 100% porosity or water point. Thus, a triangular diagram is defined.

The two component modes are the lines connecting any two points. These can be scaled in linear units of

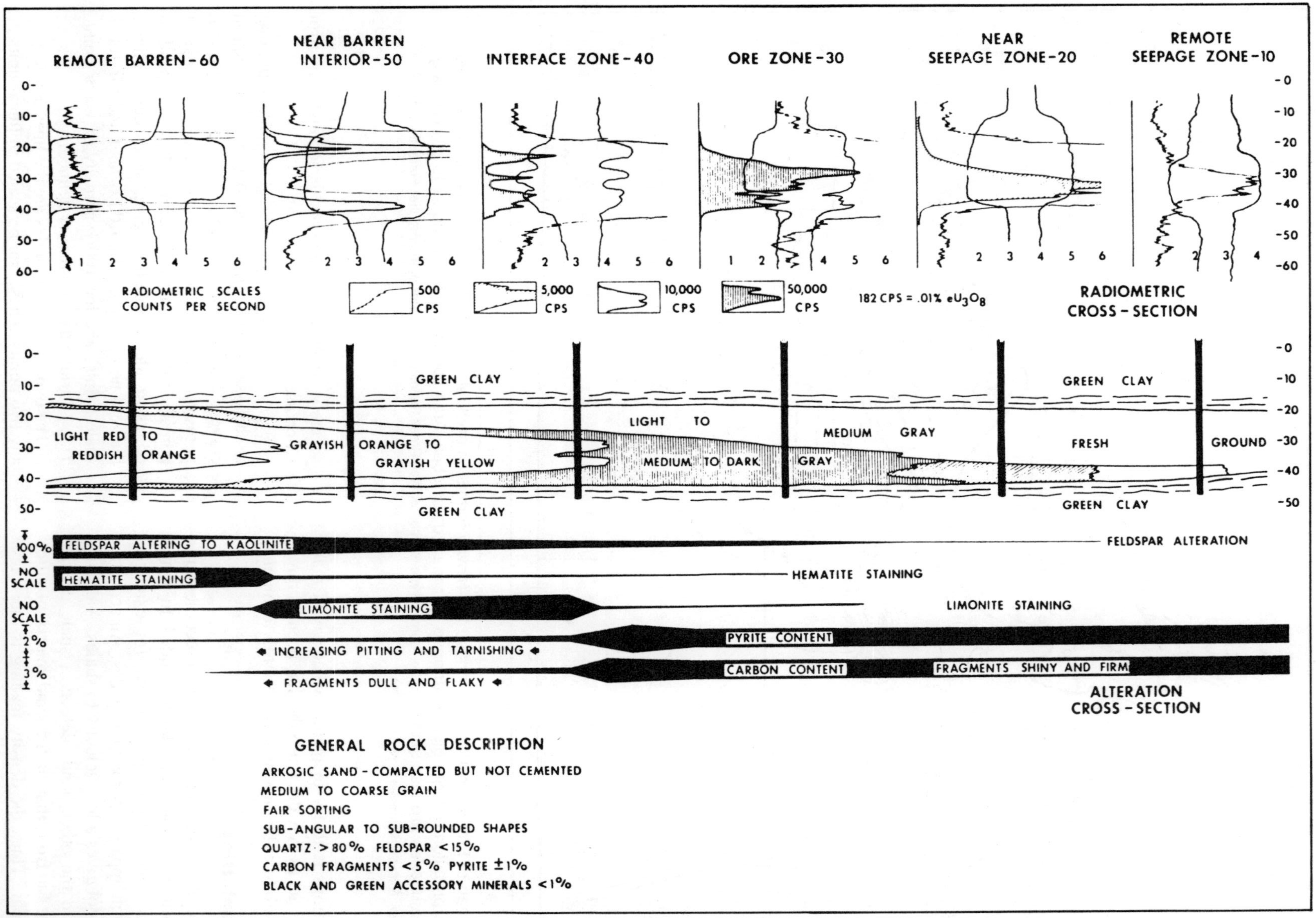

FIG. 9–20

Type of geochemical cell, Powder River basin, Wyoming (after Rubin, "Uranium Roll Front Zonation in the Southern Powder River Basin, Wyoming," WGA, *Bulletin,* December 1970)

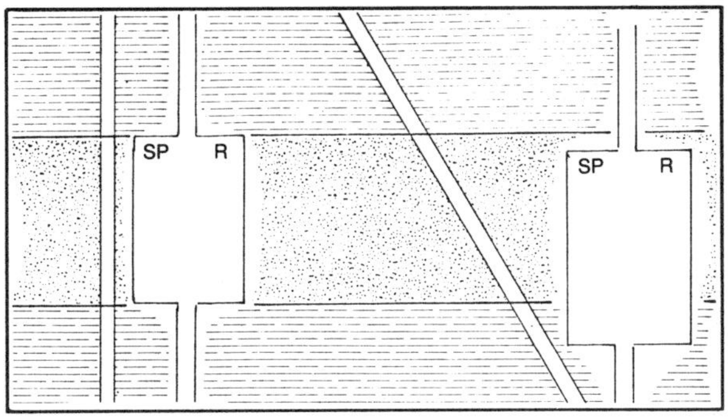

FIG. 9–21
Effect on log of a slanted borehole

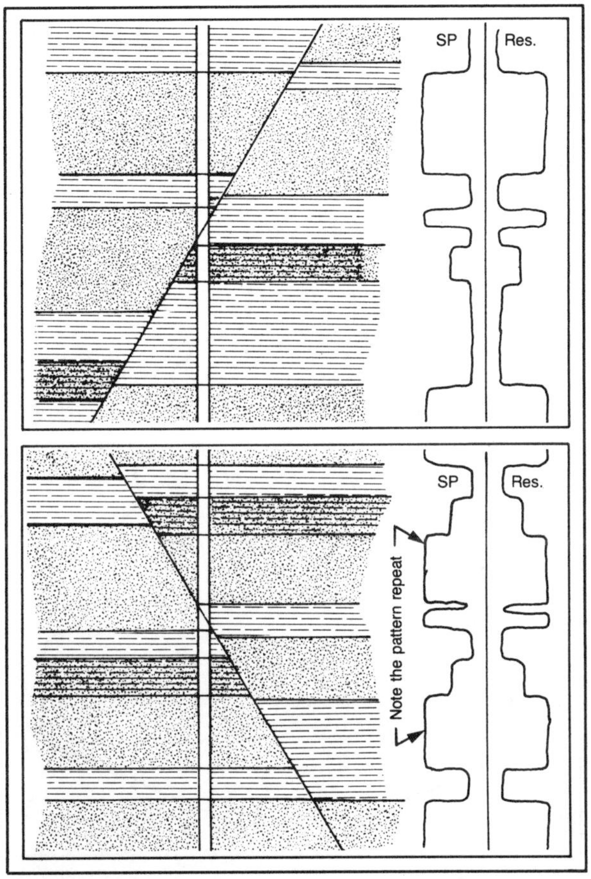

FIG. 9–22
Reverse and normal faults

percent. If we examine any clean sand point, P, and add shale to it, we find that the addition of shale increases the density value only a small amount. This is because the shale density is only slightly lighter than the sand grain (or matrix) density. On the other hand, the neutron apparent porosity is increased a much greater amount because part of the sand matrix

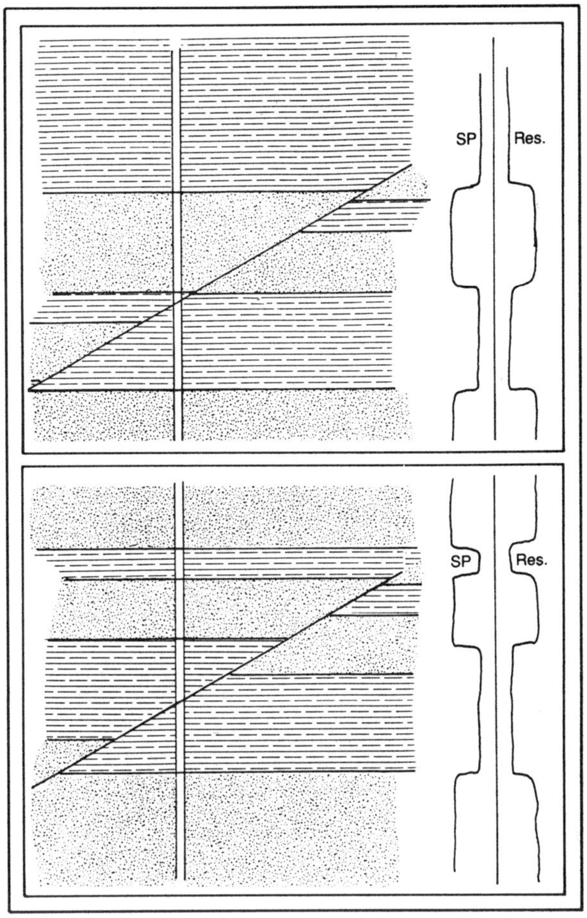

FIG. 9–23
Shallow reverse and normal faults

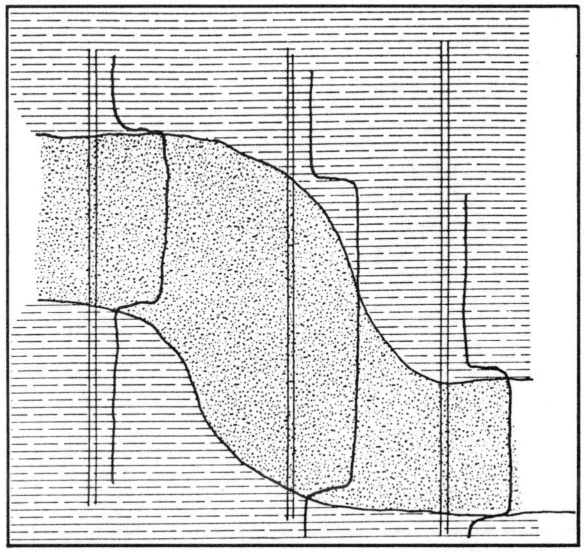

FIG. 9–24
Folded sand

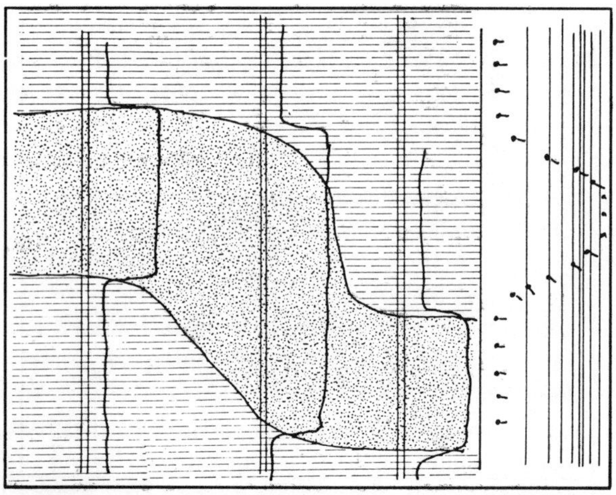

FIG. 9–25

Folded sand with dipmeter

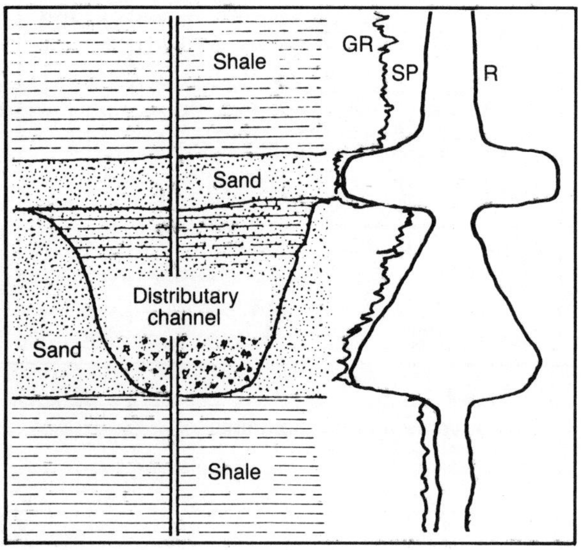

FIG. 9–26

Distributary channel pattern

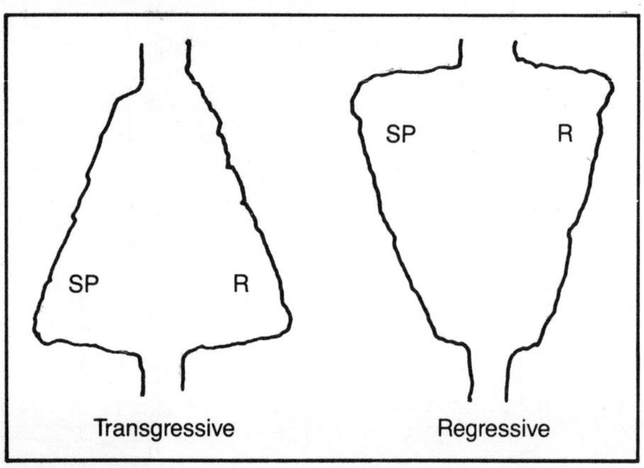

FIG. 9–27

Beach sand deposition pattern

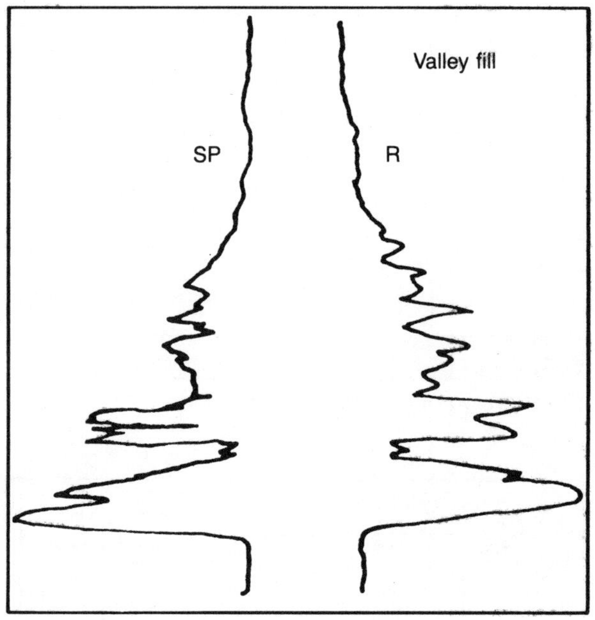

FIG. 9–28

Valley fill log pattern

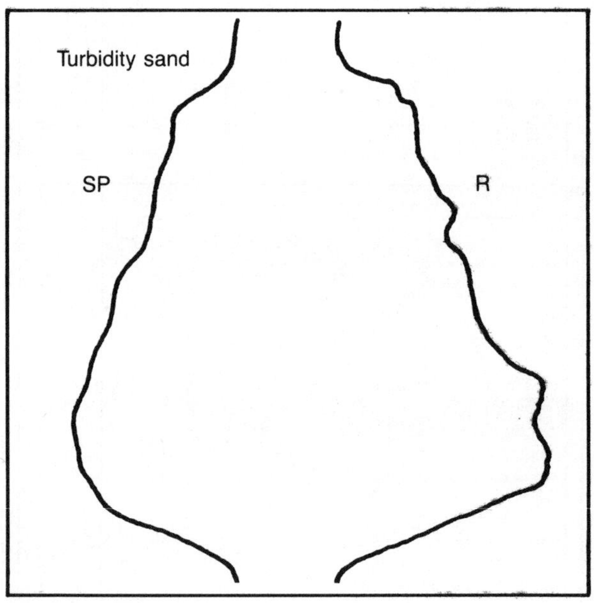

FIG. 9–29

Turbidity sand log pattern

is being displaced by a hydrogenous material, shale. The resultant value for 50% shaly sand with 18% porosity is shown on the diagram. Besides being able to determine porosity, the use of two curves allows one to tell something about composition.

Any of the crossplotting diagrams follow this general procedure. Any parameter can be chosen, and

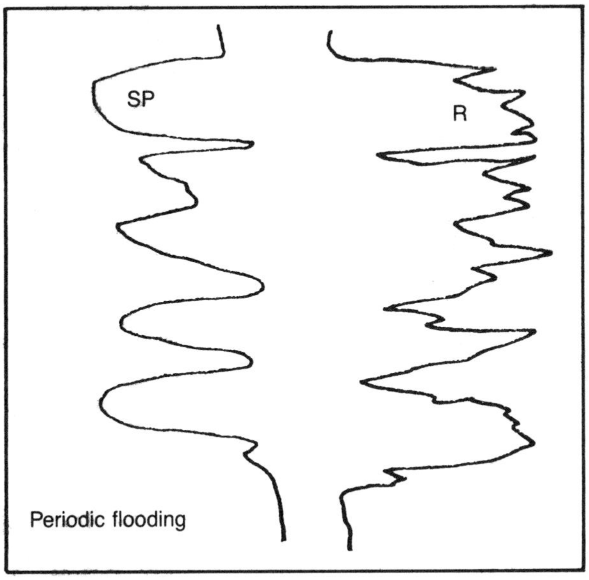

FIG. 9–30

Periodic flooding—note the fine detail

the units of the ordinate and abscissa can be any realistic ones desired. We could as easily have plotted Fig. 9–32 in counts per second, for example.

Point P is obtained from the logs. It is evident that if the porosity had been determined from the density curve alone, the apparent value would have been about 32%. If it had been obtained only from the neutron curve, the porosity would have appeared to be about 50%. The probable amounts of sand, shale, and water that make up this formation can be obtained from the diagram. Fig. 9–33 shows a porosity cross plot for density and neutron. Fig. 9–34a shows a similar one prepared for gas evaluation.

It is not necessary to restrict the crossplotting diagram to two parameters. The M – N crossplot used extensively in petroleum work is a three-element crossplot with one axis consisting of the density and acoustic velocity and the other consisting of the neutron and density.

In petroleum logging the great value of the cross plot is in obtaining a correct porosity value, although of course it is used for many other things. In non-petroleum logging the value seems to be more in the determination of lithology. Fig. 9–35 illustrates this idea qualitatively. The lithology crossplot promises to be especially valuable in determining the components of coals.

Coal is a complex material having little permeability but containing water. It has an ash content and water content that detract from the unit heat value of the coal. The heat value is a function of the carbon content and the volatile content. The form of the carbon content is unclear, but it somewhat resembles amorphous carbon in character. The volatile content depends on the method of measurement, but it appears to be largely hydrocarbons. If we determine the mean value for a large sampling of a coal when the density and conductivity are crossplotted, we can vary each of these four components and see what a crossplot would look like. If the ash content is increased, the plotted values would increase in density and in conductivity. This would cause a skewing along the line labeled A on the diagram of Fig. 9–36. Increasing moisture content would cause a decrease in density and an increase in conductivity, or a skewing along the line W. As long as the carbon particles are isolated—as they are except in better bituminous coals—increasing carbon content results in decreased conductivity and an increase in density. This is line C. Similarly, increased volatile content skews points along line V. With four parameters the crossplot

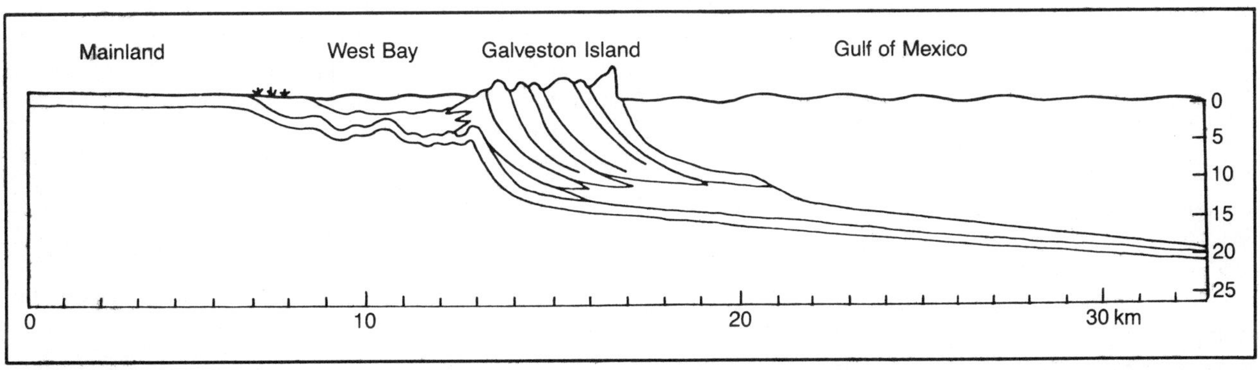

FIG. 9–31a

Cross section of a barrier island (after Rouff, "A Technique for Interpreting Depositional Environments of Sandstones from the SP Log Utilizing the Computer," *Log Analyst*, volume 17, No. 4, 1976)

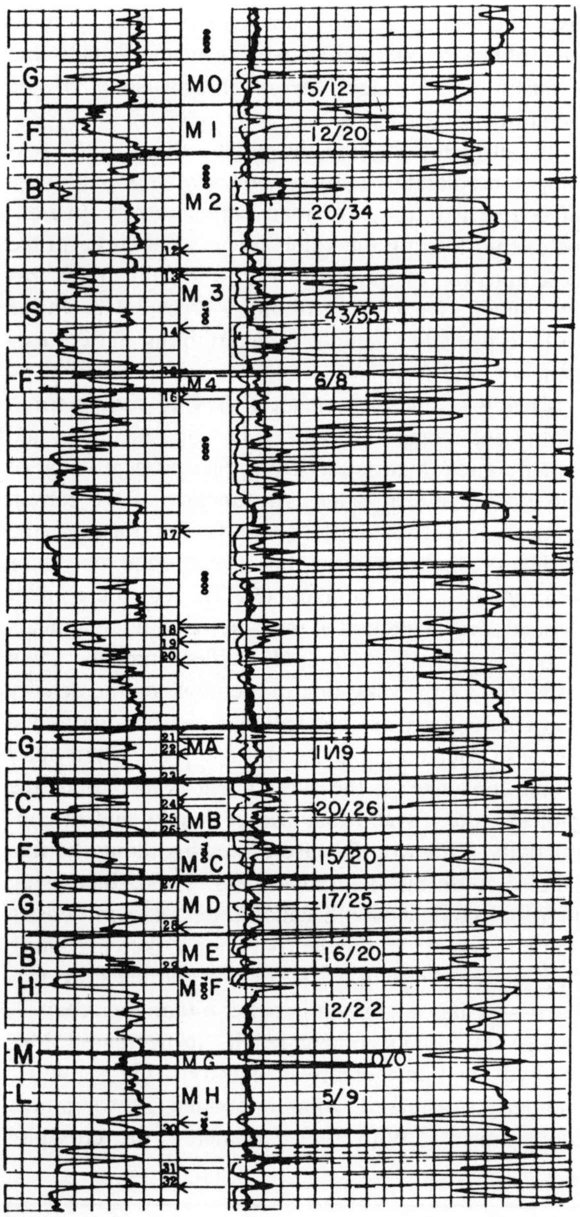

FIG. 9–31b

Drill hole log run on Galveston Island

(after Rouff, "A Technique for Interpreting Depositional Environments of Sandstones from the SP Log Utilizing the Computer," *Log Analyst*, volume 17, No. 4, 1976)

diagram is difficult to evaluate quantitatively, unless one has a gamma-ray log to evaluate the ash content or a neutron log to evaluate the moisture content.

The major problem in many crossplots is in identifying the pure components. The three-component coal diagram is a good example of this (Fig. 9–37). In this diagram the carbon and the hydrocarbon have been combined into the dry, ashless coal value. If the ash is assumed to be identical to the nearby clay or shale, there is no trouble assigning a density and a resistivity value to it. The density of any water in the coal is nearly 1.0 g/cc. The resistivity of the water, however, is difficult to determine because a sample of water is seldom available. Similarly, the resistivity of the dry, ashless coal is very high. Thus, its conductivity is nearly zero, but its density is difficult to determine because a sample of dry, ashless coal is not available.

To avoid the problem of not knowing the pure component values, first plot the 100% ash point. Then, plot a line for water whose density is 1.0 g/cc and whose conductivity is any value. Conductivity is used instead of resistivity in this plot because it will result in a linear relationship. Estimate the water conductivity on the basis of experience or by guessing. It is fairly low since the water has few dissolved solids. A reasonable value is 100–200 millimhos/m. The coal point can be handled in a like manner. While the density is unknown, the conductivity is close to zero. Therefore, draw a line at zero conductivity and a range of densities. In a lignite the density is near 1.3 g/cc. Place a point there, and connect the three points. This gives a triangular diagram. Divide each line into 4 to 10 equal parts and connect points on opposite sides.

Once the results arrive from the chemical laboratory, several points can be plotted for density and conductivity. The whole triangle can then be moved along the coal density line and the water conductivity line until the triangle is correctly placed over the laboratory points. From that time the diagram may be used quantitatively. Before that time it may be used to determine relative values.

A most valuable form of the crossplot is to plot system response against porosity. The porosity can be determined from the core determinations. It can also come from another porosity device. For example, in a 100% water-saturated horizon the resistivity-derived porosity is given by Archie's relation:

$$R_t = R_w \phi^{-m} \qquad (9.20)$$

$$\log R_t = \log R_w - m \log \phi \qquad (9.21)$$

If R_t versus ϕ is plotted, a log − log grid must be used since:

$$y = \log R_t \text{ and}$$
$$x = \log \phi$$

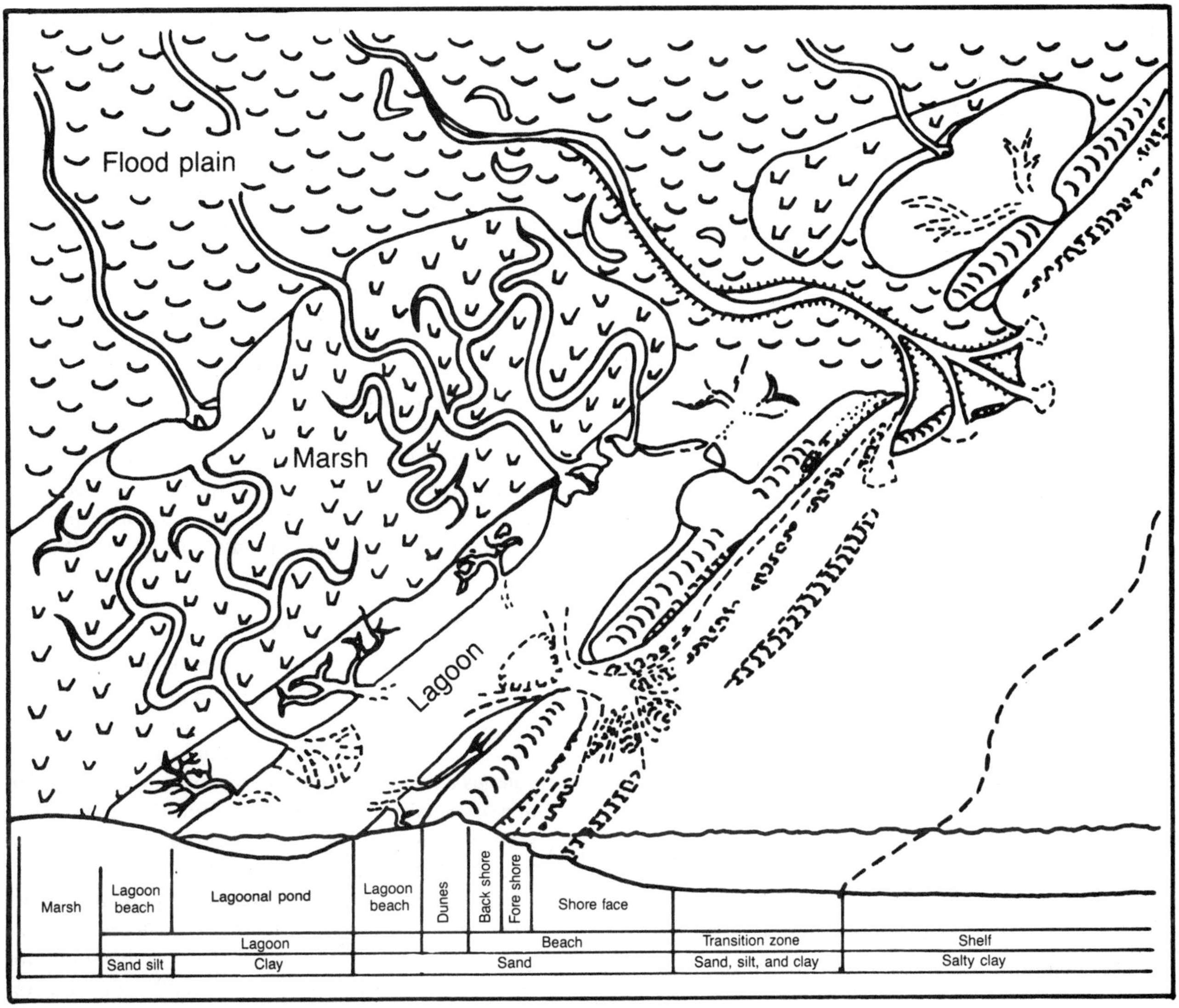

FIG. 9–31c

Reconstruction of various environments of deposition (after Rouff, "A Technique for Interpreting Depositional Environments of Sandstones from the SP Log Utilizing the Computer," *Log Analyst,* volume 17, No. 4, 1976)

In the linear equation form:

$$y = a + bx \qquad (9.22)$$

Then log R_w is the y intercept and $-m$ is the slope of the plotted line. The y intercept may be found by setting ϕ equal to zero and solving the equation. The value of the slope is the change in y divided by the change in x, or:

$$-m = \Delta y / \Delta \phi \qquad (9.23)$$

The value of ϕ can also be obtained from another porosity device. For example, the density-derived porosity is:

$$\phi_\rho = (\rho_b - \rho_m)/(\rho_f - \rho_m) \qquad (9.24)$$

If this is substituted into Archie's equation:

$$\log R_t = \log R_w + m \log(\rho_f - \rho_m) - m \log(\rho_b - \rho_m) \qquad (9.25)$$

The slope is $-m$, and the intercept is $\log R_w + m\log(\rho_f - \rho_m)$. Each rock type is defined by a different trend.

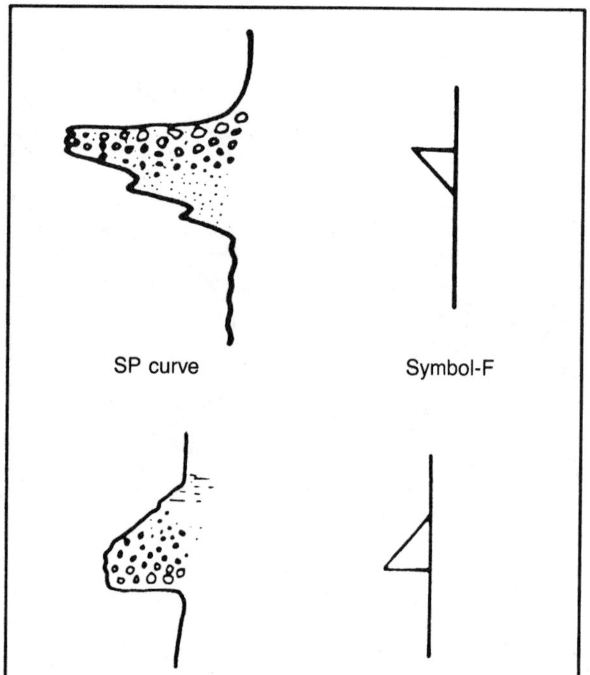

FIG. 9–31d

Log anomalies and the corresponding computer identification symbols

(after Rouff, "A Technique for Interpreting Depositional Environments of Sandstones from the SP Log Utilizing the Computer," *Log Analyst,* volume 17, No. 4, 1976)

Fig. 9–38 shows a Hingle crossplot. This plot uses a nonlinear resistivity y axis whose value is:

$$y = [(aR_w)/R_t]^{-1/m} \text{ or} \quad (9.26)$$

$$y = [(aR_{mf})/R_{xo}]^{-1/m} \quad (9.27)$$

The x axis is any function of porosity. The value of R_w may be found from another resistivity curve or from the SP. The value of R_{mf} may be found from the mud sample. Both R_w and R_{mf} remain constant for any horizon. The value of *a* may be assumed to be 1.0 as a first approximation. The trend of points of any given 100% water-saturated horizon is a straight line. If it is extrapolated to infinite resistivity, the characteristics of the solid rock matrix are indicated. If the trend is extrapolated to the function of 100% porosity, the fluid characteristics are defined.

If the acoustic travel time is plotted against the density, each lithological type is defined by a particular slope. That slope is:

$$M = (\Delta t_f - \Delta t_m)/(\rho_f - \rho_m) \quad (9.28)$$

Similarly, if the neutron-derived porosity is plotted against the density, each lithological type is defined by a particular slope. That slope is:

$$N = (\phi_{Nf} - \phi_{Nm})/(\rho_f - \rho_m) \quad (9.29)$$

The slope of each will be constant for any single rock type for all porosities. If M (x 0.01) is plotted against N, the resulting crossplot is essentially independent of porosity and indicates lithology.

There is no problem in designing crossplots for one's own use. Merely pick curves that emphasize the points to be evaluated and plot one value against another for the same depth. Be sure to stay away from bed-boundary effects. Plotting positions for pure substances, such as water, zero-porosity materials, and known materials, provides tie points for quantitative evaluation. The method can be very valuable in the mineral industry. Crossplotting can also be easily done by computers (Figs. 9–39 a–c).

■ Simultaneous Equations

Crossplotting is really a graphical solution of three or more variables as simultaneous equations. In order to solve for a number of variables, we must have one equation for each unknown. In the simultaneous equation method of log analysis, the assumption is usually made that the sum of the fractional parts, such as the volumes of the components, is equal to one. Therefore, to solve for three components, such as sand, shale, and water (or porosity), there must be measurements from two curves such as density and acoustic velocity. The combinations for equations are only limited by the number of measurements available.

If we desire to solve for the amounts of sand, shale, and water in a sand-shale sequence, for example, we must have three equations relating to the volumes of these three components. The first equation states that these are the only three components to be considered and that the sum of their fractional volumes, V, will equal the whole amount. Therefore, the equation is:

$$V_s + V_{sh} + V_w = 1 \quad (9.30)$$

Where:

s = Sand
sh = Shale
w = Water

The value of V_w can be equated with the porosity, ϕ. Since there are three unknowns, V_s, V_{sh}, and V_w, there must be two more independent equations to solve for the volumes. There are a number of choices.

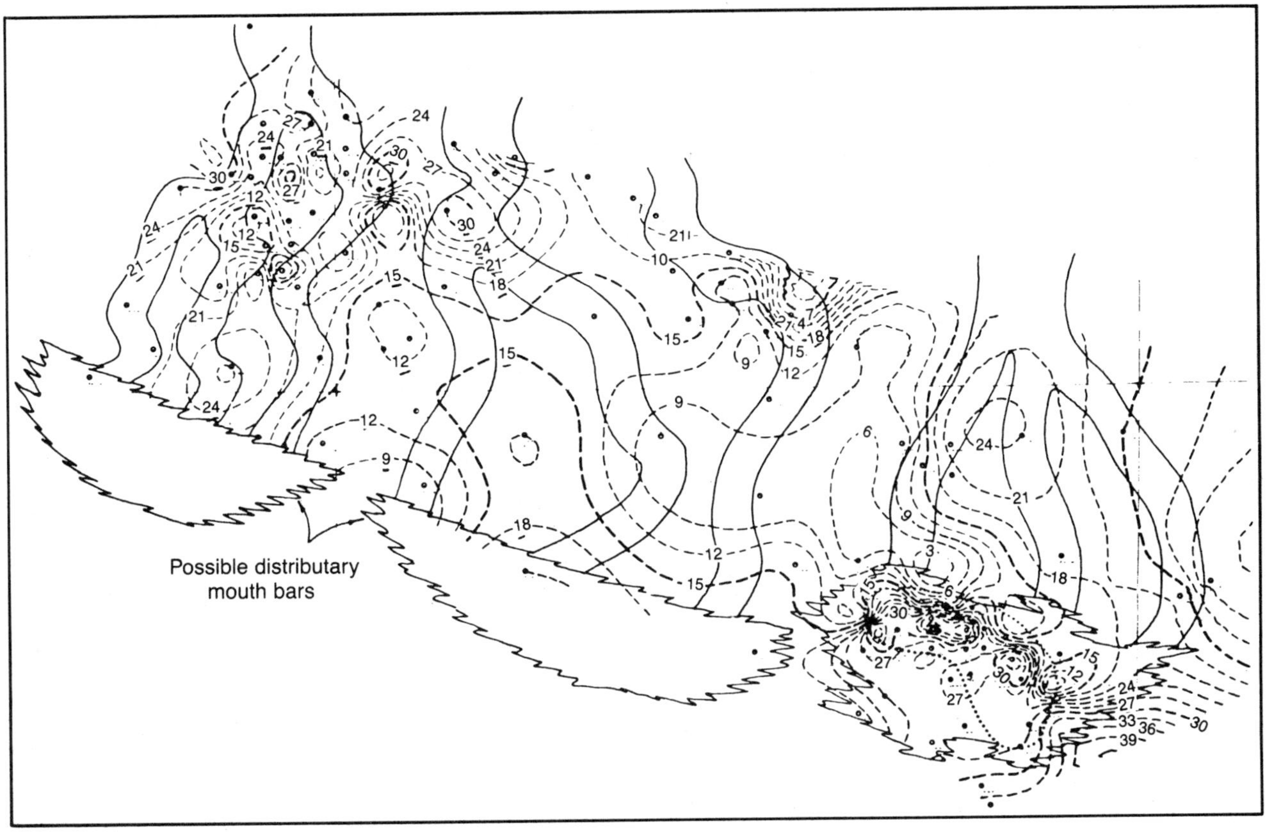

FIG. 9–31e

Pilot project paleogeologic map of MA sand (after Rouff, "A Technique for Interpreting Depositional Environments of Sandstones from the SP Log Utilizing the Computer," *Log Analyst,* volume 17, No. 4, 1976)

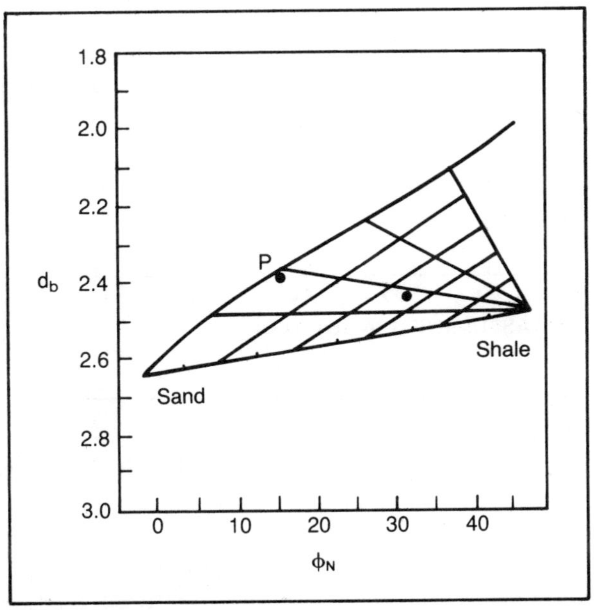

FIG. 9–32

Hypothetical density-neutron cross plot

Density:

$$\rho_b = \rho_s V_s + \rho_{sh} V_{sh} + \rho_w V_w \tag{9.31}$$

Acoustic velocity:

$$\Delta t = \Delta t_s V_s + \Delta t_{sh} V_{sh} + \Delta t_w V_w \tag{9.32}$$

Resistivity:

$$F = a V_w^{-m}, \tag{9.33}$$

Where:
a = Formation factor
m = Cementation

Gamma-ray counting rate:

$$V_{sh} = \frac{\gamma_s - \gamma}{\gamma_s \quad \gamma_{sh}} \tag{9.34}$$

Neutron:

$$ND = A + B \log V_w \tag{9.35}$$

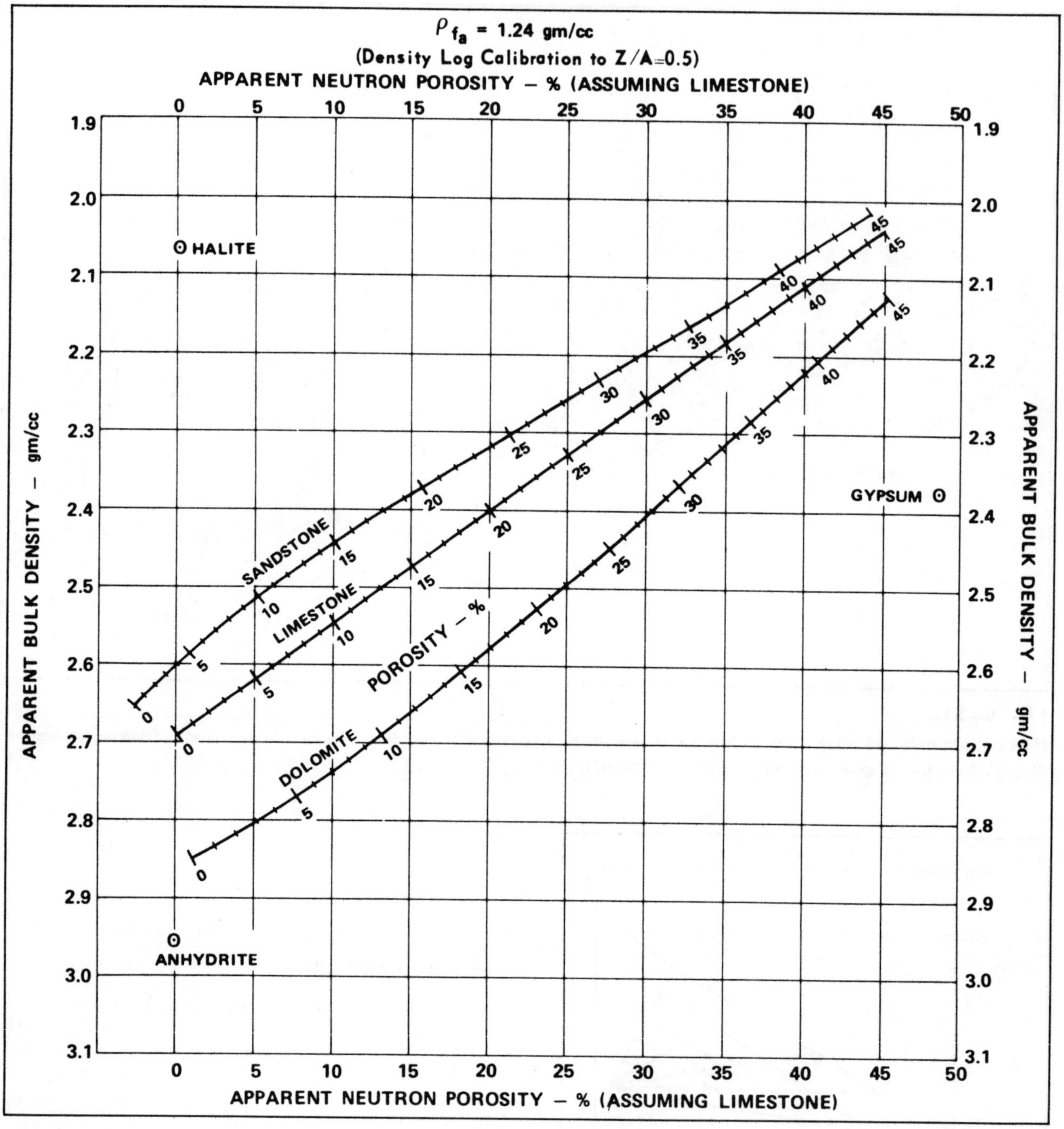

FIG. 9–33

Bulk density-neutron crossplot (courtesy Gearhart Industries Inc.)

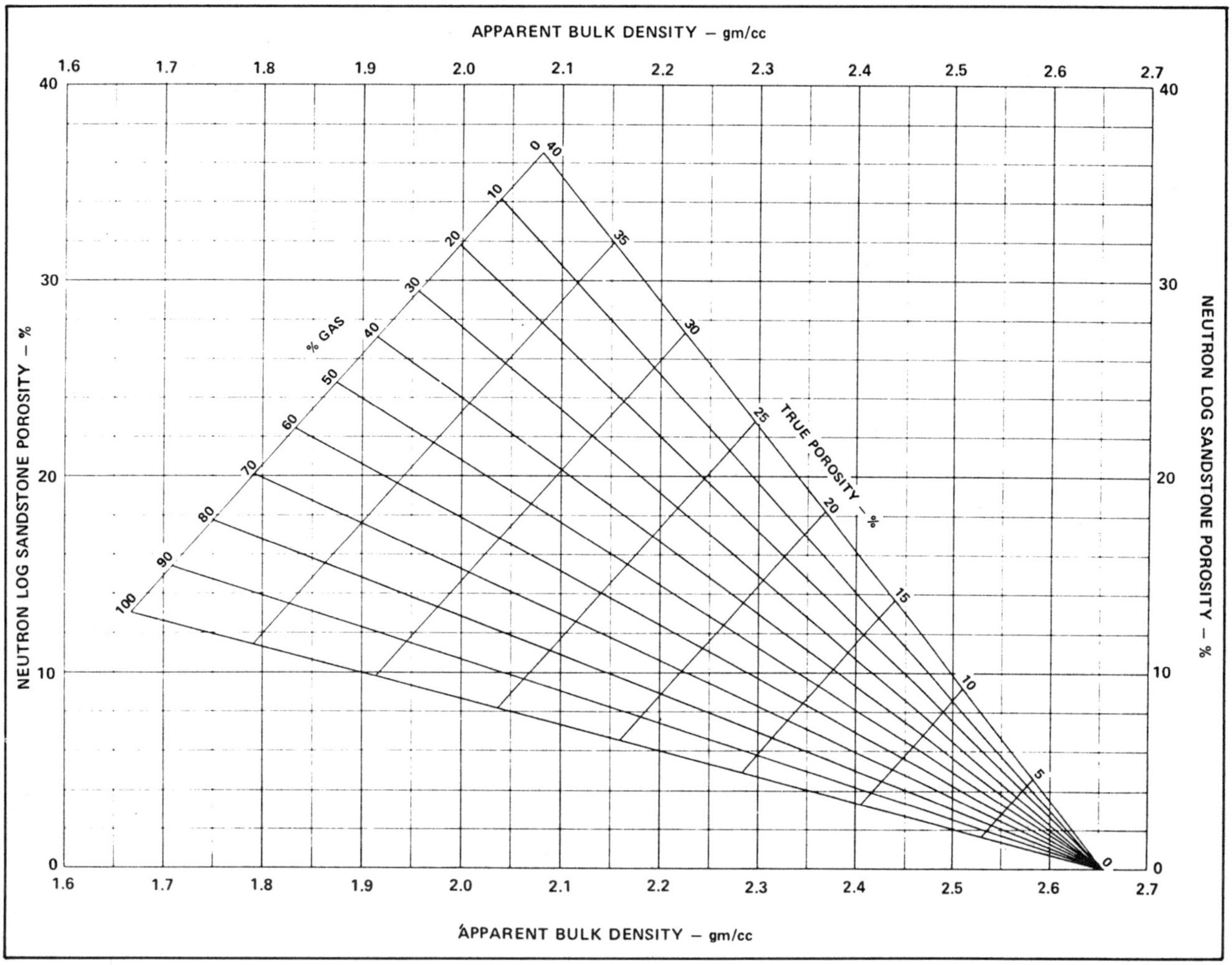

FIG. 9–34a

Gas detection and porosity determination in sands from density-neutron log crossplotting (courtesy Gearhart Industries Inc.)

Where:

ND = Neutron deflection

A and B = Constants that depend on the tool design

Thus, there are five possible equations in the listing besides Eq. 9.30. Theoretically, these could be used to solve for up to six different components.

If the logging unit being used is a computer based unit, the first equation (Eq. 9.20) can be plotted in log form to give a log of probable lithology. An example is shown as Fig. 9–40.

The logging systems must be chosen with care. A single-point resistivity would not give realistic values, for example, because the nonlinearities of the single point are too great, especially in the high resistivity of the coal. An omnidirectional density system cannot be effectively quantified and would not give a valid set of data for an equation. In some cases assigning values is difficult. The density and resistivity of the carbon component of coal are difficult to determine. One very effective computer technique, which will also work manually, is to assign values that are good estimates and correct them later when measured values are available from cores.

GENERAL CONSIDERATIONS

General considerations involve the miscellaneous interpretation and evaluation methods such as map-

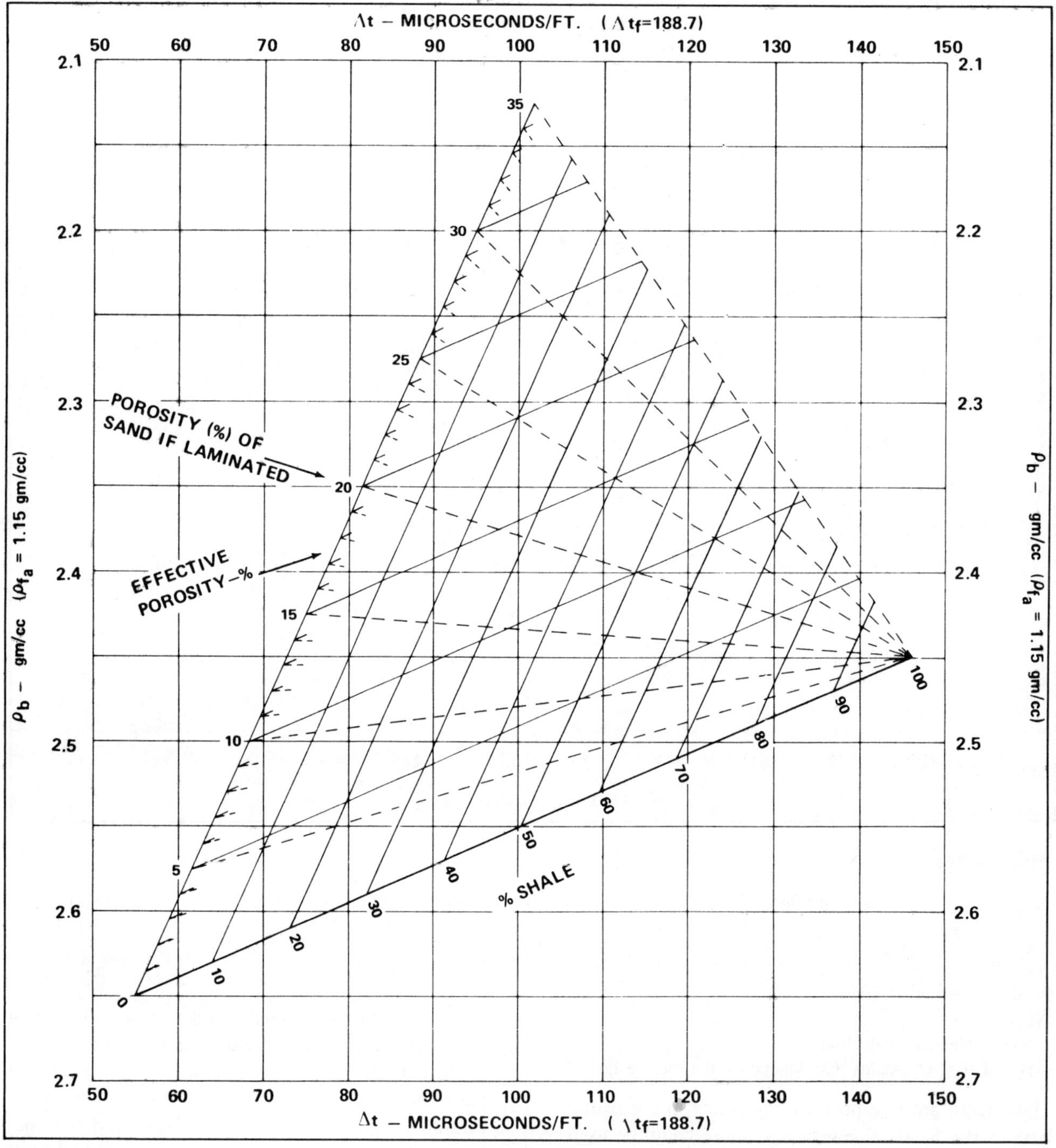

FIG. 9–34b

Shale amount and porosity determination from density and acoustic log crossplotting (courtesy Gearhart Industries Inc.)

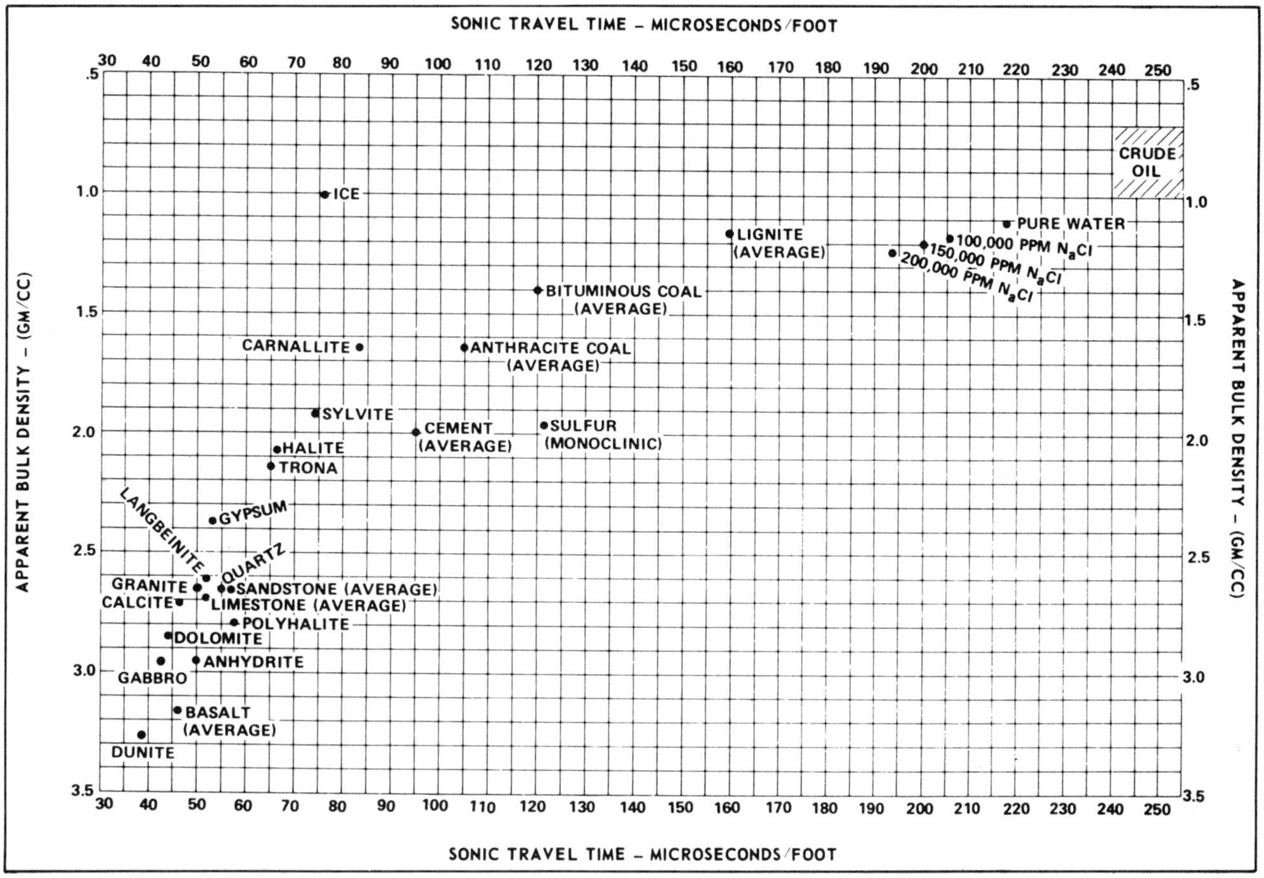

FIG. 9–35

Sonic-density cross plot (lithology), density log calibration to Z'A = 0.5 (courtesy Gearhart Industries Inc.)

ping and reserve calculations. Many such methods are available, but only these two are discussed in this text.

■ Mapping and Reserve Calculations

Mapping is a subject with which most geologists and geophysicists are familiar. Therefore, no attempt is made to go into it deeply. Rather, the discussion is limited to suggesting sources of information for several types of subsurface mapping.

Cross sections of a project, referred to elevation, are an excellent device for evolving isobase maps. These can also be used for isopach maps. The most easily used curves for these purposes are the SP, resistivity, and gamma ray. Isoclinal maps can be determined from the dipmeter logs. These can be valuable in analysis of the depositional history of an area. In fluid work, especially for petroleum reservoirs, isoclinal maps from dipmeter logs can aid in pinpointing traps. Isobathic and isopiestic information can be obtained from a combination density, neutron, and resistivity logs. Again, there are many possibilities to use logging techniques for mapping purposes. Many of them do not need highly calibrated values. Often qualitative and, thus, inexpensive logs can be effectively used.

One of the most useful and widely used techniques in the mineral industry is the use of logs to calculate the mineral reserves of a project and plan the mines, mining, and production of the ore. Probably the first step after deciding a discovery is a viable one is to make an isolith map of the mineral deposit. Initially, this shows where information is lacking. During the development stage of the project, this missing information can be obtained and filled in. At the end of the development stage, a project map showing ore body thicknesses, grades, and depths should be available. The probable amount of mineral can be calculated in

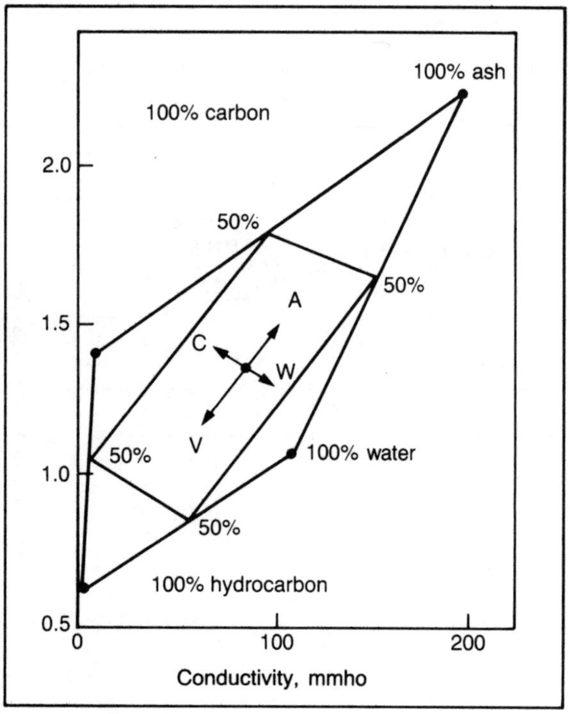

FIG. 9–36

Cross plot for four-component determination in coal

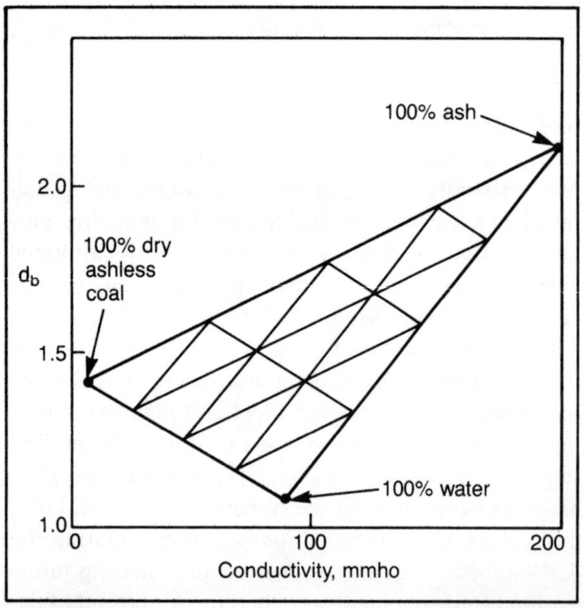

FIG. 9–37

Cross plot for three-component determination in coal

several ways. The ore body can be outlined at the cutoff-grade and grade-thickness products. An average grade for the area can be determined. Then the average thickness times the area, times the grade, times the average formation density gives the total amount of mineral.

A second technique is to use the polygon method. This method assigns a probable polygon of influence to each hole. Mineral values are determined and summed. The actual distances of the zones of influence can be varied, but customarily, a polygon around an isolated hole would be a square 50 ft on the side with the hole in the center for proven ore, 70.7 ft on a side for probable ore, and 100 ft on a side for inferred ore. This is illustrated in Fig. 9–41. If the drillholes are closer together than the radius of the polygon, the distance for a zone of influence is taken as half the distance to the next hole. The side will be normal to a line connecting the two holes. Therefore, a polygon may have more than four sides. The area of the polygon is A_p:

$$A_p = \Sigma A_T = \Sigma [s(s - a)(s - b)(s - c)]^{1/2} \tag{9.36}$$

Where:
a, b, and c = Sides of triangles illustrated in Fig. 9–41

The number of pounds of mineral available is the sum of all of the individual contributions.

The amount of ore available is the area times the thickness of the mineralized zone in cubic feet or yards. The weight of ore is the volume times the density. The volume of the overburden to be removed is the sum of the depth times the area of each hole. The amount of water to be removed by dewatering pumps is the area times the sum of thicknesses of the porous zones times the porosity of each porous zone. The permeability of the porous zone times the peripheral area of each porous zone allows an estimate of water replacement rate. The shear modulus times the shear vector times the accumulated mass above a level allows a calculation of the maximum slope of the pit wall.

Young's modulus can be used in a beam calculation to determine roof strength and the bulk modulus can be used to calculate pillar strength for underground mines. The porosity and permeability allow an estimate of the seepage into the underground mine. There are many other values that can be calculated or estimated from the logs, but those mentioned give an idea of some of the work that can be done.

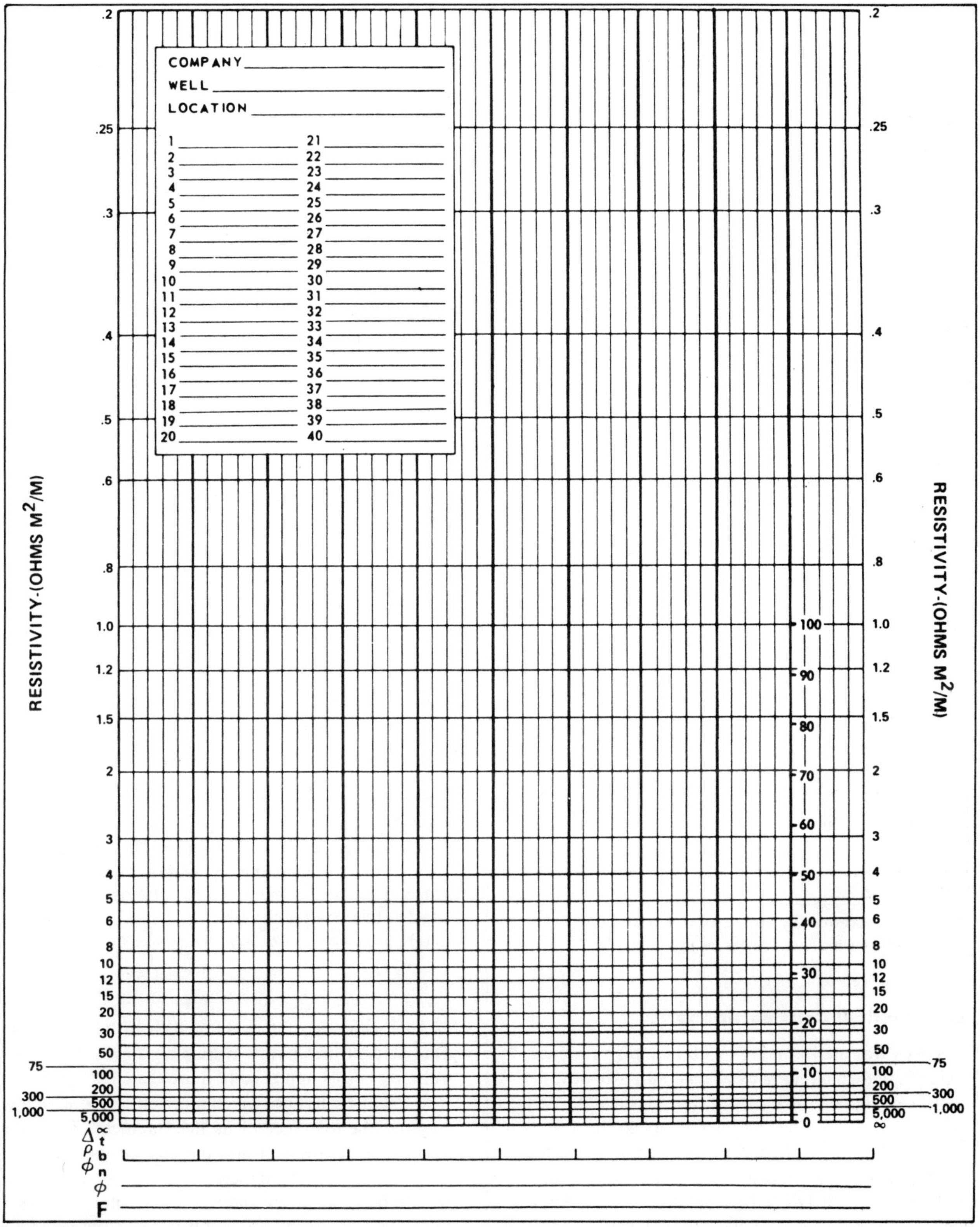

FIG. 9–38

Resistivity vs. Δ_t, ρ_b, ϕ_n, or ϕ (courtesy Gearhart Industries Inc.)

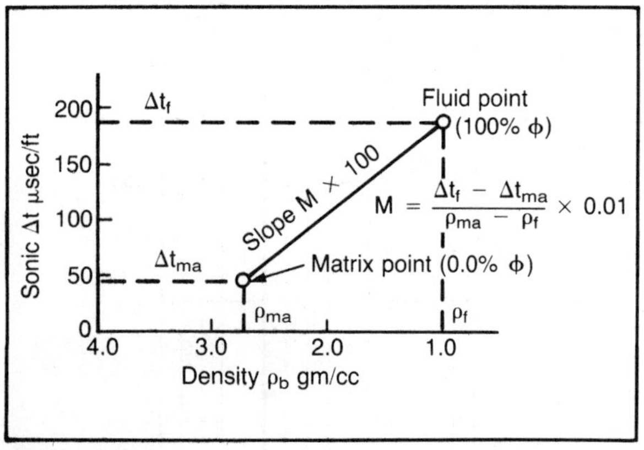

FIG. 9–39a

M from acoustic-density crossplot of mineral "A"
(courtesy Schlumberger Well Services Inc.)

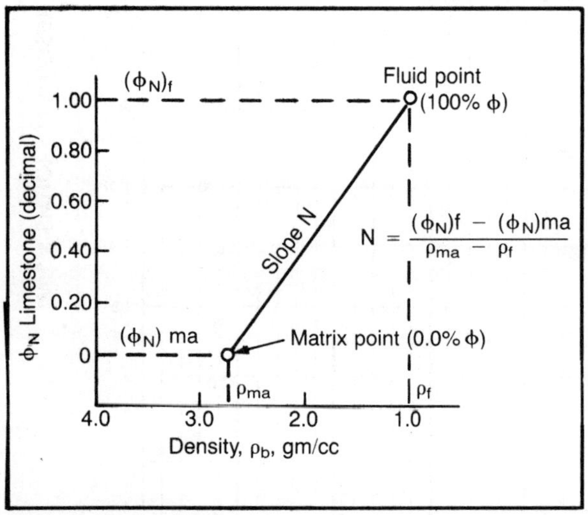

FIG. 9–39b

N from neutron-density crossplot of mineral "A"
(courtesy Schlumberger Well Services Inc.)

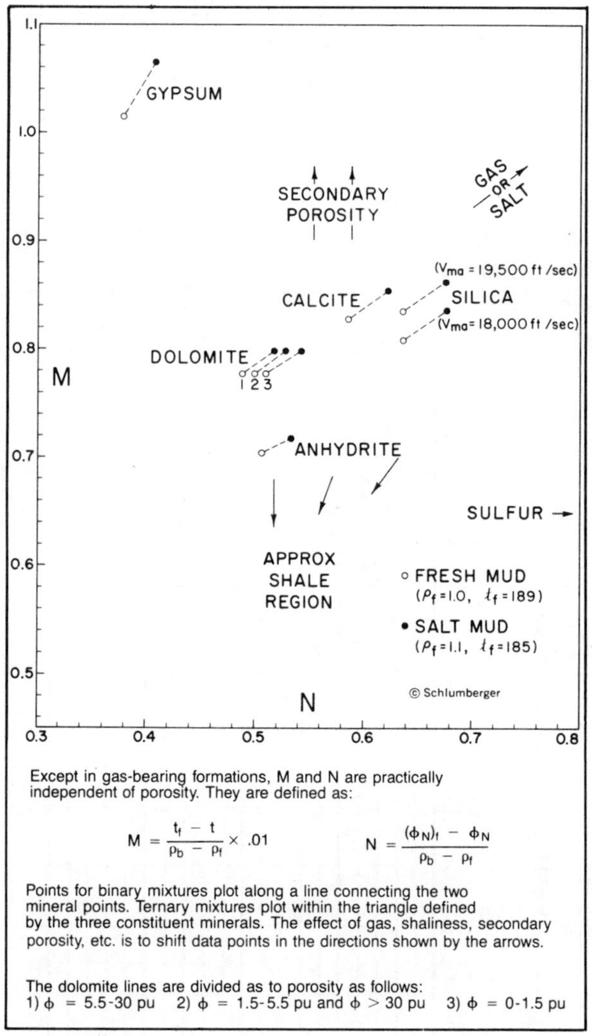

FIG. 9–39c

M-N* plot for mineral identification. This crossplot may be used to help identify mineral mixtures from sonic, density, and neutron logs. The neutron log used in this chart is the SNL*
(*mark of Schlumberger, courtesy Schlumberger Well Services Inc.)

Statistical Considerations

Since we are working with sample volumes, weights, and counting rates of random processes, the statistical characteristics must be considered. Any measurement made has an uncertainty to it. A voltmeter, for example, may have an accuracy of ±5% of full scale. If we measure the power-line voltage with that meter at 117 v, rms, we do not really know whether the real value of the line voltage is 111 v, 123 v, or something in between. The uncertainty of our measurement is ±5%.

The accuracy of the voltmeter can be improved. One-percent meters are available. However, a ±1% voltmeter costs appreciably more than a ±5% meter. The 1% meter has had much more effort spent on it to improve the tolerance of its measurement. Thus, the cost of the instrument and the cost of the measurement increases with an increase in accuracy. We must decide how much the increase in accuracy is worth to our measurement. If a motor and light bulbs with a ±10% tolerance are to be used, it is a waste of money to measure the voltage to ±1%.

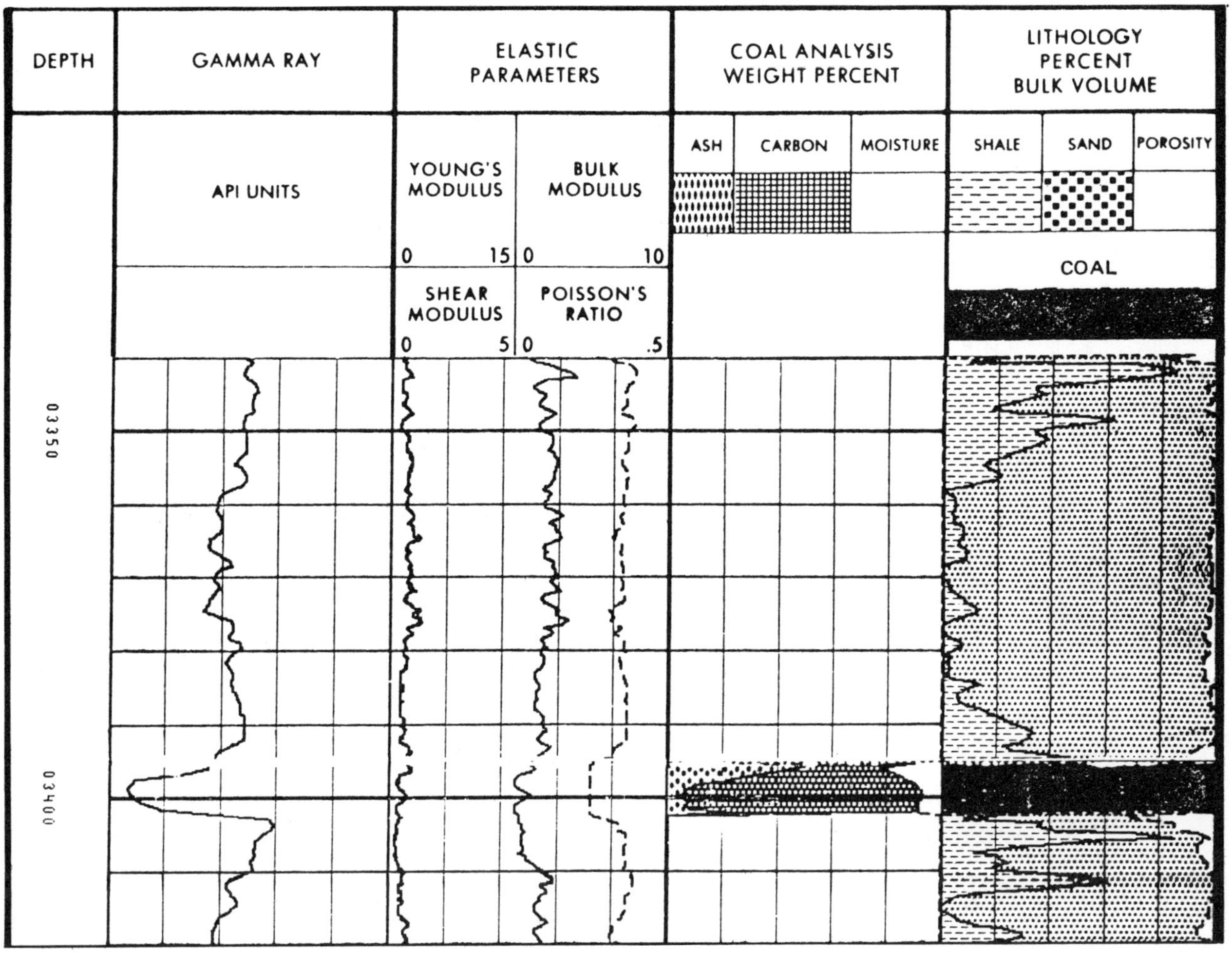

FIG. 9–40a

Computer determined lithology (after Kowalski, SPE, 5503)

This same principle is true with logging measurements. We are never able to make absolutely accurate measurements. We must determine what accuracy we need to do a particular job. Then we must determine the degree of accuracy of the measurement. If two or more measurements are to be compared, the relative accuracies must also be compared. To a great extent you must rely on the logging contractor for much of the accuracy information for individual measurements.

For radioactive measurements there are several assessments of the uncertainty we may use. Because radioactive measurements are made of random events, we must handle the information statistically. If we wish to know a counting rate under specific conditions, the arithmetic mean must be used. If individual rates, N, are taken from a curve or counter, the arithmetic mean is the sum of the rates divided by the number of entries, m:

$$\overline{N} = \frac{N_1 + N_2 + \cdots + N_m}{m} = \frac{\Sigma N}{m} \qquad (9.37)$$

The accuracy with which we can determine the value of $\overline{N}$ is the standard deviation, σ:

$$\sigma = \sqrt{\frac{\Sigma(N - \overline{N})^2}{m}} \qquad (9.38)$$

If a number of random events or particles is counted in any time interval, the expected deviation, D, from the true or mean value, $\overline{N}$ is:

$$D = (\overline{N})^{1/2} \qquad (9.39)$$

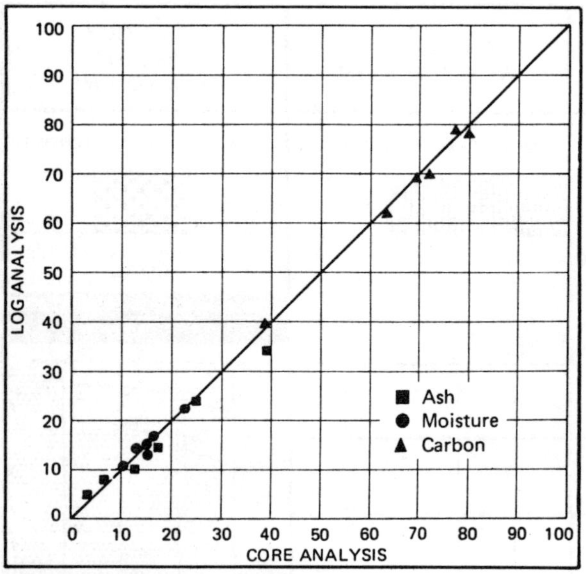

FIG. 9–40b

Computer-determined coal analysis

(after Kowalsi, SPE, 5503)

The probable error, PE, is defined as:

$$PE \simeq 0.67\,(\overline{N})^{1/2} \qquad (9.40)$$

The relative probable error, RPE, is:

$$RPE = \frac{0.67\,(\overline{N})^{1/2}}{\overline{N}} = \frac{0.67}{(\overline{N})^{1/2}} = 0.67\,(\overline{N})^{-1/2} \qquad (9.41)$$

If an analog system is scaled in counts per second, N_c, then the averaging of the time constant, T, in seconds must be accounted for:

$$RPE = 0.67\,(\overline{N}_c T)^{-1/2} \qquad (9.42)$$

The value of N, the actual number of counts, or N_c, the number of counts per second, must be corrected for the possibility that, because of their random nature, two or more events occurred closely enough together that the circuitry could not separate them:

$$N = \frac{\mu}{1 - \mu n} \qquad (9.43)$$

Where:

n = Recorded count or counting rate

μ = Dead time, sec

Sample size affects all measurements and comparisons. This is true of a comparison of log and core results. For example, a gross-count gamma-ray measurement examines a sphere about 1.5 ft in radius, r_γ. This is a volume, V_γ, of:

$$V_\gamma = 4/3\pi\, r_\gamma^3 \qquad (9.44)$$

Then V_γ is 24,429 cu in. If this is compared with a 1-in. diameter core, the comparison should be made using a core length, l, the same as the diameter of the logged volume diameter, 3 ft. This core has a volume, V, of:

$$V_c = \pi\, r_c^2 l \qquad (9.45)$$

Thus, V_c is 56.55 cu in. Since the volume of the gamma ray, V_γ is 432 times larger than the core volume, V_c, a variation of the core values of approximately $(432)^{1/2}$ greater than the gamma-ray values is to be expected. Thus, any particular core value should show a variation in porosity, uranium grade, or any other component of 21 times greater than the gamma-ray variations. For example, if the gamma-ray counting rate is 4,400 counts/sec, the relative probable error is slightly less than ±1%. If the gamma-ray uranium grade with a disequilibrium of 1.0 is 0.048% for that counting rate, it could fall between 0.047–0.049%. The core value might fall between 0.040–0.058% and still be a valid comparison. Sample size and relative sample size must be considered.

The purpose of running geophysical logs, of course, is to extract information from the subsurface environment. Like instrumentation methods in other fields, logging requires core, calibration, and interpretation. Also, as with any dynamic field, measurement and interpretation methods will change. Older, proven methods will be modified. New methods will be added. To get the maximum value from these methods, an engineer must stay up to date on current practices.

REFERENCES

"Borehole Logging for Uranium Exploration." Technical Reports Series, No. 212, IAEA, 1982.

Curry, William H. III. "Type Section of the Wall Creek Sandstone Member of the Frontier Formation." *Earth Science Bulletin,* volume 12, No. 2, June 1979.

Edmundson, H., and Raymer, L.L. "Radioactive Logging Parameters for Common Minerals." SPWLA 20th Annual Logging Symposium, 1979.

Formation Evaluation Data Handbook. Gearhart Industries Inc., 1974.

Fundamentals of Dip Meter Interpretation. Schlumberger Well Services, 1970.

Log Interpretation Charts, Dresser Atlas, Dresser Industries Inc., 1979.

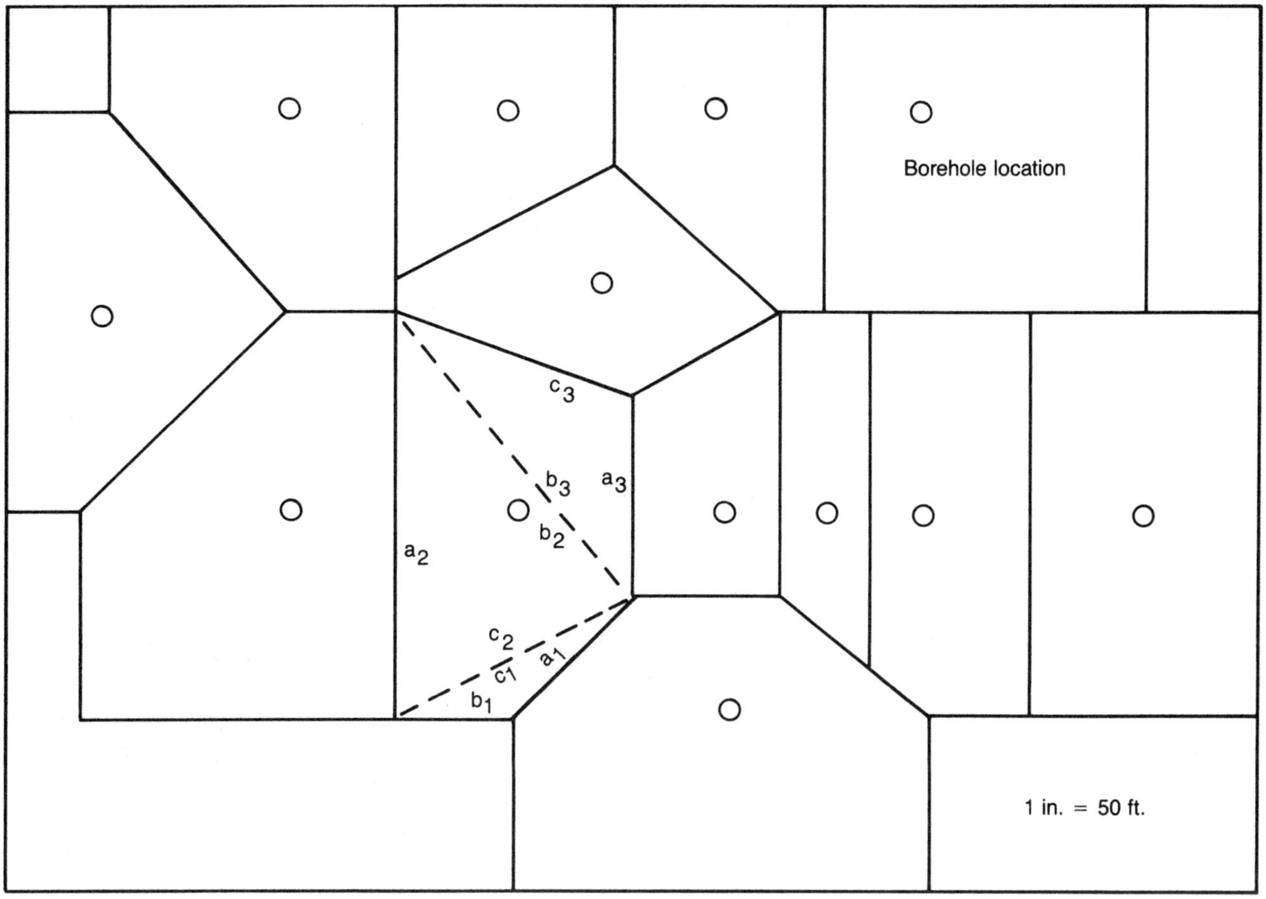

FIG. 9–41

Polygon map for reserve estimate

Mowla, Reda A., and Al-Saadoon, Faleh T. "Normalization Techniques and Interpretive Practices of Relative Permeability Curves of Reservoir Rocks." SPWLA 19th Annual Logging Symposium, 1978.

Moore, C.V., and Kaufman, R.L. "Your Unsuspected Problems: Fluid Resistivity and Water Analysis." 22nd Annual SPWLA Logging Symposium, 1981.

Pirson, Sylvain J. *Handbook of Well Log Analysis*. Prentice-Hall, 1963.

Rigby, Fred A. "Fracture Identification in an Igneous Geothermal Reservoir, Surprise Valley, California." SPWLA 21st Annual Logging Symposium, 1980.

Rubin, Bruce. "Uranium Roll Front Zonation in the Southern Powder River Basin, Wyoming," *Earth Science Bulletin,* volume 3. No. 4, December 1970.

Schlumberger Log Interpretation/Principles. Schlumberger Well Services, 1972.

Schlumberger Log Interpretation/Applications. Schlumberger Well Services, 1974.

Schlumberger Dip Meter Interpretation. Schlumberger Well Services, 1981.

Scott, James H. "Borehole Compensation Algorithms for a Small Diameter, Dual Detector Density Well Logging Probe." SPWLA 18th Annual Logging Symposium, 1977.

Scott, James H. "Pitfalls in Determining the Deadtime of Nuclear Well-Logging Probes." SPWLA 21st Annual SPWLA Logging Symposium, 1980.

Wells, L.E.; Sanyal, S.K.; and Mathews, M.A. "Matrix and Response Characteristics for Sonic, Density and Neutron." SPWLA 20th Annual Logging Symposium, 1979.

Wittstrom, Martin D. Jr., and Hagemeier, Marvin E. "A Review of Little Knife Field Development." *Earth Science Bulletin,* volume 11, No. 4, December 1978.

Zoltan, Barlai. "Some Theoretical Aspects of How to Correlate Well Log-Evaluated Data to Core-Determined Reservoir Properties," SPWLA 20th Annual Logging Symposium, 1979.

Appendix

Table A–1
Uranium Logging

Log Type	Measurement	Uses	Combinations
Natural gamma ray; gross count (GCGR)	Total radiation of gamma rays	Sand/shale sequences; effective mineral-grade correlation; lithology	Resistance; resistivity; neutron porosity; density; acoustic velocity
Spontaneous potential (SP)	Imbalances of ion concentrations in fluids	Sand/shale; permeability; relative or absolute salinity; total dissolved solids	Mud resistivity; formation resistance or resistivity; GCGR
Single-point resistance	Relative resistance of formations; sensitive to formation fluids	Sand/shale approximate lithology; stratigraphy; correlation	GCGR; SP
Focused resistivity	Absolute resistivity of formations; deep penetration plus symmetrical detailed response	Lithology; R_t stratigraphy; R_w porosity	GCGR; mud resistivity; SP; neutron porosity; density; acoustic velocity
Neutron porosity	Neutron moderation; capture cross-section of thermal neutrons	Lithology; porosity; resistivity substitute	Acoustic velocity; resistivity; density
Delayed or prompt fission neutrons (DFN or PFN)	Fission neutrons	Direct measurement of uranium content	GCGR; neutron porosity; density
Density	Scattered gamma rays; electron density	Formation bulk density, porosity, and lithology	GCGR; neutron; acoustic velocity; DFN; PFN
Deviation	Gravitational, magnetic fields; strengths and directions	Hole deviation; bed location, depth, and thickness	GCGR; resistivity; dipmeter
High-resolution dipmeter	Focused resistivity, 4 measurements at 90°	Determine bed dip and fracturing; geological history	GCGR; resistivity; deviation; acoustic velocity
Gamma-ray spectrograph	Characteristic energy peaks of gamma radiation	Determine uranium-thorium ratio; evaluate K content; depositional history; quantitative uranium	GCGR; resistivity; fission neutron

Table A–2

Coal Logging

Log Type	Measurement	Uses	Combinations
Density	Scattered gamma rays; electron density	Definitive of presence of coal; coal quality; porosity	Caliper; GCGR; resistivity; neutron; acoustic velocity
GCGR	Total gamma radiation	Lithology; ash content; shale content; correlation	Resistivity; density
Focused resistivity	Formation resistivity and symmetrical detail respectively	Formation resistivity; quality of coal; thickness; correlation	GCGR; density
Neutron porosity	Neutron moderation by hydrogen	Porosity; water content; R substance	GCGR; density; acoustic velocity
Deviation	Gravitational and magnetic field vectors	Hole deviation; bed thickness	GCGR
Acoustic velocity	Elastic properties	Definitive for presence of coal; overburden mechanical parameters; coal quality; porosity	Density; neutron; GCGR
Neutron activity	Capture gamma-ray peaks	Elemental analysis of formations	GCGR; resistivity
SP	Ion imbalances	R_w, TDS, shale content	Resistivity; GCGR

Table A–3

Engineering Logging

Log Type	Measurement	Uses	Combinations
Acoustic velocity	Elastic properties	Mechanical parameters; porosity; fracture	Density; GCGR; resistivity
Density	Electron density; scattered gamma rays	Bulk density; mechanical parameters; porosity; formation competence	Acoustic velocity; GCGR; resistivity
Resistivity	Resistivity of fluid properties	Correlation; formation competence; R_w; porosity	GCGR; neutron; acoustic velocity
GCGR	Total gamma-ray emission	Correlation; lithology; shale content	Acoustic velocity; density; resistivity
SP	Ion imbalances in fluids	Salinity; shale content	GCGR; resistivity; porosity
Deviation	Gravitational and magnetic field vectors	Depth and thickness correction; hole location	Dipmeter
High-resolution dipmeter	Focused resistivity in four directions	Formation dip; true thickness; geological history; fracturing	

Table A–4
Useful Formulas

SP:

$$SSP = -K \cdot \log_{10} (R_{mf}/R_w)$$
$$K = 0.133°F + 60°F$$
(For a clean sand; make a later correction for shale)

F:

$$F = R_o/R_w$$
(F decreases as R_w increases in the case of shaly sands and formations containing conductive solids)

$$= \phi^{-m}$$
(Archie)

$$= a/\phi^m$$

$$= R_{xo}/R_{mf}$$
(only if there is no residual hydrocarbon saturation)

ϕ_{sonic}:

$$\Delta t = \Delta t_m + B\,\phi$$

$$\Delta t = \phi \Delta t_f + (1 - \phi)\,\Delta t_m$$
(Wyllie time-average equation)

$\phi_{density}$:

$$\rho_b = \phi \cdot \rho_f + (1 - \phi)\,\rho_m$$

ϕ_{TDT}:

$$\phi = \frac{\Sigma_{ma} - \Sigma}{\Sigma_{ma} - \Sigma_f}$$

Q factor:

$$Q = \frac{\phi_{sonic} - \phi_{density}}{\phi_{sonic}}$$

Nuclear logging, analog only:

$$\text{Lag, ft} = \frac{\text{TC} \cdot \text{logging speed, ft/min}}{60}$$

$$\text{Probable error, counts} \cong 0.67(n)^{1/2}$$

Uranium logging:

$$G = KA/h$$

Table A–5
Energy Unit Conversion Chart*

Natural Gas, cu ft**	Oil, bbl	Bituminous Coal, short ton	BTU	Electricity, kwh
—	—	—	1	0.000293
1.0	0.00018	0.00004	1,000	0.293
3.41	0.00061	0.00014	3,413	1
1,000	0.18	0.04	1 million	293
3,413	0.61	0.14	3.41 million	1,000
5,600	1.0	0.22	5.6 million	1,640
25,000	4.46	1.0	25 million	7,325
1 million	180	40	1 billion	293,000
3.41 million	610	140	3.41 billion	1 million
1 billion	180,000	40,000	1 trillion	293 million
1 trillion	180 million	40 million	1 quadrillion	293 billion

*Based on the following nominal fuel heating values:
1 cu ft of natural gas = 1,000 BTU
1 bbl of crude oil = 5.6 MM BTU
1 lb of bituminous coal = 12,500 BTU

**Substitute natural gas (SNG) and liquefied natural gas (LNG) have approximately the same heating value

Glossary

absolute temperature °F plus 460.

alpha particle A particle that is a disintegration product of the nucleus of an atom that is identical to helium nuclei, being a cluster of two protons and two neutrons possessing a positive charge of two units.

API Abbreviation for American Petroleum Institute.

apparent value The uncorrected curve value read directly from the log.

atomic number Z, the number of orbital electrons surrounding the nucleus of a neutral atom of the specific type in question or the number of protons in the nucleus.

atomic weight A, the relative weight of the atom on the basis of oxygen being 16. It is equal to the total number of protons plus neutrons in the nucleus.

azimuth The number of degrees, looking from the top of the tool, measured clockwise between magnetic north and the direction of the dipmeter's #1 pad.

background radiation The radiation intensity (flux) caused by the natural environment at the surface, which is in addition to the radiation directly under consideration.

barn The customary unit for the absorption cross section equal to 10^{-24} cm^2 the order of magnitude of nuclear areas.

beta particle A negatively charged particle identical to an electron but a product of nuclear disintegration.

BTU Abbreviation for British thermal unit (252 cal).

bulk density The in situ density of a material, including any pore fluids present during the time of measurement.

calibrate To adjust the scale of a logging system into appropriate meaningful units.

calorie The amount of heat necessary to raise the temperature of 1 g of water 1°C from 14.5°C to 15.5°C (3.97×10^{-3} BTU).

centralized Positioned in the center of the hole by use of centering devices.

circulation The forced flow of fluid into a drillhole, usually down the drillpipe, casing, or tubing and up the annulus to return to the surface.

conductive solid Any material within a formation that contributes to formation conductivity without contributing to effective porosity (i.e., porosity related to permeability).

curie That quantity of a radioisotope required to supply 3.70×10^{10} disintegrations per second.

curve The continuous recording on a log mat, in the form of an unbroken or broken line of information supplied by a subsurface logging tool.

dead time The time interval, usually in microseconds, between the original ionizing event and the instant of return of the system to readiness to count another ionizing event.

decentralized Positioned against the side of the borehole wall by either gravity or a decentralizer device.

density Mass per unit volume.

density, bulk The in situ density of a material including any pore fluids present during the time of measurement.

density, matrix The density of the constituent solid particles of a substance (formation), i.e., the substance free of pore space.

departure curve Curve in graphical form that expresses the effects of nonstandard conditions on curve response (e.g., hole diameter, bed thickness, invasion, and mudcake thickness)

dielectric constant A measure of the relative ability of a material to store electrical charge for a given applied field strength.

displacement (general) The distance between the tool zeros for the various recorded parameters of a single log. Also, the correction in feet or meters necessary to record all curves at their proper depth on the log.

dose The radiation delivered to a specified area or volume or to the whole body.

dose, maximum permissible The maximum dose of radiation that is safely expected to cause no appreciable bodily injury.

drift angle The number of degrees between the vertical and the tool axis.

erg 2.39×10^{-8} calorie.

formation signal The signal caused by the formation with the tool at a given level within the hole as opposed to signals generated from the tool itself or coming from the borehole fluids only.

frequency The number of cycles of waves completed in a unit of time (generally cycles per second).

gram molecular weight Mass in grams numerically equal to the molecular weight of a substance (gram mole).

half-life The time required for a radioactive material to lose one-half of its radioactivity by decay.

half value thickness That thickness of a material that will absorb one-half of the incident rays of a stated energy level.

instrument calibration zero The chart position during calibration representing zero signal generated by the subsurface tool (may or may not be the same as curve zero).

linear absorption coefficient A factor expressing the fraction of a beam of X or gamma radiation absorbed in a unit thickness of material.

log The printed well record (log) including log heading, calibration record, and survey record. Also, the actual recording of the above information at the well.

log zero The position of the recording trace on the chart grid corresponding to no signal from the tool (zero resistivity, zero conductivity, zero radioactivity, etc.).

mass absorption coefficient The linear absorption coefficient per centimeter, divided by the density of the absorber in grams per cubic centimeter.

matrix density The density of the constituent solid particles of a substance (formation), i.e., the substance free of pore space.

maximum permissible dose The maximum dose of radiation that is safely expected to cause no appreciable bodily injury.

maximum pressure rating The maximum recommended pressure for satisfactory operation under conditions of normal wellbore temperature.

maximum temperature rating The maximum borehole temperature at which the logging tool used under normal conditions of wellbore pressure is expected to function properly.

minute mark Mark recorded on the far left side of the log at one minute intervals during the logging operation to document the logging speed.

molality The gram-molecular weight of a solute in 1,000 g of solvent; molality of a solution is the same at all temperatures, whereas concentrations on a volume basis vary.

molarity The number of moles (gram-molecular weights) of solute per liter of solution.

molecular weight The sum of the atomic weights of all the atoms in a given molecule.

molecular volume The volume occupied by 1 g molecular weight; equals gram molecular weight/density.

neutron An electrically neutral elementary nuclear particle with a mass of 1.0089.

normal curve Electric log curve that is symmetrical in response and made with a four-electrode configuration where AM is small compared to AB.

previous run Any log run made during a different trip into a well and before a subsequent log run made in the same borehole.

relative bearing The number of degrees, looking from the top of the dipmeter tool, measured clockwise between the drift angle direction and the direction of the #1 pad.

reluctance The property of a magnetic circuit that determines the total magnetic flux in the circuit when a given magnetomotive force is applied.

rem A measure of the dose to body tissue of any ionizing radiation, in terms of its estimated biological effect relative to a dose of 1 roentgen of X rays.

repeat section A short section of log, generally representing 200 ft of hole, that is recorded in addition to the main logging run to provide an interrun comparison of log repeatability (i.e., curve stability).

roentgen A unit measuring the quantity of energy transmitted by a photon to air in its path. That quantity of X or gamma radiation that under standard conditions (air at 20°C and 760 mm pressure) produces 2.083×10^9 ion pairs/cc of air; 2.083×10^9 ion pairs equals 1 electrostatic unit of charge. When two charges 1 cm apart repel with a force of 1 dyne, each charge is said to possess 1 esu of charge.

salt mud Mud in which the filtrate resistivity is not more than four times that of the formation water encountered.

shaly Including significant amounts of shale or clay.

sonde A general term used for any subsurface logging tool that carries electrodes or detectors into the borehole, e.g., induction sonde.

spacing The distance between a transmitter and detector on a given tool.

span The distance between two detectors used for a common measurement.

specific activity The amount of radioactive isotope present per unit weight of element, generally expressed as curies per gram.

specific gravity The ratio of the weight per unit volume of the material to the weight of an equal volume of pure water under standard temperature and pressure.

standard cubic foot of gas 1 cu ft of gas at 60°F and 1 atm pressure (14.7 psi).

stand-off The distance separating the tool from the borehole wall.

statistical check A recording (generally 3 min) of the environmental properties at a given depth in the hole to determine the statistical variations of the recording with time.

subsequent run Any log run made during a different trip into the hole and after a previous log run has been made in the same borehole.

survey The act of using subsurface tools to obtain formation on borehole geophysical data.

thermal conductivity The coefficient that multiplies the temperature gradient to give the rate of heat transfer by conduction expressed in heat energy crossing unit area in unit time.

thin bed A bed requiring a bed-thickness correction.

time constant The time lag between the presentation of a new environment to the tool to the recorder indication of a 63% change from the previous environmental value in the direction of the new environmental value.

total depth The total depth reached by the bottom of the logging tool during the survey as shown by the log.

Index

A

a, compaction constant, 22–23
ABS casing, 120
accelerator, 147
acoustic methods, 170–176, 220–221, 235
  caliper, 179
  travel time, 5, 170–176
  velocity, 6, 9, 28, 170–176, 391, 394, 395, 416, 425
Acoustic waves, 171–175
  holography, 176
  P-wave, compressional, pressure, 171–176
  S-wave, shear, 171, 175–176
  tube, 171
activation, gamma-ray, 144
activation, neutron, 153–159
activity, ion, 60–63
activity, radioactivity, 60
additives, mud, 12–13
adiabatic cooling, 189, 192, 194
adjacent bed effect, 39, 48
adsorption potentials, 61–64
AEC calibration, 127–130
AEC method, 212
airborne gamma ray, 125–126, 129
air drilling, 15
alpha particles, 92, 146
alpha (shale, SP), 64–66, 69
amplifiers, 106, 108
analog logging speed, 109
analog systems, 104–111, 115, 202–205
angstrom unit, 91
anions, 60–61, 69, 73–74
API, 115–116, 161–162
API, gamma ray, 115–116, 127–130, 211–212
API, neutron, 161, 162
Archie's Relation, 9, 22–23, 26, 219–220
area, cross-section, 20
arrays, electrode, 35, 42, 44, 52, 76
ash, 113, 114, 116, 209–210, 231–232, 240
assaying, 145
atomic activity, 91
atomic mass, mass units, 91–92, 98–99
atomic number, Z, 131–136, 144, 148–149
atomic weight, A, 91–93, 131–136, 147, 148–149
averaging, 111–112, 243–244
azimuth frequency polar plot, 77–78, 81

B

B, magnetic flux density, 166
barn, 149
bed boundary, 36–37, 43, 54–56, 61, 116–118, 210, 212, 216
bed boundary tool, 137–138
bed thickness, 19, 66, 212
Bernoulli effect, 16, 186
beryllium, 146–147
beta-gamma method, 145–146
beta particles, 91–93
BHC Sonic, 172
bimetallism, 74–75
Birdwell, 180
bismuth germanate, 104, 115
body dosage, 94–96, 100
borehole, 11–17, 176–186
  deviation, 182–186, 221–222
  televiewer, 176, 178–179
bottom-hole temperature, BHT, 192
bulk density, 95, 130–144
bulk modulus, 175, 243

C

cable, 34, 74, 107
cable termination, 107
calibration, 46–50, 52, 76, 115, 127–130, 143–144, 161–162, 176, 181, 193–194
caliper, 176–181, 221–222
Canada, 126–127
  Beaver Lodge, 68
  Rabbit Lake, 68
carbonates, 6, 113, 153, 175, 209
cased-hole logs, 159, 160–162
casing, 16–17, 112, 114–115, 119–121, 146, 160–161, 176
casing, ABS, 120
casing, PCV, 146, 161
catalog:
  acoustic velocity, travel time, 174
  density, 5–10, 133–136, 180
  gamma emission, 5–6
  hydrogen content, 148–149
  ion mobility, 61–62
  magnetic anomalies, 167
  magnetic susceptibilities, 165–166
  neutron cross section, thermal, 5–10, 133–136, 147–149
  potassium content, 5, 10, 114
  resistivity, 5, 21
  thermal conductivity, 188–189
cation, 60, 73
cation exchange capacity, CEC, 5, 64
cementation constant, 9, 22–23, 219, 223, 232–233
cement bond, 176
Century Geophysical, 184–185
charge, 91
chargeability, 84
charged particles, 21
chlorine log, 146
circuits, 28, 52–53, 55, 104–111, 189–191
clay, shale, 4–6, 13, 19, 21, 50–66, 112–118, 120, 142, 151, 153
closed can gamma assaying, 145
coal, 38, 116, 124, 137, 142, 209–210, 221, 231–232, 240, 243–244
collisions, 147–149
combinations, 222–239
complex resistivity. *See also* induced polarization, 45, 79–89
compressional (pressure or P wave), 170–176
compressive strength, 175
Compton scattering, 93–94, 103, 123, 131–132
computations, 98–99, 139, 142–143, 147, 155, 159, 165–166, 171–176, 179, 180–181, 185–186, 188–189, 192, 195–196, 209, 212–213, 217, 219–220, 225, 232–235, 240, 243–244, 248
computers, 77, 110, 124, 204–205, 212
concentration cell. *See also* diffusion, 60–64
conductance, 20
conductive solids, 5, 19, 21, 24
conductivity, 23–24, 60
core, coring, 13, 194–196, 232
core gamma, 195
correction factors, 16–17, 52, 63–66, 118–121, 150, 152
counting, counters, 109–110
crooked holes, 13, 185–186
crossplot, 28, 124, 227–234
cross sections, 149, 224–226

crystal, 102–104, 123, 131
cups, radon, 199
Curie, 90, 97, 137, 146
current, electrical, 19–21, 23–24, 28–30, 33–35, 40, 46, 52, 54–55, 58, 82–83
cuttings, 13–15, 196
cycle skipping, 176
Czubek, 121–122

D

Darcy's Law, 188
dead time, 101–103, 107–108, 112, 121, 136, 212, 244
delayed fission neutron, DFN, 16, 155–157
density:
- bed boundary, 137–138
- calibration, 143–144
- compensated, 138–140
- crossplotting, 142
- dual spacing, 138–140
- fluid, 140
- Litho-Density, 140–142
- omnidirectional, 136–137
- sidewall, 137–144
- single spacing, 137–138
- response, 137

density catalog, 6–9, 133–136
density log and density logging, 130–144, 235
departure curves, 32, 39–40, 48, 150, 153–154
depth of investigation, geometrical factor, 19, 30–31, 35–37, 41, 116–118
detectors, 99–104, 114–115, 123, 131, 136–139, 146, 150
deviation, 76, 182–186, 221–222
- continuous, 183–184
- inertial, 184–185
- multishot, 183
- photographic, 182–183
- radius of curvature method, 185–186
- station, 185–186
- tangential method, 185–186

dielectric constant, 71
differential SP, 75
differential temperature, 89, 190
diffusion potentials, 60–61
digital systems, 109–111, 202–205
dipmeters, 76–79, 80–83, 221–222
dipole (double dipole), 52
direct uranium measurements, 156–157
discriminator, 105–106, 131, 172
disequilibrium, 210–211
disintegration, 90
Dresser-Atlas, 77
drilling, 11–17, 194
- bits, 11–12, 194

Dunlap multiplier, 24–26

E

electric cone penetrometer, 198–200
electric induced polarization, EIP, 84
electric logging, 18–89
electrodes, 28–50, 52–53, 55–56, 68–69, 71–74
electromagnetic method, 167–168
electron, 92–95, 99
electron volt, eV, 91
elemental analysis, 146, 157–158, 159
energy discrimination, 105–106, 122–127, 157
Ertec Western, 198–200

F

factors:
- formation resistivity, F, 21–23, 28, 216–220
- casing, 120–121
- water or borehole, Fw, 118–120

Faraday's Number, 60, 70
faults, 225
ferric, ferrous iron, 66–67, 69
fission, 156–157
fluid density, 140
fluorescence, 144–145
flux gate, 167, 183–184
focused resistivity, 2–5, 37–39, 44–48, 219
formation, formation types, 4, 12–13, 21–24, 52–54, 67–69, 79, 83–84, 113–114, 173–175, 209, 227–237
Fourier's Law, 188
fractional volume, V, 114–116, 153
fracture, 6, 181
frequency, 203
- domain, 85
- modulation, 203

G

GAMLOG, 111, 121, 130, 195
gamma assaying, 145
gamma, magnetic, 167
gamma ray, 5–6, 93–98, 114–146, 150–151, 209–212, 235
gamma ray:
- activation, 144
- capture, 150–151, 157–160, 204
- intensity, 131, 144
- scattered. *See* density
- scattering, 131

gas, 23, 196–199
Geiger-Mueller detector, 101–102, 106
geological features and materials, 5–10, 23–24, 84, 142, 152–153
geometry, measurement, 19, 30–33, 36–37, 41, 44, 117–118, 151, 244
germanium detectors, 104, 124–128, 158
gold, 159–160
grain size, 4, 213, 216
grade, 110, 121–122, 195–196, 209–212
grounding resistance, 33–34
guard log, 37–39, 44–48

H

H, magnetic intensity, 166–168
half-life, 91, 97, 146, 156, 157
half value thickness, 95, 97
hard-rock environments, 38, 142
heading, 115
helium sniffer, 196–199
Helmholz' Law, 71
high resolution dipmeter, HRD, 76–79, 80–83
history, 1–3
hole size effects, 16–17, 21, 32–33, 39–40, 52, 63, 66, 150, 153

I

induced polarization, 45, 79–89
induction log, 39–41, 166, 219
inertial deviation methods, 184–185
instrumentation, 27–50, 39–41, 52–53, 55–58, 71–75, 104–111
intensity, 93–95, 116
interpretation methods, 206–245
- conventional, 207–222
- crossplotting, 207, 227–234

equation, 207, 209, 212–213, 217–220, 225, 232–237
initial data, 207–208
maps and mapping, 208, 216, 225, 227, 231, 233, 235, 239–240, 245
patterns, 224–227
single curve, 208–222
intrinsic germanium detector. *See also* germanium *and* ultrapure germanium detector, 104, 124
invasion, 11–15, 26, 38, 63
ion, 21, 24, 50, 60–64, 66, 72, 74, 87, 147, 213, 216
ion concentration, 21, 24
ionization, 147
ionization chamber, 99–101
isomagnetic lines, 167

K

K, conversion factor:
gamma-ray, 110–111, 212
IP, 86–87
relative magnetic permeability, 166
resistivity, 20–22, 24, 30–33, 37, 41, 44, 46, 50, 52
SP, 60–63, 213
Keller, 216, 220
Kelvin, 1, 189
KUT, 122–123

L

lateral resistivity, 34–37, 219
Laterolog. *See* focused resistivity
lifetime log, 146, 158–159
limestone, 38
linear absorption coefficient, 12
linear accelerator (neutron generator), 146–147
lithology, 146, 152
livetime log, 146
log suites, 2–5, 397–431
logging speed, 109
logs, 47, 51, 62, 72, 141, 184–186, 195, 199

M

m, cementation factor, 6, 9, 22–23, 219, 223, 232–234
magnetic methods, 165–169
field, 166–168
susceptibility, 165–167
magnetic permeability, 166
magnetization, 165–166
magnetotellurics, 167–168
mapping, 432–435
marine shale, 115, 124
mass absorption coefficient, 94–95
mass unit, 91
matrix, matrix density, 5–9
mechanical methods, 170–201
mercury sniffer, 196–199
metal factor, 85
microprocessors, 110
microresistivity devices, 45–46, 218
mobility, 52, 60–63
mud, 11–16, 140
mud additives, 12–13
mud cake, 12–14, 19, 45, 131, 142
mud-cake resistivity, 14, 24
mud filtrate resistivity, 14, 24–27, 60–63, 66, 68, 218, 219
mud resistivity, borehole resistivity, 19, 24, 38, 60–63
mud wave, 171–172
multiple curve methods, 222–227

N

N, counting rate, 108, 123
N, number of atoms, 91
n, recorded or uncorrected counting rate, 112, 121, 212, 244
n, saturation constant, 22, 220
Nernst potential, 52, 60–63
neutrons and neutron methods, 91–93, 98–99, 146–162, 192–193, 235
activation, 153, 155–160
detectors, 150
generator (linear accelerator), 147
logging, 146–164
porosity, 150–153, 411–417
sources, 97, 103, 146–147
normal resistivity, 28–29, 34–41, 218
nucleus, 91–92, 132

O

oil, 26
oil muds, 15
Owl Technical, 184
oxidation, 63–64, 69

P

pair production, 94, 131
patterns, 398–410
percent frequency effect, PFE, 85
permeability, 46, 71
petroleum logging, 4
PGT, Princeton Gamma Tech, 104, 124, 158
photoelectric effects, 93–94, 131
photo emission, 181
photographic-type deviation tools, 331
photomultiplier, 102–104
photons, 92–94
plateau, 101–104, 106
platinum resistors, 1, 189–190
Poisson's ratio, 175
polygon method, 239–240, 245
pore fluid, 21
porosity, 4–6, 9, 19, 21–23, 26, 46, 142, 146, 173–176, 232–234
porosity log, 220–221
potash, 6, 10, 113–114, 121, 210–212
potassium, 6, 10, 113–114, 122, 210–212
preamplifier, 105
probability, 149–150, 156
probable error, 244
prompt fission neutron, 156–157
proportional counters, 101, 150
proton procession, 167
pulses, 150
pulse shapers, 106, 108
PVC casing, 146, 161

Q

Q, charge, 108, 150
quadrature voltage, 85

R

radioactivity, 90–164
radon, 196–199
ratemeter, 108–109
raymer, 173–175
redox, 52, 66–71, 216
reserve calculations, 239–240, 245
resistance, 19–21

resistivity, 21–52, 212–220, 235
catalog, 5–10, 21
constant, geometrical, 32
flushed zone, 26, 32, 45, 218–220, 234
logging, 1–2, 216–220
mud, 14–15, 19, 32–34, 38–40, 52, 54–56, 62, 66, 68, 218
mud filtrate, 24, 26, 63, 66, 68, 218–219
true, 14–15, 22, 24, 26, 28, 38, 41, 46, 65, 219–220, 232–234
water, 24–26, 58–64, 214–216, 219, 232–234
resistivity-porosity crossplot, 232–234
resistivity, resistance logging, 18–50
return electrode, 28, 30, 34–35, 55
rugosity, 140

S
salinity, 58–64
salts, 13–14, 24
sampling, 194–199
sands and sandstones, 4–10, 50–76, 60–68, 115–118, 151, 173–175, 209–210, 213–214, 225
saturation, 1–3, 26, 234
scattered gamma-ray systems, 130–144
scattering, 93–95, 130–144
Schlumberger, 1–3, 52
scintillation detector, 102–104, 106, 150
sediments, 4–10, 21–22
selective SP, 75–76
semiconductor detector (solid state), 104
shales, 4–6, 21, 26–28, 52–66, 68, 142, 153, 173, 209–210, 213–214, 220, 225
shale, marine, 115, 124
shaly sands, 114–116, 153, 209, 211
shear modulus, 175
sidewall tools, 137–144, 151
silver, 159–160
simultaneous equations, 234–237
single curves, 208–222
single-point resistance and resistivity, 2–5, 28–34, 216–218
sniffers, 196–199
sodium chloride, 21, 24, 50, 60–62, 72, 144, 191, 213
sodium iodide, 104, 115
solid-state detectors, 104, 150
sonic logging, 170–176, 419, 420
spacing, 19, 34–41, 136–139
spectrograph, 105, 122–129, 157, 159–160
spectrum, 124–125, 158–160
spherical electrodes, 28–31
Spontaneous Potential:
differential, 75
selective, 75–76
SP, 1, 24, 51–76, 213–216, 218–219
static, 75–76
standard deviation, 108, 243
standoff, 131
statistics, 111–112, 147–150, 242–244
streaming potential, 71, 72
stripping, 123–124
sulfides, 67–68, 71, 79, 83
surface methods, 42–46, 52, 124–127, 167–168
susceptibility, 165–166
sylvite, 212

T
TDT, 146, 158–159
temperature, 24, 104, 159–160, 186–194
temperature, differential, 188–189
temperature effects, 60–63, 127
temperature logging, 1, 146, 159–160
thermal conductivity, 188–189
thermal gradient, 188–192
thermistors, 190
thermal neutrons, 148–151, 153–156, 158–159, 192–193
thermal neutron lifetime logging, 158–159
thermal neutron temperature log, 159–160, 192–193
thermocouple, 190
thermoluminescence, 193
thickness, 36–37, 116–118
thin bed, 36–37, 116–118
thorium, 6, 122–124
time averaging equation, 173
time constant, 108–109, 189
time domain, 84–85
travel time, interval, 171
true resistivity, 22

U
unconsolidated formations, 5, 11, 23
universal gas constant, R, 60, 70
uranium, 113–114, 121–124, 126–127, 146, 150, 156–157, 195, 197–198, 207–212
uranium logs and logging, 5, 127, 156–157, 166, 210–212

V
velocity, 339
velocity acoustic, 170–176
volume fractional, 153, 234–235, 243
vugular porosity, 6

W
water loss, 13
water saturation, 26, 28
weight percent, 130, 195–196
Wyllie's time average equation, 173

X
X-ray fluorescence, 144–145
X-rays, 16, 91–92, 97–98, 144–145, 205

Y
Young's modulus, 175, 243

Z
Z/A ratio, 132–136, 143
Z, atomic number, 92–93, 132–136, 144, 148–149
zeta potential, 71, 72